Lecture Notes in Computer Science 967
Edited by G. Goos, J. Hartmanis and J. van Leeuwen

Advisory Board: W. Brauer   D. Gries   J. Stoer

Springer

Berlin
Heidelberg
New York
Barcelona
Budapest
Hong Kong
London
Milan
Paris
Tokyo

Jonathan P. Bowen Michael G. Hinchey (Eds.)

ZUM '95: The Z Formal Specification Notation

9th International Conference of Z Users
Limerick, Ireland, September 7-9, 1995
Proceedings

 Springer

Series Editors

Gerhard Goos, Karlsruhe University, Germany

Juris Hartmanis, Cornell University, NY, USA

Jan van Leeuwen, Utrecht University, The Netherlands

Volume Editors

Jonathan P. Bowen
Oxford University Computing Laboratory
Wolfson Building, Parks Road, Oxford OX1 3QD, United Kingdom

Michael G. Hinchey
Real-Time Computing Laboratory, New Jersey Institute of Technology
University Heights, Newark, NJ 07102, USA and
Department of Computer Science & Information Systems
University of Limerick, Plassey Technological Park
Castletroy, Limerick, Ireland

Cataloging-in-Publication data applied for

Die Deutsche Bibliothek - CIP-Einheitsaufnahme

ZUM <9, 1995, Limerick>:
ZUM '95: the Z formal specification notation : proceedings /
9th International Conference of Z Users, Limerick, Ireland,
September 7 - 9, 1995. J. P. Bowen ; Michael G. Hinchey (ed.).
- Berlin ; Heidelberg ; New York ; Barcelona ; Budapest ;
Hong Kong ; London ; Milan ; Paris ; Tokyo : Springer, 1995
  (Lecture notes in computer science ; Vol. 967)
  ISBN 3-540-60271-2
NE: Bowen, Jonathan P. [Hrsg.]; GT

CR Subject Classification (1991): D.2, I.1.3, F.3.1, D.1, G.2

ISBN 3-540-60271-2 Springer-Verlag Berlin Heidelberg New York

© Springer-Verlag Berlin Heidelberg 1995
Printed in Germany

Typesetting: Camera-ready by author
SPIN 10485383   06/3142 – 5 4 3 2 1 0   Printed on acid-free paper

Preface

For ZUM'95, the 9th in the series, we are pleased to report that we received more than twice as many submissions as any previous Z User Meeting. The past four Z User Meeting proceedings have been published in the Springer-Verlag *Workshops in Computing* series. This year the event has been renamed the *International Conference of Z Users*, being the first in the series to be held outside the United Kingdom, although we are retaining the acronym 'ZUM'. Partly as a result of the change of name, the proceedings is now published in the Springer-Verlag *Lecture Notes in Computer Science* series instead.

The activities associated with the meeting span a whole week. The wide selection of tutorials offered for the three days immediately preceding the main meeting, organized by Norah Power, are intended to provide attendees with the opportunity to gain a grounding in the topics covered. On the day before the meeting, the Z standard committee (under ISO/IEC JTC1/SC22 and ANSI X3J21) is meeting and providing an opportunity for ZUM attendees to learn about the proposed international Z standard during part of the day.

After the 1994 Z User Meeting we held a half-day session on educational issues, organized by Neville Dean. This was very successful, and a similar session is to be held immediately after this meeting, again organized by Neville Dean. The forum gives the opportunity for some more interactive discussion in this important area.

We are fortunate to have an excellent selection of invited speakers at ZUM'95. At the last meeting, we found that inviting speakers from outside the Z community was extremely useful in generating debate and fostering cross-fertilization between different parts of the formal methods community at large. Therefore we have continued this newly-established tradition by inviting David Lorge Parnas (McMaster University, Canada), well known for his work on the certification of the Darlington nuclear reactor software using formal methods, John Rushby (SRI International, USA), originator of the PVS theorem prover based on higher-order logic, and Jeannette Wing (Carnegie Mellon University, USA), one of the developers of the Larch algebraic approach to formal specification and verification. In addition, David Gries (Cornell University, USA) is speaking at the educational issues session and giving the after-dinner speech at the conference dinner.

The meeting is organized by the Z User Group (ZUG), in association with the BCS FACS (Formal Aspects of Computing Science) special interest group. For ZUM'95 we are sponsored by BT, who provided student bursaries and some of the expenses for invited speakers, Forbairt for a donation towards the cost of the meeting, Praxis who maintain and run the Z postal mailing list, and the University of Limerick who have made all of their facilities available to us. In addition, the meeting is supported by the ESPRIT ProCoS-WG Working Group (no. 8694) of 24 European industrial and academic partners with an interest in 'Provably Correct Systems', through the provision of funding for attendance by Working Group members.

Norah Power acted as local organizer, and Gemma Ryan organized the bookings for the meeting at the University of Limerick. We are very grateful to Kevin Ryan and Séamus O'Shea for making the facilities of University of Limerick available to us, and for their invaluable support during the organization of the conference.

During the period since ZUM'94, Trevor King (Praxis, UK) has been an extremely efficient Secretary of the Z User Group. Unfortunately work pressures have meant that he must retire from these duties, but Jonathan Hammond, also of Praxis, has been co-opted as a ZUG committee member in his place. We wish Trevor well in his 'retirement'.

In May 1995, a special issue on Z of the *Information and Software Technology* journal was published. This featured revised versions of selected papers presented at the last ZUM, together with some new material. The aim was to include papers covering a balance of issues concerned with the development and use of Z, together with the more recent B-Method and its associated Abstract Machine Notation (AMN) and B-Toolkit. A copy is being issued to all ZUM'95 attendees as part of the delegate's pack.

With the location at Limerick, we decided to hold a limerick competition on Z and formal methods in general. We offer the following in the hope of inspiring a better selection by the end of the conference:

> At the Irish School of VDM
> They do it because of the TDM.
> > But if they tried Z
> > It would go to their head;
> There seems to be no happy meDM.

For US readers, we offer the following, with a non-deterministic last line:

> It's as easy as A - B - C
> But then you get to Z.
> > It has to be said
> > That those who write Z
> Don't like to say Z = B.

Further limericks submitted to the competition are held on-line as part of the World Wide Web (WWW) information on the conference to be found under the following URL (Uniform Resource Locator):

```
http://www.comlab.ox.ac.uk/archive/z/zum95.html
```

This will be kept up to date before and after the meeting with links to any information relevant to ZUM'95. More general information on Z, the B-Method, and other formal methods, maintained as part of the WWW Virtual Library, are also linked from this page.

We hope that ZUM'95 and this associated proceedings will be of interest and benefit to users of Z, and formal methods in general, both in academia and industry.

Oxford & Cambridge   Jonathan Bowen (Conference Chair)
May 1995   Mike Hinchey (Programme Chair)

Programme Committee

Jonathan Bowen, Oxford University, UK (*Conference Chair*)
Stephen Brien, Oxford University, UK
Elspeth Cusack, BT, UK
Neville Dean, Anglia Polytechnic University, UK
David Garlan, Carnegie Mellon University, USA
Howard Haughton, JP Morgan, UK
Ian Hayes, University of Queensland, Australia
Mike Hinchey, NJIT, USA & Univ. of Limerick, Ireland (*Programme Chair*)
Trevor King, Praxis, UK
Kevin Lano, Imperial College, UK
Norah Power, University of Limerick, Ireland (*Tutorial Chair*)
Gordon Rose, University of Queensland, Australia
Chris Sennett, DRA, UK
David Till, City University, UK
Sam Valentine, University of Brighton, UK
Jim Woodcock, Oxford University, UK
John Wordsworth, IBM UK Laboratories, UK

External Referees

All submitted papers were reviewed by the programme committee and/or a number of external referees. We are very grateful to these people, and apologize in advance for any names omitted from this list:

Rod Abraham
David Carrington
Jill Doakes
Andy Evans
Mark Humphrys
Ed Kazmierczak
Brendan Mahony
Steven Phillips
Liam Relihan
Lesley Semmens
John Staples
Owen Traynor
Clazien Wezeman
Pete Young

Ralph Becket
Bernie Cohen
Tim Drye
Stephen Goldsack
Stephen Jarvis
Peter Kearney
Rex Matthews
Hossein Rafsanjani
Gordon Rose
Graeme Smith
Ben Strulo
Mark Utting
Luke Wildman

Rob Booth
Mark Dawson
Roger Duke
Jonathan Hammond
Sara Kalvala
J. McDoowi
Martin Owen
Tim Regan
Paul Sanders
Jo Stanley
Paul Swatman
Jim Welsh
Jeremy Wilson

Sponsors

The 9th International Conference of Z Users greatly benefited from the support and financial assistance of the following:

BT
Forbairt
Praxis
University of Limerick

Tutorial Programme

Ten tutorials were presented on the three days prior to the main sessions (4th – 6th September); they were:

Formal Methods and Conformance with Open Distributed Processing
H. Bowman & J. Derrick (University of Kent, UK)

An Introduction to Object-oriented Formal Methods
K.C. Lano & S. Goldsack (Imperial College, UK)

Engineering of Complex Real-Time Systems
A.D. Stoyenko & P.A. Laplante (New Jersey Institute of Technology, USA)

An Introduction to Z
J. Turner (Staffordshire University, UK)

Building Models in Z: An Active Approach
N. Dean (Anglia Polytechnic University, UK)

The Rôle of Formal Specifications in Software Test
H.-M. Hörcher & J. Peleska (DST Deutsche System–Technik GmbH, Germany)

A Tutorial on Proof in Standard Z
J.C.P. Woodcock, S.M. Brien & A. Martin (Oxford University, UK)

Formal Methods, Requirements and Testing
M. Mac an Airchinnigh (University of Dublin, Ireland)

Using Z to Rigorously Review Structured Specifications
L. Semmens (Leeds Metropolitan University, UK)

Teaching Logic as a Tool
D. Gries (Cornell University, USA)

Contents

Education Session

Appendices

Methods

Language-Free Mathematical Methods for Software Design
Extended Abstract

David Lorge Parnas
Department of Electrical and Computer Engineering
CRL, McMaster University, Hamilton, Ontario, Canada L8S 4K1

One of the many mistakes made in the early days of Computer Science was using the term "language" to describe things like FORTRAN, C, Modula, ADA, or ML In normal usage, "language" denotes a set of signals, symbols and conventions for formulating, communicating, and recording facts (or ideas) *between people*. With the possible exception of ALGOL-60, which was originally designed for communication between people, that description does not fit the things that we call programming languages.

Natural languages grow by an exception handling process; when users (speakers, writers) feel that a language is incapable of expressing some thought, a new word or phrase-type is added. When the need arises, nouns are "verbed" and meanings generalised or specialised. The set of conventions and rules that results from this type of process is complex and lacking in conceptual integrity. Unfortunately, this is exactly the philosophy that has guided the design of some programming and specification "languages". They have grown like mushrooms and have become "rich", i.e. full of overlapping features, and lacking in structure.

If we look at older areas of Engineering, we will see that mathematical methods are not introduced by teaching languages. For example, Electrical Engineers learn to use mathematics when designing circuits in three quite distinct courses.

- They learn about solving sets of differential equations in mathematics classes.
- In physics classes they learn physical laws that govern the behaviour of circuits.
- In circuit design classes they learn how to use these physical models, together with their mathematical skills, to design reliable products.

I view this "separation of concerns" as extremely important. It helps to distinguish syntax from method and semantics. No Electrical Engineer confuses the "dot" notation for derivatives with the fact that an ideal resistor obeys Ohm's law. They also understand that the existence of resistors that are not ideal does not mean that the mathematics that they have learned is irrelevant. The distinction between the physics, mathematics, and design aspects has another practical advantage. The physical laws do not have to be stated for each analysis; they are implicit "axioms". The result is that engineering mathematics looks sloppy and informal to pure mathematicians but, "miraculously" yields answers that are consistent with physical reality.

A clear distinction between, fundamental semantics, notation, and design method is badly needed in the field of Software Engineering. It is particularly needed in the communities that have formed around languages such as Z, VDM, LOTOS, ESTELLE, SDL, LARCH, HOL, etc. These communities have so deeply confused syntax, semantics, and design methods that they fail to make important distinctions such as those between, models, specifications, descriptions, and prototypes. The fact that all may be described using the same syntax seems to have led mathematical formalists to use such terms interchangeably.

The first section of this talk deals with a functional model of programming. The "laws" that correspond to Kirchoff's laws for circuits have been well-known for years although they are rarely presented as "physical laws". Any computer system, or any of its components, can be represented by one or more mathematical relations. When two components are combined, we know how to compute the relations that represent their combined behaviour. These "laws" can be represented in a way that is independent of the notation used to describe the mathematical relations.

The second section of the talk deals with mathematical notation that is useful for describing these relations and functions.

The final section of the talk discusses design.

The material in this talk is taken from the following papers.

References

1.   Parnas, D.L., Madey, J., "Functional Documentation for Computer Systems Engineering (Version 2)", CRL Report 237, McMaster University, TRIO (Telecommunications Research Institute of Ontario), September 1991, 14 Pgs. (Substantially Revised Version to appear in *Science of Computer Programming, 1995*)

2.   Parnas, D.L. "Predicate Logic for Software Engineering" *IEEE Transactions on Software Engineering* Vol. 19, No. 9, September 1993

3.   Janicki, R., "Towards a Formal Semantics of Tables", Proceedings of the *17th International Conference on Software Engineering*, Seattle WA, April 23 - 30 1995, pp. 231 - 240.

4.   Parnas, D.L., "The Use of Precise Specifications in the Development of Software", Proceedings of the IFIP 1977, North-Holland Publishing Company, Amsterdam, New York, Oxford, pp. 861-867.

5.   Parnas, D.L. "Mathematical Descriptions and Specification of Software", *Proceedings of IFIP World Congress 1994, Volume I*" August 1994, pp. 354 - 359.

6.   Zucker, J.I., "Normal and Inverted Function Tables", CRL Report 265, McMaster University, Communications Research Laboratory, TRIO (Telecommunications Research Institute of Ontario), December 1993, 15 pgs.

7.   Zucker, J.I., "Transformations of Normal and Inverted Function Tables", CRL Report 291, McMaster University, Communications Research Laboratory, TRIO (Telecommunications Research Institute of Ontario), August 1994, 26 pgs.

A Formal Approach to Software Design: The Clepsydra Methodology

P. Ciaccia[1], P. Ciancarini[2], W. Penzo[1]

[1] DEIS CIOC-CNR, University of Bologna,
Viale Risorgimento, 2, 40136 Bologna, Italy,
e-mail: {pciaccia,wpenzo}@deis.unibo.it
[2] Department of Computer Science, University of Bologna,
Via Mura A. Zamboni, 7, 40127 Bologna, Italy,
e-mail: cianca@cs.unibo.it

Abstract. In order to improve software quality, specifiers can take advantage of the use of formal methods in the software development process. With regard to requirements specifications, attempts in this sense have been successfully made. We claim that also in the design phase a formal approach could lead to several benefits, such as the possibility of formally checking if the produced documents satisfy initial requirements. To this purpose we emphasize the role of tools since they provide automatic support to verification. Because of the different aims of the requirements and design specification phases, we believe appropriate the use of two different formal languages for their definition. The above considerations are gathered in the *Clepsydra* methodology, where a relationship between these early phases of the software development process is introduced, by using Z and Larch as formal languages for requirements and design specifications respectively. Moreover, verification issues are also discussed, by outlining how the introduced methodology makes easier the verifier's task.

1   Introduction

Improvement of quality is one of the most advisable goals to achieve in the software development process. By taking special care of the early phases of the process, the produced software is high-grade and coming up initial expectations. To this end, the application of formal methods can produce several benefits [13].

The specifier uses a formal notation like Z [3] or Larch [6] to write a user requirements document. This can be subsequently formally analyzed and its properties proved [11]. Several tools exist to check formal documents; for instance, parsers, type-checkers, and theorem provers are available both for Z [1, 14] and Larch [4, 5].

After a formal requirements document has been written and agreed upon, in the subsequent phase a coherent design document satisfying the specified needs has to be produced. Many questions arise at this stage of the process: How can a designer produce a design specification document from the specifier's output? How can one show that such a design document fulfills the same formal properties proved for the requirements document? Several attempts have been made aiming at offering support also at design level: these include object-oriented extensions for specification languages like Z [16], refinement [12], and animation techniques [8].

We claim that a good strategy consists in adopting a formal approach in order to yield a *formal design document* whose properties can be checked against requirements specification. Currently, it is completely ascertained that the use of programming languages for specification purposes should be avoided, because they are too "low-level" to adequately express the specifier's needs. Similarly, the use of a unique formal specification language for both requirements and design is inappropriate, since they present different features which are more properly described by means of specific notations and tools [9].

In this paper we provide a coherent formal methodology, named *Clepsydra*, in which two different formal notations are used: Z for requirements [15] and Larch [6] for design specifications. We argue that Z is adequate for formally describing a user requirements domain, since it is totally free of typical design issues such as reuse and modularity. Instead, Larch aims at guiding the designer to produce a document based on programming languages features. Our methodology provides some guidelines for the transformation of a requirements document written in Z into a design document written in Larch. The transformation process between these two levels is carried out and automatically checked by means of a set of verification tools which we have found suitable for each different phase [4, 5, 14]. In order to give a complete description of this translation process, we have established a correspondence between Z and Larch constructs. This method has been successfully used in several projects developed by our students [2].

The structure of the paper is as follows. In section 2 we discuss how Z and Larch can adequately fit the early phases of the development process and how tools can support properties verification. Section 3 deals with the presentation of the *Clepsydra* methodology. Here we study a number of strategies which help in the translation process, from Z descriptions to Larch modular structures. Furthermore, consistency maintenance considerations are introduced through the verification of properties expressed at different levels of the development process. In section 4 a classic case study is considered: the *BirthdayBook* [15], so as to provide a concrete example of the applicability of the method. Within this example, design alternatives are also discussed. Finally, in section 5 we state our conclusions on the advantages gained by the introduction of the *Clepsydra* methodology in software development.

2 Formal Requirements and Design Specifications

In order to ensure the quality of a software product, a verifiable correspondence between what has been built and its initial specification is strongly needed. To this end, an important role is played by verification tools which are able to prove properties for a specified system [1, 5]. Then, the use of a formal approach both for requirements and design specification phases would allow formal verification at both levels. Here, we present a brief view of these early phases, by outlining how languages like Z and Larch can be used as formal notations for their definition.

2.1 The Requirements Specification Phase

In this phase the accuracy in the description of the required system has great importance since requirements inconsistencies or incompletenesses found at subsequent stages in the development process could imply a redefinition of the initial specification document, thus affecting the whole development process [11]. The use of tools and formal methods during this phase ensures a rigorous approach to requirements formulation by obtaining coherent, complete, correct and unambiguous specifications. Z has been properly introduced as a formal notation for fully abstract, declarative, non-executable requirements specifications [7]. The expressiveness of Z allows for specifying requirements by clearly pointing out *what* has to be done, not *how* it should be done. In this sense, we strongly believe in the soundness and validity of the use of Z for the requirements specification phase.

2.2 The Design Phase

Once the formal requirements document has been produced, the designer takes it as the input for the subsequent phase. At this step all design choices are taken by defining and specifying the required abstract data structures and their related "methods", the emphasis being on describing the modular architecture of the system in a form suitable for programmers [13]. The result should be a quite abstract document free from implementative details.

Taking Z as a starting point for the requirements specification phase, it is not clear how a document expressed in such notation should be used in the rest of the development process. In order to adopt a formal approach covering also the design phase, two different trends appear:

1. The use of a unique formal language both for requirements and design specifications phases. Within this trend several attempts of extending Z (e.g. object-oriented versions like Z++, OOZE, Object-Z, MooZ [16]) have been made.
2. Two different formal notations are used for the above phases of the development process.

We regard the latter trend as more advisable in agreement with Hoare's position, as expressed in the preface of the VDM conference held at Kiel in 1990 [9]: the "uniform" approach can be considered improper since the use of a unique language for both phases does not adequately fit all the required purposes.

Focusing on existing formal methods we observe that some of them appear to be more suitable for formal design than for formal requirements specification, even if they have been introduced for the latter goal. In particular in this paper we consider Larch as formal design language. It is a notation used for requirements specification as well but, because of its peculiarities, it appears to offer more support than Z to the designer.

The Larch Language. Larch is a family of specification languages [6]. A Larch specification is two-tiered, that is, it includes two components written using two different notations:

- the Larch Shared Language (LSL) is used for defining a library of abstract data structures (*traits*) specified using algebraic notation [6];
- a Larch Interface Language (LIL) tailored to a specific programming language (e.g. LCL [6] for the C language) is used to specify module interfaces to be subsequently implemented [17].

While LSL is independent of any programming language, LIL syntax and semantics depend on the specific implementation language. Basically, LIL modules are used to describe possible actions on the system whose abstract state has been algebraically declared by LSL traits.

A Larch document is representable by means of a directed acyclic graph that includes two subgraphs (see Fig. 1). The top graph, said *algebra-tree*, includes several nodes, representing traits. An algebra-tree is incrementally generated in a bottom-up fashion, possibly reusing a number of predefined library traits [6]. The lower graph represents a set of procedure interfaces and specifications written in a LIL. The semantics of LIL modules is given by referring to entities and properties specified in "used" LSL traits (see Fig. 1).

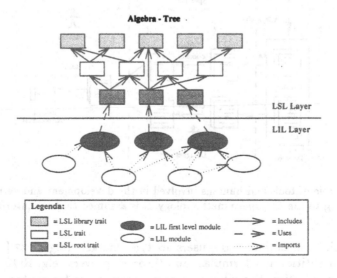

Fig. 1. General structure of a Larch specification.

2.3 The Role of Tools

Tools are of great importance in a sound and correct development process, in that they provide useful mechanisms for automatically check consistency and verify syntactic and semantic properties at each different stage.

Here, we concentrate on the process that starts from a formal requirements document and produces a design document. Ideally, a number of tools can help the designer during this phase (see Fig. 2).

- Pretty printers and text formatters, to improve readability. For instance, both Z and Larch have a LaTeX-based syntax.
- Type checkers, to improve confidence in the correctness of a specification document containing typed entities [4, 6, 10, 14].
- Theorem provers, to help in prove formal properties [1, 5].
- Animators, supporting rapid prototyping so helping in proving that a specification is really implementable [8].

Fig. 2 shows how some of the above tools are used in the development and verification process followed by the *Clepsydra* methodology.

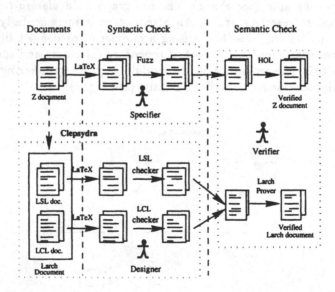

Fig. 2. The role of tools and humans involved in the development and verification process according to the *Clepsydra* methodology. It is assumed that the interface language is LCL (for *C*).

The basic tools available to Z users are type checkers (e.g. *f*UZZ [14]) to check the types of entities and formulas, and theorem provers (e.g. HOL [1]) to help in proving properties. For what concerns Larch, tools for syntactic check are available both at LSL level (LSL checker [6]) and at LIL level (e.g. LCLint [4] for LCL); then, semantic check is provided by the *Larch Prover* [5], which can help the designer to prove properties claimed for traits.

The most important role is played by theorem provers, because a formal design document must be proved correct with respect to the formal requirements specification. Then, the role of the verifier is to control that the design document satisfies the given requirements by checking that properties at the first level still hold at the subsequent phase.

3   The Clepsydra Methodology

The new methodology we introduce is named *Clepsydra* because of the form of the picture describing it (see Fig. 3). Its main characteristic stands in a clear

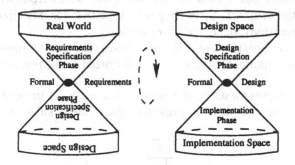

Fig. 3. The *Clepsydra* methodology.

distinction between requirements specification and design phases and in the use of a formal approach in both these phases. In fact, in *Clepsydra* we put emphasis on formal design, intended to produce a document that is possible to compare with the formal requirements. We firmly believe that the use of a formal approach even at the design phase would provide several benefits because of the possibility of using proper tools for consistency checking.

As it can be seen in Fig. 3, in *Clepsydra* the requirements analysis phase converges on a specific formal requirements document, among infinite documents which could describe user's needs. Such a document becomes the input of the design specification phase, thus defining a *design space* consisting of all the possible design choices that are compatible with the requirements specification.

Our methodology allows for translating a requirements document written in Z into a design document made of Larch modules (traits, data structures and procedures). In the *Clepsydra* methodology the designer's goal consists of defining the modular structure of the system under development by referring to two different levels, one for data objects and one for procedure interfaces. The Z distinction between schemas describing abstract states and schemas defining operations on the abstract states is kept: the abstract states are transformed into LSL traits, whereas the definition of operations is handled by a Larch Interface Language.

In the following we briefly describe the basic steps of the methodology and postpone a more detailed description to Section 4.

3.1 Step 1: Graphic Analysis of a Z Document

We introduce a graphic representation for Z specifications that is useful to concisely describe and analyze the existing interconnections among the schemas included in a Z document. Such a representation consists of a graph formed by *boxes* as nodes and *arrows* as links. The semantics to be assigned to these symbols is as follows:

- Schemas expressing abstract states of the system are represented by boxes with no arrows coming out. They are usually the destination of arrows describing abstract operations.
- Schemas describing operations on the states are boxes from which arrows leave towards the abstract states they refer to, either for initialization, modification (Δ operation), or query ($\varXi$ operation).
- Initializations, Δ operations and $\varXi$ operations stand out by means of different arrows forms, namely solid, dashed and dotted lines, respectively.

Such a graphic representation of requirements helps in understanding the connections among the schemas, and provides a guideline to subsequent analysis.

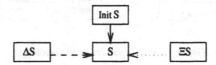

Fig. 4. Graphic representation of simple Z schemas.

3.2 Step 2: From Boxes to Modules

By examining the graphic representation of a Z specification we distinguish the components to be translated in LSL traits from those corresponding to LIL modules. The general strategy we use in the transformation process is based on the following heuristics:

- Boxes with no outgoing arrows represent data and are to be translated in LSL traits. To this end, the designer needs to consider the abstract state and related operations, as expressed in the Z document. Then, by taking advantage of available library traits, he starts to define the algebra-tree. The construction of the algebra-tree should proceed bottom-up, starting from the leaves, usually corresponding to library traits. Sometimes, renaming is advisable in order to provide a clear semantics to the set of abstract operators predefined at LSL level. Furthermore, the definition of new operators could be needed in order to cover required functionalities which cannot be expressed through library operators. These will be used in the subsequent phase to define the module interface functions.
- Boxes with outgoing arrows represent operations and are grouped and translated in LIL code expressing the operational part of the corresponding Z

schemas. The structural similarity of a LIL language (e.g. LCL) to the corresponding implementation language (e.g. C) allows the detailed aspects of the operations on the system to be almost completely described. The set of operations is partitioned into a set of LIL modules. Each of these modules imports, by means of *uses* clauses, the LSL traits defining the structures which the operations in the module apply to.

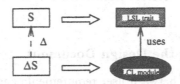

Fig. 5. Graphic description of the translation process.

Completing the Design Document. Once the design framework is obtained, in order to completely define a design document some relevant aspects have to be considered, among which:

Syntactic and semantic correspondence between Z and Larch constructs. With reference to the Z-Larch mapping, even if semantic issues are still under analysis, some basic syntactic guidelines can be given as follows:

Z Given Types are translated in LSL sorts (Larch basic types).
Other Z types can be mapped as follows:

Z notation	Larch
Enumeration	Enumeration
Z Tuple (Cartesian Product)	Larch Tuple
Free Types	Larch Union

Generic Definitions expressing polymorphic abstract states are directly transformed in LSL traits.
Axiomatic Definitions expressing abstract operations are translated in operators inside traits, because they define executable actions on abstract objects. They represent auxiliary functionalities and are considered as extensions to the trait under definition.

Different design alternatives. The transformation process does not result in a precisely determined document, neither can be automated. In fact, the description of a Z state can be described by means of several different Larch constructs, since with Z we express the user requirements, whereas with Larch we introduce possible design choices. For instance, the Z description of a database expressed by a functional operator can be indifferently translated by using Larch traits defining either functional mappings, tables, or sets.

Different expressive power between the two notations. The transformation process can be difficult basically because Z has a higher expressive power. Some Z constructs, like existential quantification, overriding and image operators, cannot be directly translated in LSL. Then, some logical transformations on expressions containing these constructs are needed. For instance, skolemization processes and extensional descriptions of elements and operations may be necessary for a correct translation of Z expressions by means of LSL structures.

3.3   Step 3: Verifying the Design Document

Consistency maintenance between user requirements and design documents is a feature that cannot be given up. Therefore, if we consider that a set of properties is shown to hold for the requirements document, then the goal is to ensure that the same properties also hold in the design document.

Besides the advantages gained by the application of our methodology as a guideline in the development process, *Clepsydra* also provides a valid support to the verification stage by easily singling out the correspondence between properties at requirements specification and design phases. In fact, since a mapping is established between Z and Larch documents (see Fig. 2), a guide for identifying properties to be checked is therefore available, so making the verifier's task easier. In *Clepsydra* we use the Larch Prover for checking LSL specifications. This tool enables the software engineer to prove equations, logical implications, and assertions contained in the design document. The Larch Prover is not completely automatic: it is an interactive theorem prover which needs a guide for the choice of proof strategies. With regard to the LIL level the proof is manual but it relies on the result of the Larch Prover.

4   Application of Clepsydra: A Simple Case Study

We have experienced our methodology in several complex projects, developed by undergraduate students, involving a large number of Z schemas, LSL traits and LCL modules. Between these projects we can mention the specification of an integrated software environment for chess playing consisting of editors, databases, automated players, etc., and of a distributed system describing a client-server organization also including communication protocols [2]. In all these cases the *Clepsydra* methodology has proved to be a sound guide for the software development process. In particular, it can be considered as a valid support for managing large specifications, since it allows the overcoming of possible disorientation due to the large amount of information to be dealt with.

Here, for lack of space, we present only a simple case study known in the literature as the *BirthdayBook* [15]. We have to specify and design a database recording relationships between a set of persons and corresponding birthdays; the database has to provide suitable data insertion and retrieval operations. Here, we assume that the target implementation language is C.

4.1 The Requirements Document

Formal specifications for the *BirthdayBook* are given in Z notation as follows:

[*NAME, DATE*]

BirthdayBook
known : **P** *NAME*
birthday : *NAME* ↦ *DATE*

dom *birthday* = *known*

Some operations are then specified, that either modify the state, by introducing some elements (*AddBirthday* schema), or maintain it unchanged, by simply retrieving the required information through proper queries on the database (*FindBirthday* and *Remind* schemas). In order to provide a complete definition of the state of the system the specification of the initial conditions, namely when the database has no elements, is needed (*InitBirthdayBook* schema) [15].

However, the above specification is not complete, because the system behavior is undefined in some situations not explicitly covered. For this purpose, the concept of output report is introduced in order to signal error conditions. Reports can be defined as enumeration of messages:

REPORT ::= *ok* | *already_known* | *not_known*

Therefore, operations will be supplied with schemas reporting proper messages signaling success (*Success* schema), or error conditions (*AlreadyKnown* and *NotKnown* schemas). Their introduction allows for the definition of "robust" specifications since all the possible behaviors of the system are specified:

RAddBirthday ≙ (*AddBirthday* ∧ *Success*) ∨ *AlreadyKnown*

RFindBirthday ≙ (*FindBirthday* ∧ *Success*) ∨ *NotKnown*

RRemind ≙ *Remind* ∧ *Success*

4.2 Translation in the Larch Formalism

We start with a graphic representation of the Z specification of the *BirthdayBook* (without error cases handling) as in Fig. 6.

A `BirthdayBook.lsl` trait is built by using library traits concerning sets (`Set.lsl`) and functional applications (`FiniteMap.lsl`). In order to define the *BirthdayBook* data structure, operators like powerset (for *known* definition) and partial function (for *birthday*) are needed. The system state definition is completed by the set of operations corresponding to those presented in Z. Hence, the LSL trait corresponding to the complete specification is defined as follows:

`BirthdayBook(BDB)`: *trait*
includes

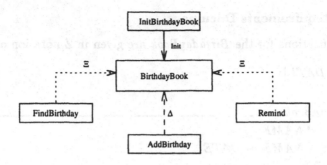

Fig. 6. Graphic description of the *BirthdayBook* in Z.

```
Set(NAME, KNOWN, emptySet for {}, addName for insert),
   FiniteMap(BIRTHDAY, NAME, DATE, emptyMap for {}, addBirthday for
update, findBirthday for apply)
BDB tuple of k: KNOWN, b: BIRTHDAY
introduces
   dom: BIRTHDAY → KNOWN
   remind: BIRTHDAY, DATE → KNOWN
asserts
BIRTHDAY partitioned by remind
∀ b: BIRTHDAY, n,n1: NAME, d,d1. DATE
   dom(emptyMap) == emptySet;
   dom(addBirthday(b, n, d)) == dom(b) ∪ {n};
   remind(emptyMap, d) == emptySet;
   remind(addBirthday(b, n, d), d1) ==
            if d = d1 then remind(b, d1) ∪ {n}
      else remind(b, d1);
implies converts dom, remind
```

The definition of the BIRTHDAY LSL sort corresponds to the Z definition of the type of *birthday* in the *BirthdayBook* schema (see Section 4.1). This is based on the assumption that this type can also be specified with a more compact definition as follows:

$$BIRTHDAY == NAME \nrightarrow DATE$$

So, the definition of *birthday* can be written as *birthday* : *BIRTHDAY*. Similarly, the definition of *known* as a set of *NAME* elements can be easily compacted in *known* : *KNOWN*, where:

$$KNOWN == \mathbf{P}\, NAME$$

These simple transformations can be regarded as slightly simplifying LSL sorts definition. The `BirthdayBook.lsl` trait reuses data structures and operations defined in pre-existing library traits (i.e. `Set.lsl` and `FiniteMap.lsl`). Moreover, the renaming of data and operations allows abstract definitions to be adapted to different cases. In the above example, besides the definition of **KNOWN** as a set of names, the statement **addName for insert** renames the **insert** operator being defined in the **Set.lsl** trait, so as to be used in the current **BirthdayBook.lsl** trait and in the related **BirthdayBook.lcl** module with a more meaningful name. Similarly we deal with the operations of the library trait **FiniteMap.lsl**, suitably renamed by means of the **addBirthday for update** and **findBirthday for apply** clauses.

While operations describing the introduction of a new birthday and the retrieval of the date corresponding to a given name have found a direct correspondence with previously defined operations, the action expressed by the *Remind* Z schema is not directly translatable in a library defined operation. To this end, the **remind** operator is introduced. The *BirthdayBook* is represented by a tuple (BDB). The **dom** function, describing the domain of a function, is needed for keeping the specification consistent with the Z invariant. In order to describe the system behavior in case of errors, we have to introduce a trait defining the structure of a report message output by the system:

RepBirth(REP): *trait*
REP *enumeration of* **Okay, Already_Known, Not_known**

LCL Modules. In the Z specification, a number of abstract operations are defined on the abstract state represented by the *BirthdayBook* schema. The schemas which specify operations are easily recognizable by the presence of Δ or Ξ letters. Correspondingly, the LCL layer consists of the following modules (see Fig. 7):

- The **BirthdayBook.lcl** module is the interface specification for the LSL trait with the same name.
- The **Error_Messages.lcl** module handles errors, and refers to the **RepBirth.lsl** trait.
- The **RBirthdayBook.lcl** module defines the interaction between the first two modules, capturing the concept of the corresponding Z *schema calculus* expressions.

Thus, to each LSL root trait we associate a corresponding LCL module describing the operational features. Modules can be furtherly combined together to give an explicit description of their interactions in the complete definition of the system. The details of the LCL modules are as follows:

BirthdayBook.lcl

The definitions of the **NAME** and **DATE** basic sorts are given as *immutable* types since they do not need an implementative representation (we assume we inherit them from the Z definition); **KNOWN** and **BDB** are instead given as *mutable* objects, since their instances can be modified.

immutable type **NAME**;
immutable type **DATE**;
mutable type **KNOWN**;

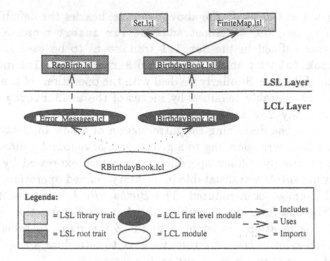

Fig. 7. Larch specification of the *BirthdayBook*.

mutable type BDB;

uses BirthdayBook;   /* refers to the BirthdayBook.lsl trait */

claims System_prop(BDB bdb){   /* system invariant */
ensures (dom(bdb°.b) = bdb°.k); }

BDB InitBDB(void) {
ensures result = [emptySet, emptyMap]; }

void Addbirthday(BDB bdb, NAME n, DATE d) {
requires ¬ (n ∈ bdb^.k);
modifies bdb;
ensures bdb'.k = addName(n, bdb^.k) ∧
        bdb'.b = addBirthday(bdb^.b, n , d); }

DATE Findbirthday(BDB bdb, NAME n) {
requires n ∈ bdb^.k;
ensures result = findBirthday(bdb^.b, n); }

KNOWN Remind(BDB bdb, DATE d) {
ensures result = remind(bdb^.b, d);
}

Error_Messages.lcl

This LCL module simply uses the definition of the error messages structure given at LSL level, then specifying functions for handling error cases.

immutable type REP;

uses RepBirth;

REP Success(void) {
ensures result = Okay; }

REP AlreadyKnown(BDB bdb, NAME n) {
requires n ∈ bdb^.k;

ensures result = Already_Known; }
REP NotKnown(BDB bdb, NAME n) {
requires n ∉ bdb^.k;
ensures result = Not_Known; }

RBirthdayBook.lcl

While the above modules (**BirthdayBook.lcl** and **Error_Messages.lcl**) define actions when given preconditions are satisfied, the **RBirthdayBook.lcl** module completely describes the behavior of the system by examining all the possible cases, i.e. this module provides a complete specification of the system's interface.

imports BirthdayBook;
imports Error_Messages;
REP RAddbirthday(BDB bdb, NAME n, DATE d) {
modifies bdb;
ensures if ¬ (n ∈ bdb^.k) then (bdb'.k = addName(n, bdb^.k) ∧
        bdb'.b = addBirthday(bdb^.b, n , d) ∧ result = Okay)
    else result = Already_Known; }
REP RFindbirthday(BDB bdb, NAME n, DATE d) {
ensures if (n ∈ bdb^.k) then (d = findBirthday(bdb^.b, n) ∧ result = Okay)
    else result = Not_Known; }
REP RRemind(BDB bdb, KNOWN k, DATE d) {
ensures k' = remind(bdb^.b, d) ∧ result = Okay; }

In general, we see that LCL modules that result from the application of the *Clepsydra* methodology can be distinguished in:

"First level" LCL modules: Each of these modules specifies the behavior of an independent part of the system, whose structure is defined in an LSL root trait.

"Structured" LCL modules: The first level LCL modules can be combined to take into consideration interrelationships between them.

With structured LCL modules we typically translate *schema calculus* formulas of the Z document. In fact, a *schema calculus* formula finds a match in the *ensures* body of a structured module, derived from first level ones through the use of *imports* clauses, where combination of logical expressions based on LSL operators is easily accomplished. This approach is especially advisable when the specification of complex systems produces several modules at first level, so defining a flat library of LCL modules. The introduction of modules encapsulating the interconnections existing among the first level modules allows the definition of interface dependencies, so providing a guide for the programmer.

Advantages resulting from this two-phase approach are:

– *Modularity* reached in the design process. Each independent part of the system is given an ad hoc specification, thus keeping a clear distinction among different modules.

- *Integrability* of modules. The global view of the system is obtained by keeping track of interconnections among modules.
- *Reusability* of the LCL first level modules.
- *Flexibility* of the global configuration. Design specifications derived with the *Clepsydra* methodology can be easily modified according to optimization needs.

Design Alternatives. There are several alternatives to the design obtained above, that should be considered in order to simplify the global structure. From the specifier's point of view, simplification means more clarity at the conceptual level, by singling out the most important parts of the system from marginal ones. From the programmer's point of view, design simplification affects the number of modules to be managed, so influencing the amount of information involved in the subsequent implementation phase. Furthermore, merging of modules eliminates redundancies, thus making easier the global system organization.

However, modification decisions of the global structure depend on several choices, based on design considerations that cannot be "a priori" fixed. In general, some basic guidelines can be applied in order to restructure interface specifications.

- One method takes advantage of simple actions whose direct integration in other modules is easy, since the definition of their first level LCL module can be considered almost trivial. In this case the considered module is removed and its operations are directly accessed at LSL level by an extended module which "uses" the corresponding trait, and therefore manages also the mentioned operations. A clear advantage of this approach is a reduced number of LCL modules.

  For instance, in the *BirthdayBook* case, the application of this method would lead to the elimination of the `Error_Messages.lcl` and `BirthdayBook.lcl` modules, by concentrating the definition of their operations inside the `RBirthdayBook.lcl` module. In fact, error messages enumeration is simply used without any considerations on preconditions, therefore making a direct reference to the related LSL trait definition. Thus, this case is particularly appropriate for compacting modules specifications, by merging the two involved modules in only one. The corresponding graphic representation is shown in Fig. 8.

  The original `RBirthdayBook.lcl` module specification is properly modified with the addition of type definitions, *claims* clauses and initialization operations previously specified in the first level LCL modules. Furthermore, two *uses* clauses are added with the aim of establishing a connection with the related `BirthdayBook.lsl` and `RepBirth.lsl` traits so that either LSL operators or enumeration values defining error messages can be used in the logical combination of expressions composing the *ensures* clauses.

- Further considerations may be related to the sharing of LCL modules on behalf of others. This is the case where module integration is not really

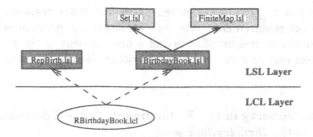

Fig. 8. Compacting LCL modules for the interface of the example in Fig. 7.

appropriate, since LCL definitions in such modules are of importance for more than one module.

In the *BirthdayBook* example this situation could occur with the Error_Messages.lcl module. It could be worth keeping it, in the view of having other modules (defining further parts of a more complex system than the one we are dealing with), which need an error management interface specification.

By assuming, on the other hand, that the original BirthdayBook.lcl module is instead integrated in the RBirthdayBook.lcl specification, the configuration of the LCL layer is depicted in Fig. 9.

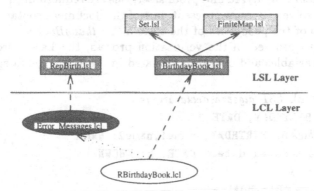

Fig. 9. Error_Messages.lcl module is kept in the view of being shared by other structured LCL modules.

The original RBirthdayBook.lcl specification is consequently modified so as to consider the *imports* clause referring to the BirthdayBook.lcl module as directly transformed in a *uses* clause of the corresponding LSL trait.

These considerations form only a part of all the possible techniques for modifying the structure of the LCL layer. Several perspectives can be taken into account, on the basis of different design features and needs.

4.3 Verification

The main property we consider in the verification process is that the set of known names belonging to the database always corresponds to the domain of the function relating to the registered dates. For space reasons, for each phase (specification of requirements, algebraic structures, and modules' interfaces) we only consider the operation of inserting a new element in the birthday-book. A similar process can be carried out for the other operations affecting the state of the system.

Verifying a Property in the Z Document. In the Z document the invariant is expressed in the *BirthdayBook* schema as:

dom *birthday* = *known*

and it is still true after the insertion of a new pair:

dom *birthday'* = dom(*birthday* ∪ { *name?* ↦ *date?* }) =
dom *birthday* ∪ { *name?* } = *known* ∪ { *name?* } = *known'*

Verifying a Property in the LSL Document. Proving consistency at LSL level is supported by an appropriate tool: the *Larch Prover*. This tool deals with properties verification both by forward and backward inference methods (by cases, normalization, induction, implication, contradiction and others [5]). A Larch document is analyzed and proof assertions are reduced until the considered conjecture is solved. Here, we present part of the document containing the Larch formalization of the properties of the system (*BirthdayBook.lp*).

In order to proceed in the verification process, the Larch Prover needs to know sorts, variables and operators declared in the LSL algebra specification:

% BirthdayBook *LSL algebra declarations*
declare sorts BIRTHDAY, DATE
declare variables b: BIRTHDAY, name1,name2: NAME
declare variables date1,date2: DATE, k: KNOWN
declare operators
   emptyMap: → BIRTHDAY
   dom: BIRTHDAY → KNOWN
   addBirthday: BIRTHDAY, NAME, DATE → BIRTHDAY
   remind: BIRTHDAY, DATE → KNOWN
   ..

Furthermore, the set of LSL axioms asserted for the introduced operators are also needed so as to give semantics to their syntactic declaration. An operational description is given with respect to the generators of the basic sorts.
asserts
   dom(emptyMap) == emptySet
   dom(addBirthday(b, name1, date1)) == dom(b) ∪ singleton(name1)
   remind(emptyMap, date1) == emptySet
   remind(addBirthday(b, name1, date1), date2) ==
     if(date1 = date2, remind(b, date1) ∪ singleton(name1),

> remind(b, date1))
>
> ..

After required algebra declarations, a part of "prove" assertions is given. Here, the verifier introduces what has to be proved. In this case we want to verify that the state invariant is maintained even after a new entry is introduced in the birthdays database.

prove (dom(b) = k) ⇒
        (dom(addBirthday(b, name1, date1)) = k ∪ singleton(name1)) by ⇒
qed

The meaning of the above assertion is as follows: if dom(b) = k holds, then, by applying the addBirthday operator, the domain of the updated database has to be the set union of the previous domain, k, and the singleton of the introduced element, name1. This has to hold whether the element is already present in the database or not. The conjecture is suggested to be proved by implication method since in the assertion the ⇒ operator appears. The result given after the *Larch Prover* analysis is the following:

prove (dom(b) = k) ⇒
        (dom(addBirthday(b, name1, date1)) = k ∪ singleton(name1)) by ⇒
        <> 1 subgoal for proof of ⇒
        [] ⇒ subgoal
        [] conjecture
qed

In order to prove the assertion, the Larch Prover tries to solve the subgoal given by the application of the ⇒-method. Once the subgoal is solved, the conjecture can be considered as proved. Sometimes, more than one subgoal can take part in the proof, as in the case of the induction method, where both basic and inductive steps are considered.

Verifying a Property in the LCL Document. At LCL level the given invariant is stated through the following expression:

claims System_prop(BDB bdb) {
ensures (dom(bdb°.b) = bdb°.k); }

Now, we show that the introduction of a new entry in the database leads to a post-state that still satisfies the invariant, provided it holds in the pre-state. For clarity, we repeat here the definition of the Addbirthday LCL function.

void Addbirthday(BDB bdb, NAME n, DATE d) {
requires ¬ (n ∈ bdb^.k);
modifies bdb;
ensures bdb'.k = addName(n, bdb^.k) ∧
        bdb'.b = addBirthday(bdb^.b, n, d); }

The proof consists of the following steps:

1. `dom(bdb'.b)` = `dom(addBirthday(bdb^.b, n, d))`
   by substitution from the second conjunct of the *ensures* clause of the `Addbirthday` function.

2. `dom(addBirthday(bdb^.b, n, d))` = `bdb^.k ∪ { n }`
   by assuming the invariant true in the pre-state and by the result obtained from the *Larch Prover*.

3. `bdb^.k ∪ { n }` = `addName(n, bdb^.k)`
   since `addName` renames the `insert` operator defined in the `Set.lsl` library trait [6].

4. `addName(n, bdb^.k)` = `bdb'.k`
   by substitution from the first conjunct of the *ensures* clause of the `Addbirthday` function.

So, also at LCL level invariant maintenance is proved.

5 Conclusions and Future Work

We have adopted a formal approach in writing a design document so as to be able to formally check it with respect to formal requirements specifications. We followed the trend which proposes the use of two different formal languages for the early phases of the development process [9].

The Z notation is appropriate for the initial phase since it is suitable to provide logic specifications and abstract descriptions free from implementative choices; moreover, its use requires only a basic knowledge of first order logic and simple skills in systems modeling [7].

We chose Larch as formal language for design. We claim that the suitability of Larch for this purpose is due to its peculiar structure divided in LSL and LIL sections. Because of its algebraic nature, LSL hierarchical and modular description of traits can be formally verified [13]. This is kept independent of the specification of the concrete operations that are introduced only at LIL level.

An option offered by Z is the possibility of expressing, by means of the *schema calculus*, complex interconnections among entities described in the requirements document. This turns to be a useful guide for establishing the interrelationships between LIL modules.

This work is a preliminary exposition of the *Clepsydra* methodology. We have presented a transformation process from a formal requirements specification to a design document that can be formally verified. Several advantages come from the application of our methodology such as formal comparisons between properties at both requirements and design specification phases, so providing a valid support to the verifier. Experiences in its use also suggest it as a good guideline in the development of complex systems, since it provides a global view at design level on structures and organization [2].

Some problems still wait to be solved. A basic one consists of developing a theory to match the semantics of Z with the semantics of Larch, in order to improve the confidence in the transformation process. To this end, we are studying a more detailed correspondence between these two specification languages, so as to provide further aids for the mapping from Z to Larch constructs.

Among the tools which could provide effective support to our approach, the following are under development: a tool to obtain a graphic representation of a Z specification, a tool to animate Z specifications, and a tool to maintain formal properties, i.e. to check the consistency of properties proved for a Larch document with respect to properties proved for a corresponding Z document. All of them are then intended to be integrated in a global environment supporting the *Clepsydra* methodology.

References

1. J. Bowen and M. Gordon. Z and HOL. 1994. URL ftp://ftp.cl.cam.ac.uk/hvg/-papers/zhol.ps.gz.
2. P. Ciaccia and P. Ciancarini. A Course on Formal Methods in Software Engineering. In A. Finkelstein and B. Nuseibeh, editors, *Proceedings ACM/IEEE International Workshop on Software Engineeering Education*, pages 97–110, Sorrento, Italy, 1994.
3. A. Diller. *Z: An Introduction to Formal Methods*. Wiley, 1990.
4. D. Evans. LCLint User's Guide. MIT/LCS Distribution Site, February 1994.
5. S. Garland and J. Guttag. An Overview of LP, the Larch Prover. In *Proc. 3rd Int. Conf. on Rewriting Techniques and Applications*, volume 355 of *LNCS*, pages 137–151. Springer-Verlag, Berlin, 1989.
6. J. Guttag and J. Horning. *Larch: Languages and Tools for Formal Specification*. Springer-Verlag, 1993.
7. I. Hayes. *Specification Case Studies*. Prentice Hall, 2 edition, 1993.
8. M. Hewitt. Automated Animation of Z Using Prolog. B.Sc. Project Report, Department of Computing, Lancaster University, UK, August 1991.
9. C. Hoare. Preface. In D. Bjorner, C. Hoare, and H. Langmaack, editors, *Proc. VDM 90: Formal Methods in Software Development*, volume 428, pages vii–x, Kiel, Germany, 1990. Springer-Verlag, Berlin.
10. D. Jordan. CADiZ - Computer Aided Design in Z. In S. Prehn and W. Toetenel, editors, *VDM 91: Formal Software Development Methods*, volume 551, pages 685–690. Springer-Verlag, Berlin, October 1991.
11. R. Kemmerer. Testing Formal Specifications to Detect Design Errors. *IEEE Transactions on Software Engineering*, 11(1):32–43, January 1985.
12. S. King. Z and the Refinement Calculus. In D. Bjorner, C. Hoare, and H. Langmaack, editors, *VDM and Z-FORMAL METHODS*, volume 428 of *LNCS*, pages 164–188. Springer-Verlag, 1990.
13. B. Liskov and J. Guttag. *Abstraction and Specification in Program Development*. MIT Press, Cambridge, 1986.
14. J. Spivey. *The ƒUZZ Manual*. 1988.
15. J. Spivey. *The Z Notation. A Reference Manual*. Prentice Hall, 2 edition, 1992.
16. S. Stepney, R. Barden, and D. Cooper. *Object-Orientation in Z*. Springer Verlag, 1990.
17. J. Wing. Writing Larch Interface Language Specifications. *ACM Transactions on Programming Languages and Systems*, 9(1):1–24, January 1987.

Refining Database Systems

David Edmond

School of Information Systems, Queensland University of Technology
GPO Box 2434, Brisbane, Queensland 4001, Australia
e-mail: davee@icis.qut.edu.au

Abstract. Most computers are used in an organisational setting and most organisational information systems are database systems. A database system is one in which the simple statements far outnumber the complex ones, to the extent that a special *database management system* is required to enable sufficiently rapid access to the data thus stored. SQL is the standard language for database access and manipulation. Z is a formal notation which has achieved some degree of acceptance – even popularity. Yet, Z and SQL are rarely considered in conjunction, which is unfortunate because *both* languages have their roots in set theory and predicate calculus. This paper examines some of the issues involved in mapping from a Z specification to an SQL-based implementation.

1 Introduction

1.1 Developing Database Systems

A database system is not an end in itself. Its purpose is to enable its users to explore, monitor and control some real-life situation. Yet, while its internal structure and workings are of relatively little interest to these people, they are of fundamental importance to the developers. For these reasons, we create such information systems in two distinct stages because, to adequately represent a situation on a computer, we must:

- *First* describe that situation in our (human) terms, and
- *then* convert that description into a machine-executable form.

In the first step, we *specify* the situation, using whatever language or languages we feel best allow us to capture and express the nature of that situation and any constraints upon it. It is *the* fundamental requirement of information systems development that any condition expressed in a specification must be properly implemented. Any constraint that we express about some real-world situation must be matched by a corresponding constraint in an information system modelling that situation. This requirement is a direct consequence of the decision to have separate specification and implementation languages. According to Figure 1, we begin the process by listening to users and capturing their viewpoints. From these we develop a so-called conceptual model which is an aggregation of the user-views. From that conceptual model, we design a relational database schema, assuming a conventional SQL-based implementation. Finally we write

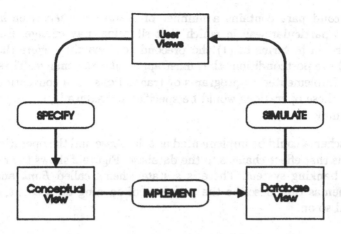

Fig. 1. User-views and program-views

programs that operate upon that database and which reconstruct or simulate the original user-views.

It is the program's purpose to sustain a mapping between the two pictures, one that simulates the external picture by suitable manipulation of the internal picture.

1.2 Z and SQL

In this paper, and for reasons to be discussed shortly, we use the Z notation as our specification language, and SQL as our implementation language. The Z language has been used to formalise a wide range of applications, such as the Unix file system, reservation systems, telephone systems and the CICS transaction processing system. These case studies were brought together in [Hay87]. The language has been standardised [Spi89] and given a formal semantics [Spi88]. Object-oriented extensions, of which the best-known is Object-Z [DK+91], are reported in [SBC92] . Z is based on typed set theory and predicate calculus. From that base are defined the commonly-used discrete mathematics of relations, functions, sequences, and so on. The unit of specification is the **schema** which is a *named* predicate that constrains some aspect of the application domain. More complex statements, again in the form of schemas, may be created using the **schema calculus**. A typical Z specification consists essentially of two parts: one part describes the structure, and the other describes the dynamics, of some situation or domain:

- The first part consists of a *state schema*, which is a static description of the application domain. It consists of a number of named observations or views. These are then related to one another in such a way as to provide a fixed picture of the situation being represented or specified. In a less rigorous specification, this would be accomplished by means of an entity-relationship diagram or equivalent.

- The second part contains a number of *operation schemas* each of which shows a particular way in which that situation may change. Each change is described in terms of (1) the preconditions existing *before* that change, and (2) the postconditions that must apply *after* the change. These schemas will be implemented as programs or transactions. In a conventional specification, these operations would be specified procedurally, using pseudocode, for example.

The state schema could be implemented as a database and the operation schemas as programs that effect changes to the database. Figure 2 shows the relationship in a small banking system. There is a state schema called *Bank* and four operation schemas that describe the processes of opening an account, depositing money, and so on.

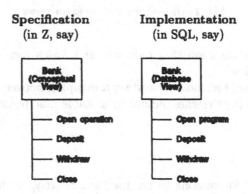

Fig. 2. Specification and implementation

The relational model [Cod70] is a data model that has the same origins as Z. A (database) relation may be viewed as a typed set. For example, a Depts(DeptNr, Name, Mgr) relation may only contain elements of a particular type. Set operations may only be applied to sets of the same type or scheme. Furthermore, the relational calculus, upon which SQL is based, is a modified form of the predicate calculus. It would seem that Z is a "natural" language for specifying relational database systems, and vice-versa, a relational database provides a straightforward target for a Z specification. But how is the link between the specification and its implementation established? Despite their common roots, Z and SQL are different languages. Z is a rich and varied language with an extensive vocabulary of operators, SQL is comparatively simple. This is not a coincidence. We judge a specification language by its expressive power. We look for *economy of expression* and operators that help us economise. Conversely, SQL was originally designed as an end-user query language. One approach is to *refine* the original specification. This requires that we rewrite that specification, still in Z, but using data structures that are close to those available in the target programming language. A typical refinement might involve the replacement of a sequence by an array.

Of course, the operations upon that sequence must be replaced by appropriate array operations. This process of simulation entails a number of proof obligations. These are designed to guarantee that the two specifications are equivalent. Once this has taken place, we may "convert" the concrete specification into a program. This paper uses the concrete data structure most commonly used for database systems – namely (database) relations.

The paper has the following structure: in Section 2, we present a small example which we specify and then implement. In Section 3, we discuss what it means to "implement" by introducing data refinement. In Section 4, some laws are given: these concern the use of relational calculus in data refinement proofs. In Section 5, we examine the **insert**, **update** and **delete** statements of SQL. Each of these is given an equivalence in terms of the relational calculus. The particular syntax used for the relational calculus is adapted from the Z notation. Finally, in Section 6, some worked examples that use these techniques and laws, in the data refinement of a simple database application, are given.

2 The Savings Model

2.1 User Views

Suppose we are required to develop a specification for a (very) small savings bank. The manager "sees" the bank and its relationships with customers in two ways:

1. How much they have saved:

    Bill   100
    Bob   20
    Sue   100

2. Where they live:

    Bill   Ashgrove
    Bob   Springhill
    Sue   Ashgrove

We can abstract these views:

    balView::={Person+Number}
    locView::={Person+Address}

Based on these view descriptions, we may decide that the following basic types should be used to represent the savings bank:

[*Address*] The set of addresses.
[*N*]   The set of integers $0, 1, 2, \ldots$
[*Person*] The set of *possible* account holders.

We can now further abstract the views into Z declarations:

$bal : \mathbf{P}(Person \times N)$
$loc : \mathbf{P}(Person \times Address)$

However, both these relationships are many-to-one. So we would declare them
as functions:

$bal : Person \nrightarrow N$
$loc : Person \nrightarrow Address$

2.2 The Savings Bank State Schema

Both user views have now been expressed formally, but in isolation. Now we
may form a global view – a single picture of the bank as a whole. To do this, we
combine the individual views and say how they relate to one another.

```
Savings
bal : Person ⇸ N
loc : Person ⇸ Address
────────────────
dom loc = dom bal
```

The state schema incorporates the two views and its predicate requires that at
all times:

$dom\ loc = dom\ bal$
Everybody with a balance has an address, and vice-versa.

Things that can be said about a state schema are that:

- It is as if the information system, at the user-level, is just one big record –
  with two fields or components, *bal* and *loc*.
- Each component provides one particular *view* or *observation*.
- Good design suggests that we look for components that are as unconnected
  as possible, that is, have as little overlap as possible.
- It is an expression of the user's concept of the application domain. It is the
  user's database.
- It will be replaced by an *equivalent* relational database.

2.3 The Database Schema

A relational database is a collection of relations, each of which is restricted in its
structure. A relation is a set of composite objects or **records**. The components
of each record are limited to simple atomic values, giving effect to the "flat-file"
or tabular appearance associated with database relations.

The information on any given account may be kept in a record of the following
type:

```
AccountRecord
Name : Person
Bal : N
Loc : Address
```

The savings bank may be captured by a database consisting of just one relation:

```
┌─SavingsDB─────────────────────────────────────────────────
│  Accounts : P AccountRecord
│ ─────────────────────────────────────────────────────────
│  #{A : Accounts • A.Name} = # Accounts
└───────────────────────────────────────────────────────────
```

The account *Name* attribute is the relation *key*. The number of account records (*#Accounts*) equals the number of account names, that is, every account has a different name. Relation keys are used to enforce **functional dependencies**.

The language most commonly used to implement database systems is SQL which is based on the **relational calculus**, an amalgam of the predicate calculus and set comprehension. If we use the Z notation for set comprehension, then we can say that the general form of an expression in the relational calculus is:

{*Declaration* | *Predicate* • *Term*}

Where:

- The *Declaration* allows us to introduce the sets, in this case database relations, used as a basis for the query.
- The optional *Predicate* allows us to express constraints that elements of these sets must satisfy. In relational calculus, the elements are tuples of some kind. If no predicate is supplied then all tuples are selected.
- The optional *Term* allows us to identify the exact nature of the new set that is to be formed. If no term is supplied then tuples from the base relations are to be used. Any term specified must conform to the requirements of the relational model, that is, it must have simple-valued components.

A simple example is:

{*A* : *Accounts* | *A.Bal* < 1000 • *A.Name*}

This statement returns the names of those account holders with a balance less than $1 000. The tuple variable *A* ranges over the entire relation, taking in turn the value of each account record. We can see that SQL is based on relational calculus if we "read" the expression above as:

```
From    Accounts A
Where   A.Bal < 1000
Select  A.Name
```

2.4 The State Schema and the Database

The state schema and the database schema are simply two different pictures of the same situation. The database is a machine-oriented realisation of the state schema. We can show this relationship between the database and the state schema by means of a **mapping** schema.

```
┌─ Mapping ─────────────────────────────────────────────
  Savings
  SavingsDB
 ├────────────────────────────────────────────────────
  bal = {A : Accounts • (A.Name, A.Bal)}
  loc = {A : Accounts • (A.Name, A.Loc)}
```

This schema constructs and defines the components of *Savings* in terms of *SavingsDB* components. From these equations, we can see that:

$$dom\ loc = \{A : Accounts \bullet A.Name\}$$
$$= dom\ bal$$

Thus the predicate *dom loc = dom bal* of the *Bank* schema is enforced in the database by representing them as the same column of the *Accounts* relation. The functional nature of *bal* and *loc* is guaranteed because the *Name* component is the relation key.

2.5 Opening an Account

A static database is of little interest to a bank. We need to be able to handle change. What happens when an account is to opened? What effect does that operation have upon the database?

The *Open* Operation This schema describes what happens when a new customer *p?* who lives at *home?* opens an account with a deposit of *d?* dollars.

```
┌─ Open ────────────────────────────────────────────────
  ΔSavings
  p? : Person
  home? : Address
  d? : N
 ├────────────────────────────────────────────────────
  p? ∉ dom bal
  bal' = bal ∪ {(p?, d?)}
  loc' = loc ∪ {(p?, home?)}
```

The person must not already have an account. Entries for the new customer's balance and address are made.

The *OpenEXE* Operation Now, *without altering the user interface*, we will redo this operation, but this time the savings bank will be represented by the database. This is what a programmer does when implementing a specification. The program that implements the *Open* operation will involve two segments of SQL:

1. It will check that the person does not already have an account:

```
p? not in (Select Name
            From    Accounts)
```

2. It will insert the new customer:

```
Insert
Into  Accounts
Values(p?,d?,home?)
```

To compare this program with the original specification, we (1) view the situation as a relational database, and (2) ensure that the language used is restricted to the relational calculus. That language is as close as we can get, in Z, to SQL.

$$\begin{array}{l}
\underline{\quad OpenEXE \quad} \\
\Delta SavingsDB \\
p? : Person \\
home? : Address \\
d? : N \\
\hline
p? \notin \{a : Accounts \bullet a.Name\} \\
Accounts' = Accounts \cup \{AccountRecord(p?, d?, home?)\}
\end{array}$$

The expression:

$$AccountRecord(p?, d?, home?)$$

constructs a new account record from the tuple $(p?, d?, home?)$

3 Refinement

3.1 Three Obligations

So now we have two versions of the operation:

- An *Open* operation that modifies the situation as represented by the abstract *Savings* state.
- An *OpenEXE* that modifies the situation as represented by the concrete *SavingsDB* state.

How do we know that the program, as described by *OpenEXE*, is correctly implemented? How do we verify that it is a valid **refinement** of the original? There are three proof obligations [Spi89]. All of these assume that the abstract and the concrete states are related in the way specified by the mapping schema.

I. Applicability:
If the pre-conditions of the abstract operation are met and then the concrete operation should go ahead, that is, its pre-conditions should also be satisfied.

$$pre \; Op \wedge Mapping \vdash pre \; OpEXE$$

If we assume that (1) the preconditions for the abstract operation (*pre Op*) are satisfied, and (2) that the mapping is in place, then we must show ($\vdash$) that the preconditions for the concrete operation (*pre OpEXE*) are satisfied.

II. Correctness:

Once we have demonstrated that the concrete version of the operation is still applicable, then we must show that, if the concrete operation (ie, program) goes ahead, then its effect is exactly what the users would have expected.

$$pre\ Op \land Mapping \land OpEXE \land Mapping' \vdash Op$$

If the pre-conditions are met and the concrete operation is satisfied, then the abstract operation should also be satisfied.

III. The Initial State:

The third condition requires that we establish that the initial state of the database satisfies the initial requirements of the abstract state:

$$SituationDBinit \land Mapping \vdash SituationInit$$

The applicability and correctness criteria must be demonstrated for every operation. The final requirement need only be demonstrated once.

3.2 *Open* vs *OpenEXE*

In this section we will examine the applicability and correctness requirements within the context of the savings bank situation and the operation of opening an account.

Applicability: Here, we must establish that the precondition of *Open* leads to or implies the precondition of *OpenEXE*.

$$p? \notin dom\ bal \dotfill \text{[pre Open}$$
$$\Rightarrow p? \notin \{A : Accounts \bullet A.Name\} \dotfill \text{[Mapping}$$
QED

The justification comes from the *Mapping* schema (and its consequences as discussed in Section 2.4). **Correctness:** Here, we must establish that, given the changes to the database described in *OpenEXE* we can establish the postconditions of *Open*. There are two conditions we must establish:

$$bal' = bal \cup \{(p?, d?)\}$$
$$loc' = loc \cup \{(p?, home?)\}$$

We will look at the second of these.

$$loc'$$
$$= \{A : Accounts' \bullet (A.Name, A.Loc)\} \dotfill \text{[Mapping'}$$
$$= \{A : (Accounts$$
$$\cup \{AccountRecord(p?, d?, home?)\})$$
$$\bullet (A.Name, A.Loc)\} \dotfill \text{[OpenEXE}$$

$$= \{A : Accounts \bullet (A.Name, A.Loc)\}$$
$$\cup\{A : \{AccountRecord(p?, d?, home?)\}$$
$$\bullet (A.Name, A.Loc)\} \dots\dots\dots\dots\dots\dots\dots\dots\dots\dots\dots\dots\dots\dots\dots\dots\dots\dots\dots [S3$$
$$= loc \dots\dots\dots\dots\dots\dots\dots\dots\dots\dots\dots\dots\dots\dots\dots\dots\dots\dots\dots\dots\dots\dots\dots\dots [Mapping$$
$$\cup\{(p?, home?)\} \dots\dots\dots\dots\dots\dots\dots\dots\dots\dots\dots\dots\dots\dots\dots\dots\dots\dots [R3$$
QED

The justifications S3 and R3 are described in Section 4.

4 Refinement Laws

Data refinement is the process of ensuring that operations on some relatively concrete data structures are equivalent to other operations on more abstract data structures. In this paper, the concrete data structures are the relations of a relational database, and the abstract data structures are those used in some specification written in the Z notation. In this section, we look at three collections of laws that may be useful in developing refinement proofs.

4.1 Mapping Laws

We begin by describing, in schematic form, how a situation expressed in a state schema *Situation* is to be mapped to a relational database expressed in a schema *SituationDB*. Suppose we have a state schema:

```
┌─ Situation ─────────────────────────────────
│  f : KT ↦ AT
│
│  ⋮
└─────────────────────────────────────────────
```

In particular, it has an observation f in the form of a function. Suppose also that the corresponding database has, as one of its record types, the following:

```
┌─ RT ────────────────────────────────────────
│  Key : KT
│  A : AT
│
│  ⋮
└─────────────────────────────────────────────
```

The database includes a relation *Rel*:

```
┌─ SituationDB ───────────────────────────────
│  Rel : P RT
│
│  ⋮
│─────────────────────────────────────────────
│  #{r : Rel • r.Key} = #Rel
│
│  ⋮
└─────────────────────────────────────────────
```

The attribute *Key* is the key of the relation.

The observation f in the state schema is represented, according to the mapping schema, in the following way:

Mapping
Situation
SituationDB

$f = \{r : Rel \bullet (r.Key, r.A)\}$

$\vdots$

If we project the *Key* and *A* attributes from the concrete relation *Rel* then we form the abstract observation f. Here are some laws arising from this mapping. The symbol t_K represents a term of type KT, that is, of the same type as the relation key; the symbol t_A is a term of type AT.

M1 – an equivalence:

$$\frac{t_K \in dom\, f \wedge t_A = f(t_K)}{\exists\, r : Rel \bullet r.Key = t_K \wedge t_A = r.A}$$

M2 – an equality:

$$\frac{r : Rel \mid r.Key = t_K}{r.A = f(t_K)}$$

M3 – an equality:

$$\frac{t_K \in dom\, f}{\{r : Rel \mid r.Key = t_K \bullet (r.Key, r.A)\} = \{(t_K, f(t_K))\}}$$

M4 – an equality:

$$\frac{t_A \in ran\, f}{\{r : Rel \mid r.A = t_A \bullet r.Key\} = \{x : dom\, f \mid f(x) = t_A\}}$$

4.2 Set Laws

These laws allow us to simplify a set comprehension expression that, in its declaration, contains either (1) a set operation such as set union, or (2) further set comprehension. The first group of laws involve set operations:

S1 Distribution:
$$\{x : (A \cup B)\} = \{x : A\} \cup \{x : B\}$$
S2 Distribution with predicate.
$$\{x : (A \cup B) \mid P\} = \{x : A \mid P\} \cup \{x : B \mid P\}$$
S3 Distribution with term.

$$\{x : (A \cup B) \bullet t\} = \{x : A \bullet t\} \cup \{x : B \bullet t\}$$

Similar rules or equations will apply with set intersection and set subtraction instead of set union.

The second group apply when the declaration itself involves set comprehension:

S4 Simplification:
$$\{x : \{a : A\} \bullet t\} = \{x : A \bullet t\}$$
S5 When there is a predicate involved:
$$\{x : \{a : A \mid P\} \bullet t\}$$
$$= \{x : A \mid P[x/a] \bullet t\}$$
$$= \{a : A \mid P \bullet t[a/x]\}$$

4.3 Record Laws

These laws simplify expressions involving records. Suppose we have a schema or record type:

┌─ *RT* ─────────────────────────────
│ $a : A$
│ $b : B$
│ $c : C$
└─────────────────────────────────────

We are also given:

$$R : RT$$
$$Rel : \mathbb{P}\, RT$$

and that t_A, t_B and t_C are terms of types A, B and C respectively. Suppose also that we assume that we can use the type name as a constructor. The expression $RT(t_A, t_B, t_C)$ creates an object of type RT with its a component equal to t_A, its b component set to t_B and so on.

R1 Projection:
$$RT(t_A, t_B, t_C).c$$
$$= [a : A;\ b : B;\ c : C](t_A, t_B, t_C).c$$
$$= t_C$$
R2 In set comprehension:
$$\{x : \{R\} \bullet x.a\} = \{R.a\}$$
$$\{x : \{R\} \bullet (x.a, x.b)\} = \{(R.a.R.b)\}$$
R3 Set comprehension over a constructed record:
$$\{x : \{RT(t_A, t_B, t_C)\} \bullet x.a\} = \{t_A\}$$
$$\{x : \{RT(t_A, t_B, t_C)\} \bullet (x.a, x.b)\} = \{(t_A, t_B)\}$$
R4 Constructing "after" versions of a relation:
$$\{r : Rel;\ r' : RT \mid r'.a = t_A \wedge r'.b = t_B \wedge r'.c = t_C \bullet r'\}$$
$$=$$
$$\{r : Rel \bullet RT(t_A, t_B, t_C)\}$$

R5 Simplifying set comprehension:
$$\{x : \{r : Rel \bullet RT(t_A, t_B, t_C)\} \bullet (x.a, x.b)\}$$
$$=$$
$$\{r : Rel \bullet (t_A, t_B)\}$$

5 Mapping from Z to SQL

In this section, we examine the `insert`, `update` and `delete` statements of SQL. Each of these is given an equivalence in terms of the relational calculus.

5.1 A Sample Database

Suppose we have a personnel database with two tables, one for permanent members of staff and one for trainees.

Staff

Id	Loc	Salary
Alan	IS	100000
Dave	CS	50000
Kate	IS	60000
:	:	:

Trainees

Id	Loc	Level
Jill	CS	6
Bill	IS	7
Ian	IS	6
:	:	:

Let the `Id` attribute be the key of both these relations. We define each relation by first introducing a record type that matches the tuples of the relation. There are two corresponding record types required:

$$StaffRT \;\widehat{=}\; [Id : Person; \; Loc : Dept; \; Salary : N]$$
$$TraineeRT \;\widehat{=}\; [Id : Person; \; Loc : Dept; \; Level : N]$$

The database may be defined as follows:

```
┌─ PersonnelDB ──────────────────────────────
│ Staff : P StaffRT
│ Trainees : P TraineeRT
├─────────────────────────────────────────────
│ #{s : Staff • s.Id} = #Staff
│ #{t : Trainees • t.Id} = #Trainees
└─────────────────────────────────────────────
```

5.2 The Insert Statement

The purpose of this statement is to enable one or more rows to be added to a table (or view). There are two forms:

```
insert_statement:
    INSERT
    INTO table_name [(list_of_column_names)]
    VALUES (list_of_constants)

or  INSERT
    INTO table_name [(list_of_column_names)]
    select_statement
```

Example 1: Suppose the IS department hires a new staff member, Kevin, at a salary of \$45000. We might write:

```
Insert
Into   Staff
Values('Kevin','IS',45000)
```

In Z we could write:

$$Staff' = Staff \cup \{StaffRT(Kevin, IS, 45000)\}$$

In general terms, the insertion of a new row would be written as:

$$Rel' = Rel \cup \{RT(t_1, \ldots)\}$$

where RT is some record type, Rel is a set of records of that type, and $t_1, \ldots$ are terms consistent with the attributes of RT. The second form of the insert statement allows the insertion of rows extracted from other table(s) in the database, or even from the same table.

Example 2: Suppose we want to promote all trainees in the IS department who have reached level 7 or above. The starting salary is to be \$30000.

```
Insert
Into   Staff
Select Id, Loc, 30000
From   Trainees t
Where  t.Loc = 'IS'
  and  t.Level >= 7
```

In Z we could write:

$$
\begin{aligned}
Staff' = Staff \cup \\
\{t : Trainees \\
\mid t.Loc = IS \wedge t.Level \geq 7 \\
\bullet StaffRT(t.Id, t.Loc, 30000)\}
\end{aligned}
$$

The syntax of the insert statement may be rewritten as:

```
INSERT
INTO  Rel
SELECT SelectList
FROM  FromTable f
WHERE WhereCondition
```

In general terms, the insertion of several new rows could be written as:

$Rel' = Rel \cup$
    $\{f : FromTable$
         $| \ WhereCondition$
         $\bullet \ RT(SelectList)\}$

5.3   The Update Statement

This has the form:

```
update_statement:
    UPDATE table_name
    SET list_of_assignments
    [where_clause]
assignment:
    column_name = expression
```

Example 3: Suppose we were to give Kate a $1000 raise. We would update her row in the Staff table:

```
Update Staff s
Set    s.Salary = s.Salary + 1000
Where  s.Id = 'Kate'
```

Because the **where** clause equates the relation key with some value, this update will affect only one row in the table. We imagine that we are removing that row and replacing it with a modified version.

$\exists s : Staff;\ s' : StaffRT \bullet$
        $s.Id = Kate \wedge$
        $s'.Id = s.Id \wedge$
        $s'.Loc = s.Loc \wedge$
        $s'.Salary = s.Salary + 1000 \wedge$
        $Staff' = Staff - \{s\} \cup \{s'\}$

We establish the existence of a staff record s with an Id component equal to *Kate*. We then construct an "after" version of that record s' with the same Id and Loc components, but with a $Salary$ component equal to 1000 more than the original. The $Staff$ relation is reconstructed by taking away the old record and inserting the new. Generally, the **update** statement allows a set of rows to be updated. In set theoretical terms, the operation involves the subtraction of a set of rows, followed by the insertion of a modified version of that set. The set to be removed is determined by the **where** clause. the modifications are determined by the **set** clause.

Example 4: Suppose we wanted to give everybody in the IS department a $5000 raise. We would write:

```
Update Staff s
Set    s.Salary = s.Salary + 5000
Where  s.Loc = 'IS'
```

This is equivalent to:

$$\exists\, whereset : \mathbf{P}\ Staff;\ setset : \mathbf{P}\ StaffRT \bullet$$
$$whereset = \{s : Staff \mid s.Loc = IS\}$$
$$setset = \{s : whereset;\ s' : StaffRT \mid$$
$$s'.Id = s.Id \wedge$$
$$s'.Loc = s.Loc \wedge$$
$$s'.Salary = s.Salary + 5000$$
$$\bullet\ s'\} \wedge$$
$$Staff' = Staff - whereset \cup setset$$

The *whereset* is the set of staff records to be updated; the *setset* is the modified version of these records. The syntax of the **update** statement may be written as:

```
UPDATE Rel r
SET    SetCondition
[WHERE WhereCondition]
```

The SetCondition is a list of assignments, but these may be viewed as a condition relating the after-version of each attribute to its before-version. The corresponding Z template is:

$$\exists\, whereset : \mathbf{P}\ Rel;\ setset\ \mathbf{P}\ RT \bullet$$
$$whereset = \{r : Rel \mid \text{WhereCondition}\}$$
$$setset = \{r : whereset;\ r' : RT \mid$$
$$\text{SetCondition}$$
$$\bullet\ s'\} \wedge$$
$$Rel' = Rel - whereset \cup setset$$

5.4 The Delete Statement

This has the form:

```
delete_statement:
    DELETE
    FROM table_name
    [where_clause]
```

Example 5: Suppose that, having promoted certain trainees, we want to remove them from the Trainees table:

```
Delete
From  Trainees t
Where t.Loc = 'IS'
  and  t.Level >= 7
```

The corresponding Z would be:

$$Trainees' = Trainees$$
$$-\{t : Trainees$$
$$\mid t.Loc = IS \wedge t.Level \geq 7\}$$

The syntax of the `delete` statement may be rewritten as:

```
DELETE
FROM Rel r
[WHERE WhereCondition]
```

The corresponding Z template is:

$$Rel' = Rel - \{r : Rel \mid WhereCondition\}$$

6 Complete Examples

In this section, we specify some further bank operations and their implementations. Verifying these implementations makes use of the material introduced in Sections 4 and 5.

6.1 Depositing Some Money

The *Deposit* Operation This schema describes what happens when an existing customer p? makes a deposit of d? dollars.

$$
\begin{array}{l}
\hline
\quad Deposit \underline{\qquad\qquad\qquad\qquad\qquad\qquad\qquad\qquad} \\
\Delta Savings \\
p? : Person \\
d? : N \\
\hline
p? \in dom\ bal \\
d? > 0 \\
bal' = bal \oplus \{(p?, bal(p?) + d?)\} \\
loc' = loc \\
\hline
\end{array}
$$

The person must have an account and something in that account. If so, the new balance is calculated by adding the amount deposited to the current balance. The customer's address is not changed.

The *DepositEXE* Operation The SQL for this operation would be:

```
Update Accounts a
Set    a.Bal = a.Bal + d?
Where  a.Name = p?
```

```
┌─ DepositEXE ──────────────────────────────────────────────
│ ΔSavingsDBp? : Person
│ d? : N
├───────────────────────────────────────────────────────────
│ p? ∈ {a : Accounts • a.Name}
│ d? > 0
│ ∃ a : Accounts;  a' : AccountRecord •
│         a.Name = p? ∧
│         a'.Name = a.Name ∧
│         a'.Bal = a.Bal + d? ∧
│         a'.Loc = a.Loc ∧
│         Accounts' = Accounts − {a} ∪ {a'}
└───────────────────────────────────────────────────────────
```

The rules for rewriting an SQL **update** in Z terms are discussed in Section 5.3.

Deposit vs *DepositEXE* Correctness:

bal'
$= \{A : Accounts' • (A.Name, A.Bal)\}$ ....[Mapping'
$= \{A : (Accounts − \{a\} ∪ \{a'\}) • (A.Name, A.Bal)\}$
   .... [DepositEXE
$= \{A : Accounts • (A.Name, A.Bal)\}$
   $−\{A : \{a\} • (A.Name, A.Bal)\}$
   $∪\{A\{a'\} • (A.Name, A.Bal)\}$ .... [S3
$= bal$ ....[Mapping
   $−\{(a.Name, a.Bal)\}$ .... [R2
   $∪\{(a'.Name, a'.Bal)\}$ .... [R2
$= bal$
   $−\{(p?, a.Bal)\}$ .... [DepositEXE
   $∪\{(p?, a'.Bal)\}$ .... [DepositEXE
$= bal ⊕ \{(p?, a'.Bal)\}$ ....[⊕ property
$= bal ⊕ \{(p?, a.Bal + d?)\}$ .... [DepositEXE
$= bal ⊕ \{(p?, bal(p?) + d?)\}$ .... [M2
QED

6.2 Closing an Account

The *Close* Operation This schema describes what happens when customer
p? closes his or her account.

```
┌─ Close ───────────────────────────────────────────────────
│ ΔSavings
│ p? : Person
├───────────────────────────────────────────────────────────
│ p? ∈ dom bal
│ bal' = { p? } ◁ bal
│ loc' = { p? } ◁ loc
└───────────────────────────────────────────────────────────
```

The SQL would be:

```
Delete
From  Accounts a
Where a.Name = p?
```

$$\begin{array}{|l}
\underline{\quad CloseEXE \quad}\hfill\\
\Delta SavingsDB\\
p? : Person\\
\hline
p? \in \{a : Accounts \bullet a.Name\}\\
Accounts' = Accounts\\
\qquad\qquad -\{a : Accounts \mid a.Name = p?\}\\
\end{array}$$

Close vs **CloseEXE** Correctness:

bal'
$= \{A : Accounts' \bullet (A.Name, A.Bal)\}$ ....[Mapping'
$= \{A : (Accounts - \{a : Accounts \mid a.Name = p?\})$
$\quad \bullet (A.Name, A.Bal)\}$ ....[CloseEXE
$= \{A : Accounts \bullet (A.Name, A.Bal)\}$
$\quad -\{A : \{a : Accounts \mid a.Name = p?\}$
$\quad \bullet (A.Name, A.Bal)\}$ ....[S2
$= bal$ ....[Mapping
$\quad -\{A : Accounts \mid A.Name = p?$
$\quad \bullet (A.Name, A.Bal)\}$
....[R3
$= bal - \{(p?, bal(p?))\}$ ....[M3
$= \{p?\} \lhd bal$ ....[property of $\lhd$
QED

7   Summary and Conclusions

The creation of a database system is an activity that, *almost invariably*, involves the use of different kinds of notation at different stages of the process. Each notation will be chosen because it is particularly suitable for expressing or describing some part of the system and its domain. In using these notations, we may choose to represent different parts differently, or the same part in different ways. As an example of the latter, we may specify a calculation using summation (Σ) and "program" the same calculation by means of a loop. We take a multi-notation approach because it leads to a more comprehensible and maintainable information system.

In this paper, we have deliberately restricted ourselves to a specification language for conceptual descriptions, and an implementation language for implementation or computational purposes. However, even this restriction obliges

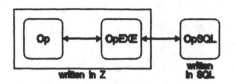

Fig. 3. Refining database systems

us to demonstrate the equivalence of the two representations. We have discussed how we might perform such a demonstration: see Figure 3.

The transition from an abstract description *Op* of some operation to the equivalent program *OpSQL* cannot be performed directly. As proposed by this paper, several stages are required:

1. Using the specification language, a relational schema must be defined.
2. Still using the specification language, the operation *Op* is restated as an operation *OpEXE* acting on the database. In this latter form, the language is restricted to the relational calculus.
3. The validity of the *OpEXE* is checked by against the applicability and correctness criteria. This validation is supported by the various laws discussed in Section 4.
4. Once validated, the operation *OpEXE* may be transliterated into a skeleton program *OpSQL* using the templates of Section 5 as a guide.

Acknowledgements

I would like to thank Athman Bouguettaya and Peter Bancroft for comments on earlier drafts of this paper.

References

[Cod70]   Codd E.F. (1970). "A Relational Model of Data for Large Shared Data Banks" *CACM*, Vol 13, No 6 (June 1970).

[DK+91]   Duke R., King P., Rose G. and Smith G. (1991). "The Object-Z Specification Language: Version 1" *Technical Report 91-1, Dept of Computer Science, University of Queensland.*

[Hay87]   Hayes I.J. (ed.) (1987). *Specification Case Studies.* London, England: Prentice Hall International.

[SBC92]   Stepney S., Barden R. and Cooper D. (eds.) (1992). *Object-orientation in Z.* London; New York: Springer-Verlag.

[Spi88]   Spivey J.M. (1988). *Understanding Z.* Cambridge, England: Cambridge University Press.

[Spi89]   Spivey J.M. (1989). *The Z Notation: A Reference Manual.* Hemel Hempstead, England: Prentice Hall International.

Applications I

Structuring a Z Specification to Provide a Formal Framework for Autonomous Agent Systems

Michael Luck[1] and Mark d'Inverno[2]

[1] Department of Computer Science, University of Warwick, Coventry, CV4 7AL, UK.
Email: mikeluck@dcs.warwick.ac.uk
[2] School of Computer Science, University of Westminster, London, W1M 8JS, UK.
Email: dinverm@westminster.ac.uk

Abstract. This paper describes a project which is using Z in the field of artificial intelligence (AI) to provide a defining framework for agency and autonomy. Specifically, the use of Z has provided a means for escaping from the terminological chaos surrounding agency and autonomy that is prevalent not just in the AI community, but also in other areas of computer science. We outline how we have developed a Z specification which serves as a framework that satisfies three distinct requirements. First, a framework should be defining in the sense that it must precisely and unambiguously provide meanings for the common concepts and terms. Second, it should be designed in such a way as to enable alternative models of particular classes of system to be explicitly described, compared and evaluated. Third, the framework should be sufficiently well-structured to provide a foundation for subsequent development of increasingly more refined concepts. The state based specification language Z is accessible to researchers from a variety of different backgrounds and allows us to provide a consistent unified formal account of an abstract agent system.

1 Introduction

1.1 Agency and Autonomy

The use of *agents* of many different kinds in a variety of fields of computer science and artificial intelligence is increasing rapidly. In artificial intelligence, the introduction of the notion of agents is partly due to the difficulties that have arisen when attempting to solve problems without regard to a real external environment or to the entity involved in that problem-solving process. Thus, though the solutions constructed to address these problems are significant, they are limited and inflexible in not coping well in real-world situations. In response, agents have been proposed as *situated* and *embodied* problem-solvers capable of functioning effectively in their environment.

It is now common for people to speak of *software agents*, *intelligent agents*, *interface agents*, *autonomous agents* and so on. The richness of the agent metaphor that leads to such different uses of the term is both a strength and a weakness. Its strength lies in the fact that it can be applied in very many different ways in many situations for different purposes. The weakness, however, is that the term *agent* is now used so frequently that there is no commonly accepted notion of what it is that constitutes an agent. For example, *agents* are often taken to be the same as *autonomous agents*, and the two terms are used interchangeably, without regard to their relevance or significance. The difference between these related, but distinct, notions is both important and useful in considering aspects of intelligence. Given the range of areas in which the notions and terms are applied, this lack of consensus over meaning is not surprising. As Shoham [7] points out, the number of diverse uses of the term *agent* are so many that it is almost meaningless without reference to a particular notion of agenthood.

It is now generally recognised that there is no agreement on what it is that makes something an agent, and it is standard, therefore, for many researchers to provide their own definition. In a recent collection of papers, for example, several different views emerge. Smith [8], takes an agent to be a "persistent software entity dedicated to a specific purpose." Selker [6] views agents as "computer programs that simulate a human relationship by doing something that another person could do for you." More loosely, Riecken [5] refers to "integrated reasoning processes" as agents. Others avoid the issue completely and leave the interpretation of their agents to the reader. In this paper, we report on a formal framework that we have constructed for *autonomy* and *agency* that attempts to bring together such disparate notions [4].

This section considers why we need a formal framework and what should be required of such formalisms. The second section introduces the key ideas contained in the framework and provides as overview of the specification which follows in the next section. Then we discuss the framework specification in relation to its usefulness in artificial intelligence and consider how it can be extended to incorporate architecture-specific ideas, illustrating this with the particular example of *knowledge interchange protocols*.

1.2 Formal Specification

In the current work, we have adopted the specification language Z [9] for two major reasons. First, it provides modularity and abstraction and is sufficiently expressive to allow a consistent, unified and structured account of a computer system and its associated operations. Such structured specifications enable the description of systems at different levels of abstraction, with system complexity being added at successively lower levels. Second, we view our enterprise as that of building programs. Z schemas are particularly suitable in squaring the demands of formal modelling with the need for implementation by allowing transition between specification and program. Thus our approach to formal specification is pragmatic — we need to be formal to be precise about the concepts we discuss, yet we want to remain directly connected to issues of implementation.

Furthermore, Z is gaining increasing acceptance as a tool within the artificial intelligence community (e.g. [3], [2]) and is therefore appropriate in terms of standards and dissemination capabilities.

Our concern has been to develop well-defined formal concepts that can be used both as the basis of an implementation, and also as a precise but general framework for further research.

1.3 Requirements of Formal Frameworks

We argue that a *formal framework* must satisfy three distinct requirements:

1. A formal framework must precisely and unambiguously provide meanings for common concepts and terms and do so in a readable and understandable manner. The availability of readable explicit notations allows a movement from vague and conflicting informal understandings of a class of models towards a common conceptual framework. A common conceptual framework will exist if there is a generally held understanding of the salient features and issues involved in the relevant class of models.

2. A framework should enable alternative designs of particular models and systems to be explicitly presented, compared and evaluated. It must provide a description of the common abstractions found within that class of models as well as a means of further refining these descriptions to detail particular models and systems.

3. The framework should be sufficiently well-structured to provide a foundation for subsequent development of new and increasingly more refined concepts. In other words, a practitioner should be able to choose the level of abstraction suitable for their purpose.

The use of abstraction renders prejudice about design unnecessary and enables a specification of a general system to be written. Z schema boxes are ideal for manipulation in the design process since by viewing the design process as a constraint of possible states, design strategies can be presented as further predicates in an abstract state schema.

2 A Framework for Agency and Autonomy: Overview

It was stated earlier that there is a lack of consensus over the meaning of the term agent and that there exist many diverse notions of agency. In this section, we introduce terms and concepts which are used to explicate our understanding of *agents* and *autonomous agents* and then developed into formal definitions.

Shoham [7] takes an *agent* to be any entity to which mental state can be ascribed. Such mental state consists of components such as beliefs, capabilities and commitments, but there is no unique correct selection of them. This is sensible, and we too do not demand that all agents necessarily have the same set of mental components. Indeed, we recognise the limitations associated with assuming an environment comprising homogeneous agents and consequently include in this discussion heterogeneous agents with varying capabilities. However, we do specify what is minimally *required* of an entity for it to be considered an agent in our framework.

We require a model that initially describes the environment and then, through increasingly detailed description, defines objects, agents and autonomous agents to provide an account of a general multiagent system. (This is what we mean by decreasing the level of abstraction in our descriptions of the world.) The definition of agency that follows is intended to subsume existing concepts as far as possible. In short, we propose a three-tiered hierarchy of entities comprising *objects*, *agents* and *autonomous agents*. The basic idea underlying this hierarchy is that all known entities are objects. Of this set of objects, some are agents, and of these agents, some are autonomous agents.

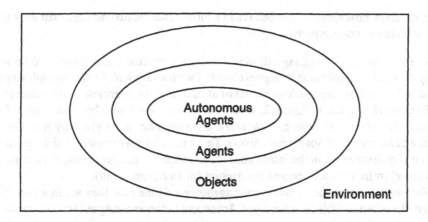

Fig. 1. The Entity Hierarchy

The specification had to be structured so that it reflected our view of the world as shown in the Venn diagram of Figure 1. It was required that the specification be built up in such a way that, starting from a basic description of an environment, each succeeding definition could be a refinement of the previously described entity. In this way, an object would be a refinement of the environment, an agent a refinement of an object, and an autonomous agent a refinement of an agent. Accordingly, the specification is thus structured into four parts.

Environment Our most abstract description of the world describes an environment simply as a collection of attributes.

Object Next, we cluster together some of the attributes in the environment and consider each such cluster as an object. Naturally, these objects are still collections of attributes, but we also give a more detailed description of these entities by describing their capabilities. The capabilities of an object are defined by a set of action primitives which can theoretically be performed by the object in some environment and, consequently, change the state of that environment.

Agent If we consider objects more closely, we can distinguish some objects which are serving some purpose or, equally, can be attributed some set of goals. This then becomes our definition of an agent, namely, an object with goals. With this increased level of detail in our description, we can define the greater functionality of agents over objects.

Autonomous Agent Further refinement of this description enables us to distinguish a subclass of the previously defined class of agents which are those agents that are autonomous. These autonomous agents are self-motivated agents in the sense that they follow their own agendas as opposed to functioning under the control of another agent. We thus define an autonomous agent as an agent with motivations and, in turn, show how these agents behave in a more sophisticated manner than their non-autonomous counterparts.

The Z specification language allowed just such a structured specification to be written by describing a system at its highest level of abstraction with further complexity being added at each successive lower level of abstraction. An overview of the structure of our framework is given in Figure 2, where the arrows represent schema inclusion. Definitions of *environment*, *object*, *agent* and *autonomous agent* are given by Env, Object, Agent and AutonomousAgent respectively. This shows how we constructed the most detailed entity description in the framework (of an autonomous agent situated in some environment) from the least detailed description (of an environment).

For objects, agents and autonomous agents, we define how they act in an environment in the schemas ObjectAction, AgentAction and AutonomousAgentAction respectively. For agents and autonomous agents, we detail how they perceive in a given environment in AgentPerception and AutonomousAgentPerception. For objects, agents and autonomous agents we define their state when situated in an environment in ObjectState, AgentState and AutonomousAgentState.

The framework is specified below. Whenever a new concept is introduced, a textual definition is given before its formal specification.

3 The Framework Specification

3.1 Environment

The first primitive that we need to define is an attribute. Attributes are simply features of the world and are the only characteristics which are manifest. They need not be perceived by any particular entity, but must be potentially perceivable in an omniscient sense.

Definition: An *attribute* is a perceivable feature.

[ATTRIBUTE]

Definition: An *environment* is some collection of attributes.

```
┌─ Env ─────────────────────────────
│ Environment : ℙ ATTRIBUTE
└───────────────────────────────────
```

For the purposes of readability, we define a new type ENVIRONMENT as the power set of attributes.

ENVIRONMENT == ℙ ATTRIBUTE

3.2 Objects

At a basic level, an object can be defined in terms of its abilities and its attributes. An object, in this sense, is just a 'thing' or entity with no further defining characteristics. This provides us with the basic building block to develop our notion of agency.

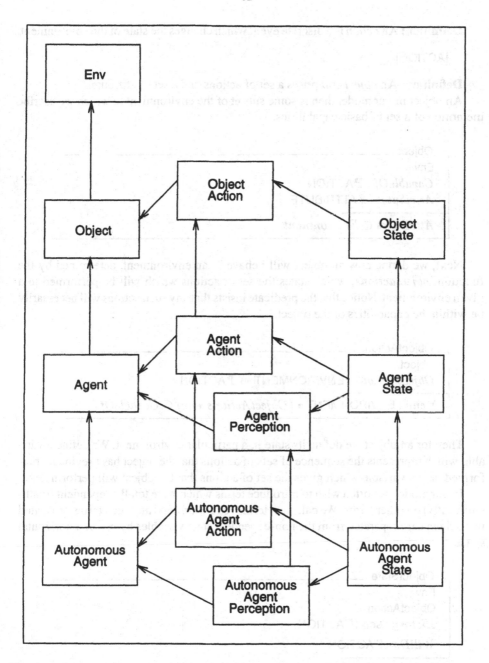

Fig. 2. Bottom-up View of the Use of Schema Inclusion in the Framework for Agency and Autonomy

Definition: An *action* is a discrete event which changes the state of the environment.

[ACTION]

Definition: An *object* comprises a set of actions and a set of attributes.

An object in our model then is some subset of the environment to which we ascribe the notion of a set of basic capabilities.

```
┌─Object─────────────────────────────────────────────
│ Env
│ CapableOf : ℙ ACTION
│ Attributes : ℙ ATTRIBUTE
│────────────────────────────────────────────────────
│ Attributes ⊂ Environment
└────────────────────────────────────────────────────
```

Next, we define how an object will behave in an environment, determined by the function, *objectactions*, which states the set of actions which will be performed in a given environment. Notice that the predicate insists that any such actions will necessarily be within the capabilities of the object.

```
┌─ObjectAction──────────────────────────────────────
│ Object
│ ObjectActions : ENVIRONMENT → ℙ ACTION
│────────────────────────────────────────────────────
│ ∀ env : ENVIRONMENT • (ObjectActions env) ⊆ CapableOf
└────────────────────────────────────────────────────
```

Then for an object we define its state in a particular environment. We define a variable which represents the sequence of sets of actions that the object has previously performed, and a variable which gives the set of actions that the object will perform next.

In our model, we often wish to introduce terms which were totally dependent (mathematically) on other terms. We call such terms *redundant* and any such terms presented in a schema are separated from the non-dependent state variables by use of a horizontal space.

```
┌─ObjectState───────────────────────────────────────
│ Env
│ ObjectAction
│ History : seq (ℙ ACTION)
│ WillDo : ℙ ACTION
│────────────────────────────────────────────────────
│ ∀ as : ran History • as ⊆ CapableOf
│ WillDo = ObjectActions Environment
└────────────────────────────────────────────────────
```

Lastly, we show how an object actually behaves in an environment. When an object performs some set of actions, neither the basic capabilities, nor the action selection function will be changed.

```
┌─ΔObjectState────────────────────────────────────────
│ ObjectState
│ ObjectState′
├──────────────────────────────────────────────────────
│ CapableOf′ = CapableOf
│ ObjectActions′ = ObjectActions
└──────────────────────────────────────────────────────
```

An *interaction* is simply that which happens when actions are performed in an environment. The effects of an interaction on the environment are determined by applying the *EffectInteraction* function in the axiom definition below to the current environment and the actions taken. An axiom definition was chosen since in our model, all actions will result in the same change to an environment whether taken by an object, agent or autonomous agent.

$$EffectInteraction : \text{ENVIRONMENT} \rightarrow \mathbb{P}\,\text{ACTION} \nrightarrow \text{ENVIRONMENT}$$

When an object interacts in its environment and performs an action, both the state of the object and the environment change. The history variable is updated by concatenating the current set of actions to the end of the sequence, and the new environment is given by applying *EffectInteraction* to the old environment and the current actions. The last line states that the actions which follow the current actions are found by applying *ObjectActions* to the new environment.

```
┌─ObjectEnvInteract────────────────────────────────────
│ ΔObjectState
├──────────────────────────────────────────────────────
│ History′ = History ⌢ ⟨WillDo⟩
│ Environment′ = EffectInteraction Environment WillDo
│ WillDo′ = ObjectActions Environment′
└──────────────────────────────────────────────────────
```

3.3 Agents

We proceed to define agents in much the same way as for objects. There are many dictionary definitions for an agent. A recent paper by Wooldridge and Jennings [10] quotes *The Concise Oxford Dictionary* [11] definition of an agent as "one who, or that which, exerts power or produces an effect." However, they omitted the second sense of agent which is given as "one who acts for another ...". This is important, for it is not the acting alone that defines agency, but the acting for *someone or something* that is the defining characteristic.

In this sense, agents are just objects with certain dispositions. They may always be agents, or they may revert to being objects in certain circumstances. An object is an agent if it serves a useful purpose either to a different agent, or to itself, in which case the agent is *autonomous*. This latter case is discussed further later. Specifically, an agent is something that 'adopts' or satisfies a goal or set of goals (often of another). Thus, if I want to store coffee in a cup, then the cup is my agent for storing coffee. It has been *ascribed* or has *adopted* my goal to have the coffee stored. It is, therefore, the goals of an agent which are its defining characteristics. We take a traditional AI view of goals as describable environmental states.

Definition: A *goal* is a state of affairs to be achieved in the environment.

[GOAL]

Definition: An *agent* is an instantiation of an object together with an associated goal or set of goals.

The schema for an agent is simply that of an object but with the addition of goals.

```
┌─Agent─────────────────────────────────────────────
│ Object
│ Goals : ℙ GOAL
│ ──────────────────────────────────────────────────
│ Goals ≠ { }
└───────────────────────────────────────────────────
```

Thus an agent has, or is *ascribed*, a set of goals which it retains over any instantiation (or lifetime). The same object may give rise to different instantiations of agents. An agent is instantiated from an object in response to another agent. Thus agency is *transient*, and an object may become an agent at some point in time, but may subsequently revert to being an object.

We now introduce perception. An agent in an environment may have a set of percepts available. These are the possible attributes that an agent could perceive, subject to its capabilities and current state. However, due to limited resources, an agent will not normally be able to perceive all those attributes possible, and bases action on a subset, which we call the *actual* percepts of an agent. Some agents will not be able to perceive at all. In the case of a cup, for example, the set of possible percepts will be empty and consequently the set of actual percepts will also be empty. The robot, however, may have several sensors which allow it to perceive. Thus it is not a requirement of an agent that it is able to perceive.

In our model we wish to distinguish between sets of attributes which represent a mental model of the world and those which represent the 'actual' environment. For clarity of exposition, we choose to define a type VIEW to be the perception of an Environment by an agent. This has an equivalent type to that of ENVIRONMENT, but now we can distinguish between physical and mental components of type ℙ ATTRIBUTE.

VIEW == ℙ ATTRIBUTE

It is also important to note that it is only meaningful in our model to consider perceptual abilities in the context of goals. Thus when considering objects which have no goals, perceptual abilities are not relevant. Objects respond directly to their environments and make no use of percepts even if they are available. We say that perceptual capabilities are *inert* in the context of objects.

In the schema for agent perception, AgentPerception, we add further detail to the definition of agency and so include the schema Agent. An agent has: a set of perceiving actions, *PerceivingActions*, which are a subset of the capabilities of an agent; a function, *CanPerceive*, which determines the attributes potentially available to an agent through its perception capabilities; and a function, *WillPerceive*, which describes those attributes which are actually perceived by an agent.

```
┌─ AgentPerception ─────────────────────────────────────────
│ Agent
│ PerceivingActions : ℙ ACTION
│ CanPerceive : ENVIRONMENT → ℙ ACTION ⇸ VIEW
│ WillPerceive : ℙ GOAL → ENVIRONMENT → VIEW
├───────────────────────────────────────────────────────────
│ PerceivingActions ⊆ CapableOf
│ ∀ env : ENVIRONMENT; as : ℙ ACTION •
│         as ∈ dom (CanPerceive env) ⇒ as = PerceivingActions
│ dom WillPerceive = {Goals}
└───────────────────────────────────────────────────────────
```

Directly corresponding to the goal or goals of an agent is an action-selection function, dependent on the goals, current environment and the actual perceptions. This is built up from the Agent and ObjectAction schemas.

```
┌─ AgentAction ─────────────────────────────────────────────
│ Agent
│ ObjectAction
│ AgentActions : ℙ GOAL → VIEW → ENVIRONMENT → ℙ ACTION
├───────────────────────────────────────────────────────────
│ ∀ gs : ℙ GOAL; env1, env2 : ENVIRONMENT •
│         (AgentActions gs env1 env2) ⊆ CapableOf
│ dom AgentActions = {Goals}
│ ObjectActions = AgentActions Goals Environment
└───────────────────────────────────────────────────────────
```

The state of an agent includes two variables which describe those percepts possible in the current environment and a subset of these which are the current (actual) percepts of the agent in the environment.

```
┌─ AgentState ──────────────────────────────────────────────
│ AgentPerception
│ AgentAction
│ ObjectState
│ PossiblePercepts : VIEW
│ ActualPercepts : VIEW
├───────────────────────────────────────────────────────────
│ ActualPercepts ⊆ PossiblePercepts
│ PossiblePercepts = CanPerceive Environment PerceivingActions
│ ActualPercepts = WillPerceive Goals PossiblePercepts
│ (PerceivingActions = { }) ⇒ (PossiblePercepts = { })
│ ObjectActions = AgentActions Goals ActualPercepts
└───────────────────────────────────────────────────────────
```

We now define which of the *AgentState* variables remain unchanged after a set of actions has been performed by that agent. If any of these variables ever did change, a different agent schema would have to be instantiated.

```
┌─ΔAgentState──────────────────────────────────────────
│ AgentState
│ AgentState′
├──────────────────────────────────────────────────────
│ CapableOf′ = CapableOf
│ Goals′ = Goals
│ PerceivingActions′ = PerceivingActions
│ CanPerceive′ = CanPerceive
│ WillPerceive′ = WillPerceive
│ AgentActions′ = AgentActions
└──────────────────────────────────────────────────────
```

The history and environment are altered in exactly the same way as previously described in the ObjectEnvInteract schema.

```
┌─AgentEnvInteract─────────────────────────────────────
│ ΔAgentState
│ ObjectEnvInteract
└──────────────────────────────────────────────────────
```

3.4 Autonomous Agents

The definition of agency that we have developed above relies upon the existence of other agents which provide goals that are adopted in order to instantiate an agent. In order to ground the chain of goal adoption, to escape what could be an infinite regress, and also to bring out the notion of *autonomy*, we introduce *motivation*.

Definition: A *motivation* is any desire or preference that can lead to the generation and adoption of goals and which affects the outcome of the reasoning or behavioural task intended to satisfy those goals.

[MOTIVATION]

Definition: An *autonomous agent* is an instantiation of an agent together with an associated set of motivations.

```
┌─AutonomousAgent──────────────────────────────────────
│ Agent
│ Motivations : ℙMOTIVATION
├──────────────────────────────────────────────────────
│ Motivations ≠ { }
└──────────────────────────────────────────────────────
```

The cup of the previous example cannot be considered autonomous because it cannot generate its own goals. By contrast, the robot is potentially autonomous in the sense that it may have a mechanism for internal goal generation, depending on its environment. Suppose the robot has motivations of achievement, hunger and self-preservation, where achievement is defined in terms of fixing tyres onto a car on a production line, hunger is defined in terms of maintaining power levels, and self-preservation is defined in terms of avoiding system breakdowns. In normal operation, the robot will generate goals to attach tyres to cars through a series of subgoals. If its power levels are low, however, it may replace the goal of attaching tyres with a newly-generated goal of recharging its batteries. A third possibility is that in satisfying its achievement motivation, it works for too long and is in danger of overheating. In this case, the robot can generate a goal of pausing for an appropriate period in order to avoid any damage to its components. Such a robot is autonomous because its goals are not imposed, but are generated in response to its environment.

The motivations, as well as the goals, of autonomous agents determine the way in which they perceive their environment. In the schema given below, the function, *Auto-WillPerceive*, is then a more complex version of an agent's *WillPerceive*, but they are related — and must be since an autonomous agent is still an agent — as shown in the schema. However, that which an autonomous agent is *capable* of perceiving at any time is independent of its motivations. Indeed, it will always be independent of goals and motivations and there is consequently no equivalent increase in functionality to *Can-Perceive*.

AutonomousAgentPerception
AutonomousAgent
AgentPerception
AutoWillPerceive : $\mathbb{P}$ MOTIVATION $\rightarrow$ $\mathbb{P}$ GOAL $\rightarrow$
                              ENVIRONMENT $\rightarrow$ VIEW

$WillPerceive = AutoWillPerceive\ Motivations$
dom $AutoWillPerceive = \{Motivations\}$

We build up our schemas as follows:

AutonomousAgentAction
AutonomousAgent
AgentAction
AutoActions : $\mathbb{P}$ MOTIVATION $\rightarrow$ $\mathbb{P}$ GOAL $\rightarrow$
                        VIEW $\rightarrow$ ENVIRONMENT $\rightarrow$ $\mathbb{P}$ ACTION

dom $AutoActions = \{Motivations\}$
$AgentActions = AutoActions\ Motivations$

```
┌─AutonomousAgentState──────────────────────
│ AutonomousAgentPerception
│ AutonomousAgentAction
│ AgentState
├───────────────────────────────────────────
│ WillPerceive = AutoWillPerceive Motivations
│ AgentActions = AutoActions Motivations
└───────────────────────────────────────────
```

Lastly, we specify the operation of an autonomous agent performing its next set of actions in its current environment. Notice that no explicit mention is made of any change in motivations, they may change in response to changes in the environment. If they do change, then the agent functions *WillPerceive* and *AgentActions* will also change. Further, motivations may generate new and different goals for the agent to pursue. In any of these cases, the characterizing features of an agent are in flux so that an autonomous agent can be regarded as a continually re-instantiated non-autonomous agent. In this sense, autonomous agents are permanent as opposed to transient non-autonomous agents.

```
┌─ΔAutonomousAgentState─────────────────────
│ AutonomousAgentState
│ AutonomousAgentState'
├───────────────────────────────────────────
│ CapableOf' = CapableOf
│ PerceivingActions' = PerceivingActions
│ CanPerceive' = CanPerceive
│ AutoWillPerceive' = AutoWillPerceive
│ AutoActions' = AutoActions
└───────────────────────────────────────────
```

A description of an autonomous agent acting in the environment is built from that of an agent acting in the environment.

```
┌─AutonomousAgentEnvInteract────────────────
│ ΔAutonomousAgentState
│ AgentEnvInteract
└───────────────────────────────────────────
```

4 Evaluating the Framework

We have formal definitions for agents and autonomous agents which are clear, precise and unambiguous, but which do not specify a prescribed internal architecture for agency and autonomy. This is exactly right, since it allows a variety of different architectural and design views to be accommodated within a single unifying structure. All that is required by our specification is a minimal adherence to features of, and relationships between, the entities described therein.

Thus we allow a cup to be viewed as an object or an agent depending on the manner in which it functions or is used. Similarly, we allow a robot to be viewed as an object, an agent or an autonomous agent depending on the nature of its control structures. We do not specify here how those control structures should function, but instead how the control is directed.

The framework is suitable for reasoning both *about* entities in the world, and *with* entities in the world. That is to say that an agent itself can also use the entity hierarchy as a basis for reasoning about the functionality of other agents and the likelihood, for example, that they may or may not be predisposed to help in the completion of certain tasks.

4.1 A Foundation for Further Work

The framework described and specified here is intended to stand by itself, but also to provide a base for further development of agent architectures in an incremental fashion through refinement and schema inclusion. In this context, we are currently extending the framework to incorporate notions and mechanisms of communication through *knowledge interchange protocols*.

Knowledge interchange protocols (KIPs) provide scripts for communication between autonomous agents, first introduced by Campbell and d'Inverno [1]. While they have been used in many implemented systems, they seem to have been used in conflicting ways, possibly because a formal description has never been constructed. Essentially, KIPs provide a well-defined structure within which both agents and autonomous agents may communicate freely. If the communicating agents can agree on the reason for the communication, their dialogue follows a pre-compiled script or protocol, governed by the agreed purpose of the particular dialogue. Thus agents' contributions to a dialogue will depend on what is regarded as the purpose or intention of an exchange of information.

Each protocol can be seen as a sequence of general statements by agents taking part in the information exchange. For a particular instantiation of a protocol in a communication, a sequence of well-defined slots will be filled by the particular information that an agent has to contribute. Thus the actual amount of information communicated is minimised. There is no need for an agent to say things that are redundant because all agents know which protocol is being used, and the protocol itself ensures that risks of loops and long meaningless interactions are avoided. Communication overheads are thus minimised.

Here we give a very brief description of the specification of these protocols in relation to the framework described above.

In order to define protocols, we must show how actions can be structured and must therefore model the world at the *object* level. First, we describe an object with the ability to communicate in that some subset of its capabilities are potentially *communicating actions*.

CommunicateObject
Object
$CommunicatingActions : \mathbb{P}\ ACTION$
$CommunicatingActions \subseteq CapableOf$

The set of protocols is given by:

[PROTOCOL]

A protocol object is defined as an object with communication capabilities, which has at its disposal a set of protocols which may be followed. Each protocol has associated with it a set of basic communicating actions.

```
┌─ ProtocolObject ─────────────────────────────────────────
│ CommunicateObject
│ Protocols : ℙ PROTOCOL
│ ActionsOfProtocol : PROTOCOL ⇸ ℙ ACTION
├───────────────────────────────────────────────────────────
│ dom ActionsOfProtocol = Protocols
│ ∀ p : Protocols • ActionsOfProtocol p ⊆ CommunicatingActions
└───────────────────────────────────────────────────────────
```

An autonomous agent with a set of protocols is then as follows:

```
┌─ ProtocolAutonomousAgent ────────────────────────────────
│ AutonomousAgent
│ ProtocolObject
└───────────────────────────────────────────────────────────
```

5 Conclusions

We have constructed a formal specification which identifies and characterises those entities that are called agents and autonomous agents. The work is not based on any existing classifications or notions because there is no consensus. Recent papers define agents in wildly different ways, if at all, and this makes it extremely difficult to be explicit about their nature and functionality. The taxonomy given here provides clear and precise definitions for objects, agents and autonomous agents that allow a better understanding of the functionality of different systems. It explicates those factors that are necessary for agency and autonomy, and is sufficiently abstract to cover the gamut of agents, both hardware and software, intelligent and unintelligent, and so on.

Z has enabled us to produce a specification that is generally accessible to researchers in AI, as well as practitioners of formal methods. Through the use of schema inclusion, we are able to describe our framework at the highest level of abstraction and then, by incrementally increasing the detail in the specification, we add system complexity at appropriate levels. Our use of Z does not restrict us to any particular mathematical model, but instead provides a general mathematical framework within which different models, and even particular systems, can be defined and contrasted.

In particular, the nature of Z allows us to extend the framework and to refine it further to include a more varied and more inclusive set of concepts. The particular case of *knowledge interchange protocols*, by which the original framework was extended through new schemas and schema inclusion, indicates just how we intend to proceed in this respect, and how appropriate Z is for this task. It enables a practitioner to choose the level of detail required to present a particular design, and further provides an environment in which the design itself can be presented in increasing levels of detail.

Acknowledgements

Thanks to John Campbell, Jennifer Goodwin, Paul Howells, Colin Myers, Mark Priestley and John Wolstencroft for comments on earlier versions of this paper. The specification contained in this paper was checked for correctness using the ƒUZZ package.

References

1. J. A. Campbell and M. d'Inverno. Knowledge interchange protocols. In Y. Demazeau and J. P. Muller, editors, *Decentralized Artificial Intelligence*. Elsevier North Holland, 1989.
2. I. D. Craig. The formal specification of ELEKTRA. Research Report RR261, Department of Computer Science, University of Warwick, 1994.
3. R. Goodwin. Formalizing properties of agents. Technical Report CMU-CS-93-159, Carnegie-Mellon University, 1993.
4. M. Luck and M. d'Inverno. A formal framework for agency and autonomy. In *Proceedings of the First International Conference on Multi-Agent Systems*, 1995.
5. D. Riecken. An architecture of integrated agents. *Communications of the ACM*, 37(7):107–116, 1994.
6. T. Selker. A teaching agent that learns. *Communications of the ACM*, 37(7):92–99, 1994.
7. Y. Shoham. Agent-oriented programming. *Artificial Intelligence*, 60:51–92, 1993.
8. D. C. Smith, A. Cypher, and J. Spohrer. Programming agents without a programming language. *Communications of the ACM*, 37(7):55–67, 1994.
9. J. M. Spivey. *The Z Notation*. Prentice Hall, Hemel Hempstead, 2nd edition, 1992.
10. M. J. Wooldridge and N. R. Jennings. Agent theories, architectures, and languages: A survey. In *Proceedings of the 1994 Workshop on Agent Theories, Architectures, and Languages*, 1994.
11. *The Concise Oxford Dictionary of Current English*. Oxford University Press, 7th edition, 1988.

On the Use of Formal Specifications in the Design and Simulation of Artificial Neural Networks

Patrícia Duarte de Lima Machado
Silvio Lemos Meira

Federal University of Pernambuco
Department of Informatics
PO Box 7851, Recife, PE, Brazil, 50.732-970
Tel.: +55 (81) 271 8430, Fax: +55 (81) 271 4925
E-mail: {pdlm, srlm}@di.ufpe.br

Abstract. This paper presents an experience on the use of object-oriented formal specifications in the process of software development for artificial neural networks. A formal specification of artificial neural networks using the *MooZ* language is presented. This specification of class hierarchies shows the gradual inclusion of neural network concepts such that new models or paradigms are easily incorporated by reusing previous definitions. *EASY*, a neural network simulation environment which was developed using the formal specification, is also described.
Keywords: Neural networks, formal specifications, object orientation, neural network simulation.

1 Introduction

Artificial neural networks [23, 19, 4] (*ANN*) are abstract models of known properties of information processing in biologic neural systems. They are composed of a set of elements, the artificial neurons, which simulate some features of the biologic neuron. Such elements are organized in a network of connections (an architecture). *ANN* learn by experience, generalize and abstract essential features of input patterns which has relevant and irrelevant data. *ANN* learning is done by the application of a systematic method which adjusts internal information in order to modify each neuron's individual behavior. This process is defined by a learning algorithm.

Computer simulation is the usual way adopted by *ANN* designers to specify their architectures and to simulate their behavior. The lack of a formalism which allows for formal proofs to verify properties of *ANN* is the major factor that contributes to simulation continues to be a dominant activity.

Some formal notations for specifying *ANN* have already been proposed [20, 18, 2]. They are used to specify *ANN* applications, where a fixed number of neurons and a fixed pattern of connections are defined. They cannot be used to formalize *ANN* concepts in a generic way. There are other informal notations based on programming languages, like *AXON* [4]. The specification methodology presented in this paper and first introduced in [10, 9] is an attempt to formalize *ANN* concepts and paradigms. This specification can be used in the process of *ANN* software development and in the formal verification of *ANN* properties. In particular, in this paper, we present the application of this specification in the development of an environment for *ANN* simulation.

Descriptions of neuron and *ANN* paradigms [23, 19, 4] are usually informal, in a natural language, with the transfer and learning functions specified using a mathematical notation. Many other aspects are specified informally, like the neurons' connection pattern, the effect of the execution of transfer and learning functions in the network, and so on. Meyer [14] presents seven classes of deficiency found in natural language specifications that make them unacceptable for rigorous software development. These deficiencies are redundancy, omission, overspecification, contradiction, ambiguity, forward reference and wishful thinking.

ANN software implementations lose in velocity when compared with their hardware counterparts. However, they are more easily modified and expanded. In general, they are constructed using a general purpose programming language like C, C++ or Smalltalk. When programming is the only phase of the whole software development process, it becomes an artisanal task, in which no formal technique nor method is used to avoid ambiguous, inconsistent and incomplete definitions, compromising the final product's reliability and quality. The possibility of the program being reused by other people is very small, due to the difficulties in understanding conventional programming language code.

The programming task is very time consuming. An *ANN* simulator is a software package with the purpose of reducing the time and effort involved in problem solution using *ANN* [15]. Normally, simulator users can specify *ANN* architectures using a description language which has a higher level of abstraction than conventional programming languages. Simulators may provide the processing power necessary to investigate complex paradigms and they may allow new ideas to be tested in an efficient way.

EASY (An [E]nvironment for [A]rtificial Neural [SY]stems Simulation) [7, 8] was designed to support all steps in a problem solving process by using *ANN*. In order to achieve this main objective, requirement analysis was performed to identify services, properties and restrictions of this kind of software and a model for the process of *ANN* simulation was proposed. Such is explained in detail in [6]. Formal specification and object orientation were used to develop *EASY*, in order to generate a reliable product, without ambiguity or inconsistence, and that would make easy the incorporation of new *ANN* paradigms.

The next sections present the *MooZ* language, the formal specification of families of artificial neurons and neural networks, some aspects of *EASY*'s design and, finally, some conclusions.

2 The *MooZ* Language

MooZ (Modular Object-Oriented Z) [12] is an object-oriented extension of the Z specification language [21] aimed at the specification of large software systems. A system specification in *MooZ* is represented by a class whose definition may depend on subsystem or subclass specifications. The complete language definition is given in [13].

The general structure of a *MooZ* class is shown in figure 1. The order of the clauses is mandatory and all of them are optional. The *givensets* clause introduces the given sets which are used to represent an object that does not require a model at a given abstraction level. Given sets can also be used to parameterize a class. A class definition containing the *givensets* clause is generic with its given sets as parameters. The hierarchy structures among classes is established in the *superclasses* clause. Public and private definitions to class clients are introduced in the *private or public* clause. The *constants* clause is used in the definition of global constants. The state components and invariant of the class are those specified in the superclasses and those introduced in the *state* clause. Operations for object initialization are specified in the *initialstates* clause. Class operations are specified in the *operations* clause.

```
Class ⟨Class_Name⟩
givensets ⟨type_name_list⟩
superclasses ⟨class_reference_list⟩
    ⟨auxiliary_definitions⟩
private ⟨definition_name_list⟩
or
public ⟨definition_name_list⟩
constants
⟨axiomatic_description_list⟩
    ⟨auxiliary_definitions⟩
state
⟨anonymous_schema⟩ or ⟨constraint⟩
    ⟨auxiliary_definitions⟩
initialstates
⟨schema⟩
    ⟨auxiliary_definitions⟩
operations
⟨definition_list⟩
EndClass ⟨Class_Name⟩
```

Fig. 1. General Form of a Class.

Comments may be included at any point of the class specification. Any Z paragraph (*given sets, schemas, axiomatic descriptions, abbreviations, free types, constraints*) may be used. *MooZ* also includes two new syntatic structures, the anonymous schema and the semantic operation. The anonymous schema is a nameless schema which can be used only to introduce state components and invariants. Semantic operations allow operations to be defined in a way similar to axiomatic descriptions. However, while the latter can introduce only constants and, in particular, functions, the former defines operations with state components as formal parameters.

3 The Formal Specification

In this section, we present a formal specification of *ANN*. The specification has two class hierarchies: neuron (figure 2) and neural network (figure 3). Classes at the higher levels of these hierarchies are abstract and their purpose is to introduce general concepts of neuron and *ANN* paradigms. These paradigms have similar features which can be encapsulated in abstract classes such that distinct properties may be gradually added, resulting in a hierarchy of classes where the most specialized represents the actual paradigms.

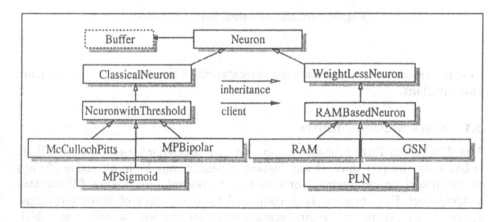

Fig. 2. Hierarchy of neuron classes.

A preliminary version of this specification is described in [10], where a different hierarchy is used and only the abstract classes like *NeuralNetwork*, *ClusteredNeuralNetwork* and *LayeredNeuralNetwork* are presented. A more complete version of that specification, which uses the class hierarchies presented in figures 2 and 3 and which includes neuron specifications (*Neuron*, *WeightedNeuron* and *WeightLessNeuron*) and further concepts of *ANN* can be found in [9].

In the works mentioned above, abstract classes only are specified. Here, we illustrate how classes that model real paradigms are added to the hierarchies shown in figures 2 and 3. The complete specification of all classes can be found in [6].

The next subsections present the specification of the classes *McCullochPitts* and *MLPBackpropagation*. The specification of some operations is omitted due

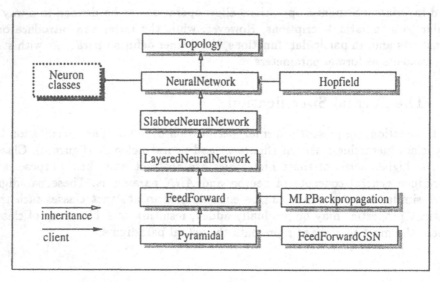

Fig. 3. Hierarchy of neural network classes.

to space reasons. We emphasize the presentation of the class' state components and invariant.

3.1 Class McCullochPitts

The *McCulloch-Pitts* neuron was proposed by *McCulloch and Pitts* in [11]. It is based on the fact that, at any instant, a neuron may be in an active (firing) or inactive state, producing 1 or 0 as output, respectively. Figure 4 illustrates a *McCulloch-Pitts* neuron. It is composed by a sequence of input terminals, $x_1, x_2, \ldots, x_n$, where each input terminal is associated with a weight w_j which represents the power of a synaptic connection. The activation level, *activation*, is computed by the weighted sum of input terminals and weight values. The output of the neuron is computed by an activation function F. The process of computing the neuron's activation level and output is called *neuron activation*.

Each neuron has a particular activation function. *McCulloch-Pitts* activation function is specified by:

$$o = \begin{cases} 1 \text{ iff } z = \sum_j x_j w_j \geq \Theta \\ 0 \text{ iff } z = \sum_j x_j w_j < \Theta \end{cases} \tag{1}$$

where x_j is the value of the input terminal j, w_j is the value of its weight, o is the neuron's output, *activation* is the neuron's activation level, and Θ_i is the neuron's threshold level.

In order to specify the *McCullochPitts* class, the essence of its superclasses (which are illustrated in figure 2) is briefly described here. The formal specification of the abstract classes *Neuron*, *WeightedNeuron* and *WeightLessNeuron* is

68

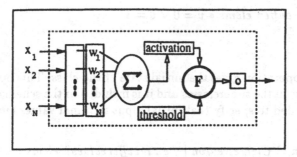

Fig. 4. *McCulloch-Pitts* neuron.

introduced in [9]. The *Neuron* class introduces the most generic features of an artificial neuron, including input terminals, activation level and output, along with basic operations to include and remove input terminals, to change input terminal values, and to compute the neuron's activation level and output. Although the functions to compute the neuron's activation level and output are not supplied by this class, we specify the effect of changing such values in the state of the class.

The *WeightedNeuron* class specifies the neuron models which has weights between connections with other neurons. Weights are similar to biologic neuron synapses. This class includes a state component to specify weights and its operations. The *WeightLessNeuron* class represents another basic type of neuron model which does not have weights.

The *NeuronwithThreshold* class introduces a state component to model the neuron's threshold. It also includes operations related to changing this component.

As can be perceived, in the inclusion of a concrete class which specifies a real neuron model like *McCulloch-Pitts*, it is only necessary to specify its activation function and maybe some constraints over input terminals. The specification of the class *McCullochPitts* is given below.

Class *McCullochPitts*

superclasses *NeuronwithThreshold*

private *ActivationFunction*

state

There is no need to specify any state components because they were specified in the superclasses. Only a constraint over the input values of the neuron is introduced. Input terminals are represented by the state component *inbuffer*. *McCulloch-Pitts* neurons receive either 0 or 1 at their input terminals.

⊢ ∀ v : inbuffer buffer elems • v = 0 ∨ v = 1

initialstates

CreateNeuron operation gives the initial state of the class. This operation is previously specified in the superclasses and it is included in the schema below. A new constraint is added to specify that initial input terminal values (*inbuffervalues?*) must be 0 or 1.

$$CreateNeuron \; \widehat{=} \; CreateNeuron \mid \forall v : inbuffervalues? \; elems \bullet v = 0 \lor v = 1$$

operations

ActivationFunction is an axiomatic description specifying the *McCulloch-Pitts* activation function, defined by equation 1 above.

$$ActivationFunction : \; \mathbb{R} \times \mathbb{R} \to \mathbb{R}$$
$$\forall a : \mathbb{R}; \; t : \mathbb{R} \bullet$$
$$a \geq t \Rightarrow ActivationFunction(a, t) = 1 \land$$
$$a < t \Rightarrow ActivationFunction(a, t) = 0$$

The *ActivateNeuron* operation specifies neuron activation. It associates the value of the activation function to the neuron's output. The value of this state component is given, in the superclasses, by the value of an auxiliary component *o*. The activation function receives the actual values of the activation level and threshold as parameters. In the superclass *WeightedNeuron*, the function to compute the activation level is included in the operation *ActivateNeuron*.

$$ActivateNeuron \; \widehat{=} \; ActivateNeuron \mid o = $$
$$ActivationFunction(activation', threshold)$$

$$ActivateNeuron \; \widehat{=} \; ActivateNeuron \setminus (o)$$

EndClass *McCullochPitts*.

3.2 Class *MLPBackpropagation*

Perceptrons are neural networks composed by *McCulloch-Pitts* neurons, which normally have just one layer of neurons *Single Layer Perceptrons*. Some constraints were proved over *perceptrons*, like their limitation to compute only a single class of problems where the set of output points can be geometrically separated. The exclusive-or function is an example of a problem that cannot be computed using a single layer perceptron.

Until the mid-80's, the only known learning algorithms were for *Single Layer Perceptrons*. Rumelhart, Hinton and Williams [17] proposed the *Backpropagation Error Correction Algorithm* for *Multi-Layer Perceptron* (MLP) neural networks. Despite its limitations, this algorithm largely expanded the class of problems solved using neural networks.

Backpropagation is a supervised learning algorithm where, during its application, each input pattern is presented to the neural network with the target outputs it should produce. The neural network computes the error between the outputs that it had produced and the target outputs and propagates this error to adjust the neuron' weights. The set of pairs (input pattern, target output) is called *training set*.

The architecture of an MLP network is composed by one or more layers of neurons, as can be seen in figure 5. The first, or input layer, buffers the received external impulses and distributes them to other neurons in the network. The following layers are composed by neurons of the type specified in the class *MPSigmoid*[1]. An extra layer of buffers is needed to receive, during the learning process, the target outputs.

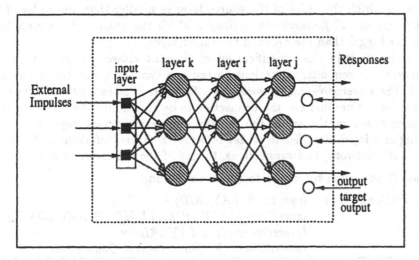

Fig. 5. *Multi Layer Perceptron* architecture with 3 layers.

Before presenting the specification of the class *MLPBackpropagation*, which introduces MLP neural networks using the *Backpropagation* algorithm, we give a brief description of its superclasses (which are illustrated in figure 3). The formal specification of the abstract classes *Topology* and *NeuralNetworks* is introduced in [9]. The *Topology* class specifies a topology of nodes and directed arcs. It has operations to include and remove nodes and arcs. Statically, a neural network may be seen as a topology. The *NeuralNetwork* class introduces general neural network concepts. In this class, we specify which target neuron input terminal is used in a connection. It is also necessary to associate each node identifier in the topology to an object of the *Neuron* class or its subclasses. This mapping

---

[1] MPSigmoid specifies a type of *McCulloch-Pitts* neuron which uses a sigmoid as its activation function.

is introduced by the state component *processing*. Operations to activate a set of neurons and to propagate a neuron' output values through the network of connections are included too.

The *SlabbedNeuralNetwork* class is created to specify neural networks whose neurons are grouped into clusters called slabs. Operations to create and remove slabs and to insert and remove neurons from a slab are defined in this class. When slabs are linearly ordered, they are called layers. The first layer is generally called input layer and the last layer is called output layer. The class *LayerNeuralNetwork* introduces the concept of layer.

A neural network with layers may have three types of connection: 1) *interlayer connection*, a connection between neurons in adjacent layers; 2) *intralayer connection*, between neurons in the same layer; and 3) *supralayer connection*, between neurons that do not belong to the same or adjacent layers.

Another classification defines two types of connection: 1) *feedforward connection*, in which the order of the source layer is smaller than the order of the target layer; and 2) *feedback connection*, in which the order of the source layer is equal or bigger than the order of the target layer.

In order to help the specification of different classes of layered neural networks, some semantic operations to test the type of a connection are specified in the *LayeredNeuralNetwork* class. They are illustrated here. All operations receive, as parameters, the connection to be tested and the following state components: *layers*, the set of the network's layer identifiers, *layerNeurons*, a mapping of a layer identifier to its set of neurons, *layersOrder*, the order of the layers in the network, and *connections*, the set of the network's connections.

isForwardConnection specifies a forward connection.

---

isForwardConnection : *layers* : ($\mathbb{P}$ *LAYERID*)$\times$
        layerNeurons : (*MAP LAYERID* ($\mathbb{P}$ *NEURONID*))$\times$
        layersOrder : (*Seq LAYERID*)$\times$
        connections : ($\mathbb{P}$ *Connection*) $\times$ *Connection* $\nrightarrow$ $\mathbb{B}$

$\forall c$: *Connection* | $c \in$ *connections* •
isForwardConnection(*layers*, *layerNeurons*, *layersOrder*, *connections*, *c*) $\Leftrightarrow$
 $\exists$ *slayer*, *tlayer* : *layers* |
  c.source $\in$ *layerNeurons*(*slayer*) $\wedge$ *c.target* $\in$ *layerNeurons*(*tlayer*) •
 $\exists s, t : \mathbb{N}$ | *layersOrder*(*s*) = *slayer* $\wedge$ *layersOrder*(*t*) = *tlayer* • $t > s$

---

isFeedBackConnection specifies a feedback connection.

---

isFeedBackConnection : *layers* : ($\mathbb{P}$ *LAYERID*)$\times$
        layerNeurons : (*MAP LAYERID*($\mathbb{P}$ *NEURONID*))$\times$
        layersOrder : (*Seq LAYERID*)$\times$
        connections : ($\mathbb{P}$ *Connection*) $\times$ *Connection* $\nrightarrow$ $\mathbb{B}$

$\forall c$: *Connection* •
isFeedBackConnection(*layers*, *layerNeurons*, *layersOrder*, *connections*, *c*) $\Leftrightarrow$
 $\exists$ *slayer*, *tlayer* : *layers* |
  c.source $\in$ *layerNeurons*(*slayer*) $\wedge$ *c.target* $\in$ *layerNeurons*(*tlayer*) •
 $\exists s, t : \mathbb{N}$ | *layersOrder*(*s*) = *slayer* $\wedge$ *layersOrder*(*t*) = *tlayer* • $t <= s$

isInterLayerConnection specifies an interlayer connection.

$$
\begin{array}{|l}
isInterLayerConnection:\quad layers:(\mathbb{P}\ LAYERID)\times \\
\qquad\qquad\qquad\qquad layerNeurons:(MAP\ LAYERID \\
\qquad\qquad\qquad\qquad\qquad\qquad (\mathbb{P}\,NEURONID))\times \\
\qquad\qquad\qquad\qquad layersOrder:(Seq\ LAYERID)\times \\
\qquad\qquad\qquad\qquad connections:(\mathbb{P}\ Connection)\times Connection \nrightarrow \mathbb{B} \\
\hline
\forall\,c:Connection\mid c\in connections\ \bullet \\
\ isInterLayerConnection(layers,layerNeurons,layersOrder,connections,c)\Leftrightarrow \\
\quad \exists\,slayer,tlayer:layers\mid \\
\qquad c.source\in layerNeurons(slayer)\wedge c.target\in layerNeurons(tlayer)\ \bullet \\
\quad \exists\,s,t:\mathbb{N}\mid layersOrder(s)=slayer\wedge layersOrder(t)=tlayer\ \bullet \\
\qquad s-t=1\vee t-s=1
\end{array}
$$

isIntraLayerConnection specifies an intralayer connection.

$$
\begin{array}{|l}
isIntraLayerConnection:\quad layers:(\mathbb{P}\ LAYERID)\times \\
\qquad\qquad\qquad\qquad layerNeurons:(MAP\ LAYERID \\
\qquad\qquad\qquad\qquad\qquad\qquad (\mathbb{P}\ NEURONID))\times \\
\qquad\qquad\qquad\qquad layersOrder:(Seq\ LAYERID)\times \\
\qquad\qquad\qquad\qquad connections:(\mathbb{P}\ Connection)\times Connection \nrightarrow \mathbb{B} \\
\hline
\forall\,c:Connection\mid c\in connections\ \bullet \\
\ isIntraLayerConnection(layers,layerNeurons,layersOrder,connections,c)\Leftrightarrow \\
\quad \exists\,slayer,tlayer:layers\mid \\
\qquad c.source\in layerNeurons(slayer)\wedge c.target\in layerNeurons(tlayer)\ \bullet \\
\quad slayer=tlayer
\end{array}
$$

isSupraLayerConnection specifies a supralayer connection.

$$
\begin{array}{|l}
isSupraLayerConnection:\quad layers:(\mathbb{P}\ LAYERID)\times \\
\qquad\qquad\qquad\qquad layerNeurons:(MAP\ LAYERID \\
\qquad\qquad\qquad\qquad\qquad\qquad (\mathbb{P}\ NEURONID))\times \\
\qquad\qquad\qquad\qquad layersOrder:(Seq\ LAYERID)\times \\
\qquad\qquad\qquad\qquad connections:(\mathbb{P}\ Connection)\times Connection \nrightarrow \mathbb{B} \\
\hline
\forall\,c:Connection\mid c\in connections\ \bullet \\
\ isSupraLayerConnection(layers,layerNeurons, \\
\qquad\qquad\qquad\qquad\qquad layersOrder,connections,c)\Leftrightarrow \\
\quad \neg\ isInterLayerConnection(layers,layerNeurons, \\
\qquad\qquad\qquad\qquad\qquad layersOrder,connections,c)\wedge \\
\quad \neg\ isIntraLayerConnection(layers,layerNeurons, \\
\qquad\qquad\qquad\qquad\qquad layersOrder,connections,c)
\end{array}
$$

Feedforward neural networks have *feedforward* connections only. *Feedback* neural networks can have both *feedback* or *feedforward* connections. So, it is not necessary to specify a new class for this kind of neural network. The MLPs are feedforward neural networks. In addition, their connections are of the *interlayer* type. The *FeedForward* class specification is briefly illustrated below.

Class *FeedForward*

superclasses *LayeredNeuralNetwork*

state

A new constraint is added to the state invariant to say that all connections must be classified as *feedforward*. The semantic operation *isForwardConnection* of the superclass is used to test the type of the connections of the state component *connections*, which was introduced in the superclass *Topology*.

$\vdash \forall c : connections \bullet$
$\qquad isForwardConnection(layers, layerNeurons, layersOrder, connections, c)$

operations

InsertConnections, a superclass operation to include new connections, is extended to specify that new connections must be classified as *feedforward*.

ForwardLayer operation specifies the synchronous activation of a layer's set of neurons. *GetNeuronInputs* apdates the input terminal values of the neurons and *ActivateLayerNeurons* activates the neurons.

$$ForwardLayer \mathbin{\hat{=}} GetNeuronInputs \mathbin{\text{\small 9}} ActivateLayerNeurons$$

EndClass *FeedForward*.

It can be noticed that all of the support needed to build an MLP neural network architecture is given by its superclasses. In the *MLPBackpropagation* class specification, we just introduce the definitions related to its learning algorithm. The specification of this class is given below.

Class *MLPBackpropagation*

superclasses *FeedForward*

private *WeightsAdjust, ThresholdAdjust, HiddenErrors*

state

Three new state components are included in the class state. The target outputs are specified by the *targetOutputs* component, which represents a mapping of neurons of the output layer (last layer of the network) to target output values. The actual values to adjust weights of the neurons are specified by the *weightAdjust* component, and the actual values to adjust threshold values of the neurons are specified by the *thresholdAdjust* component. Each neuron must have the same quantity of weights and values to adjust weights. With the exception of the neurons of the input layer, all others in the network are instances of the class *MPSigmoid*. The set of neuron identifiers of the input layer are specified by the state component *inputneurons* which is introduced in the superclass *NeuralNetwork*. Input neurons do not have weight or threshold adjust because they are instances of the class *InputNeuron* and their only function is to distribute external impulses to the network.

$targetOutputs : MAP \ NEURONID \ \mathbb{R}$
$weightsAdjust : MAP \ NEURONID \ (Seq \ \mathbb{R})$
$thresholdAdjust : MAP \ NEURONID \ \mathbb{R}$

$targetOutputs \ dom = layerNeurons(outputLayer)$
$weightsAdjust \ dom = neurons \setminus inputneurons$
$thresholdAdjust \ dom = neurons \setminus inputneurons$
$\forall \ n : weightsAdjust \ dom \ \bullet$
   $weightsAdjust(n) \ len = processing(n) \ inbuffer \ length$
$\forall \ c : connections \ \bullet$
   $isInterLayerConnection(layers, layerNeurons,$
                                $layersOrder, connections, c)$
$\forall \ n : neurons - inputneurons \ \bullet \ processing(n) \in MPSigmoid$

---

initialstates

The new components are mappings whose domain is defined directly or indirectly in terms of the component *neurons* (set of all neurons identifiers). So these components are implicitly initialized.

operations

We now illustrate some primitive operations used by the *Backpropagation* learning algorithm, where the following steps are required.

1. Select a pair of (input pattern, target output) and apply it to the neural network.

GetInNet is an operation of the superclass which specifies the reception of external impulses by some input neurons. These external impulses are then propagated to the neurons which are target of connections from the corresponding input neurons. In this class, this operation is redefined to specify that all input neurons will receive external impulses at the same time.

The *GetTargetNet* operation specifies the reception of new values of target outputs.

2. Compute neural network outputs, beginning at the next layer after the input until the output layer. When the outputs of a layer are computed, they become the inputs of the next. The operation *ForwardLayer* which is introduced in the superclass may be used to activate neurons of a layer, after changing their input terminal values to the actual outputs of the previous layer.

3. Compute the error between the output produced by the network and the target outputs. This error may be computed by equation 2:

$$\delta_j = o_j(d_j - o_j)(1 - o_j) \tag{2}$$

where d_j is neuron j's target output and o_j is neuron j's output.

OutputErrors is defined to compute the error between outputs produced by the network and target outputs based on equation 2. These values are received as arguments.

OutputErrors : *MAP NEURONID* $\mathbb{R} \times$
 targetOutputs : (*MAP NEURONID* $\mathbb{R}$) $\rightarrowtail$
 MAP NEURONID $\mathbb{R}$

$\forall$ *outputs* : *MAP NEURONID* $\mathbb{R}$ | *outputs dom* = *targetOutputs dom* •
 OutputErrors(*outputs*, *targetOutputs*) =
 λ j : *outputs dom* •
 outputs(j) * (*targetOutputs*(j) − *outputs*(j)) * (1 − *outputs*(j))

---

4. From the last to the first layer, adjust weights in order to minimize error. This step is subdivided into to smaller steps: adjust weights of the output layer and adjust weights of the hidden layers. Both steps use equation 3:

$$\Delta w_{ij} = \eta \delta_j o_i \tag{3}$$

which represents the variation to be applied in the weight associated to the connection between neuron j and i (which belongs to the preceding layer). η is the learning rate, o_i is neuron i's output e δ_j is neuron j's error. If j's layer is the output one, then the error is given by equation 2. Otherwise, the error it is given by another equation which is described in [6].

Rumelhart, Hinton e *Williams* [17] described the momentum method to minimize the time spent by *Backpropagation* algorithm to train an MLP neural network. Equation 3 is changed to include this method, giving:

$$\Delta w_{ij}(n + 1) = \eta \delta_j o_i + \alpha \Delta w_{ij}(n) \tag{4}$$

where η is the learning rate, n is the number of iteractions[2], and α is the momentum constant which represents the effect of the previous adjusts in the current adjusts.

The neuron's *threshold* or *bias* is considered as a weight of a connection whose source and target are the neuron to which it belongs. So, the threshold adjust value is also necessary and is computed by equation 5:

$$\Delta\Theta_i(n + 1) = \eta \delta_i o_i + \alpha \Delta\Theta_i(n) \tag{5}$$

---

[2] An iteraction is one pass forward to compute neural network outputs followed by one pass backward adjusting weights to minimize the error between outputs produced by the network and target outputs.

WeightsAdjust returns the weight adjust values for a neuron n, using equation 4. Its arguments are the neuron (n) identifier, the weight adjusts in the previous state, the learning rate, the error to be applied and the momentum constant, along with some state components.

$$
\begin{array}{|l}
WeightsAdjust: \quad NEURONID \times Seq\ \mathbb{R} \times \mathbb{R} \times \mathbb{R} \times \mathbb{R} \times \\
\qquad processing: (MAP\ NEURONID\ Sigmoid) \times \\
\qquad sourcesOfTarget: (MAP\ NEURONID\ (Seq\ NEURONID)) \\
\qquad \to Seq\ \mathbb{R} \\
\hline
\forall n: NEURONID;\ oldAdjusts: Seq\ \mathbb{R};\ r, e, m: \mathbb{R} \bullet \\
\quad WeightsAdjust(n, oldAdjusts, r, e, m, processing, sourcesOfTarget) = \\
\quad (\ \lambda\, i: sourcesOfTarget(n)\ dom \bullet \\
\qquad r * e * processing(sourcesOfTarget(n)(i))\ currentOut+ \\
\qquad m * oldAdjust(i)\ )\ squash
\end{array}
$$

ThresholdAdjust returns the adjust value for the threshold of a neuron n, following equation 5.

$$
\begin{array}{|l}
ThresholdAdjust: \quad NEURONID \times \mathbb{R} \times \mathbb{R} \times \mathbb{R} \times \mathbb{R} \times \\
\qquad processing: (MAP\ NEURONID\ Sigmoid) \to \mathbb{R} \\
\hline
\forall n: NEURONID;\ oldAdjust, r, e, m: \mathbb{R} \bullet \\
\quad ThresholdAdjust(n, oldAdjust, r, e, m, processing) = \\
\qquad r * e * processing(n)\ currentOut + m * oldAdjust
\end{array}
$$

AdjustOutputLayer is an operation to adjust weights of the output layer. In this operation, the output errors, the weights and threshold adjusts are computed using the previous semantic operations. The adjusts are applied to each neuron of the output layer. It is also specified that the other state components of the network are not changed.

HiddenErrors is a semantic operation to compute the error propagated to a hidden layer and *AdjustHiddenLayers* is an operation to adjust weights of a hidden layer.

5. Repeat steps 1 to 4 for each pair (input pattern, target output) of the training set until the error becomes as low as desired.

hasConverged verifies if the error between outputs produced by the network and target outputs is less than or equal to a global error which is received as argument.

$$
\begin{array}{|l}
hasConverged: \quad MAP\ NEURONID\ \mathbb{R} \times \\
\qquad targetOutputs: (MAP\ NEURONID\ \mathbb{R}) \times \mathbb{R} \nrightarrow \mathbb{B} \\
\hline
\forall outputs: MAP\ NEURONID\ \mathbb{R};\ globalError: \mathbb{R}\ | \\
\qquad outputs\ dom = targetOutputs\ dom \bullet \\
\quad hasConverged(outputs, targetOutputs, globalError) \Leftrightarrow \\
\qquad \forall n: outputs\ dom \bullet (outputs(n) - targetOutputs(n) \leq globalError) \vee \\
\qquad\qquad (targetOutputs(n) - outputs(n) \leq globalError)
\end{array}
$$

After the learning stage, the weights and threshold are frozen and the neural network enters the operation stage.

EndClass *MLPBackpropagation*.

4 *EASY*

The biggest problem of most simulators is that they do not support entire experiments. They do not cover all steps of the process of simulating a neural network. In addition, in most cases, they have unfriendly interfaces, and users spend, with a simulator, almost the same time as they would spend while writing their own programs. Such simulators are, in practice, more used by students for small experiments. As mentioned in the introduction, the research area of neural networks is largely dependent on the use of computer simulation. However, a part of research software and even software simulator packages available today [16, 3] look as if their development process was composed basically by the programming step, in an artisanal way. So, the results produced by such tools cannot be quite relied upon.

EASY [7, 8] is a proposal for a simulation environment whose objective is to give support to all steps of problem solving using neural networks, offering the following resources: 1) multiple paradigms, which can be used to generate different types of applications; 2) flexibility to modify and/or incorporate new paradigms; 3) interactive definitions of network configurations and execution parameters; 4) presentation of the interface options at different levels of abstraction; 5) interaction during simulation, so that process evolution can be followed or interrupted in order to change parameters.

Object-oriented formal specifications in (*MooZ*), as shown before, were used to develop *EASY*; object orientation has been also adopted by other simulator projects [3] as a natural path to neural network modelling for it promotes modularity and reusability. *EASY* has a basic hierarchy of neuron and neural network classes (figures 2 and 3) which may be reused in the inclusion of new models. Formal specifications made possible a precise and non-ambiguous description of *EASY*'s properties, avoiding erroneous interpretations that could lead to error propagation in subsequent stages of the development process.

The *EASY* prototype has been implemented in *Smalltalk* (*VisualWorks* 1.0 [22], which runs in SUNs, PCs and Macs). From the formal specifications of neurons, architectures and learning algorithms, the classes in the implementation language were easily derived. Although the transformation was informal, a systematic development was adopted, since the implementation of formally specified classes was almost a direct mapping of the specification' classes to the classes in the implementation language.

The architecture of the prototype is shown in figure 6. Classes in the low level layers supply services to classes at the high level layers using a method protocol. The *Neural Networks* layer has class hierarchies that implement neural network paradigms (illustrated in figures 2 and 3). The *Configuration and Execution* layer is responsible for the control of architecture definitions, simulation execution and result analysis. The *Session and Presentation* layer is responsible for the user interface.

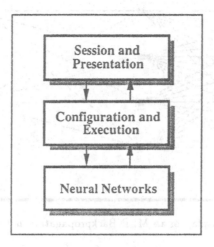

Fig. 6. EASY's prototype architecture.

All layers of the architecture presented in figure 6 were implemented in 2 person/months, achieved due to the previous work on formal specification in which 3 person/months were spent and the facilities of interface building provided by *VisualWorks*. A total of 10.000 lines of Smalltalk code was produced, originating from 3.000 lines of *MooZ* specification. The time spent in the implementation of all neuron and neural network classes illustrated in figures 2 and 3 was 7 days. The prototype has a number of tools to specify architectures, to control the process of learning and testing of the neural network paradigms, along with tools for graphically viewing architectures, input patterns, and internal information like weights, dynamic view of neural network outputs, and so on. It supports multiple specification/training/testing runs at the same time. Figure 7 illustrates the graphical network architecture viewer. The view is changed automatically to reflect new definitions made by the user. The network architecture illustrated has 8 positions in its input layer because this is the length of the input pattern. There is a hidden layer with 4 neurons and an output layer with 2 neurons. The neurons in each layer are fully connected with the neurons in the previous layer. *EASY*'s user interface to specify *MLP* network architectures is described in [7]. *EASY*'s user interface to control a *MLP* network learning stage is shown in figure 8. All learning parameters are specified in an interactive way.

In many simulators the specification of a neural network experiment is done by writing code in a low level, conventional, programming language like C. One of

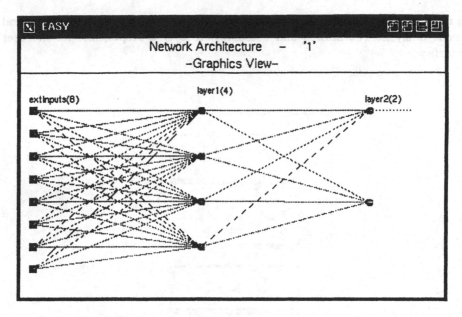

Fig. 7. EASY's display of an MLP Backpropagation network architecture.

EASY's objectives is to avoid this. When writing code, one basically "programs" the experiment, thereby risking the introduction of many bugs. A formal language is more appropriate for the inclusion of new functions in a software simulator package.

Some case studies were simulated using the prototype, including *QRS* complex recognition (an electrocardiogram processing problem), rebuilding of degenerated images of characters, off-line signature recognition and verification. These case studies are presented in [6]. When simulating the three case studies, it was clear that only a minimum effort is needed; there is no need to program in low level languages and the user can follow the simulation process with a number of visualization tools. The prototype allows for a better interaction of the user with the experiment itself.

5 Summary

In this work, we presented an object-oriented formal specification of artificial neural networks used in the development of *EASY*, a simulation environment. The inclusion of new concepts and paradigms was done via the reuse of abstract class definitions in a formal way. This avoids the respecification of all basic features of neural networks and, indeed, reduces the development effort.

Neural networks are fault tolerant, because the failure of one or more neurons or connections may not result in a loss of knowlegde. However, software implementations of neural networks must be reliable, because bugs may change, for example, the behavior of a learning algorithm, compromising the final results. *MooZ* made possible a precise and non-ambiguous description of artificial neural network properties.

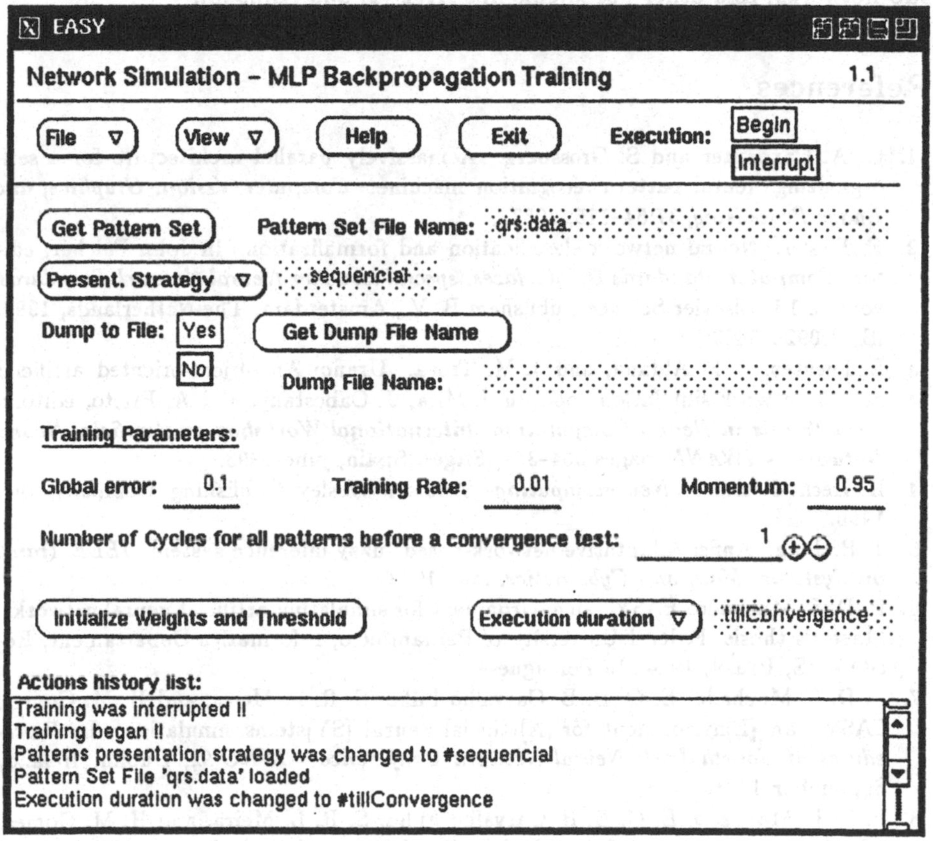

Fig. 8. EASY's tool for controlling MLP network learning.

The theory of neural networks has had a horizontal development so far, as a great number of researchers around the world have been extending or creating new concepts and paradigms. Only very little effort has been done to establish standards or to formalize concepts in order to promote a more vertical growth. The specification presented in this work represents a formalization of some of the basic neural network concepts, including real paradigms like *MLP Backpropagation*.

In the development of *EASY*, a small part of the effort corresponded to the programming. Formal specifications, which were adopted in the process of development, helped to expose ambiguities and inconsistencies which normally would not be detected in an informal development.

EASY was tested in three case studies including real world applications. It is now being used by the UFPE neural networks group and also by cooperating groups. The formal specification is being extended to include other concepts, including different kinds of learning, macro-structures or networks composed by sub-networks, and other paradigms like ART [1] and Anfis [5].

References

1. G. A. Carpenter and S. Grossberg. A massively parallel architecture for a self-organizing neural pattern recognition machine. *Computer Vision, Graphics, and Image Processing*, 37:54–115, 1987.
2. E. Fiesler. Neural network classification and formalization. In John Fulcher, editor, *Computer Standards & Interfaces, special issue on Neural Network Standards*, volume 16. Elsevier Science Publishers B. V., Amsterdam, The Netherlands, 1994. ISSN 0920-5489.
3. L. Fuentes, J. F. Aldana, and J. M. Troya. Urano: An object-oriented artificial neural network simulation tool. In J. Mira, J. Cabestany, and A. Prieto, editors, *New Trends in Neural Computation: International Workshop on Artificial Neural Networks - IWANN*, pages 364–369, Sitges, Spain, june 1993.
4. R. Hecht-Nielsen. *Neurocomputing*. Addison-Wesley Publishing Company, Inc, 1990.
5. J. R. Jang. Anfis: Adaptative-network-based fuzzy inference system. *IEEE Trans. on Systems, Man, and Cybernetics*, May 1993.
6. P. D. L. Machado. EASY - an environment for simulating artificial neural networks. Master's thesis, Federal University of Pernambuco, Informatics Departament, Recife - PE, Brazil, 1994. In Portuguese.
7. P. D. L. Machado, E. C. D. B. Carvalho Filho, S. R. L. Meira, and H. M. Gomes. EASY - an [E]nvironment for [A]rtificial neural [SY]stems simulation. In *Proceedings of Fourth Irish Neural Network Conference - INNC'94*, Dublin, Ireland, September 1994.
8. P. D. L. Machado, E. C. D. B. Carvalho Filho, S. R. L. Meira, and H. M. Gomes. EASY - an environment for modelling, simulation and analysis of artificial neural networks. In *Proceedings of I Brazilian Symposium on Neural Networks*, Caxambu, Minas Gerais, Brazil, August 1994. In Portuguese.
9. P. D. L. Machado, S. R. L. Meira, E. C. D. B. Carvalho Filho, and H. M. Gomes. A formal object-oriented specification of artificial neural networks. In *Proceedings of VIII Brazilian Symposium on Software Engineering*, Curitiba, PR, Brazil, October 1994. In Portuguese.
10. P. D. L. Machado, S. R. L. Meira, E. C. D. B. Carvalho Filho, and H. M. Gomes. Specifying artificial neural networks in MooZ. In *Proceedings of XX Conferencia Latino Americana de Informatica - PANEL'94*, Mexico, September 1994. In Portuguese.
11. W. S. McCulloch and W. H. Pitts. A logical calculus of the ideas immanent in nervous activity. *Bull Math Biophys*, 5:115–133, 1943. formal neuron.
12. S. R. L. Meira and A. L. C. Cavalcanti. Modular Object-Oriented Z Specifications. In Prof. C. J. van Rijsbergen, editor, *Workshop on Computing Series*, pages 173 – 192, Oxford - UK, December 1990. Springer-Verlag.

13. S. R. L. Meira and A. L. C. Cavalcanti. The MooZ Specification Language. Technical report, Universidade Federal de Pernambuco, Departamento de Informática, Recife - PE, 1992. Available via anonymous ftp from rosa.cr-pe.rnp.br, file pub/MooZ/MooZ.ps.Z.

14. B. Meyer. On formalism in specifications. *IEEE Software*, pages 6–26, january 1985.

15. J. M. J. Murre. Neurosimulators. In M. A. Arbib, editor, *HandBook of Brain Research and Neural Networks*. MIT Press, MRC APU, Cambridge, 1994.

16. H. R. Myler, A. R. Weeks, R. K. Gillis, and G. W. Hall. Object-oriented neural simulation tools for a hypercube parallel machine. *Neurocomputing*, 4(5):235–248, 1992.

17. D. E. Rumelhart, G. E. Hinton, and R. J. Williams. Learning internal representations by error propagation. In *Parallel Distributed Processing*, chapter 8, pages 318–362. MIT Press, Cambridge, 1986.

18. D. A. Santos. A formal model for the specification of neural networks - MOFEU. Master's thesis, Federal University of Pernambuco, Informatics Departament, Recife - PE, Brazil, 1994. In Portuguese.

19. P. K. Simpson. *Artificial Neural Systems: Foundations, Paradigms, Aplications and Implementations*. Pergamon Press, Inc, 1990.

20. L. S. Smith. A framework for neural net specification. *IEEE Transactions on Software Engineering*, 18(7):601–612, july 1992.

21. J. M. Spivey. *The Z Notation: A Reference Manual*. Prentice Hall, 1989.

22. ParcPlace Systems. *VisualWorks* ^{TM} *Release 1.0 - User's Guide*. 999 E. Arques Ave., Sunnyvale, California 94086-4593, 1992.

23. P. D. Wasserman. *Neural Computing: Theory and Practice*. ANZA Research, Inc, New York, 1989.

Structuring Specification in Z to Build a Unifying Framework for Hypertext Systems

Mark d'Inverno and Mark Priestley

School of Computer Science, University of Westminster,
115 New Cavendish Street, London, W1M 8JS, UK.
Email {dinverm,priestm}@westminster.ac.uk

Abstract. A report is given on work undertaken to produce a structured specification in Z of a model which aims to capture the essential abstractions of hypertext systems. The specification is presented in part and the potential value of this specification to the hypertext community is explored and discussed. We argue that this specification provides a framework for hypertext systems in that it provides: explicit and unambiguous definitions of hypertext terms, an explicit environment for the presentation, comparison and evaluation of hypertext systems and a foundation for future research and development in the field. Although there are many formal reference models of hypertext, we have found Z expressive enough to allow a unified account of a system and its operations. Our model does not restrict the specifier to any particular design, but provides a mathematical framework within which different models may be compared. Further, we were able to structure the specification in order that the model could be described initially at the highest level of abstraction with complexity added at increasingly lower levels of abstraction. This structured specification allows the functionality of hypertext systems to be considered at different levels of granularity which, we argue, gives rise to a well-defined robust model and a beneficial environment within which to reason about hypertext design. The use of this model in presenting and comparing existing models, as well as its use in developing a new hypertext learning model, is briefly discussed.

1 Introduction

Many formal reference models of hypertext have been presented in the literature [1, 12, 17, 20, 21], and whilst these models give valuable theoretical insights into certain aspects of the structure of hypertext, they are not by themselves adequate vehicles for the presentation, evaluation and comparison of different systems. In this paper we describe an approach to the formal specification of hypertext systems which allows the development of a common conceptual framework and provides an environment in which to discuss, design, develop and evaluate hypertext systems.

The language Z is based upon basic mathematical ideas. This means that it is accessible to many hypertext practitioners, and unlike many models [20, 21], is expressive enough to allow a consistent formal unified account of a system and its associated operations. We claim further that a well-structured Z specification built up from basic mathematical ideas can provide the following for the hypertext community:

Clarity. The use of formal concepts allows explicit and unambiguous descriptions of terms and complex systems to be given.

Common Conceptual Framework. The availability of explicit notations allows a movement from informal mutually inconsistent descriptions of systems towards a common understanding of the basic features and concerns of a particular class of systems.

Design Evaluation. This common framework enable alternative designs of particular systems to be explicitly presented, compared and evaluated.

Reliability. A formal specification language provides a proof system and a set of proof obligations which enables a reliable and robust model of systems to be built.

A well-structured specification in this context is one which first describes a system at its highest level of abstraction, with complexity added at each successive lower level of abstraction, allowing irrelevant information to be removed from consideration. Further, different modular components of a system can be isolated and described separately and commonalities in different parts of a system can be recognised and presented as such. Abstraction renders prejudice about design unnecessary; consequently, a specification of a "general system" can be written. Indeed, Z schema boxes have been ideal for manipulation in the design process since, viewing in many cases the design process as a constraint of possible states, design strategies can be presented as predicates in the appropriate state schemas. Such a structured specification, we argue, is a tool which enables a more considered hypertext analysis.

1.1 Motivation

This paper is a consequence of two separate ongoing areas of work. The first area involved the writing of programs to translate between various hypertext systems currently used at University of Westminster. In order to ensure that the structure of a document is preserved in its translation, it was decided to produce full formal specifications of the University of Westminster systems [9, 10]. This activity produced great benefits in understanding the differences and similarities between these systems and indicated that a suitable formal specification of a general hypertext system would be of great value in comparing different hypertext systems. The second major influence is the writing and presentation of formal specifications (in Z) with similar 'unifying' motivations in certain areas of computer science including the fields of interactive conferencing systems [8], distributed artificial intelligence [7] and multi-agent systems [18]. These papers demonstrate the need to provide further suitable formal specifications of complex systems in diverse application areas.

1.2 Related Work

There is another attempt in the hypertext literature to provide a formal specification of a 'general' hypertext system known as the Dexter Model of Hypertext [15]. Interestingly, the authors also chose Z with the motivation - to capture formally and informally the important abstractions found in a wide range of existing and future hypertext systems - similar in some respects to our own. The model essentially comprises a collection of components - links and nodes - with an accessor function which maps a unique identifier to a node and an resolver function which maps descriptions of components to the components themselves. The operations specified are that of adding, modifying and retrieving components.

However the specification is often obtuse and over-complicated, and only the most experienced Z practitioner with a good knowledge of hypertext would be able to gain much from the specification. Part of the problem is that no structuring of the specification takes place, rather it starts with a large collection of given sets and introduces many concepts and functions before the first state schema is actually presented. In this sense, the specification is very 'flat' and does not aid the reader in building up a picture of their model of a hypertext. Further, the specification describes hypertext at a very low level of detail and hence is much more orientated towards implementation concerns.

We argue that the immediate complexity will not serve the standard hypertext practitioner in providing an accessible model which can be commonly adopted by the hypertext community, and that the lack of abstraction mechanisms within the 'flat' specification does not provide a framework in which to present and develop ideas in the design of hypertext systems.

In addition, there is a system known as HAM - A General-Purpose Hypertext Abstract Machine [3], which is a general purpose server for a hypertext storage system. HAM has several general features similar to our own including nodes, links, graphs and attributes and describes the way information is represented before it is used in Information Retrieval. The motivation is essentially to lay the foundations for a standard terminology for the development of hypertext technology. Important as this work is, the model is only concerned with problems of storage, and not with representing an Information Retrieval session. In addition, the model is not formal, and subsequently does not provide the precision of a mathematical specification. Our work on the other hand is both formal and, we argue, sufficiently expressive to provide a framework within which to detail all aspects of hypertext.

1.3 An Overview of the Paper

The specification of our framework is split into two parts, and these are defined in sections 2 and 3. The first part, given in section 2, presents what we believe is the most straightforward and intuitive description of a model of hypertext systems where nodes are treated as given sets and links as a pair of system nodes. The second part, given in section 3, builds on the basic model and increases the level of specification detail in order to describe the internal details of a node. Each of these two sections is divided into three subsections: the first defines the structure of the system, the second the state of the system as it is being read and the third presents a description of the basic applications of hypertext, namely how it facilitates structured movement through a particular information space. Section 4 outlines how the model can be used to detail other features and applications of hypertext. Section 5 provides a summary of the paper and details current and future work.

2 The Basic Hypertext

2.1 Structure

If we consider a hypertext system at its highest level of abstraction, it consists of a collection of basic elements. These elements are typically called nodes, but the name can vary from system to system: for example, they are called *cards* in NoteCards [14], *frames* in KMS [2], *documents* in Intermedia [22], and *statements* in Augment [11]. In this specification we chose to use the word *node*, being the most common.

[NODE]

```
┌─ CONTENTS ─────────────────────────────────────
│  Nodes : ℙ NODE
└────────────────────────────────────────────────
```

However, if we take a closer look at a hypertext system, we find that the structure is more sophisticated and that between nodes there exist certain connections, known as links, each suggesting some relationship between the nodes they connect. A link is directional, pointing from one node (sometimes referred to as the parent node) to another node (sometimes referred to as the child node). A link is therefore characterised by the nodes it connects.

LINK == NODE × NODE

```
┌─ LINKS ─────────────────────────────────────────
│  Links : ℙ LINK
└─────────────────────────────────────────────────
```

We define a hypertext system as a collection of nodes and links, where links must point *from* an existing system node. However, it is not the case that a link must necessarily point to a system node: many hypertext systems include the notion that some links only have the potential to point to such a node (e.g. [9]).

```
┌─ HYPERTEXT ─────────────────────────────────────
│  CONTENTS
│  LINKS
│  ────────────
│  dom Links ⊆ Nodes
└─────────────────────────────────────────────────
```

Button Nodes A more detailed investigation of a hypertext system will reveal that some nodes are special in that they may be reached without using a link. We call such nodes *button* nodes. Further, there may or may not be a default starting node when a hypertext system is first used in an IR session. (For definitions of optional and related concepts, please consult the Appendix).

```
┌─ButtonHYPERTEXT──────────────────────────────────
│ HYPERTEXT
│ Buttons : ℙ NODE
│ StartNode : optional [NODE]
├──────────────────────────────────
│ StartNode ⊆ Nodes
│ Buttons ⊆ Nodes
│ StartNode ⊆ Buttons
└──────────────────────────────────
```

Typed Links In certain hypertext systems [9], links may be grouped into *Link Functions*.

LINKFUNCTION == NODE ↦ NODE

```
┌─TypedLINKS──────────────────────────────────
│ LINKS
│ LinkFunctions : ℙ LINKFUNCTION
├──────────────────────────────────
│ ⋃ LinkFunctions ⊆ Links
└──────────────────────────────────
```

A particular system might insist that a link could not belong to more than one function and that every link should belong to a function. In which case, we would simply include the following predicate in the above schema.

setdisjoint *LinkFunctions* ∧ ⋃ *LinkFunctions* = *Links*

2.2 The State of the Hypertext

The next aspect of this simple hypertext model is to specify the state of the model as it is used in an Information Retrieval (IR) session. In all systems that we have investigated, there is a notion of the position of a user within the information space, and the history of that user's IR session. A history provides a record of the nodes visited by a user in a session, and possibly the way in which they were visited.

Our general model of a hypertext session history is, then, a set of sequences of nodes, where each of these sequences is updated in one of four ways depending on what type of move is made: the sequence can remain unchanged, the last-visited node can be appended, the sequence can be truncated at the first occurrence of the last-visited node, or fourthly, at the last occurrence of the last-visited node. For the sake of brevity, we do not present this mechanism in this paper, but for the purposes of exposition, we will show how two commonly-found browsing histories are used in IR. We define *StandardHistory* as a sequence of all the nodes visited, and *Visited* to be the set of nodes which have been visited.

```
┌─HISTORY──────────────────────────────────
│ StandardHistory : seq NODE
│ Visited : ℙ NODE
├──────────────────────────────────
│ Visited = ran StandardHistory
└──────────────────────────────────
```

We represent a user session by the hypertext, their history and their current position within the information space.

```
┌─ HYPERTEXTState ─────────────────────────────────
│ HYPERTEXT
│ HISTORY
│ CurrentNode : NODE
├──────────────────────────────────────────────────
│ CurrentNode ∈ Nodes
└──────────────────────────────────────────────────
```

As mentioned, the buttons which become available during a session might be dependent on the session itself. Here the variable *RunButtons*, a superset of *Buttons*, represents those nodes which can currently be visited without the use of a link. In particular, it is typical that any previously visited node can be re-visited without using a link.

```
┌─ ButtonHYPERTEXTState ───────────────────────────
│ HYPERTEXTState
│ ButtonHYPERTEXT
│ RunButtons : ℙNODE
├──────────────────────────────────────────────────
│ (defined StartNode) ∧ ( Visited ≠ { }) ⇒
│                         head StandardHistory = the StartNode
│ Buttons ⊆ RunButtons
│ RunButtons ⊆ Nodes
│ Visited ⊆ RunButtons
└──────────────────────────────────────────────────
```

2.3 Applications

One of the benefits of Z is that the operations that a system provides can be specified within the same formal framework. This property is not shared by many of the mathematical models presented in the literature; for example, [21] uses hypergraphs to give a formal account of the structure of a hypertext, but then specifies the operations of reading the hypertext using a mixture of pseudocode and informal English description. A particular advantage of a unified specification, as provided by Z, is that the properties of operations, and their effects on the state of the system, can be explored and reasoned about formally.

Any operation in an IR session will not alter the actual linked structure of the hypertext.

```
┌─ ΔHYPERTEXTState ────────────────────────────────
│ HYPERTEXTState
│ HYPERTEXTState'
│ ΞHYPERTEXT
└──────────────────────────────────────────────────
```

Starting a hypertext session changes the state of the hypertext by resetting the history.

```
┌─Login───────────────────────────────────────────
│ ΔHYPERTEXTState
├──────────────────────────────────────────────────
│ StandardHistory' = ⟨⟩
└──────────────────────────────────────────────────
```

We now show how the hypertext is used in an IR session by moving through an information space using the hypertext system. Essentially a hypertext system supports two types of moves: first, a user may move from one node to another node by use of a link from their current node to another related node; second, if they have some knowledge of a node - because it is a button node or a previously visited node, for example - they may move directly to that node without using a link.

We structure the specification by first describing the properties that we want a general move operation to have before giving details of a particular move. This provides an example of how Z has enabled us to modularise this specification and so present the model in levels of increasing detail. In particular it shows how the small schemas defined in our specification can be combined to define more complex states and operations. In general, a move operation will change the state of the hypertext: the current node may alter and the history may alter, but the actual linked structure of the hypertext will remain unchanged. Further, a move operation may return a message to the user.

```
┌─Move────────────────────────────────────────────
│ ΔHYPERTEXTState
│ message! : optional [ERROR]
└──────────────────────────────────────────────────
```

We can next distinguish between successful and unsuccessful attempts to move. A successful move will update the history list, appending the node that has just been visited. Further, there will be no message.

```
┌─MoveOk──────────────────────────────────────────
│ Move
├──────────────────────────────────────────────────
│ Visited' = Visited ∪ { CurrentNode }
│ StandardHistory' = StandardHistory ⌢ ⟨CurrentNode⟩
│ undefined message!
└──────────────────────────────────────────────────
```

Failed moves leave the hypertext state unchanged. An error message is given.

```
┌─MoveFail────────────────────────────────────────
│ Move
│ ΞHYPERTEXTState
├──────────────────────────────────────────────────
│ defined message!
└──────────────────────────────────────────────────
```

Now we describe using a link to move from the current node to another.

```
┌─ FollowLinkOk ─────────────────────────────────────────
│ link : LINK
│ MoveOk
├────────────────────────────────────────────────────────
│ CurrentNode' = second link
└────────────────────────────────────────────────────────
```

This move can be made in one of two ways. The user can select either the link to be used, or, a link function intended to isolate an appropriate link.

```
┌─ UserChosesLinkOk ─────────────────────────────────────
│ FollowLinkOk[link?/link]
├────────────────────────────────────────────────────────
│ link? ∈ Links ∧ first link? = CurrentNode ∧ second link? ∈ Nodes
└────────────────────────────────────────────────────────
```

```
┌─ FollowLinkFunctionOk ─────────────────────────────────
│ LinkFunction? : LINKFUNCTION
│ FollowLinkOk
│ TypedLINKS
├────────────────────────────────────────────────────────
│ LinkFunction? ∈ LinkFunctions
│ CurrentNode ∈ (dom LinkFunction?)
│ link = (CurrentNode, LinkFunction? CurrentNode)
└────────────────────────────────────────────────────────
```

UserChoosesLinkFunctionOk ≙ FollowLinkFunctionOk \ (link).

Moves can be made using buttons by supplying the node to be visited.

```
┌─ MoveButtonOk ─────────────────────────────────────────
│ MoveOk
│ ΔButtonHYPERTEXTState
│ ΞButtonHYPERTEXT
│ button? : NODE
├────────────────────────────────────────────────────────
│ button? ∈ RunButtons
│ CurrentNode' = button?
└────────────────────────────────────────────────────────
```

Moving back to the previously visited node requires no input node.

```
┌─ MoveBack ─────────────────────────────────────────────
│ MoveOk
├────────────────────────────────────────────────────────
│ StandardHistory ≠ ⟨⟩
│ CurrentNode' = last StandardHistory
└────────────────────────────────────────────────────────
```

The history may be up updated in a different way as follows. This justifies the need to model a more general history mechanism as mentioned previously.

```
┌─ MoveBackAlternative ──────────────────────────────────
│ Move
├──────────────
│ StandardHistory ≠ ⟨⟩
│ undefined message!
│ CurrentNode' = last StandardHistory
│ StandardHistory = front StandardHistory
└──────────────────────────────────────────────────────
```

3 The Highlight Hypertext

We now lower the level of description of a hypertext system, in order to model the fact that a node contains certain hypertext elements. These elements are references, typically taking the form of highlighted text, which can then serve as the destination of, or source for, hypertext links. Many hypertext systems facilitate not only links connecting nodes, but regions within nodes.

3.1 Structure

Each node has a - possibly empty - set of internal hypertext references which we call highlights. Many hypertext systems support the operations of scrolling backwards and forwards through these highlights. Assuming that the highlights are unique within a node, we represent them using an injective sequence.

[HIGHLIGHT]

```
┌─ HighlightNODE ───────────────────────────────────────
│ Highlights : iseq HIGHLIGHT
└──────────────────────────────────────────────────────
```

The inside of each node then consists of highlights.

```
┌─ HighlightNODES ──────────────────────────────────────
│ CONTENTS
│ GetHighlights : NODE ↠ HighlightNODE
├──────────────
│ dom GetHighlights = Nodes
└──────────────────────────────────────────────────────
```

Next, we extend the notion of a link so that they can point to highlights within a node. We use the categories of Conklin [6] who differentiates two categories of link by drawing a distinction between *organisational* and *referential* links. We use these categories and introduce a third type which can be found in current hypertext systems [10] known as *span* links.

Organisational links Many hypertexts have an underlying structure, either as a consequence of the information space itself, or the way that the information space is required to be presented to a user. Organisational links capture this underlying structure. For example, these may be hierarchical in nature so that there might be a standard way within the hypertext of moving from a given node to a "parent", "child" or "sibling" node.

Referential links Referential links are typically non-hierarchical. They connect a *highlight*, which can be a point or a region within a node, to a another node. Referential links are motivated by the *content* of a node, rather than by the underlying structure of the hypertext or information space.

Span links We define Span links to be links which connect a *highlight* within a node to a *highlight* within another node. The notion of a cross-reference, for example, could be modelled in this fashion.

In this specification, we differentiate between these three types of link: we call organisational links *orglinks*, referential links *reflinks* and span links *spanlinks*. Some other mathematical models have had problems defining different kinds of link. In [13], this ability is described as an "innovative feature". In order to specify this more detailed hypertext, we must define a new type to represent links between highlights, and define each of the three link categories as subtypes.

HighlightLINK _____

$From, To$: NODE
$FromHighlight, ToHighlight$: optional [HIGHLIGHT]

OrgLINK _____

HighlightLINK

undefined $FromHighlight \land$ undefined $ToHighlight$

RefLINK _____

HighlightLINK

defined $FromHighlight \land$ undefined $ToHighlight$

SpanLINK _____

HighlightLINK

defined $FromHighlight \land$ defined $ToHighlight$
$(From \neq To) \land (FromHighlight \neq ToHighlight)$

We may wish to reason about these links in terms of the nodes they connect without concern about the kind of link they are. In order to do this we introduce a function which maps our new representation of links to our old representation.

$RecoverLink$: HighlightLINK $\twoheadrightarrow$ LINK
$RecoverLinks$: $\mathbb{P}$ HighlightLINK $\twoheadrightarrow$ $\mathbb{P}$ LINK

$\forall c$: HighlightLINK; cs : $\mathbb{P}$ HighlightLINK $\bullet$
$\qquad RecoverLink\ c = (c.From, c.To) \wedge$
$\qquad RecoverLinks\ cs = RecoverLink (\!|\ cs\ |\!)$

We can now give the set of all links of the hypertext and relate the two representations of organisational link within the model.

HighlightLINKS

LINKS
$OrgLinks$: $\mathbb{P}$ OrgLINK
$RefLinks$: $\mathbb{P}$ RefLINK
$SpanLinks$: $\mathbb{P}$ SpanLINK
$HighlightLinks$: $\mathbb{P}$ HighlightLINK

$HighlightLinks = OrgLinks \cup RefLinks \cup SpanLinks$
$Links = RecoverLinks\ OrgLinks$

Our new model of hypertext is then given by the following schema which ensures that all links are well defined.

HighlightHYPERTEXT

HYPERTEXT
HighlightLINKS
HighlightNODES

$\forall l$: $HighlightLinks \bullet (l.From \in Nodes) \wedge$
$\qquad (l.FromHighlight \subseteq$ ran $(GetHighlights\ l.From).Highlights)$

Typing the links is similar to that given in the basic model, but the definition of what constitutes a link function is slightly different. A link function is any set of links for which no two links have the same 'from-node' and 'from-highlight'. In other words, there is only one way to leave a given position given a particular link function. Further, we assert that a typed link function will only contain links which are either all organisational, referential or span.

$$HlightLINKFUNCTION == \{xs : \mathbb{P} HighlightLINK;\ x, y : HighlightLINK\ |$$
$$((x \in xs) \wedge (y \in xs) \wedge (x \neq y)) \Rightarrow$$
$$(x.From, x.FromHighlight) \neq (y.From, y.FromHighlight) \bullet xs\}$$

TypedHighlightLINKS

HighlightLINKS
$OrgLinkFuns, RefLinkFuns, SpanLinkFuns$: $\mathbb{P}$ HlightLINKFUNCTION

$\bigcup OrgLinkFuns \subseteq OrgLinks$
$\bigcup RefLinkFuns \subseteq RefLinks$
$\bigcup SpanLinkFuns \subseteq SpanLinks$

3.2 State

We now define the position of a user within the hypertext. This will not only include the current node and history from the state of the basic model, but also the position of a user within a node. This position will either be defined, in which case the user will be positioned at some highlight, or undefined, which occurs, for example, when a node has no highlights or an organisational link has just been used to move to the current node.

```
┌─ HighlightHYPERTEXTState ─────────────────────────────────
│ HYPERTEXTState
│ HighlightHYPERTEXT
│ Position : optional [HIGHLIGHT]
│ HighlightNODE
├───────────────────────────────────────────────────────────
│ θHighlightNODE = GetHighlights CurrentNode
│ Position ⊆ (ran Highlights)
└───────────────────────────────────────────────────────────
```

3.3 Applications

A change in the state will not affect the structure.

```
┌─ ΔHighlightHYPERTEXTState ─────────────────────────────────
│ HighlightHYPERTEXTState
│ HighlightHYPERTEXTState'
│ ΞHighlightHYPERTEXT
└───────────────────────────────────────────────────────────
```

Any move using highlights may affect the state.

```
┌─ HighlightMove ────────────────────────────────────────────
│ ΔHighlightHYPERTEXTState
│ Move
└───────────────────────────────────────────────────────────
```

We distinguish between two types of move. Internal moves involve the user scrolling through or selecting one of the highlights of the current node. External moves, on the other hand, involve the user taking a link, or moving to a button node. An internal move will not affect the state of the basic hypertext - the current node and history are not altered - and can only be made if the current node actually contains highlights.

```
┌─ InternalMoveOk ──────────────────────────────────────────
│ HighlightMove
│ ΞHYPERTEXTState
│ MoveOk
├───────────────────────────────────────────────────────────
│ Highlights ≠ ⟨⟩
└───────────────────────────────────────────────────────────
```

There are three basic internal moves within a node: moving to the next highlight, moving to the previous highlight - both which necessarily entail that the current position is defined - and moving to a chosen highlight. For a definition of CycleNext and CyclePrevious please see the appendix.

```
┌─ NextHighlight ─────────────────────────────────────
│ InternalMoveOk
├─────────────────────────────────────────────────────
│ defined Position
│  the Position' = CycleNext (( the Position), Highlights)
└─────────────────────────────────────────────────────
```

```
┌─ PreviousHighlight ─────────────────────────────────
│ InternalMoveOk
├─────────────────────────────────────────────────────
│ defined Position
│  the Position' = CyclePrevious (( the Position), Highlights)
└─────────────────────────────────────────────────────
```

```
┌─ SelectHighlight ───────────────────────────────────
│ InternalMoveOk
│ highlight? : HIGHLIGHT
├─────────────────────────────────────────────────────
│ highlight? ∈ (ran Highlights)
│  the Position' = highlight?
└─────────────────────────────────────────────────────
```

The general external move may affect the position and the current node. Note that the use of buttons is not changed in any way in this more sophisticated model.

```
┌─ ExternalMoveOk ────────────────────────────────────
│ HighlightMove
│ MoveOk
│ ΞHighlightHYPERTEXT
└─────────────────────────────────────────────────────
```

To use a link successfully, the parent node of the link must necessarily be the current node. If we use an organisational link then this is sufficient; they can be used from any position within a node. However, if we use a span or referential link then the link must have a from-highlight equal to the current position. We re-use our basic model definition as follows:

```
┌─ FollowHighlightLinkOk ─────────────────────────────
│ ExternalMoveOk
│ FollowLinkOk
│ highlightlink : OrgLINK
├─────────────────────────────────────────────────────
│ highlightlink.From = CurrentNode
│ highlightlink ∈ OrgLinks ∨ (highlightlink ∈ (RefLinks ∪ SpanLinks)
│                          ∧ Position = highlightlink.FromHighlight)
│ link = (highlightlink.From, highlightlink.To)
│ Position' = highlightlink.ToHighlight
└─────────────────────────────────────────────────────
```

UserChoosesHighlightLinkOk $\widehat{=}$ FollowHighlightLinkOk \ (*link*)

Again the user either chooses the link, or a link function intended to isolate the required link.

4 Extensions

Although many formal models of hypertext have been proposed in the literature, there is still little consensus about what a definitive model should be. Indeed one might argue that, in view of the rapid progress of the technology, it is probably premature to attempt a definitive formalisation. However, a significant benefit of Z that we have discovered in this respect is that it does not restrict the specifier to any particular mathematical model; rather it provides a general mathematical framework within which different models, and even particular systems, can be defined and contrasted.

We now justify this claim here by considering a number of more sophisticated features of hypertext and show how the model defined in the previous sections can be elaborated to define and describe features and extensions found in a variety of hypertext systems.

4.1 Values and Attributes

A node or link may have a collection of types with associated values which may be used to structure the state space. In this case, the hypertext can be structured so that only certain types of links have access to certain types of nodes. In the following schema, we assert that for any two nodes connected by a link, the link and the nodes must have an associated type in common.

[TYPE, VALUE]
TYPEVALUEPAIRS $==$ $\mathbb{P}$(TYPE $\times$ VALUE)

TypedHYPERTEXT
HighlightHYPERTEXT
NodeType : NODE $\rightarrow$ TYPEVALUEPAIRS
HighlightLinkTypes : HighlightLINK $\rightarrow$ TYPEVALUEPAIRS

---

dom *NodeType* $=$ *Nodes*
dom *HighlightLinkTypes* $=$ *HighlightLinks*
$\forall c : HighlightLinks \mid c.\,To \in Nodes \bullet$
    $(\exists t : \text{TYPE} \bullet (t \in \bigcap \{first(\!\mid HighlightLinkTypes\ c\ \mid\!),$
        $first(\!\mid NodeType\ (c.\,To)\ \mid\!), first(\!\mid NodeType\ (c.From)\ \mid\!) \}))$

4.2 Specifying Different Topologies

The *topology* of a hypertext describes the way in which the nodes are connected. In the simplest case, a hypertext is seen simply as being a directed graph, but other organisations are possible to facilitate the successful movement though an information space. In

[19] a survey is given of possible topologies and, as an example, we define a hypertext which has a *hierarchical* structure. This is defined in terms of the organisational link function *Parent* and a set of organisational link functions called *Children*.

```
┌─ HierarchicalHYPERTEXT ─────────────────────────────
│ HighlightHYPERTEXT
│ TypedHighlightLINKS
│ ButtonHYPERTEXT
│ Parent : HlightLINKFUNCTION
│ Children : ℙ HlightLINKFUNCTION
├─────────────────────────────────────────────────────
│ Parent ∈ OrgLinkFuns
│ Children ⊆ OrgLinkFuns
│ defined StartNode
│ ran (RecoverLinks (⋃ Children)) = Nodes \ StartNode
│ (RecoverLinks Parent)~ = RecoverLinks (⋃ Children)
└─────────────────────────────────────────────────────
```

4.3 User Navigation

As a mechanism to help the user navigate through a document, a number of proposals have been made for defining *paths* through the document [20]. A path would offer a reader a pre-defined route through a subset of the document, thus enabling an overview of the hypertext, or of a particular subject to be presented. A path may be just a set of nodes. Equally it may take the form of a sequence of links which actually take the user through particular highlights in the hypertext.

```
SIMPLEPATH == seq NODE
PATH == seq HighlightLINK
```

```
┌─ SimplePATHS ───────────────────────────────────────
│ CONTENTS
│ SimplePaths : ℙ SIMPLEPATH
├─────────────────────────────────────────────────────
│ ran (⋃ SimplePaths) ⊆ Nodes
└─────────────────────────────────────────────────────
```

```
┌─ PATHS ─────────────────────────────────────────────
│ HighlightLINKS
│ paths : ℙ PATH
├─────────────────────────────────────────────────────
│ ∀ p : paths • (∀ l, m : HighlightLINK | ⟨l, m⟩ in p •
│                (l.To, l.ToHighlight) = (m.From, m.ToHighlight))
└─────────────────────────────────────────────────────
```

4.4 Content of Nodes

In this specification we have not considered the text that is stored at each node. In the following, *Text* is defined as a schema since we wish to allow for the possibility of different nodes sharing the same content, cited as an advantage of the hypergraph model in [21].

```
[CHAR]
STRING == seq CHAR
TEXT ≙ [text : STRING]
```

```
__ TextHYPERXTET _____
  ΞHYPERTEXT
  Text : NODE ↦ TEXT
 _____
  dom Text = Nodes
```

Once a notion of content has been defined, it is possible to define a further class of
move operations which Conklin [6] calls *keyword links*. A simple example would be to
search for all nodes containing a given string. The schema which describes the successful
operation is given below. The input, *keyword*?, is the search string and the set of nodes
in the document containing the string is returned in the set *found*!. One of these nodes
may become the current node.

```
__ KeywordSearch _____
  TextHYPERXTET
  keyword? : STRING
  found! : ℙNODE
 _____
  found! = {n : Nodes | keyword? in (Text n).text • n}
```

4.5 Properties of Hypertext

It is straightforward in the Z framework to define additional properties of hypertext. We
give two examples here. The first states that every node in the hypertext can be reached
from the start node of a user session (whether user or system defined) using organisational links only. The second that no nodes are dead ends, or equivalently, that there is
always at least one organisational, referential or span link out of any node.

```
__ Accessibility _____
  ButtonHYPERTEXT
 _____
  defined StartNode ⇒ Nodes ⊆ ran (StartNode ◁ Links*)
  undefined StartNode ⇒ (∀n : Nodes • Nodes ⊆ ran ({n} ◁ Links*))
```

```
__ NoDeadEnds _____
  HighlightHYPERTEXT
  TypedHighlightLINKS
 _____
  Nodes ⊆ dom (RecoverLinks (OrgLinks ∪ RefLinks ∪ SpanLinks))
```

4.6 Authoring

In addition to operations for reading an information space using hypertext, many systems also provide *authoring* operations by means of which nodes and links can be added to a hypertext to represent the information space in a more effective way. Adding a new node is known as indexing, adding a new link is known as hyperization. In the following schema we describe the hyperization operation of creating a new organisational link in the highlight hypertext.

```
AddHighlightLink
ΔHighlightHYPERTEXT
hlink? : HighlightLINK
─────────────────────────
hlink? ∉ HighlightLinks
hlink?.From ∈ Nodes
hlink? ∈ OrgLINK
OrgLinks' = OrgLinks ∪ {hlink?}
RefLinks' = RefLinks
SpanLinks' = SpanLinks
```

As a further example, we define the operation of removing a node from the basic model where any links which point to or from that node are also deleted.

```
RemoveNode
node? : NODE
ΔHYPERTEXT
─────────────────────────
node? ∈ Nodes
Nodes' = Nodes \ {node?}
Links' = {node?} ⊲ Links ⊳ {node?}
```

5 Summary and Future Work

In this paper we have presented part of the formal specification we have developed in order to define a framework for hypertext systems. This, we argue, provides an environment in which to discuss, design, develop and evaluate such systems. Z has enabled us to produce a well structured specification accessible to researchers from a non-formal background. We have been able to describe hypertext at the highest level of abstraction and then, using increasingly detailed specification, we have been able to add necessary system complexity at appropriate levels. We have shown how constructing Z in such a way does not restrict a specifier to any particular mathematical model; rather it provides a general mathematical framework within which different models, and even particular systems, can be defined and contrasted.

The framework has provided the foundation upon which to build a formal model of a new learning system in hypertext [16]. This system uses statistical information collected in Information Retrieval sessions over a period of time to learn how best to aid the user in navigating through a given information space. Since our framework specification is well structured, we were able to choose the appropriate level of abstraction relevant to our purpose of modelling this new system. In this case, the learning model is concerned only with organisational links between nodes, and so we have developed the formalisms of the learning techniques within our most basic model of hypertext. Next we intend to formalise certain hypertext systems within this framework in order that the new learning model of hypertext may be incorporated into these systems.

Further work continues in developing a more general and generic version of the specification and in applying it to existing hypertext systems. We have specified several existing hypertext systems including HyperCard [4] and the World Wide Web [5] within our framework. Further, using the specifications of the two systems used by Westminster University [9, 10], we are now able to investigate a means of providing automatic translation rules between the two.

Acknowledgements

Thanks to Claire Cohen, Jennifer Goodwin, Paul Howells, Michael Hu and Michael Luck and the anonymous referees for comments on earlier versions of this paper. The specification contained in this paper was checked for correctness using the *f*UZZ package.

References

1. F. Afrati and C. Koutras. A hypertext model supporting query mechanisms. In *Hypertext: Concepts, Systems and Applications. Proceedings of the European Conference on Hypertext*, pages 52–66, 1990.
2. R. Akscyn, D. McCracken, and E. Yoder. KMS: A distributed hypertext for managing knowledge in organizations. *Communications of the ACM*, 31(7), July 1988.
3. B. Campell and J. Goodman. HAM: A General Purpose Hypertext Abstract Machine. *Communications of the ACM*, 31(7):856–861, 1988.
4. C. Cohen, M. d'Inverno, and M. Priestley. Z Specification of HyperCard. Technical report, Software Engineering Division, University of Westminster, 1995.
5. C. Cohen, M. d'Inverno, and M. Priestley. Z Specification of the World Wide Web. Technical report, Software Engineering Division, University of Westminster, 1995.
6. J. Conklin. Hypertext: An Introduction and Survey. *Computer*, 20(9):17–41, 1987.
7. M. d'Inverno. Using Z to Capture the Essence of a Contract Net. Master's thesis, Programming Research Group, Oxford University, 1988.
8. M. d'Inverno and J. Crowcroft. Design, Specification and Implementation of an Interactive Conferencing System. In *Proceedings of IEEE Infocom*, 1991.
9. M. d'Inverno and M. Priestley. Z specification of GNU Emacs info system. Technical report, Software Engineering Division, University of Westminster, 1994.
10. M. d'Inverno and M. Priestley. Z specification of IDEAs system. Technical report, Software Engineering Division, University of Westminster, 1994.
11. D. Englebart. Authorship provisions in Augment. In *Proceedings of the IEEE COMPCON*, Spring 1984.
12. P. Garg. Abstraction Mechanisms in Hypertext. *Communications of the ACM*, 31(7):862–870, 1988.

13. F. Garzotto, P. Paolini, and D. Schwabe. HDM–A Model-Based Approach to Hypertext Application Design. *ACM Transactions on Information Systems*, 11(1):27–50, 1993.
14. F. Halasz. Reflections on NoteCards: Seven Issues for the Next Generation of Hypermedia Systems. *Communications of the ACM*, 31(7), July 1988.
15. F. Halasz and M. Schwartz. The Dexter Hypertext. *Communications of the ACM*, 37(2):30–39, 1994.
16. M. J. Hu. *An Intelligent Information System*. PhD thesis, UCL, 1994.
17. D. Lange. A formal model for hypertext. In *Proceedings of the NIST Hypertext Standardization Workshop*, 1990.
18. M. Luck and M. d'Inverno. A formal framework for agency and autonomy. In *Proceedings of the First International Conference on Multi-Agent Systems*, 1995.
19. H. Van Dyke Parunak. Hypermedia Topologies and User Navigation. In *Hypertext '89 Proceedings*, 1989.
20. D. Stotts and R. Furata. Petri-Net-Based Hypertext: Document Structure with Browsing Semantics. *ACM Transactions on Information Systems*, 7(1):3–29, 1989.
21. F. Tompa. A Data Model For Flexible Hypertext Database Systems. *ACM Transactions on Information Systems*, 7(1):85–100, 1989.
22. N. Yankelovich, B. Haan, N. Meyrowitz, and S. Drucker. Intermedia: The concept and the construction of a seamless information environment. *IEEE Computer*, 1988.

Appendix: Z Extensions

We have found it useful in a specification to be able to assert that an element is optional. For example, in the specification given in this paper, the error message returned by a hypertext move is optional. If the move is unsuccessful, there is an error message, if the move is successful then there is no error message. The following definitions provide for a new type, optional T, for any existing type, T, along with the predicates defined and undefined which test whether an element of optional T is defined or not. The function, the, extracts the element from a defined member of optional T. We further define a prefix relation, setdisjoint, which holds for a set of sets if all the members of that set of sets are pairwise disjoint. Lastly we define two functions which cycle forwards and backwards through non-empty injective sequences.

$$\text{optional}\,[X] == \{xs : \mathbb{P}\,X \mid \#\,xs \leq 1\}$$

$$
\begin{array}{l}
\llbracket X \rrbracket \\
\hline
\text{defined}\_,\text{undefined}\_ : \mathbb{P}(\text{ optional}\,[X]) \\
\text{the: optional}\,[X] \nrightarrow X \\
\hline
\forall xs : \text{optional}\,[X] \bullet \text{defined}\,xs \Leftrightarrow \#\,xs = 1 \wedge \\
\qquad\qquad\qquad\qquad \text{undefined}\,xs \Leftrightarrow \#\,xs = 0 \\
\forall xs : \text{optional}\,[X] \mid \text{defined}\,xs \bullet \\
\qquad\qquad \text{the}\,xs = (\mu\,x : X \mid x \in xs)
\end{array}
$$

$=[X]\!=\!=$

setdisjoint $\_$: $\mathbb{P}(\mathbb{P}(\mathbb{P}\,X))$

$\forall xss : \mathbb{P}(\mathbb{P}\,X) \bullet$ setdisjoint $xss \Leftrightarrow (\forall xs, ys : \mathbb{P}\,X \bullet$
$\qquad ((xs \in xss) \wedge (ys \in xss) \wedge (xs \neq ys)) \Rightarrow (xs \cap ys) = \emptyset)$

$=[X]\!=\!=$

CycleNext, CyclePrevious: $(X \times \text{iseq}\ X) \twoheadrightarrow X$
Index: $(X \times \text{iseq}\ X) \twoheadrightarrow \mathbb{N}$

$\forall s : \text{iseq}\ X;\ x : X \mid x \in (\text{ran}\ s) \bullet \text{Index}\,(x, s) = s^{\sim}x \wedge$
$\qquad \text{Index}\,(x, s) \neq \#s \Rightarrow \text{CycleNext}\,(x, s) = s\,(\text{Index}\,(x, s) + 1) \wedge$
$\qquad \text{Index}\,(x, s) = \#s \Rightarrow \text{CycleNext}\,(x, s) = head\ s \wedge$
$\qquad \text{Index}\,(x, s) \neq 1 \Rightarrow \text{CyclePrevious}\,(x, s) = s\,(\text{Index}\,(x, s) - 1) \wedge$
$\qquad \text{Index}\,(x, s) = 1 \Rightarrow \text{CyclePrevious}\,(x, s) = last\ s$

Proof

Mechanizing Formal Methods: Opportunities and Challenges

John Rushby

Computer Science Laboratory, SRI International,
Menlo Park, CA 94025, USA

Abstract. Mechanization makes it feasible to calculate properties of formally specified systems. This ability creates new opportunities for using formal methods as an exploratory tool in system design. Achieving enough efficiency to make this practical raises challenging problems in automated deduction. These challenges can be met only by approaches that integrate consideration of its mechanization into the design of a specification language.

1 Introduction

All formal methods rest on the conviction that requirements and designs for computer systems and software can be modeled mathematically, and that many questions concerning properties of those requirements and designs can then be settled by calculation. Advocates of formal methods differ, however, in the extent to which they stress the conceptual and methodological aspects of these methods, as opposed to their calculational aspects.

While appreciating the methodological benefits of mathematical concepts and notations, I believe that the distinctive merit of specifically *formal* methods is that they support calculation. Using automated deduction (i.e., theorem proving) and related techniques (e.g., model checking) it is possible to calculate whether a formally-described design satisfies its specification or possesses certain properties. To be useful, it must be possible to perform these calculations for specifications and properties of practical interest with reasonable ease and efficiency.

It takes a lot of theorem-proving power to do this, and mechanized specification languages must be designed to mesh well with the most effective theorem-proving techniques. For example, reasoning about equality in the presence of uninterpreted function symbols is crucial to most applications of mechanized formal methods, and efficient techniques for achieving this (such as congruence closure [14]) require that functions are total. However, it is draconian and rather unnatural to force all functions (including, for example, division) to be total, so an effectively mechanized formal method requires careful and integrated design choices to be made for both the specification language and its supporting mechanization. In this particular case, it is necessary to find some reasonably attractive treatment for partial functions that does not compromise the efficiency of equality reasoning.

While developing such a treatment poses a challenge, the hypothesized availability of a powerful theorem prover creates new opportunities: for example, the typechecker of our specification language could use the theorem prover to check that partial functions are never applied outside their domains. This balancing of challenges and opportunities, and the corresponding need for integrated design decisions, arises again and again when mechanizing formal methods. In the following sections, I will briefly describe the main opportunities created by mechanized formal methods, and the technical challenges in achieving effective mechanization. My perspective on these topics is influenced by experiences in the development and use of PVS [9].

2 Opportunities for Making Effective Use of Mechanized Formal Methods

It is often assumed that the main goal of mechanized formal methods is "proving correctness" of programs or detailed hardware designs, and this assumption may be reinforced by the term "verification system" that is commonly used to describe mechanized tools for formal methods. In fact, however, this assumption is wrong on both counts: most advocates of mechanized formal methods consider such early-lifecycle products as requirements, architectural designs, and algorithms to be more attractive targets for their tools than finished programs or gate layouts, and their focus is at least as much on finding faults and on design exploration as it is on verifying correctness.

Preference for early-lifecycle applications of mechanized formal methods is partly a consequence of the strength of traditional methods for late-lifecycle activities. Traditional methods for the development and quality assurance of program code and detailed hardware designs are sufficiently effective that very few significant faults are introduced and remain undetected at these late stages of the lifecycle (for example, of 197 critical faults found during integration testing of two JPL spacecraft, only three were programming mistakes [8]), and only a small fraction of overall development costs (typically, less than 10%) are incurred here. In contrast, mechanized formal methods are quite expensive to apply at these stages. Primarily, this is because the products of the late lifecycle are usually large—typically, hundreds of thousands, or even millions, of gates or lines of code—and their sheer size makes formal verification costly.

But if traditional methods are effective in the later stages of the lifecycle, the same cannot be said of the earlier stages. Natural-language, diagrams, pseudocode, and other conventional ways for describing requirements and preliminary designs do not support calculation, so the main way to deduce their properties and consequences is through the fallible processes of inspection and review. The fallibility of these processes is illustrated by the JPL data cited earlier: since only three of the 197 critical faults were due to programming errors, it follows that the other 194 were introduced at earlier stages. Lutz reports that 50% of these faults were due to flawed requirements (mainly omissions) for individual components, 25% were due to flawed designs for these components, and the remaining 25% were due to flawed interfaces between components and incorrect interactions among them [8].

Rapid prototyping and simulation provide more repeatable and systematic examination of these issues, but often force premature consideration of implementation questions and thereby divert attention from the most important topics. Mechanized formal methods, on the other hand, can support direct analysis and exploration of the products of the early lifecycle: as soon as we have written down a few logical formulas that describe some aspect of the system, so we can begin to check their consequences—such as whether they entail some expected property, or are mutually contradictory.

The purposes for which mechanized formal methods may be used in the early lifecycle are as much those of validation and exploration as verification: the opportunities are to validate requirements specifications, to explore different system architectures and the interactions of their components, to debug critical algorithms and to understand their properties and assumptions, and to cope with changes.

Mechanized formal methods can assist requirements validation by checking whether a formal statement of requirements entails other expected properties: intuitions such as "if I've got that right, then this ought to follow," can be examined in a formal manner using theorem proving or model checking. While confirmation of an expected property is gratifying, a more common outcome—at least in the early stages—is the discovery that the requirements must be revised, or that some new assumption must be adopted. The rigor of mechanized analysis renders the discovery of such oversights far more systematic than is the case for informal reviews. Furthermore, mechanization allows us to check, rapidly and reliably, that previously examined properties remain true following each revision to the requirements.

The assurance derived by checking that expected properties are entailed by a requirements specification may be specious if the requirements are inconsistent (i.e., mutually contradictory): every property is entailed by an inconsistent specification. Consistency of formal specifications can be demonstrated by exhibiting a model; an equivalent demonstration can be checked mechanically using theory interpretations: the basic idea is to establish a translation from the types and constants of the "source" specification (the one to be shown consistent) to those of a "target" specification and to show that the axioms of the source specification, when translated into the terms of the target specification, become provable theorems of that target specification. If this can be done, then we have demonstrated *relative* consistency: the source specification is consistent if the target specification is. Generally, the target specification is one that is specified definitionally, or one for which we have some other good reason to believe in its consistency.

Following requirements validation, mechanized formal methods can be used to explore candidate system architectures. Architectures consist of interacting components; the concerns at this stage are generally to verify that the properties assumed of the components and of their interaction are sufficient to ensure satisfaction of the overall requirements. The chief benefits of applying mechanically checked formal verification to this task are the ability to explore alternative designs and assumptions, and to prune unnecessary assumptions. For example, formal examination of an architecture for fault masking and transient recovery in flight control systems reveals the need for interactive consistency on sensor inputs [11]. This can be achieved by a Byzantine Agreement algorithm [5]. Inputs to the majority vote function must also satisfy interactive consistency and it may therefore appear as if these, too, need to be run through a Byzantine Agreement algorithm. In fact, this is not required: it is possible to prove that interactive consistency of voter inputs is an inherent property of the architecture. Mechanized formal analysis allows this attribute of the architecture to be determined with certainty, and it also allows determination of the exact circumstances under which a modified architecture provides recovery from transient faults [11].

As with requirements validation, exploration of architectural design choices is an iterative process that is greatly facilitated by the rigor and repeatability of mechanized formal methods. Simulation and direct execution share the repeatability of formal methods but can examine only a few cases, whereas deductive formal methods allow consideration of *all* cases. But, unfortunately, the price of this generality can be less automated and more difficult analysis: we may need to invent invariants and to undertake proofs by induction in order to cover an infinite state space. Often, however, an effective alternative is available: we may be able to abstract or "downscale" a specification to a finite state space that can be examined, exhaustively and automatically, by model checking or by explicit state exploration. The great automation of formal finite-state methods creates new opportunities for applying formal methods in the design loop. Experience indicates that examining *all* the cases of such an abstracted system description is generally more effective at finding faults than testing or simulating *some* of the cases of the full description [3].

Once a preferred system architecture has been selected we may, recursively, explore architectures for its components or—once a sufficiently detailed level has been reached—investigate algorithms for those components. As with the earlier stages, the great benefits of using mechanized formal methods to examine algorithms are the abilities to explore alternatives, to prune assumptions, and to adapt to design changes. For example, the journal presentation of the interactive convergence clock synchronization algorithm [4] has an assumption that all initial clock adjustments are zero. Friedrich von Henke and I retained this assumption when we formally verified the algorithm [13]. Subsequently, when contemplating design of a circuit to implement part of the algorithm, it became clear that this assumption is exceedingly inconvenient. I explored the conjecture that it is unnecessary by simply striking it out of our formal specification and rerunning the proofs of all the lemmas and theorems that constitute the verification. (There are about 200 proofs in the full verification and it takes about 10 minutes for the theorem prover to check them all.) It turned out that the proofs of a few lemmas failed without the assumption, but examination showed that those lemmas could easily be restated, or given different proofs. A few hours of work were sufficient to make these adjustments to the formal specification and mechanically checked verification.

In other revisions to this algorithm and its verification, I have tightened the bound on the achieved clock skew, and extended the fault model so that the algorithm tolerates larger numbers of simple faults, without compromising its ability to resist arbitrary (i.e., Byzantine) faults [12]. In each case, the effort required to investigate the proposed revision and to rework the formal specification and verification was on the order of a day or two. In another example, Lincoln and I developed the formal specification and verification of a Byzantine Agreement algorithm for an asymmetric architecture in less than a day by modifying an existing treatment for a symmetric architecture [7].

The ability to make these enhancements to complex algorithms, rapidly and reliably, is an opportunity created by mechanized formal methods. Informal methods of proof are unreliable in these domains (see [6, 9, 13] for examples) and it requires superhuman discipline to bring the same level of care and skepticism to the scrutiny of a modified algorithm as to the original. A formal specification and verification, on the other hand, is a reusable intellectual resource: its properties can be calculated, and those calculations can be mechanized.

In summary, the opportunities created by mechanized formal methods are similar to those offered by mechanized calculation in other fields, such as computational fluid dynamics: exploration of design alternatives, early detection of design errors, identification of assumptions, the ability to analyze the consequences of changes and responses to them, and the acquisition of an enhanced understanding that can lead to further improvements. In the next section, I will briefly examine some of the challenges in developing mechanizations of formal methods that can make these opportunities reality.

3 Technical Challenges in Mechanizing Formal Methods

The primary challenge to achieving the benefits described above is the difficulty of mechanizing formal deduction in a way that is both efficient and enlightening. Since automated deduction—i.e., theorem proving—is a fairly well-developed field, applying this technology to formal methods might seem to be simply a matter of engineering. In fact, most of the challenges are those of engineering, but they are not simple. Theorem proving in support of formal methods raises issues that are quite different from those that have traditionally been of interest in the theorem-proving community. Most notably, candidate theorems generated by formal methods in their exploratory and debugging roles are very likely to be false. In these cases, mechanization is expected to quickly reveal the falsehood, and to help identify its causes. The traditional concern of theorem proving, however, has been to prove true theorems, and most off-the-shelf theorem provers are therefore ill-suited to the needs of formal methods.

But although an existing theorem prover is unlikely to be useful in support of formal methods, we must not ignore the component techniques developed for automated theorem proving. One of the principal reasons that many attempts at mechanizing formal methods have failed is that their developers did not appreciate the raw power and speed that is needed from their theorem-proving components, and did not make use of the relevant techniques. The most important of these techniques are decision procedures for specialized, but ubiquitous, theories such as arithmetic, equality, function updates (i.e., overriding), and propositional calculus. Decision procedures are helpful in discovering false theorems (especially if they can be extended to provide counterexamples) as well as in proving true ones, and their automation dramatically improves the efficiency of proof.

There are decision procedures for many useful theories, but few problems fall precisely in the domain of any single one of them, so one of the big engineering challenges in mechanizing formal methods is to develop effective combinations of decision procedures. This requires very careful selection of the individual procedures. For example, the decision procedure for Presburger arithmetic (i.e., the first order theory of linear arithmetic with relation symbols such as $<$) does not consider uninterpreted function symbols. Since function symbols are pervasive in formal specifications, a better choice than true Presburger arithmetic is the theory of ground (i.e., unquantified) linear arithmetic, which can be combined with the theory of equality over uninterpreted function symbols [15].

Small extensions to decidable theories can have considerable value. For example, it is not possible to add full nonlinear multiplication to the decision procedures for linear arithmetic, but it is possible to add the ability to reason about the signs of products (e.g., "a minus times a minus is a plus") and this proves to be a significant benefit in practice.

Rewriting is another technique that is essential to efficient mechanization of formal methods. Unrestricted rewriting provides a decision procedure for theories axiomatized by terminating and confluent sets of rewrite rules, but few such theories arise in practice. Consequently, rewriting cannot be unrestricted, in general, but must be performed under some control strategy. For example, one of the control strategies used in PVS will rewrite a definition whose body involves a top-level *if-then-else* only if the condition to the *if* can be reduced to *true* or *false*. This reduction may involve use of decision procedures and further rewriting, and it is possible to expend considerable resources on a search that is ultimately unsuccessful (because it does not succeed in reducing the condition). These resources will not have been wasted, however, if they allow the theorem prover to avoid making an unprofitable rewrite: in most contexts, the heuristic effectiveness of a good control strategy is likely to be more beneficial than the raw speed of a blind rewriter. As this description suggests, rewriting and decision procedures cannot stand apart: truly effective theorem provers must integrate them very tightly. A classic account of the issues in such integration is given by Boyer and Moore [1].

Integration is a pervasive theme in the effective mechanization of formal methods: many individual techniques work well on selected examples, but fail in more realistic contexts because problems seldom fall exactly within the scope of one method. Sometimes the integration must be tight, as in cooperating decision procedures, and the integration of decision procedures with rewriting. In other cases, the integration can be less tight; a good example is model checking within a theorem-proving context. Whereas theorem proving attempts to show that a formula follows from given premises, model checking attempts to show that a given system description is a model for the formula. As noted in the previous section, an advantage of model checking is that, for certain finite state systems and temporal logic formulas, it is much more automatic and efficient than theorem proving. The additional benefits of a system that provides both theorem proving and model checking are that model checking can be used to discharge some cases of a larger proof or, dually, that theorem proving can be used to justify the reduction to finite state that is required for automated model checking [10].

Model checking can provide a further benefit: before undertaking a potentially difficult and costly proof, we may be able to use model checking to examine some restricted or special cases. Any errors that can be discovered and eliminated in this way will save time and effort in theorem proving. This is a particular case of a more general desideratum: mechanized formal methods should provide a graduated collection of tools and techniques that apply increasingly strict scrutiny at correspondingly increasing cost. Representative techniques include typechecking, animation or direct execution, model checking, and theorem proving. These techniques become more effective if they are integrated so that each can use capabilities of the others. For example, typechecking can be made more strict if it is allowed to use theorem proving, rather than being restricted to trivially decidable properties; certain specifications can be executed on test cases by using the theorem prover to perform rewriting (in this case, fast "blind" rewriting may be desirable); and theorem proving can be more efficient if it makes use of type information provided by the typechecker.

Achieving the necessary integrations involves more than careful engineering of theorem proving and support tools: it extends to the design of the specification language itself. As noted in the introduction, efficient equality reasoning, for example, requires that functions are total. If we allow the concerns of mechanization to dominate language design, we may then decide that our specification language should provide only total functions. Similar considerations may lead us to restrict quantification to first-order (or to eliminate explicit quantification altogether), to restrict recursive definitions to the syntactic form of primitive recursion, or to require all formulas to be equations. In the limit, we may provide a raw logic devoid of the features expected of a specification language. Conversely, concerns for an expressive notation may lead us to provide a specification language that cannot be mechanized effectively. This is not to say that expressiveness cannot be combined with mechanization, but that expressiveness must not be considered in isolation from mechanization. For this reason, I consider it dangerous to look to the classical foundations of mathematics for guidance when designing a specification language. These formal systems (notably, first-order logic with axiomatic set theory) were created in order to be studied, not in order to be used—the"...interest in formalized languages being less often in their actual and practical use as languages than in the general theory of such use and its possibilities in principle" [2, page 47]. Unsurprisingly, therefore, set theory has characteristics that pose difficulty for mechanization—for example, as already noted, functions are inherently partial in set theory (they are sets of pairs). Also, it is difficult to provide really strict typechecking (and hence, early error detection) for set theory without sacrificing some of its flexibility: for example, a function is a set, so it can (sometimes) make sense to form its union with another set, and it is therefore not clearcut whether type restrictions should prohibit or allow this sort of construction.

If mechanization is a goal, then design of the specification language and selection of its underlying logic should not be undertaken independently of consideration of its mechanization. This does not mean that concern for mechanization must inevitably restrict or impoverish a notation—rather, I believe that availability of powerful mechanization creates new opportunities for the design of expressive, yet mechanically tractable, notations. An example is provided by the treatment of partial functions such as division in PVS, whose specification language supports the notion of *predicate subtypes*. These allow the nonzero real numbers to be defined as follows.

```
nonzero_real: TYPE = { r: real | r ≠ 0 }
```

We can then give the signature of division as

```
/: [real, nonzero_real → real]
```

and the function is total on this precisely specified domain. We can then state and prove the following result.

```
inverse_sum: LEMMA
    ∀ (a, b: real):
        a ≠ 0 ∧ b ≠ 0 ⊃ (1/a + 1/b) = (a + b)/(a × b)
```

PVS allows a value of a supertype to appear where one of a subtype is required, provided the value can be proved, in its context, to satisfy the defining predicate of the subtype concerned. In this case, PVS will generate the following three proof obligations, called Type Correctness Conditions (TCCs), that must be discharged before inverse_sum is considered fully type-correct. The three TCCs correspond to the three appearances of division in the formula inverse_sum and collectively ensure that the value of the formula does not require division by zero. The first two TCCs can be discharged automatically by a decision procedure for linear arithmetic; the third requires extensions mentioned earlier that reason about the signs of nonlinear products.

tcc1: OBLIGATION $\forall$ $(a, b:$ real$): a \neq 0 \wedge b \neq 0 \supset a \neq 0$

tcc2: OBLIGATION $\forall$ $(a, b:$ real$): a \neq 0 \wedge b \neq 0 \supset b \neq 0$

tcc3: OBLIGATION $\forall$ $(a, b:$ real$):$
$$a \neq 0 \wedge b \neq 0 \supset (a \times b) \neq 0$$

Predicate subtypes in PVS provide many more capabilities than are suggested by this simple example: in particular, injections and surjections are defined as predicate subtypes of the functions, and state-machine invariants can be enforced by the same mechanism. The source of these conveniences and benefits is allowing typechecking to require theorem proving (i.e., to become algorithmically undecidable), which is only feasible if a powerful and automated theorem prover is assumed to be available. Conversely, the power of the theorem prover is enhanced by the precision of the type information that is provided by the language and its typechecker.

4 Conclusion

Mechanization creates new opportunities for formal methods: by making it feasible to calculate properties of formally specified designs, mechanization allows exploration of alternative designs, examination of assumptions, adaptation to changed requirements, and verification of desired properties. These opportunities are likely to have maximum benefit when applied early in the development lifecycle, and to the hardest and most important problems of design.

To realize these benefits, mechanizations of formal methods must provide several capabilities ranging from very strict typechecking, to powerfully automated theorem proving. These individual capabilities need to be closely integrated with each other, and with the specification language. Because of the integration required, consideration of its mechanization must be factored into the design of a specification language.

References

1. R. S. Boyer and J S. Moore. Integrating decision procedures into heuristic theorem provers: A case study with linear arithmetic. In *Machine Intelligence*, volume 11. Oxford University Press, 1986.

2. Alonzo Church. *Introduction to Mathematical Logic, Volume 1*. Princeton University Press, Princeton, NJ, 1956. (Volume 2 never appeared.)

3. David L. Dill, Andreas J. Drexler, Alan J. Hu, and C. Han Yang. Protocol verification as a hardware design aid. In *1992 IEEE International Conference on Computer Design: VLSI in Computers and Processors*, pages 522–525. IEEE Computer Society, 1992. Cambridge, MA, October 11-14.

4. L. Lamport and P. M. Melliar-Smith. Synchronizing clocks in the presence of faults. *Journal of the ACM*, 32(1):52–78, January 1985.

5. Leslie Lamport, Robert Shostak, and Marshall Pease. The Byzantine generals problem. *ACM Transactions on Programming Languages and Systems*, 4(3):382–401, July 1982.

6. Patrick Lincoln and John Rushby. A formally verified algorithm for interactive consistency under a hybrid fault model. In *Fault Tolerant Computing Symposium 23*, pages 402–411, Toulouse, France, June 1993. IEEE Computer Society.

7. Patrick Lincoln and John Rushby. Formal verification of an interactive consistency algorithm for the Draper FTP architecture under a hybrid fault model. In *COMPASS '94 (Proceedings of the Ninth Annual Conference on Computer Assurance)*, pages 107–120, Gaithersburg, MD, June 1994. IEEE Washington Section.

8. Robyn R. Lutz. Analyzing software requirements errors in safety-critical embedded systems. In *IEEE International Symposium on Requirements Engineering*, pages 126–133, San Diego, CA, January 1993.

9. Sam Owre, John Rushby, Natarajan Shankar, and Friedrich von Henke. Formal verification for fault-tolerant architectures: Prolegomena to the design of PVS. *IEEE Transactions on Software Engineering*, 21(2):107–125, February 1995.

10. S. Rajan, N. Shankar, and M.K. Srivas. An integration of model-checking with automated proof checking. In Pierre Wolper, editor, *Computer-Aided Verification, CAV '95*, Springer-Verlag *Lecture Notes in Computer Science*, Liege, Belgium, June 1995. To appear.

11. John Rushby. A fault-masking and transient-recovery model for digital flight-control systems. In Jan Vytopil, editor, *Formal Techniques in Real-Time and Fault-Tolerant Systems*, Kluwer International Series in Engineering and Computer Science, chapter 5, pages 109–136. Kluwer, Boston, Dordecht, London, 1993. An earlier version appeared in *Formal Techniques in Real-Time and Fault-Tolerant Systems*, volume 571 of Springer-Verlag *Lecture Notes in Computer Science*, pages 237–257, Nijmegen, The Netherlands, January 1992.

12. John Rushby. A formally verified algorithm for clock synchronization under a hybrid fault model. In *Thirteenth ACM Symposium on Principles of Distributed Computing*, pages 304–313, Los Angeles, CA, August 1994. Association for Computing Machinery.

13. John Rushby and Friedrich von Henke. Formal verification of algorithms for critical systems. *IEEE Transactions on Software Engineering*, 19(1):13–23, January 1993.

14. Robert E. Shostak. An algorithm for reasoning about equality. *Communications of the ACM*, 21(7):583–585, July 1978.

15. Robert E. Shostak. Deciding combinations of theories. *Journal of the ACM*, 31(1):1–12, January 1984.

An Algebraic Proof in $VDM^{\clubsuit}$

A. P. Hughes   A. A. Donnelly

Department of Computer Science,
Trinity College Dublin, Dublin 2, Ireland
e-mail: (Arthur.P.Hughes , Alexis.Donnelly)@cs.tcd.ie

Abstract. We present a simple example to illustrate the algebraic, constructive style of specification and proof used in the Irish School of the VDM ($VDM^{\clubsuit}$). The example exploits a new and fundamental result which we have used frequently in $VDM^{\clubsuit}$. Our illustration also demonstrates a novel approach within in the VDM tradition which harnesses additional branches of abstract algebra for use in the specification and development of software. This innovation offers, we believe, several benefits: (i) an extension of the mathematical foundation of the emerging software engineering discipline, (ii) a more appropriate and abstract mathematical level at which the designer may work which abbreviates and simplifies proof and consequently (iii) a mathematical setting where insights into designs are easier to make and on foot of which designs may be confidently modified or improved. As a representative example, we also offer a proof,using the same approach, which we believe to be new, of one of the theorems used in the example.

1 Introduction

The VDM specification notation [5] has been used widely in the development and construction of software systems. Its mathematical foundations are largely set theory and logic, a foundation which is also shared with the Z notation [10]. However it is also possible to bring other results of abstract algebra to bear in the development process, a preference of the Irish School of the VDM ($VDM^{\clubsuit}$) [6], [8]. Another feature of this school is the classical engineering mathematics style of proof in which equals are substituted for equals. The mathematics used is also constructive. We believe this style of VDM offers several advantages:

- The mathematical foundations of the engineering discipline are extended into another useful branch of discrete mathematics.

- The additional mathematics is at a higher, more abstract level than logic and set theory and offers the designer more appropriate and powerful tools. This results shorter proofs focussed on the central abstraction/model being described. To use an electrical engineering metaphor, an impedance mismatch problem is solved.

- A slightly more subjective, but we believe crucial, advantage is that by working in the algebraic, constructive style of $VDM^{\clubsuit}$ the designer acquires a view of his model that is at once conveniently high level and mathematically insightful. The point can be illustrated with another electrical engineering metaphor. Assuming identical graphical presentation modules, a symbolic (*Mathematica*-style) circuit analysis/simulator tool can provide sound mathematical reasons (from equational data) for the presence of troublesome poles or zeroes in the frequency characteristics of a circuit, not just merely indicate their presence. The advantage to the designer is that the mathematical insight makes it easier to modify the design and do so with greater confidence that it will perform as required — the equations reveal the *origin* of the troublesome pole, not merely indicate its presence.

The benefit of using $VDM^{\clubsuit}$ is that the designer is now free to work at a higher, more appropriate and insightful level of abstraction than logic and set theory. More generally, a practitioner's increased familiarity with this appropriate abstract mathematics, emphasised in the Irish School, may serve to deepen his understanding of commonly used models in software designs and thus appreciate quickly the implications of possible modifications. Such insights are not as easy to make at the lower level of logic and set theory.

We illustrate our claims above with a simple example — a specification of a structured map $X \to (Y \to Z)'$ which is used to model some of the workings of a parliament (the Irish Dáil). The centrepiece of the example is the surprisingly short proof of an invariant which states that a member of parliament cannot be a member of two political parties (the so-called "servant of one master" invariant). The invariant is proved in the context of removal and insertion operations which model the death or resignation of an M.P. and his/her replacement by the winner of the resulting by-election. Pre-conditions (with natural interpretations) for the operations arise naturally in the proof. A new proof of one of theorems (used frequently by us in this and other examples) is also presented using the characteristic constructive engineering style in the same mathematical setting.

The rest of this paper is organised as follows: Section 2 briefly presents some mathematical definitions and preliminaries used in the $VDM^{\clubsuit}$ style which may be unfamiliar to some readers. The central model and invariant proof of the paper are dealt with in Section 3. Space permits only one of theorems used in this proof to be outlined and proved in Section 4. Finally Section 5 summarises our conclusions.

2 Mathematical Definitions and Preliminaries

The mathematics used in $VDM^{\clubsuit}$ is described at length elsewhere [6], [7], [8] and an example of the notation and proof style can be found in [1]. However we briefly introduce some notations and state some important definitions and results used in our example for the benefit of readers not familiar with this style.

We use priming in the $VDM^{\clubsuit}$ to denote the absence of an obvious null element from a structure. Thus $\mathcal{P}'X$ denotes the space of sets of X (the power set of X) with the empty set removed. Hence $x \in \mathcal{P}'X$ means that x is some non-empty set of elements of X. Similarly, $\rho \in (X \to \mathcal{P}'Y)'$ states that ρ is some non-empty relation from X to (non-empty) sets of Ys.

For two functions $f : X \to A$ and $g : Y \to B$ we define a functor called a map iterator, written $(f \to g)$ which can be applied to a map $\rho : X \to Y$. The result is a map from A to B inheriting the associations of ρ and constructed from it by applying f to elements of the domain of ρ and applying g to corresponding elements of the range of ρ.

A *Monoid*, denoted $(X, \bullet, \iota)$, is an algebraic structure consisting of a base set, X, and a binary operator, $\bullet$, for which there exists a unique identity element, ι. The base set

X, must be closed under the operator $\bullet$ which must also be associative. If in addition, $\bullet$ is also commutative the structure is known as a *commutative* monoid. Examples of Monoids include the natural numbers under addition, $(\mathcal{N}, +, 0)$ and the natural numbers excluding zero under multiplication $(\mathcal{N}', \times, 1)$.

A *homomorphism* $h : M \rightarrow P$ between two monoids $(M, \oplus, 0)$ and $(P, \otimes, 1)$ is a function h which renders the effects of the $\oplus$ operation in M as the effects of the $\otimes$ operation in P. The identity elements must also be mapped. Thus:

$$h(m_1 \oplus m_2) = h(m_1) \otimes h(m_2) \tag{1}$$

$$h(0) = 1 \tag{2}$$

If it happens that the sets from the domain and range monoids of h above coincide (i.e. $M = P$) then h is known as an *endomorphism*.

A structure used frequently in the example presented here and in others in the Irish School of *VDM* is that of the *Indexed Monoid* ([8], page 29). An indexed monoid, $I \bigcirc (X, \bullet, \iota)$, can be constructed from a base monoid, say $(X, \bullet, \iota)$ and an indexing set I. Elements of the new monoid are simply maps from I to X' in $(I \rightarrow X')$ where $X' = \{\iota\} \triangleleft X$, the identity element is the null map, θ, and the associative binary operator $\odot$ is an indexed version of the base operator $\bullet$. Thus, in this case, the indexed monoid is $(I \rightarrow X', \odot, \theta)$. If $\bullet$ is set union for example, then we usually construct $\odot$ to be a map-building operator defined in terms of map extend and map override (see [8] page 15, and equation (3) below for examples).

Additional results used in the course of proof are stated in the text.

3   The Dáil Model, Resignations and Elections

The Irish parliament, or Dáil , is composed of T.D.s (*teachta dála* = member of parliament) and each T.D. has declared financial interests. The financial interests of a T.D. can be represented and built by the monoid of sets of financial interests $(\mathcal{P}FI, \cup, \emptyset)$. If we now take this monoid and index it with respect to the space TD of T.D.s we obtain an indexed structure $TD \bigcirc (\mathcal{P}FI, \cup, \emptyset)$ which is also a monoid of binary relations $(TD \rightarrow \mathcal{P}FI, \bigcirc, \theta)$. This relational monoid is used to represent and construct the T.D.-financial interest relationship. Note that a null financial interest is impossible in this structure by definition. For a pair of T.D.-financial interest relationships ρ and $[td \mapsto s]$ in $TD \rightarrow \mathcal{P}FI$ we can define a relational union in terms of map extend ($\sqcup$) and map override ($\dagger$):

$$\rho \,\textcircled{\cup}\, [td \mapsto S] = \begin{cases} \rho \sqcup [td \mapsto S] & \neg\chi[\![td]\!]\rho \\ \rho \dagger [td \mapsto \rho(td) \cup S] & \text{otherwise} \end{cases} \tag{3}$$

Where $\chi[\![td]\!]\rho$ denotes the characteristic function of the relation ρ. Thus, a typical relation ρ of T.D.s and their financial interests may be represented by

$$\rho = \begin{bmatrix} td_1 \mapsto \{fi_{11}, \cdots, fi_{1w}\} \\ \vdots \;\; \mapsto \;\;\; \vdots \\ td_i \mapsto \{fi_{i1}, \cdots, fi_{ix}\} \end{bmatrix}$$

The Dáil is also composed of political parties. Each T.D. is a member of one and only one party (ideally). If we take the monoid of relations of T.D.-financial interest relationships $(TD \rightarrow \mathcal{P}'FI, \textcircled{\tiny U}, \theta)$ and index it with respect to the space P of political parties obtaining the structure $P \bigcirc (TD \rightarrow \mathcal{P}'FI, \textcircled{\tiny U}, \theta)$ we have another monoid, specifically, $(P \rightarrow (TD \rightarrow \mathcal{P}'FI)', \textcircled{\tiny U}\theta)$ which can represent and construct a Dáil structure. The notation $(TD \rightarrow \mathcal{P}'FI)' = \theta \triangleleft (TD \rightarrow \mathcal{P}'FI)$ denotes the space of T.D.-financial interest relations with the empty relation deleted.

If δ and $[p \mapsto \rho]$ are Dáils in $P \rightarrow (TD \rightarrow \mathcal{P}'FI)'$ then we define the double-indexed union operator, $\textcircled{\tiny U}$ in a similar fashion to equation (3):

$$\delta \, \textcircled{\tiny U} [p \mapsto \rho] = \begin{cases} \delta \sqcup [p \mapsto \rho] & \neg \chi[\![p]\!]\delta \\ \delta \dagger [p \mapsto \delta(p) \, \textcircled{\tiny U} \, \rho] & \text{otherwise} \end{cases} \tag{4}$$

A typical δ in $P \rightarrow (TD \rightarrow \mathcal{P}'FI)'$ may be represented by

$$\delta = \begin{bmatrix} p^1 \mapsto & \begin{bmatrix} td_1^1 \mapsto \{fi_{11}^1, \cdots, fi_{1w}^1\} \\ \vdots \; \mapsto \; \vdots \\ td_i^1 \mapsto \{fi_{i1}^1, \cdots, fi_{ix}^1\} \end{bmatrix} \\ \vdots \; \mapsto \; \vdots \\ p^n \mapsto & \begin{bmatrix} td_1^n \mapsto \{fi_{11}^n, \cdots, fi_{1y}^n\} \\ \vdots \; \mapsto \; \vdots \\ td_j^n \mapsto \{fi_{j1}^n, \cdots, fi_{jz}^n\} \end{bmatrix} \end{bmatrix}$$

Sadly, not every δ in $P \rightarrow (TD \rightarrow \mathcal{P}'FI)'$ will represent a Dáil . This is because there exists a δ in $P \rightarrow (TD \rightarrow \mathcal{P}'FI)'$ where where a T.D. can be a member of more than one political party. To identify precisely that subspace of $P \rightarrow (TD \rightarrow \mathcal{P}'FI)'$ which models a Dáil we supply a partition invariant.

3.1 The Dáil Partition Invariant — Party Loyalty

In the Dáil each T.D. must be a member of exactly one political party. Thus the political parties partition the Dáil into groups of T.D.s belonging to the same party. More formally, for some non-null δ in $P \rightarrow (TD \rightarrow \mathcal{P}'FI)'$ we can split δ according to political party as follows

$$\delta = [p^1 \mapsto \rho^1] \sqcup \ldots \sqcup [p^n \mapsto \rho^n] \tag{5}$$

where the $[p^1 \mapsto \rho^1], \ldots, [p^n \mapsto \rho^n]$ are in $P \rightarrow (TD \rightarrow \mathcal{P}'FI)'$. We can apply the map iterator $(I \rightarrow dom)$ to δ in order to extract just the party-T.D.s information from the structure. The identity function I in the iterator is applied to each element of the domain of δ and the dom function is applied to the corresponding element of the range of δ. The association between the outputs of these two functions is inherited from δ. The application of the map iterator proceeds as follows:

$$(I \rightarrow dom)\delta = (I \rightarrow dom)([p^1 \mapsto \rho^1] \sqcup \ldots \sqcup [p^n \mapsto \rho^n])$$
$$= (I \rightarrow dom)([p^1 \mapsto \rho^1]) \, \textcircled{\tiny U} \ldots \textcircled{\tiny U} (I \rightarrow dom)([p^n \mapsto \rho^n])$$

$$= [p^1 \mapsto dom\,\rho^1] \,\textcircled{\tiny U}\, \cdots \,\textcircled{\tiny U}\, [p^n \mapsto dom\,\rho^n]$$

$$= \begin{bmatrix} p^1 \mapsto \{td_1^1, \cdots, td_i^1\} \\ \vdots \quad \mapsto \qquad \vdots \\ p^n \mapsto \{td_1^n, \cdots, td_j^n\} \end{bmatrix}$$

The first line above simply substitutes the partitioned version of δ from equation (5). The next line derives from the fact that the map iterator, $(I \to dom)$, is actually a homomorphism between two indexed monoid structures sharing a common index and "front end" under relational union. The relevant results are:

$$(I \to dom)(\delta_1 \,\textcircled{\tiny U}\, \delta_2) = (I \to dom)\delta_1 \,\textcircled{\tiny U}\, (I \to dom)\delta_2$$
$$(I \to dom)(\theta) = \theta$$

but more detail can be found in [4]. Finally, the last line of the calculation pictorially represents the party-T.D. relation so derived.
The partition property is such that $dom\,\rho^i \cap dom\,\rho^j = 0$ if $i \neq j$ for all i, j in $\{1, \cdots, n\}$ i.e. no two parties have T.D.s in common.
Next we apply relational inversion $(-)^{-1}$ to $(I \to dom)\delta$ giving us the T.D.-party relationship in $TD \to \mathcal{P}'P$

$$((I \to dom)\delta)^{-1} = ([p^1 \mapsto dom\,\rho^1] \,\textcircled{\tiny U}\, \cdots \,\textcircled{\tiny U}\, [p^n \mapsto dom\,\rho^n])^{-1} \qquad (6)$$

$$= [p^1 \mapsto dom\,\rho^1]^{-1} \,\textcircled{\tiny U}\, \cdots \,\textcircled{\tiny U}\, [p^n \mapsto dom\,\rho^n]^{-1} \qquad (7)$$

$$= [td \mapsto \{p^1\} : td \in dom\,\rho^1] \,\textcircled{\tiny U}\, \cdots$$
$$\cdots \,\textcircled{\tiny U}\, [td \mapsto \{p^n\} : td \in dom\,\rho^n] \qquad (8)$$

where $[td \mapsto \{p^1\} : td \in dom\,\rho^1]$, ... , $[td \mapsto \{p^n\} : td \in dom\,\rho^n]$ are elements of $TD \to \mathcal{P}'P$. The above expressions make use of the following theorem which is taken from [8] page 24.

Theorem 3.1 (Relational Union & Relational Inversion). *Relational inversion is a homomorphism from the monoid* $(X \to \mathcal{P}'Y, \textcircled{\tiny U}, \theta)$ *to the monoid* $(Y \to \mathcal{P}'X, \textcircled{\tiny U}, \theta)$

$$(-)^{-1} : (X \to \mathcal{P}'Y, \textcircled{\tiny U}, \theta) \longrightarrow (Y \to \mathcal{P}'X, \textcircled{\tiny U}, \theta)$$
$$(\rho_1 \,\textcircled{\tiny U}\, \rho_2)^{-1} = \rho_1^{-1} \,\textcircled{\tiny U}\, \rho_2^{-1} \qquad (9)$$
$$(\theta)^{-1} = \theta$$

Most generally this relation will be of the form

$$((I \to dom)\delta)^{-1} = \begin{bmatrix} td_1^1 \mapsto \{\cdots, p^1, \cdots\} \\ \vdots \quad \mapsto \qquad \vdots \\ td_i^1 \mapsto \{\cdots, p^1, \cdots\} \\ \vdots \quad \mapsto \qquad \vdots \\ td_1^n \mapsto \{\cdots, p^n, \cdots\} \\ \vdots \quad \mapsto \qquad \vdots \\ td_j^n \mapsto \{\cdots, p^n, \cdots\} \end{bmatrix}$$

However, if the partition property holds for δ then all the sets in the range of $((I \to dom)\delta)^{-1}$ will be singletons. An expression denoting this condition can be constructed by applying another map iterator, $(I \to card)$, to the relation $((I \to dom)\delta)^{-1}$:

$$((I \to dom)\delta)^{-1} =$$
$$[td \mapsto \{p^1\} : td \in dom\, \rho^1] \oplus \ldots \oplus [td \mapsto \{p^n\} : td \in dom\, \rho^n]$$
$$(I \to card)((I \to dom)\delta)^{-1} =$$
$$[td \mapsto 1 : td \in dom\, \rho^1] \sqcup \ldots \sqcup [td \mapsto 1 : td \in dom\, \rho^n]$$

where each T.D. is mapped (in a bag structure in $TD \to N'$) to the number of parties of which he is a member. Thus if δ is partitioned then this bag will be of the form

$$(I \to card)((I \to dom)\delta)^{-1} = \begin{bmatrix} td_1^1 \mapsto 1 \\ \vdots \mapsto \vdots \\ td_i^1 \mapsto 1 \\ \vdots \mapsto \vdots \\ td_1^n \mapsto 1 \\ \vdots \mapsto \vdots \\ td_j^n \mapsto 1 \end{bmatrix}$$

It simply remains to take the range of this bag to check that each T.D. is a member of exactly one party:

$$rng(I \to card)((I \to dom)\delta)^{-1} = \{1\} \cup \ldots \cup \{1\}$$
$$= \{1\}$$

We must not forget the case where δ is θ in which we have

$$rng(I \to card)((I \to dom)\delta)^{-1} = \emptyset$$

At last the required invariant for δ in $P \to (TD \to \mathcal{P}'FI)'$ to be a Dáil is

$$rng(I \to card)((I \to dom)\delta)^{-1} = \begin{cases} \emptyset & \delta = \theta \\ \{1\} & otherwise \end{cases} \tag{10}$$

In passing we note that this invariant guarantees that our Dáil structure is in fact a fibre bundle ([3], page 88), i.e. that there is no sharing between range elements of the structured map. The constructive style in which both the model and invariant were developed have been at a level appropriate to our example, with little mental effort devoted to lower level details. We now proceed to consider the removal and insertion of a T.D. from a typical Dáil structure.

3.2 Removal — Resignation or Death of a T.D.

Sometimes a T.D. must be removed from the Dáil structure, δ, due to death or resignation. As the death and resignation events modify the structure identically we implement both by a single removal operator, $\textcircled{\scriptsize $\ni$}'$, as follows:

$$R[\![p,td]\!]\delta \triangleq [p \mapsto \{td\}] \; \textcircled{\scriptsize $\ni$}' \; \delta \tag{11}$$

$$[p \mapsto \{td\}] \; \textcircled{\scriptsize $\ni$}' \; \delta \triangleq \begin{cases} \delta & \neg\chi[\![p]\!]\delta \\ \delta \dagger [p \mapsto \{td\} \blacktriangleleft \delta(p)] & \text{otherwise} \end{cases} \tag{12}$$

where $[p \mapsto \{td\}]$ is in $P \to \mathcal{P}'TD$ and δ is in $P \to (TD \to \mathcal{P}'FI)'$. The relational removal operator, $\textcircled{\scriptsize $\ni$}'$, is defined (see [4], page 4) in terms of the map override operator, $\dagger$, and set removal, $\blacktriangleleft$, from a map (the prime denotes removal of entries of the form $p \mapsto \theta$).

An important algebraic property of the relational removal operation on indexed structures states that the removal operation is an endomorphism of the indexed structure ([4], page 5). The significant results, which follow from removal with respect to a relation ρ, written $\rho \; \textcircled{\scriptsize $\ni$}' \; (-)$, being an outer law of the indexed structure (see Equation (1)), are:

$$\rho \; \textcircled{\scriptsize $\ni$}' \; (\iota_1 \; \circledast \; \iota_2) = (\rho \; \textcircled{\scriptsize $\ni$}' \; \iota_1) \; \circledast \; (\rho \; \textcircled{\scriptsize $\ni$}' \; \iota_2) \tag{13}$$

$$\rho \; \textcircled{\scriptsize $\ni$}' \; (\theta) = \theta \tag{14}$$

The indexed structures might be the Dáil monoid, $(P \to (TD \to \mathcal{P}'FI)', \textcircled{\scriptsize Q}\theta)$, or the party-T.D. monoid, $(P \to \mathcal{P}'TD, \textcircled{\scriptsize Q}, \theta)$, a substitution which we exploit later. We now show that no pre-condition is necessary for the removal operation by demonstrating that the modified Dáil structure (after removal) also satisfies the partition invariant.

3.3 Calculation of Removal Pre-Condition

We now apply the partition invariant of Equation (10) to the modified Dáil to see if the invariant still holds. The first step is to iterate over the altered structure by applying the domain operator to its range giving us

$$(I \to dom)R[\![p,td]\!]\delta = (I \to dom)([p \mapsto \{td\}] \; \textcircled{\scriptsize $\ni$}' \; \delta)$$
$$= [p \mapsto \{td\}] \; \textcircled{\scriptsize $\ni$}' \; ((I \to dom)\delta)$$

The first equation above is simply the definition of the removal operation. The second relies on the fact that the map iterator, $(I \to dom)$, is an admissible homomorphism. Any homomorphism between two related structures which allows us to commute the application of the homomorphism with a removal operator defined upon the structure is known as an *admissible* homomorphism. This property may be written:

$$\text{ah}(\rho \; \textcircled{\scriptsize $\ni$}' \; \delta) = \rho \; \textcircled{\scriptsize $\ni$}' \; \text{ah}(\delta) \tag{15}$$

and is explained in more detail in [6], page 128. Next we apply relational inversion resulting in

$$((I \to dom)R[\![p,td]\!]\delta)^{-1} = ([p \mapsto \{td\}] \; \textcircled{\scriptsize $\ni$}' \; ((I \to dom)\delta))^{-1}$$
$$= [p \mapsto \{td\}]^{-1} \; \textcircled{\scriptsize $\ni$}' \; ((I \to dom)\delta)^{-1}$$
$$= [td \mapsto \{p\}] \; \textcircled{\scriptsize $\ni$}' \; ((I \to dom)\delta)^{-1}$$

The passage from the first to the second line is accomplished using the following theorem on relational removal and relational inversion.

Theorem 3.2 (Relational Removal & Relational Inversion). *For any pair of relations* ρ_1 *and* ρ_2 *in* $X \to \mathcal{P}'Y$,

$$(\rho_1 \circledcirc' \rho_2)^{-1} = \rho_1^{-1} \circledcirc' \rho_2^{-1} \qquad (16)$$

where $(\rho_1 \circledcirc' \rho_2)^{-1}, \rho_1^{-1}, \rho_2^{-1}$ *in* $Y \to \mathcal{P}'X$.

We now iterate over the previous result by applying the cardinality function to the range using the map iterator, $(I \to card)$, as before:

$$
\begin{aligned}
&(I \to card)((I \to dom)R[\![p,td]\!]\delta)^{-1}\\
&= (I \to card)([td \mapsto \{p\}] \circledcirc' ((I \to dom)\delta)^{-1})\\
&= (I \to card)([td \mapsto \{p\}] \circledcirc' \tau)\\
&= (I \to card)\tau \ominus' (I \to card)([td \mapsto \{p\}] \circledcirc' \tau)
\end{aligned}
$$

where for brevity we have replaced $((I \to dom)\delta)^{-1}$ by τ. The step between the last two lines above appeals to the fact that for a pair of relations ρ_1 and ρ_2 in $X \to \mathcal{P}'Y$

$$(I \to card)(\rho_1 \circledcirc' \rho_2) = (I \to card)\rho_2 \ominus' (I \to card)(\rho_1 \circledcirc' \rho_2) \qquad (17)$$

where $\ominus'$ denotes bag subtraction which takes special care of the null cases, and $\circledcirc'$ denotes relational intersection which is defined as follows. For a pair of relations $[x \mapsto s]$ and ρ in $X \to \mathcal{P}'Y$ we define relational intersection as

$$[x \mapsto S] \circledcirc' \rho = \rho \circledcirc' [x \mapsto S] = \begin{cases} \theta & \neg\chi[\![x]\!]\,dom\,\rho \\ \rho \dagger [x \mapsto S \cap \rho(x)] & otherwise \end{cases} \qquad (18)$$

where the prime denotes the removal of entries of the form $x \mapsto \emptyset$.

We must consider three cases of the expression $[td \mapsto \{p\}] \circledcirc' \tau$ and deal with them separately. The first case is where the T.D. we are removing is not in the Dáil structure δ.

Case A : $\neg\chi[\![td]\!]\tau$

$$
\begin{aligned}
&(I \to card)\tau \ominus' (I \to card)([td \mapsto \{p\}] \circledcirc' \tau)\\
&= (I \to card)\tau \ominus' (I \to card)(\theta)\\
&= (I \to card)\tau \ominus' \theta\\
&= (I \to card)\tau
\end{aligned}
$$

The first step follows from the fact that td is not in τ and the rest from the resulting null map. Applying the range operator we find that

$$
\begin{aligned}
rng(I \to card)((I \to dom)R[\![p,td]\!]\delta)^{-1} &= rng(I \to card)\tau\\
&= rng(I \to card)((I \to dom)\delta)^{-1}
\end{aligned}
$$

which follows simply from replacing τ. We are left with the original partition invariant which is assumed to be true. Thus in the case where td is not in τ the removal operation satisfies the partition invariant for the Dáil. The next case is where the T.D. is in the Dáil structure δ but is *not* in the party from which we are removing him or her.

Case B : $\chi[\![td]\!]\tau \wedge \tau(td) \cap \{p\} = \emptyset$

$$(I \rightarrow card)\tau \ominus' (I \rightarrow card)([td \mapsto \{p\}] \,\widehat{\ominus}'\, \tau)$$
$$= (I \rightarrow card)\tau \ominus' (I \rightarrow card)(\theta)$$
$$= (I \rightarrow card)\tau \ominus' \theta$$
$$= (I \rightarrow card)\tau$$

using a similar argument to Case A. Applying the range operator we find that

$$rng(I \rightarrow card)((I \rightarrow dom)R[\![p,td]\!]\delta)^{-1}$$
$$= rng(I \rightarrow card)\tau$$
$$= rng(I \rightarrow card)((I \rightarrow dom)\delta)^{-1}$$

We are again left with the original partition invariant which is assumed to be true. Thus in the case where $\chi[\![td]\!]\tau \wedge \tau(td) \cap \{p\} = \emptyset$ the T.D. removal operation satisfies the partition invariant for the Dáil . The final case is where the T.D. is in the Dáil structure δ and *is* in the party we are removing him or her from.

Case C : $\chi[\![td]\!]\tau \wedge \tau(td) \cap \{p\} = \{p\}$

$$(I \rightarrow card)\tau \ominus' (I \rightarrow card)([td \mapsto \{p\}] \,\widehat{\ominus}'\, \tau)$$
$$= (I \rightarrow card)\tau \ominus' (I \rightarrow card)[td \mapsto \{p\}]$$
$$= (I \rightarrow card)\tau \ominus' [I(td) \mapsto card(\{p\})]$$
$$= (I \rightarrow card)\tau \ominus' [td \mapsto 1]$$
$$= \beta \dagger [td \mapsto (\beta(td) - 1)]$$
$$= \beta \dagger [td \mapsto 0]$$
$$= \{td\} \blacktriangleleft \beta$$

where β denotes $(I \rightarrow card)\tau$, $\blacktriangleleft$ denotes set removal from a bag and $\dagger$ denotes override. The final steps follow from the fact that $\chi[\![td]\!]\tau$ implies $\chi[\![td]\!]\beta$ and because δ is assumed to be partitioned prior to the removal operation (i.e. $\beta(td) = 1$). Finally we apply the range operator:

$$rng(I \rightarrow card)((I \rightarrow dom)R[\![p,td]\!]\delta)^{-1} = rng(\{td\} \blacktriangleleft \beta)$$

This expression expands to one of two cases depending on whether td is the last T.D. being deleted from the Dáil . The expression below is a standard expansion in $VDM^{\clubsuit}$ and evaluates to the null set if td is the last T.D. and to $\{1\}$ otherwise.

$$rng(\{td\} \blacktriangleleft \beta) = \begin{cases} \{\beta(td)\} \blacktriangleleft (rng\,\beta) = \{1\} \blacktriangleleft \{1\} = \emptyset & |\beta^{-1} \circ \beta(td)| = 1 \\ rng\,\beta = \{1\} & otherwise \end{cases}$$

where $\blacktriangleleft$ here denotes set removal. Both cases satisfy the Dáil invariant. So, in the case where $\chi[\![td]\!]\tau \wedge \tau(td) \cap \{p\} = \{p\}$, the T.D. removal operation satisfies the Dáil partition invariant. As the T.D. removal operation satisfies the Dáil partition invariant for all three cases it requires no pre-condition.

3.4 Insertion — Installing a By-election Winner

We now define an insertion operation, $A[\![p,td,s]\!]$, on a well-structured Dáil , δ, in terms of the (double-indexed) insertion/union operator, $\bigcirc\!\!\!\!\bigcirc$ of the Dáil monoid as follows:

$$A[\![p,td,s]\!]\delta \triangleq \delta \,\bigcirc\!\!\!\!\bigcirc[p \mapsto [td \mapsto s]] \tag{19}$$

where δ is in $P \to (TD \to \mathcal{P}'FI)'$, p is in P, td is in TD and s is a non-empty set of financial interests in $\mathcal{P}'FI$. We wish to know whether this insertion operation preserves the Dáil partition invariant, i.e. does the invariant hold for $A[\![p,td,s]\!]\delta$ assuming it holds for δ? Also what interpretation should we give to $\delta\,\bigcirc\!\!\!\!\bigcirc[p \mapsto [td \mapsto s]]$ with respect to the model of the Dáil? The first step is to iterate over the updated structure by applying the domain operator to the range giving us

$$
\begin{aligned}
(I \to dom)A[\![p,td,s]\!]\delta &= (I \to dom)(\delta\,\bigcirc\!\!\!\!\bigcirc[p \mapsto [td \mapsto s]]) \\
&= (I \to dom)\delta \,\bigcirc\!\!\!\!\bigcirc (I \to dom)[p \mapsto [td \mapsto s]] \\
&= (I \to dom)\delta \,\bigcirc\!\!\!\!\bigcirc [I(p) \mapsto dom\,[td \mapsto s]] \\
&= (I \to dom)\delta \,\bigcirc\!\!\!\!\bigcirc [p \mapsto \{td\}]
\end{aligned}
$$

The second step derives from the fact that the map iterator $(I \to dom)$ is a homomorphism from the monoid $(P \to (TD \to \mathcal{P}'FI)', \bigcirc\!\!\!\!\bigcirc \theta)$ to the monoid $(P \to \mathcal{P}'TD, \bigcirc\!\!\!\!\bigcirc, \theta)$. This fact was already exploited in constructing the partition invariant. Next we apply relational inversion to the result to give:

$$
\begin{aligned}
((I \to dom)A[\![p,td,s]\!]\delta)^{-1} &= ((I \to dom)\delta \,\bigcirc\!\!\!\!\bigcirc [p \mapsto \{td\}])^{-1} \\
&= ((I \to dom)\delta)^{-1} \,\bigcirc\!\!\!\!\bigcirc [p \mapsto \{td\}]^{-1} \\
&= ((I \to dom)\delta)^{-1} \,\bigcirc\!\!\!\!\bigcirc [td \mapsto \{p\}]
\end{aligned}
$$

The first step appeals to the fact that relational inverse is a homomorphism from the monoid $(P \to \mathcal{P}'TD, \bigcirc\!\!\!\!\bigcirc, \theta)$ to the monoid $(TD \to \mathcal{P}'P, \bigcirc\!\!\!\!\bigcirc, \theta)$. We now iterate over the previous result by applying the cardinality function to the range:

$$
\begin{aligned}
&(I \to card)(((I \to dom)A[\![p,td,s]\!]\delta)^{-1} \\
&= (I \to card)(((I \to dom)\delta)^{-1} \,\bigcirc\!\!\!\!\bigcirc [td \mapsto \{p\}]) \\
&= (I \to card)(\tau \,\bigcirc\!\!\!\!\bigcirc [td \mapsto \{p\}]) \\
&= (I \to card)\tau \oplus (I \to card)[td \mapsto \{p\}] \ominus' (I \to card)(\tau \,\widehat{\bigcirc}'\,[td \mapsto \{p\}])
\end{aligned}
$$

where τ is again an abbreviation for $((I \to dom)\delta)^{-1}$. We take the final step using the fact that for a pair of relations ρ_1 and ρ_2 in $X \to \mathcal{P}'Y$

$$
\begin{aligned}
&(I \to card)(\rho_1 \,\bigcirc\!\!\!\!\bigcirc \rho_2) = \\
&(I \to card)\rho_1 \oplus (I \to card)\rho_2 \ominus' (I \to card)(\rho_1 \,\widehat{\bigcirc}'\,\rho_2)
\end{aligned}
$$

where $\oplus$ and $\ominus'$ denote bag addition and bag subtraction respectively, the latter with special removals in the case of null entries. To continue we must consider (as before) the three cases of the expression $\tau \odot' [td \mapsto \{p\}]$ and deal with them separately. In the first case the T.D. is not in the Dáil.

Case A : $\neg\chi[\![td]\!]\tau$

$$(I \to card)\tau \oplus (I \to card)[td \mapsto \{p\}] \ominus' (I \to card)(\tau \odot' [td \mapsto \{p\}])$$
$$= (I \to card)\tau \oplus [I(td) \mapsto card(\{p\})] \ominus' (I \to card)(\theta)$$
$$= (I \to card)\tau \oplus [td \mapsto 1] \ominus' \theta$$
$$= (I \to card)\tau \oplus [td \mapsto 1]$$
$$= (I \to card)\tau \sqcup [td \mapsto 1]$$

The final step appeals to the definition of bag addition in terms of bag extend and bag override, the former applying in this case. Applying the range operator we find that

$$rng(I \to card)((I \to dom)A[\![p,td,s]\!]\delta)^{-1}$$
$$= rng((I \to card)\tau \sqcup [td \mapsto 1])$$
$$= rng(I \to card)\tau \cup rng[td \mapsto 1]$$
$$= rng(I \to card)\tau \cup \{1\}$$
$$= rng(I \to card)((I \to dom)\delta)^{-1} \cup \{1\}$$
$$= \{1\}$$

The final step follows since the first term in the union reduces to the empty set if δ was empty to start with and to the singleton $\{1\}$ otherwise. This result satisfies the partition invariant. Thus in the case $\neg\chi[\![td]\!]\tau$ the addition operation satisfies the Dáil partition invariant. In this case the addition operation can be interpreted as the winner of a by-election entering Dáil.

In the next case the T.D. is in the Dáil and is not in the party p specified in the operation. We expect that this case should invalidate the Dáil partition invariant.

Case B : $\chi[\![td]\!]\tau \wedge \tau(td) \cap \{p\} = \emptyset$

$$(I \to card)\tau \oplus (I \to card)[td \mapsto \{p\}] \ominus' (I \to card)(\tau \odot' [td \mapsto \{p\}])$$
$$= (I \to card)\tau \oplus [I(td) \mapsto card(\{p\})] \ominus' (I \to card)(\theta)$$
$$= (I \to card)\tau \oplus [td \mapsto 1] \ominus' \theta$$
$$= (I \to card)\tau \oplus [td \mapsto 1]$$
$$= \beta \oplus [td \mapsto 1]$$
$$= \beta \dagger [td \mapsto \beta(td) + 1]$$
$$= \beta \dagger [td \mapsto 1 + 1]$$
$$= \beta \dagger [td \mapsto 2]$$
$$= \{td\} \blacktriangleleft \beta \sqcup [td \mapsto 2]$$

where β is again an abbreviation for $(I \to card)\tau$ and $\dagger$, $\blacktriangleleft$ and $\sqcup$ here denote bag override, set removal from a bag and bag extend respectively. Applying the range operator we find:

$$rng(I \to card)((I \to dom)A[\![p,td,s]\!]\delta)^{-1} = rng(\{td\} \blacktriangleleft \beta \sqcup [td \mapsto 2])$$

$$= rng(\{td\} \triangleleft \beta) \cup rng\,[td \mapsto 2]$$
$$= rng(\{td\} \triangleleft \beta) \cup \{2\}$$

This expression expands to either of two cases

$$rng(\{td\} \triangleleft \beta) \cup \{2\} = \begin{cases} \{\beta(td)\} \triangleleft (rng\,\beta) \cup \{2\} \\ = \{1\} \triangleleft \{1\} \cup \{2\} = \{2\} & |\beta^{-1} \circ \beta(td)| = 1 \\ rng\,\beta \cup \{2\} = \{1\} \cup \{2\} = \{1,2\} & \text{otherwise} \end{cases}$$

Neither of the above expressions satisfies the Dáil partition invariant since td is related to two parties. So when $\chi[\![td]\!]\tau \wedge \tau(td) \cap \{p\} = \emptyset$, the addition operation violates the Dáil partition invariant and there is no interpretation for the addition operation.

The final case is where the T.D. is in the Dáil and is already in the party specified in the operation. This case should not invalidate the Dáil partition invarant. Case C :
$\chi[\![td]\!]\tau \wedge \tau(td) \cap \{p\} = \{p\}$

$$(I \to card)\tau \oplus (I \to card)[td \mapsto \{p\}] \ominus' (I \to card)(\tau \ominus' [td \mapsto \{p\}])$$
$$= (I \to card)\tau \oplus (I \to card)[td \mapsto \{p\}] \ominus' (I \to card)[td \mapsto \{p\}]$$
$$= (I \to card)\tau \oplus [I(td) \mapsto card(\{p\})] \ominus' [I(td) \mapsto card(\{p\})]$$
$$= (I \to card)\tau \oplus [td \mapsto 1] \ominus' [td \mapsto 1]$$
$$= (I \to card)\tau$$

where $\oplus$ and $\ominus'$ denote bag addition and bag subtraction respectively as before. Applying the range operator we find

$$rng(I \to card)((I \to dom)A[\![p,td,s]\!]\delta)^{-1} = rng(I \to card)\tau$$
$$= rng(I \to card)((I \to dom)\delta)^{-1}$$

This is the original Dáil partition invariant which is assumed to be true. Thus in the case where $\chi[\![td]\!]\tau \wedge \tau(td) \cap \{p\} = \{p\}$, the addition operation satisfies the Dáil partition invariant as we expected. This case may be interpreted as the addition of new financial interests for a T.D. who is already in the Dáil.

As the addition operation violates the Dáil partition invariant in one case a pre-condition is required for the operation. This can be constructed by combining, using logical or, the permitted cases (A and C):

$$\text{pre-A}[\![p,td,s]\!]\delta \triangleq$$
$$\neg\chi[\![td]\!]dom((I \to dom)\delta)^{-1} \vee$$
$$\chi[\![td]\!]dom((I \to dom)\delta)^{-1} \wedge ((I \to dom)\delta)^{-1}(td) \cap \{p\} = \{p\}$$

The above proof has illustrated some features of the constructive style of $VDM^{\clubsuit}$. Most of the work remains at an appropriately high level of abstraction, with little effort spent on low-level details. The cases considered in the proof (and the pre-conditions derived from them) have natural and intuitive interpretations in the example. We have several times appealed to some general theorems (which we have also used in other examples with similar catalogue structures [6], [8] and [4]). Our experience suggests that these results extend the mathematical foundations of software engineering in a direction that appears generally useful.

4 Outer Laws & Relational Inversion

In this section we present an outline proof of a representative theorem used in the course of the Dáil example. However we first establish a result concerning relational removal operators required for the proof.

4.1 Monoid of Relational Removal Endomorphisms

The space $X \to \mathcal{P}'Y$ is a general space of relations which can be used to construct $⊖'$ $[\![X \to \mathcal{P}'Y]\!]$, the space of removal operators defined with respect to elements of the relation space. We now show that these relational removal operators form a commutative monoid, $(⊖' [\![X \to \mathcal{P}'Y]\!], \circ, ⊖' [\![\theta]\!])$, where the binary operation is functional composition.

$⊖' [\![X \to \mathcal{P}'Y]\!]$ is closed under $\circ$:
$\forall\ ⊖' [\![\rho_1]\!], ⊖' [\![\rho_2]\!] \in ⊖' [\![X \to \mathcal{P}'Y]\!]$,

$$⊖' [\![\rho_1]\!] \circ ⊖' [\![\rho_2]\!] = ⊖' [\![\rho_1 \, ⓤ \, \rho_2]\!] \in ⊖' [\![X \to \mathcal{P}'Y]\!]$$

$\circ$ is associative:
$\forall\ ⊖' [\![\rho_1]\!], ⊖' [\![\rho_2]\!], ⊖' [\![\rho_3]\!] \in ⊖' [\![X \to \mathcal{P}'Y]\!]$,

$$(⊖' [\![\rho_1]\!] \circ ⊖' [\![\rho_2]\!]) \circ ⊖' [\![\rho_3]\!] = ⊖' [\![\rho_1 \, ⓤ \, \rho_2]\!] \circ ⊖' [\![\rho_3]\!]$$
$$= ⊖' [\![\rho_1 \, ⓤ \, \rho_2 \, ⓤ \, \rho_3]\!]$$
$$= ⊖' [\![\rho_1]\!] \circ ⊖' [\![\rho_2 \, ⓤ \, \rho_3]\!]$$
$$= ⊖' [\![\rho_1]\!] \circ (⊖' [\![\rho_2]\!] \circ ⊖' [\![\rho_3]\!])$$

$⊖' [\![\theta]\!]$ is the unique identity element for $\circ$:
$\forall\ ⊖' [\![\rho]\!], ⊖' [\![\theta]\!] \in ⊖' [\![X \to \mathcal{P}'Y]\!]$,

$$⊖' [\![\rho]\!] \circ ⊖' [\![\theta]\!] = ⊖' [\![\rho \, ⓤ \, \theta]\!]$$
$$= ⊖' [\![\rho]\!]$$
$$= ⊖' [\![\theta \, ⓤ \, \rho]\!]$$
$$= ⊖' [\![\theta]\!] \circ ⊖' [\![\rho]\!]$$

Since we have shown that $⊖' [\![\theta]\!]$ is both a left and a right identity it follows trivially that it is unique.

$\circ$ is commutative:
$\forall\ ⊖' [\![\rho_1]\!], ⊖' [\![\rho_2]\!] \in ⊖' [\![X \to \mathcal{P}'Y]\!]$,

$$⊖' [\![\rho_1]\!] \circ ⊖' [\![\rho_2]\!] = ⊖' [\![\rho_1 \, ⓤ \, \rho_2]\!]$$
$$= ⊖' [\![\rho_2 \, ⓤ \, \rho_1]\!]$$
$$= ⊖' [\![\rho_2]\!] \circ ⊖' [\![\rho_1]\!]$$

thus we obtain a composition law of relational removal endomorphisms:

$$⊖' [\![\rho_1]\!] \circ ⊖' [\![\rho_2]\!] = ⊖' [\![\rho_1 \, ⓤ \, \rho_2]\!]$$
$$\rho_1 \, ⊖' \, (\rho_2 \, ⊖' \, -) = (\rho_1 \, ⓤ \, \rho_2) \, ⊖' \, -$$

4.2 Theorem of Relational Removal & Inversion

We now state and prove the following theorem on the interchangeability of relational removal and relational inversion.

Theorem 4.1 (Relational Removal & Relational Inversion). *For any pair of relations* ρ_1 *and* ρ_2 *both in* $X \to \mathcal{P}'Y$,

$$(\rho_1 \circledast' \rho_2)^{-1} = \rho_1^{-1} \circledast' \rho_2^{-1} \tag{20}$$

where $(\rho_1 \circledast' \rho_2)^{-1}, \rho_1^{-1}, \rho_2^{-1}$ *are all in* $Y \to \mathcal{P}'X$.

Proof. By induction.
Base Case. Prove true for,

$$(\theta \circledast' \rho)^{-1} = \theta^{-1} \circledast' \rho^{-1} \tag{21}$$

$$(\theta \circledast' \rho)^{-1} = \rho^{-1}$$
$$= \theta \circledast' \rho^{-1}$$
$$= \theta^{-1} \circledast' \rho^{-1}$$

These steps follow from the definition of the relational removal operator, $\circledast'$, and from the null map being its own inverse (see [8], page 24).
General Case. We assume true for,

$$(\rho_1 \circledast' \rho_2)^{-1} = \rho_1^{-1} \circledast' \rho_2^{-1} \tag{22}$$

and attempt to prove true for,

$$(([x \mapsto \{y\}] \circledcirc \rho_1) \circledast' \rho_2)^{-1} = ([x \mapsto \{y\}] \circledcirc \rho_1)^{-1} \circledast' \rho_2^{-1} \tag{23}$$

The proof follows in a slightly different format to permit annotations to be attached to each line:

1   $(([x \mapsto \{y\}] \circledcirc \rho_1) \circledast' \rho_2)^{-1}$
.1 $= ([x \mapsto \{y\}] \circledast' (\rho_1 \circledast' \rho_2))^{-1}$
.2 $= [x \mapsto \{y\}]^{-1} \circledast' (\rho_1 \circledast' \rho_2)^{-1}$
.3 $= [x \mapsto \{y\}]^{-1} \circledast' (\rho_1^{-1} \circledast' \rho_2^{-1})$
.4 $= ([x \mapsto \{y\}]^{-1} \circledcirc \rho_1^{-1}) \circledast' \rho_2^{-1}$
.5 $= ([x \mapsto \{y\}] \circledcirc \rho_1)^{-1} \circledast' \rho_2^{-1}$

Annotations:

(1.1) Using the composition law of the relational removal endomorphisms.

(1.2) By the constructive step proven in the Appendix below. We invite the interested reader to examine the construction and cases used in the proof.

(1.3) Using the inductive hypothesis.

(1.4) Again using the composition law of the relational removal endomorphisms.

(1.5) From [8], page 24, relational inversion is a homomorphism from the monoid, $(X \rightarrow \mathcal{P}'Y, \mathbb{O}, \theta)$ to the monoid, $(Y \rightarrow \mathcal{P}'X, \mathbb{O}, \theta)$.

♣

5 Summary, Conclusions and Future Work

In this paper we have presented an example to illustrate the algebraic, constructive style of specification and proof used in the Irish School of the *VDM*.

Several observations can be made. Firstly, in our proofs of the Dáil model, we have done most of the work by exploiting the algebraic properties of the indexed operators. Only in the later stages have we used the definitions (in terms of override and extend). In the presented theorem proof however we had to appeal to the definitional details more frequently. It is our contention that work at this detailed level should be performed once for a general result and then (frequently) re-used. To the best of our knowledge this is perhaps the first publication of such an approach using these results. Secondly, the particular cases which we examined in both the addition and removal operations on the Dáil model were the same. This followed from the mathematical similarity of the basic operation definitions — hinting that we may have happened upon a more fundamental result which might be exploited again. The calculation of pre-conditions for the operations closely followed the cases considered and the latter always had a natural and intuitive interpretation in the example. Thirdly, the theorems used here form part of a collection of related results (some of them mutual duals) which we have also found useful in other examples with related structure such as simple doctor-patient database, a blood-test database and other catalogue structures (see [6], [8], [4]). Our illustration demonstrates a novel approach within in the *VDM* tradition which harnesses additional branches of abstract algebra for use in the specification and development of software. This innovation offers, we believe, several benefits: (i) the extension of the mathematical foundation of the emerging software engineering discipline, (ii) a more appropriate and abstract mathematical level at which the designer may work which abbreviates and simplifies proof and consequently (iii) a mathematical setting in which insights into designs are easier to make and on foot of which designs may be confidently modified or improved. We have also given a detailed presentation of one of the theorems found useful in this and other examples.

Although we are not as familiar with the Z notation [10], we see no reason why our approach cannot also be taken within the Z tradition. One such attempt [2], [11] is constructive in style, though based on Geometric Logic and not algebra. Nevertheless, indexed algebra techniques such as those we have demonstrated might be incorporated into Z schemas.

Much work remains to be done in two important directions. Other interesting algebraic structures and properties might be developed from the theoretical results presented here. The monoid may not be the only algebraic structure which could be usefully indexed in a similar way. Secondly, the indexed monoid theorems must be applied more widely to assess their utility to the discipline. Another line of work, also reported in this volume [9], is the construction of a *Mathematica*-based tool to support experimentation with examples defined in $VDM^{\clubsuit}$.

If we must choose between a tool or a methodology which provides such a convenient and powerful mathematical context or remain confined to using the lower-level, tedious (but nonetheless useful) tools of logic and set theory, we know what our choice will be.

The authors would like to acknowledge their debt to Dr. Mícheál Mac an Airchinnigh for his early work on indexed monoids and for encouraging us to experiment with them. We would also like to thank other members of the Formal Methods Working Group and the anonymous referees whose comments helped us improve the presentation of this paper.

References

1. Andrew Butterfield. A VDM$^{\clubsuit}$ Study of Fault-Tolerant Stable Storage: Towards a Computer Engineering Mathematics. In J. Woodcock and P. G. Larsen, editors, *FME'93: Industrial Strength Formal Methods, Odense, Denmark, Lecture Notes in Computer Science, Volume 670*, pages 216–223. Springer-Verlag, 1993.

2. Mark Dawson and Steven Vickers. Towards a GeoZ Toolkit. Unpublished Draft Technical Report, Revision 1.3, Department of Computing, Imperial College of Science and Technology, 1994.

3. R. Goldblatt. *Topoi: The Categorical Analysis of Logic*, volume 98 of *Studies in Logic and the Foundations of Mathematics*. North-Holland, 1984.

4. Arthur P. Hughes. Outer Laws for the Indexed Monoid. Final Year B.A. (Mod) Computer Science Project Report, Department of Computer Science, Trinity College Dublin, June 1994.

5. C. B. Jones. *Systematic Software Development Using VDM*. Prentice Hall International Series in Computer Science, 2nd edition, 1990.

6. Mícheál Mac an Airchinnigh. *Conceptual Models and Computing*. PhD thesis, Department of Computer Science, Trinity College Dublin, 1990.

7. Mícheál Mac an Airchinnigh. Tutorial Lecture Notes on the Irish School of the VDM. In S. Prehn and W. J. Toetenel, editor, *VDM'91: Formal Software Development Methods, Volume 2: Tutorials, Lecture Notes in Computer Science, Volume 552*, pages 141–237. Springer-Verlag, 1991.

8. Mícheál Mac an Airchinnigh. Formal Methods and Testing. In *Tutorial Notes: 6th International Software Quality Week*. Software Research Institute, 625 Third Street, San Fancisco, CA 94107-1997., 1993.

9. Colman Reilly. Exploring Specifications with Mathematica. In J.P. Bowen & M.G. Hinchey, editors, *ZUM'95: 9th International Conference of Z Users* (this volume), Lecture Notes in Computer Science, Springer-Verlag, 1995.

10. J. M. Spivey. *The Z Notation: A Reference Manual*. Prentice Hall International Series in Computer Science, 2nd edition, 1992.

11. Steven J. Vickers. Geometric Logic as a Specification Language. Unpublished Draft Technical Report, Revision 1.2.1 Theory and Formal Methods 1994: Proceedings of the Second Imperial College, Department of Computing, Workshop on Theory and Formal Methods, Moller Centre, Cambridge, UK., September 1994.

A Appendix: Constructive Step of Theorem 4.2

This appendix presents the details of the Constructive Step used in the proof of Theorem 4.2. We approach this step by considering cases — whether x is in the domain of ρ already or not. In the former case we consider three sub-cases depending in whether y occurs in $\rho(x)$ as part of a larger set, as a singleton or not at all. The construction used here is to split ρ into three parts, that element in which x appears in the domain, a relation, α, in whose range y appears and a third relation, β, in which y does not appear in the range. Thus four cases must be considered. We must show

$$([x \mapsto \{y\}] \circledcirc' \rho)^{-1} = [x \mapsto \{y\}]^{-1} \circledcirc' \rho^{-1} \tag{24}$$

Case A: $\neg\chi[\![x]\!] \, dom \, \rho$

$$
\begin{aligned}
2 \quad & ([x \mapsto \{y\}] \circledcirc' \rho)^{-1} \\
.1 \quad & = \rho^{-1} \\
.2 \quad & = (\alpha \circledcirc \beta)^{-1} \\
.3 \quad & = \alpha^{-1} \circledcirc \beta^{-1} \\
.4 \quad & = [y \mapsto \{x\}] \circledcirc' \alpha^{-1} \circledcirc [y \mapsto \{x\}] \circledcirc' \beta^{-1} \\
.5 \quad & = [y \mapsto \{x\}] \circledcirc' (\alpha^{-1} \circledcirc \beta^{-1}) \\
.6 \quad & = [x \mapsto \{y\}]^{-1} \circledcirc' (\alpha \circledcirc \beta)^{-1} \\
.7 \quad & = [x \mapsto \{y\}]^{-1} \circledcirc' \rho^{-1}
\end{aligned}
$$

Annotations:

(2.1) As $\neg\chi[\![x]\!] \, dom \, \rho$.

(2.2) Splitting ρ up where $\alpha = (\{y\} \lhd \rho^{-1})^{-1}$ ($\lhd$ denotes restriction) and where $\beta = \alpha \circledcirc' \rho$. Note that $\beta = \alpha \circledcirc' \rho \Rightarrow \neg\chi[\![y]\!]^{\cup}/\circ \, rng \, \beta$ i.e. y does not occur in any element of the range of β.

(2.3) Using the relational inversion homomorphism.

(2.4) As $\neg\chi[\![x]\!] \, dom \, \rho \Rightarrow \neg\chi[\![x]\!] \, dom \, \alpha \Rightarrow \neg\chi[\![x]\!]^{\cup}/\circ \, rng \, \alpha^{-1}$ (x cannot occur in the range of α^{1}). Also $\neg\chi[\![y]\!]^{\cup}/\circ \, rng \, \beta$ by construction. Thus removal of $[y \mapsto \{x\}]$ has no effect on α^{-1} or on β^{-1}.

(2.5) Using the composition law of relational removal proved earlier.

(2.6) Again using the relational inversion homomorphism. The left hand operand has also been re-written.

Case B: $\chi[\![x]\!] \, dom \, \rho \wedge \neg\chi[\![y]\!]\rho(x)$

$$
\begin{aligned}
3 \quad & ([x \mapsto \{y\}] \circledcirc' \rho)^{-1} \\
.1 \quad & = ([x \mapsto \{y\}] \circledcirc' ([x \mapsto \rho(x)] \circledcirc \alpha \circledcirc \beta))^{-1} \\
.2 \quad & = ([x \mapsto \{y\}] \circledcirc' [x \mapsto \rho(x)] \circledcirc [x \mapsto \{y\}] \circledcirc' \alpha \circledcirc [x \mapsto \{y\}] \circledcirc' \beta)^{-1} \\
.3 \quad & = ([x \mapsto \rho(x)] \dagger [x \mapsto \{y\} \lhd \rho(x)] \circledcirc \alpha \circledcirc \beta)^{-1} \\
.4 \quad & = ([x \mapsto \{y\} \lhd \rho(x)] \circledcirc \alpha \circledcirc \beta)^{-1}
\end{aligned}
$$

.5 $= ([x \mapsto \rho(x)] \; \textcircled{0} \; \alpha \; \textcircled{0} \; \beta)^{-1}$

.6 $= [x \mapsto \rho(x)]^{-1} \; \textcircled{0} \; \alpha^{-1} \; \textcircled{0} \; \beta^{-1}$

.7 $= [y' \mapsto \{x\} | y' \in \rho(x)] \; \textcircled{0} \; \alpha^{-1} \; \textcircled{0} \; \beta^{-1}$

.8 $= [y \mapsto \{x\}] \; \textcircled{\blacktriangleleft}' \; [y' \mapsto \{x\} | y' \in \rho(x)] \; \textcircled{0}$

                 $[y \mapsto \{x\}] \; \textcircled{\blacktriangleleft}' \; \alpha^{-1} \; \textcircled{0} \; [y \mapsto \{x\}] \; \textcircled{\blacktriangleleft}' \; \beta^{-1}$

.9 $= [y \mapsto \{x\}] \; \textcircled{\blacktriangleleft}' \; ([y' \mapsto \{x\} | y' \in \rho(x)] \; \textcircled{0} \; \alpha^{-1} \; \textcircled{0} \; \beta^{-1})$

.10 $= [x \mapsto \{y\}]^{-1} \; \textcircled{\blacktriangleleft}' \; ([x \mapsto \rho(x)]^{-1} \; \textcircled{0} \; \alpha^{-1} \; \textcircled{0} \; \beta^{-1})$

.11 $= [x \mapsto \{y\}]^{-1} \; \textcircled{\blacktriangleleft}' \; ([x \mapsto \rho(x)] \; \textcircled{0} \; \alpha \; \textcircled{0} \; \beta)^{-1}$

.12 $= [x \mapsto \{y\}]^{-1} \; \textcircled{\blacktriangleleft}' \; \rho^{-1}$

Annotations:

(3.1) The element involving x in $dom\,\rho$ is removed and the resulting relation is split into two components: α which involves only $\{y\}$ in its range and β which does not involve y. More formally, splitting up ρ into $[x \mapsto \{y\}]$, $\alpha = (\{y\} \triangleleft (\{x\} \blacktriangleleft \rho)^{-1})^{-1}$ and $\beta = \alpha \; \textcircled{\blacktriangleleft}' \; (\{x\} \blacktriangleleft \rho)$ (again $\triangleleft$ is restriction). Note that $\alpha = (\{y\} \triangleleft (\{x\} \blacktriangleleft \rho)^{-1})^{-1}$ implies $\neg\chi[\![x]\!]dom\,\alpha$ and $\beta = \alpha \; \textcircled{\blacktriangleleft}' \; (\{x\} \blacktriangleleft \rho)$ implies $\neg\chi[\![x]\!]dom\,\beta \wedge \neg\chi[\![y]\!]^{\cup}/\circ\,rng\,\beta$.

(3.2) As relational removal distributes over relational union (i.e. relational removal is an outer law of the relational monoid, $(X \to \mathcal{P}'Y, \textcircled{0}, \theta)$).

(3.3) α and β are unaffected by removal of $[x \mapsto \{y\}]$ since $\neg\chi[\![x]\!]dom\,\alpha$ and $\neg\chi[\![x]\!]dom\,\beta$. A standard expansion for $\textcircled{\blacktriangleleft}'$ is also used.

(3.5) As $\neg\chi[\![y]\!]\rho(x)$.

(3.6) Using the relational inversion homomorphism.

(3.7) By inverting $[x \mapsto \{y\}]$.

(3.8) Each term is unchanged after removal of $[y \mapsto \{x\}]$. For $\neg\chi[\![x]\!]\rho(x)$ and $\neg\chi[\![y]\!]^{\cup}/\circ\,rng\,\beta$ and also $\neg\chi[\![x]\!]dom\,\alpha$ implies $\neg\chi[\![x]\!]^{\cup}/\circ\,rng\,\alpha^{-1}$.

(3.9) By the outer relational removal law, $\textcircled{\blacktriangleleft}'$, for the monoid $(X \to \mathcal{P}'Y, \textcircled{0}, \theta)$).

(3.11) Using the relational inversion homomorphism again.

Case C: $\chi[\![x]\!]dom\,\rho \wedge \rho(x) = \{y\}$

4   $([x \mapsto \{y\}] \; \textcircled{\blacktriangleleft}' \; \rho)^{-1}$

.1 $= ([x \mapsto \{y\}] \; \textcircled{\blacktriangleleft}' \; ([x \mapsto \{y\}] \; \textcircled{0} \; \alpha \; \textcircled{0} \; \beta))^{-1}$

.2 $= ([x \mapsto \{y\}] \; \textcircled{\blacktriangleleft}' \; [x \mapsto \{y\}] \; \textcircled{0} \; [x \mapsto \{y\}] \; \textcircled{\blacktriangleleft}' \; \alpha \; \textcircled{0} \; [x \mapsto \{y\}] \; \textcircled{\blacktriangleleft}' \; \beta)^{-1}$

.3 $= (\{x\} \blacktriangleleft [x \mapsto \{y\}] \; \textcircled{0} \; \alpha \; \textcircled{0} \; \beta)^{-1}$

.4 $= (\theta \; \textcircled{0} \; \alpha \; \textcircled{0} \; \beta)^{-1}$

.5 $= \theta^{-1} \; \textcircled{0} \; \alpha^{-1} \; \textcircled{0} \; \beta^{-1}$

.6 $= \theta \; \textcircled{0} \; \alpha^{-1} \; \textcircled{0} \; \beta^{-1}$

.7 $= [y \mapsto \{x\}] \; \textcircled{\blacktriangleleft}' \; [y \mapsto \{x\}] \; \textcircled{0} \; [y \mapsto \{x\}] \; \textcircled{\blacktriangleleft}' \; \alpha^{-1} \; \textcircled{0} \; [y \mapsto \{x\}] \; \textcircled{\blacktriangleleft}' \; \beta^{-1}$

.8 $= [y \mapsto \{x\}] \; \textcircled{\blacktriangleleft}' \; ([y \mapsto \{x\}] \; \textcircled{0} \; \alpha^{-1} \; \textcircled{0} \; \beta^{-1})$

.9 $= [x \mapsto \{y\}]^{-1} \; \textcircled{\blacktriangleleft}' \; ([x \mapsto \{y\}]^{-1} \; \textcircled{0} \; \alpha^{-1} \; \textcircled{0} \; \beta^{-1})$

.10 $= [x \mapsto \{y\}]^{-1} \; \textcircled{\blacktriangleleft}' \; ([x \mapsto \{y\}] \; \textcircled{0} \; \alpha \; \textcircled{0} \; \beta)^{-1}$

.11 $= [x \mapsto \{y\}]^{-1} \; \textcircled{\blacktriangleleft}' \; \rho^{-1}$

Annotations:

(4.1) Splitting ρ up into $[x \mapsto \{y\}]$, $\alpha = (\{y\} \triangleleft (\{x\} \blacktriangleleft \rho)^{-1})^{-1}$ and $\beta = \alpha \oplus' (\{x\} \blacktriangleleft \rho)$. Note that $\alpha = (\{y\} \triangleleft (\{x\} \blacktriangleleft \rho)^{-1})^{-1}$ implies $\neg\chi[\![x]\!]\,dom\,\alpha$, and $\beta = \alpha \oplus' (\{x\} \blacktriangleleft \rho)$ implies $\neg\chi[\![x]\!]\,dom\,\beta \wedge \neg\chi[\![y]\!]^{\cup}/\circ\,rng\,\beta$.

(4.2) By the relational removal endomorphism of the monoid, $(X \to \mathcal{P}'Y, \odot, \theta)$.

(4.3) As $\neg\chi[\![x]\!]\,dom\,\alpha$ and $\neg\chi[\![x]\!]\,dom\,\beta$.

(4.5) Using the relational inversion homomorphism.

(4.7) By definition of $\oplus'$, and the removal of $[y \mapsto \{x\}]$ does not affect α^- or β^{-1} since $\neg\chi[\![x]\!]\,dom\,\alpha$ implies $\neg\chi[\![x]\!]^{\cup}/\circ\,rng\,\alpha^{-1}$ and $\neg\chi[\![x]\!]^{\cup}/\circ\,rng\,\beta^{-1}$.

(4.10) Using the relational inversion homomorphism again.

Case D: $\chi[\![x]\!]\,dom\,\rho \wedge \rho(x) \supset \{y\}$

5 $([x \mapsto \{y\}] \oplus' \rho)^{-1}$

.1 $= ([x \mapsto \{y\}] \oplus' ([x \mapsto \rho(x)] \odot \alpha \odot \beta))^{-1}$

.2 $= ([x \mapsto \{y\}] \oplus' [x \mapsto \rho(x)] \odot [x \mapsto \{y\}] \oplus' \alpha \odot [x \mapsto \{y\}] \oplus' \beta)^{-1}$

.3 $= ([x \mapsto \rho(x)] \dagger [x \mapsto \{y\} \blacktriangleleft \rho(x)] \odot \alpha \odot \beta)^{-1}$

.4 $= ([x \mapsto \{y\} \blacktriangleleft \rho(x)] \odot \alpha \odot \beta)^{-1}$

.5 $= [x \mapsto \{y\} \blacktriangleleft \rho(x)]^{-1} \odot \alpha^{-1} \odot \beta^{-1}$

.6 $= [y' \mapsto \{x\}|y' \in \{y\} \blacktriangleleft \rho(x)] \odot \alpha^{-1} \odot \beta^{-1}$

.7 $= [y \mapsto \{x\}] \oplus' [y' \mapsto \{x\}|y' \in \{y\} \blacktriangleleft \rho(x)]$
$$\odot [y \mapsto \{x\}] \oplus' \alpha^{-1} \odot [y \mapsto \{x\}] \oplus' \beta^{-1}$$

.8 $= [y \mapsto \{x\}] \oplus' ([y' \mapsto \{x\}|y' \in \{y\} \blacktriangleleft \rho(x)] \odot \alpha^{-1} \odot \beta^{-1})$

.9 $= \theta \odot [y \mapsto \{x\}] \oplus' ([y' \mapsto \{x\}|y' \in \{y\} \blacktriangleleft \rho(x)] \odot \alpha^{-1} \odot \beta^{-1})$

.10 $= \{y\} \blacktriangleleft [y \mapsto \{x\}] \odot [y \mapsto \{x\}] \oplus' ([y' \mapsto \{x\}|y' \in \{y\} \blacktriangleleft \rho(x)]$
$$\odot \alpha^{-1} \odot \beta^{-1})$$

.11 $= [y \mapsto \{x\}] \oplus' [y \mapsto \{x\}] \odot [y \mapsto \{x\}] \oplus' ([y' \mapsto \{x\}|y' \in \{y\} \blacktriangleleft \rho(x)]$
$$\odot \alpha^{-1} \odot \beta^{-1})$$

.12 $= [y \mapsto \{x\}] \oplus' ([y \mapsto \{x\}] \odot [y' \mapsto \{x\}|y' \in \{y\} \blacktriangleleft \rho(x)] \odot \alpha^{-1} \odot \beta^{-1})$

.13 $= [x \mapsto \{y\}]^{-1} \oplus' ([x \mapsto \{y\}]^{-1} \odot [x \mapsto \{y\} \blacktriangleleft \rho(x)] \odot \alpha^{-1} \odot \beta^{-1})$

.14 $= [x \mapsto \{y\}]^{-1} \oplus' ([x \mapsto \{y\}] \odot [x \mapsto \{y\} \blacktriangleleft \rho(x)] \odot \alpha \odot \beta)^{-1}$

.15 $= [x \mapsto \{y\}]^{-1} \oplus' ([x \mapsto y] \dagger [x \mapsto \{y\} \cup \{y\} \blacktriangleleft \rho(x)] \odot \alpha \odot \beta)^{-1}$

.16 $= [x \mapsto \{y\}]^{-1} \oplus' ([x \mapsto \{y\} \cup \{y\} \blacktriangleleft \rho(x)] \odot \alpha \odot \beta)^{-1}$

.17 $= [x \mapsto \{y\}]^{-1} \oplus' ([x \mapsto \{y\} \cup \rho(x)] \odot \alpha \odot \beta)^{-1}$

.18 $= [x \mapsto \{y\}]^{-1} \oplus' ([x \mapsto \rho(x)] \odot \alpha \odot \beta)^{-1}$

.19 $= [x \mapsto \{y\}]^{-1} \oplus' \rho^{-1}$

Annotations:

(5.1) Splitting ρ up into $[x \mapsto \rho(x)]$, $\alpha = (\{y\} \lhd (\{x\} \blacktriangleleft \rho)^{-1})^{-1}$ and $\beta = \alpha \oslash' (\{x\} \blacktriangleleft \rho)$. Note that $\alpha = (\{y\} \lhd (\{x\} \blacktriangleleft \rho)^{-1})^{-1}$ implies $\neg\chi[\![x]\!]\,dom\,\alpha$, $\beta = \alpha \oslash' (\{x\} \blacktriangleleft \rho)$ implies $\neg\chi[\![x]\!]\,dom\,\beta \wedge \neg\chi[\![y]\!]^{\cup}/\circ\,rng\,\beta$.

(5.2) By the relational removal endomorphism of the monoid, $(X \to \mathcal{P}'Y, \oplus, \theta)$.

(5.3) Using the definition of $\oslash'$. Also α and β are unaffected by the removal since $\neg\chi[\![x]\!]\,dom\,\alpha$ and $\neg\chi[\![x]\!]\,dom\,\beta$.

(5.5) Using the relational inversion homomorphism.

(5.7) As $\neg\chi[\![x]\!]\,dom\,\alpha$ implies $\neg\chi[\![x]\!]^{\cup}/\circ\,rng\,\alpha^{-1}$ and $\neg\chi[\![x]\!]^{\cup}/\circ\,rng\,\beta^{-1}$.

(5.8) Using the relational removal endomorphism.

(5.12) Again using the relational removal endomorphism.

(5.14) Using again the relational inversion homomorphism.

(5.17) By a set identity on removal and union of the same set. For any pair of sets s_1 and s_2 in $\mathcal{P}X$, $\quad s_1 \cup s_1 \blacktriangleleft s_2 = s_1 \cup s_2 \quad$ where $s_1 \cup s_1 \blacktriangleleft s_2$, $s_1 \cup s_2$ in $\mathcal{P}X$.

(5.18) As $\rho(x) \supset \{y\} \Rightarrow \chi[\![y]\!]\rho(x)$.

Testing

Testing as Abstraction

Susan Stepney

Logica UK Ltd.
Cambridge Division, Betjeman House,
104 Hills Road, Cambridge, CB2 1LQ, UK

Abstract. The PROST-Objects project has developed a method for formally specifying tests. The method is based on systematic abstraction from a 'state-plus-operation' style specification. It is explained here, and illustrated with a small example. Test developers can use this method, along with their own skills for choosing good tests, to produce a suite of formal test specifications. The project has also developed a prototype tool, which provides organisational support for the (potentially large) collection of test specifications as they are generated.

1   Introduction

On the one hand we have a Z specification of the behaviour of a system; on the other hand we have a purported implementation of that system. We want to *test* that the implementation *conforms to* the specification. To do this we need a test suite whose individual tests have good coverage and little overlap. The original specification is the obvious place to start in order to derive such a test suite.

The PROST-Objects project has developed a method for taking the formal specification of a system and systematically deriving from it formal specifications of conformance tests. This paper describes our method by using a small worked example; it also describes a prototype tool we have been building to support the method.

PROST-Objects is applying this method to the specifications of OSI Managed Objects, in order to derive specifications of tests that can be used to show that implementations of Managed Objects conform to their formal specifications.

The project is using ZEST, an object oriented extension to Z being developed at British Telecommunications' Research Labs [Cusack & Rafsanjani 1992]. However, the underlying method is not specific to object oriented specifications, or to Managed Object specifications. It is a general method applicable to plain Z specifications written in a 'state-and-operations' style, and is described here in those terms.

2   Test Suites

Testing is an essential part of the software development lifecycle. The existence of a formal specification can help reduce the burden of defining the tests, because it can be used to help derive a good test suite.

Here, we discuss a methodical (but not automatic) approach to deriving Z specifications of tests from Z specifications of operations.

We want to test that some system correctly implements the specification of some operation, *Op*. By 'correctly implements', we mean that for every specified before-state (including inputs), the implementation produces at least one of the specified after-states (including outputs), and not some other state or output.

In general, it is not possible to test exhaustively the entire state space, due to combinatorial explosion. Often it is not useful to test everything; many possible before-states are qualitatively similar to others, and so tests for these tell us nothing new. Also, it is not always useful to test that the entire after-state has been established as specified. For example, consider the specification of a query operation that outputs some value and leaves the state unchanged. It might be that only the value of the output is checked, and not that the state has been unaffected.

The process of building a test suite comprises selecting a 'useful' subset of the interesting before-states, and selecting those parts of the corresponding after-state to check.

There are various criteria for such a selection: it should give the best test coverage given limited testing resources; it should concentrate on the important parts of the implementation; it should identify bugs; .... These are all informal criteria, and it does not seem sensible to attempt to provide some automatic method to perform this selection. Indeed, attempts at automatic test generation can easily succeed in generating a copious number of trivial tests, but automatically generating pertinent tests has proved much more difficult. Rather than try to devise an automatic test generation method, our approach in PROST-Objects has been to develop a framework method within which test suite developers can apply their own skills, and to build prototype tool support to assist this process.

The method is described in the following sections, and the tool support in section 9.

3 The Test Method

Using our test method, a tester starts with the Z specification of an operation, *Op* say, and systematically manipulates it to produce a set of specifications of the relevant tests. Each test is itself specified by a Z operation schema.

Our method is based on the following observations:

- We test only a subset of the interesting inputs and before-states. Such a test is captured by a specification *OpTest* that is less deterministic than the original specification of *Op*, because *OpTest* does not constrain the behaviour of states starting outside the tested subset.

- We test only some of the consequences of the state transition specified by *Op*, that is, examining only some outputs and a part of the after-state. Such a test is also captured by a specification that is less deterministic than *Op*'s, because *OpTest* does not constrain the untested part of the after-state.

Our method produces test specifications by systematically introducing non-determinism into the original specification.

For example, consider the following file system specification, which has a state consisting of a mapping from file names to file contents, and the set of files open for reading.

$$[NAME, FILE]$$

__ FileSys _____
$fsys : NAME \nrightarrow FILE$
$open : \mathbb{P}\, NAME$

$open \subseteq \text{dom}\, fsys$
_____

This file system has a read operation that takes a list of *NAME*s, and returns a corresponding list of *FILE*s. For each file, if it is open, the operation gives the actual contents, but if it is closed, or does not exist, the operation gives some error flag. (In order to keep this example simple, these error results have the same type as the file; this is not necessarily the best way to specify such a system for real.)

| $closed, doesNotExist : FILE$

__ Read _____
$\Xi FileSys$
$n? : \text{seq}\, NAME$
$f! : \text{seq}\, FILE$

$\#f! = \#n?$

$\forall i : 1\,..\,\#n? \bullet$
$\quad (n?\, i \in open \Rightarrow f!\, i = fsys(n?\, i))$
$\quad \wedge\ (n?\, i \in \text{dom}\, fsys \setminus open \Rightarrow f!\, i = closed)$
$\quad \wedge\ (n?\, i \notin \text{dom}\, fsys \Rightarrow f!\, i = doesNotExist)$
_____

Consider how we might want to test a purported implementation of *Read*. Based on our (presumably short?) experience as a builder of file system test suites, we decide that we will test only open files, and only those file name lists of length one, $\#n? = 1$, and check that we get the correct single output in response. We decide not to test that the entire state remains unchanged, but we do want to check that the chosen file itself remains open after the read, because we have some suspicion that this might not be the case. So we specify our test as:

__ ReadTest _____
$\Delta FileSys$
$n? : \text{seq}\, NAME$
$f! : \text{seq}\, FILE$

$n?\, 1 \in open \Rightarrow$
$\quad n?\, 1 \in open'$
$\quad \wedge\ (\#n? = 1 \Rightarrow f! = \langle fsys\ (n?\, 1)\rangle)$
_____

ReadTest is a less deterministic specification than *Read*. The after-state is much less constrained than before: $\theta FileSys'$ is constrained only to obey the predicates on $FileSys'$, plus the openness condition on n? 1, whereas before it was completely determined to have the same value as $\theta FileSys$. Also, f! is constrained to have a particular value only when the implication's antecedent, $\#n? = 1 \wedge n? 1 \in open$, is true, not for every value of n? as before.

Putting this a more familiar way, *Read* is more deterministic than *ReadTest*, and hence *Read* is a *refinement* of *ReadTest*. Equivalently, *ReadTest* is an *abstraction* of *Read*. We use this insight of the existence of an abstraction relation between operations and their tests in our method for structuring the test discovery process.

4   Preamble and Checking Phases

There is one more step to be taken after specifying a test like *ReadTest*, before we can perform the test. How do we know that the file exists, and is open, before the operation? How do we know that the f! output from our *ReadTest* is the correct value? And how do we know that the file is still open after the operation?

We know because during the *preamble* to the test, before we perform (our implementation of) *ReadTest*, we open the file and set its contents to some known value to compare against f!. And later, during the *checking* phase of the test, we interrogate the open status of the file.

So a complete specification for performing and checking a test has the following structure:

$$ReadTest_C \;\hat{=}\; Preamble \,\fatsemi\, ReadTest \,\fatsemi\, Check$$

Here *Preamble* and *Check* are combinations of operations that put the system into some known state, and query the system's state, respectively.

The rest of this paper discusses the building of *ReadTest*. Specifying *Preamble* and *Check* is also being investigated by PROST-Objects.

5   Structuring the Test Suite

Although for the simple example given above it is possible to write down the final version of the desired test, the task is not so simple for a real world specification. Also, a single operation will likely have many different tests defined, testing different aspects of its behaviour. We would like some systematic way for deriving and documenting these tests, and we would like to get some idea of test coverage; some idea of what is *not* being tested.

This is what the test method that we have developed in the PROST-Objects project does. The method provides a set of well-defined actions, which can be performed at each stage of the process, building successively more abstract operations. These operations are structured as a *tree* of specifications, with the original operation at the root, and the tests at the leaves. Starting from the specification of a single operation, test developers working with the method select an appropriate action to perform at each stage, in order to reach a good set of tests. Which action to choose is part of the tester's skill, and is not the subject of this paper.

The method defines various actions to be used in developing the tests: partitioning, weakening, and simplification, described below. Our prototype tool, described later, provides support for each of these actions. The method also provides heuristics that reduce the need to discharge various proof obligations.

5.1 Proof obligations

Once the tree of specifications has been built in accordance with the method, the set of all the leaves comprises the specifications of the tests. Each time we apply an action, we divide an operation into several parts, each of which will either form a test, or be further divided. The allowed division is not arbitrary, but must satisfy the following correctness conditions (here C is the current specification being abstracted from; T_i are the tests produced at this step:

- **Soundness:** As noted above, the test method is based on a refinement/abstraction relation holding between a specification and each of its tests. For total operations, this reduces to the set of conditions:

$$\forall i \bullet C \Rightarrow T_i$$

- **Completeness:** Nothing must be lost when deriving the specification of the tests. The conjunction of all the leaves must be equivalent to the original specification. (For an explanation of why the tests are conjoined, rather than disjoined, see appendix A.)

$$C = \bigwedge_i T_i$$

Each leaf is the specification of a single test. We have some 'niceness' conditions that allow us to ensure we do not specify too many tests, and allow us to interpret the meaning of failing a test, or of not performing a test. These criteria can be understood by considering the set of specifications $\{\neg T_i\}$. Any system that implements $\neg T_i$ fails to implement the specification T_i.

- **Independence:** No test should be redundant: it should test at least something not covered somewhere in another test. Test j is independent of all the other tests if the specification $\neg T_j$ is not wholly contained in the union of the other $\neg T_i$ specifications:

$$\forall j \bullet \neg (\neg T_j \Rightarrow \bigvee_{i \neq j} \neg T_i)$$

This simplifies to

$$\forall j \bullet \neg (\bigwedge_{i \neq j} T_i \Rightarrow T_j)$$

- **Orthogonality:** Ideally, we would like the tests to have no overlap at all. This would mean that an implementation could fail at most one test, and that any test designated as a residual would specify precisely what is not being tested (would be 'honest'). The tests are orthogonal if the specifications of $\neg T_i$ are disjoint:

$$\forall i \neq j \bullet \neg (\neg T_i \wedge \neg T_j)$$

which simplifies to

$$\forall i \neq j \bullet T_i \vee T_j$$

Orthogonality is a very strong condition. It is convenient to relax it to orthogonal 'up to' a predicate, for example, up to (the implicit predicate in) a declaration.

In general, discharging such proof obligations would be quite onerous. However, in practice the choice of division is not arbitrary, but governed by the form of the specification being divided. Certain choices can reduce the complexity of the required proofs. In addition, the method describes certain special cases, which automatically satisfy some proof obligations. Use of these special cases substantially reduces the proof burden to a manageable level.

5.2 Partitioning

A set of less deterministic operations is derived from the current version of the operation by partitioning the input space and before-state. This gives a set of operation definitions each covering part of the before-state.

$$C \mathrel{\widehat{=}} T_1 \wedge \ldots \wedge T_n$$

For example, the first step in defining the tests for *Read* could be a partitioning of the input $n?$ on the length of the sequence:

1. $\#n? = 0$

2. $\#n? = 1$

3. $\#n? > 1$

Other partitions on the length of the sequence might be chosen. For example, if sequences of length two were thought to be deserving of special attention, this might be included explicitly as an extra partition $\#n? = 2$ (and changing the last partition to $\#n? > 2$).

Once a suitable n-way partition has been decided, in the form of n predicates H_i, n sound abstract operations are defined, each of the form

$$C \mathrel{\widehat{=}} [\, D \mid P \,]$$
$$T_i \mathrel{\widehat{=}} [\, D \mid H_i \Rightarrow P \,]$$

So, for the case of $\#n? = 1$, we get

$$
\begin{array}{|l}
\hline
\quad ReadOne \\
\hline
\Xi FileSys \\
n? : \text{seq } NAME \\
f! : \text{seq } FILE \\
\hline
\#n? = 1 \\
\quad \Rightarrow \#f! = \#n? \\
\quad \land (\, \forall i : 1..\#n? \bullet \\
\qquad\quad (n?\ i \in open \Rightarrow f!\ i = fsys(n?\ i)) \\
\qquad\quad \land (n?\ i \in \text{dom } fsys \setminus open \Rightarrow f!\ i = closed) \\
\qquad\quad \land (n?\ i \notin \text{dom } fsys \Rightarrow f!\ i = doesNotExist)\,) \\
\hline
\end{array}
$$

The choice of the term 'partition' for this action gives a clue to a heuristic for satisfying the proof obligations: the tests are sound by construction, and if the predicates H_i do partition the space, then the operations combine to form the original, and do not overlap. That is, the completeness proof obligation is satisfied if $\bigvee_i H_i$, and additionally the tests are orthogonal (up to the declaration) if

$$
\forall i \neq j \bullet \neg (H_i \land H_j)
$$

These conditions hold for the choice of partition in our example. So we have

$$
Read \equiv ReadNone \land ReadOne \land ReadMany
$$

5.3 Simplification

The action of partitioning by textually replacing P by $H \Rightarrow P$ usually leads to a specification that can be written in a simpler form (since in practice H is not an arbitrary predicate, but governed by the form of the declaration or of P). H may be assumed in P and used to simplify it. For example, assuming $\#n? = 1$ in $ReadOne$ enables us to remove the quantifier and give a more explicit form for the sequence $f!$:

$$
\begin{array}{|l}
\hline
\quad ReadOneSimp \\
\hline
\Xi FileSys \\
n? : \text{seq } NAME \\
f! : \text{seq } FILE \\
\hline
\#n? = 1 \\
\quad \Rightarrow (n?\ 1 \in open \Rightarrow f! = \langle fsys(n?\ 1) \rangle) \\
\quad \land (n?\ 1 \in \text{dom } fsys \setminus open \Rightarrow f! = \langle closed \rangle) \\
\quad \land (n?\ 1 \notin \text{dom } fsys \Rightarrow f! = \langle doesNotExist \rangle) \\
\hline
\end{array}
$$

We have a proof obligation to show that the simplification does not alter the meaning of the schema.

Here we have

$$
ReadOneSimp \equiv ReadOne
$$

Having simplified the schema, we look to see which action to perform next. Another partitioning of the input, this time on whether the name corresponds to an open, closed, or unknown file, seems a good choice.

1. $n? \, 1 \in open$

2. $n? \, 1 \in \mathrm{dom}\, fsys \setminus open$

3. $n? \, 1 \notin \mathrm{dom}\, fsys$

The *open* case gives us, remembering that $A \Rightarrow (B \Rightarrow C) \equiv A \wedge B \Rightarrow C$,

---
ReadOneOpen

$\Xi FileSys$
$n? : \mathrm{seq}\, NAME$
$f! : \mathrm{seq}\, FILE$

$n? \, 1 \in open$
$\wedge \, \#n? = 1$
$\quad\quad \Rightarrow (n? \, 1 \in open \Rightarrow f! = \langle fsys(n? \, 1) \rangle)$
$\quad\quad \wedge (n? \, 1 \in \mathrm{dom}\, fsys \setminus open \Rightarrow f! = \langle closed \rangle)$
$\quad\quad \wedge (n? \, 1 \notin \mathrm{dom}\, fsys \Rightarrow f! = \langle doesNotExist \rangle)$

Simplifying this, using $(A \wedge B) \Rightarrow (B \Rightarrow C) \equiv (A \wedge B) \Rightarrow C$, gives

---
ReadOneOpenSimp

$\Xi FileSys$
$n? : \mathrm{seq}\, NAME$
$f! : \mathrm{seq}\, FILE$

$n? \, 1 \in open$
$\wedge \, \#n? = 1$
$\quad\quad \Rightarrow f! = \langle fsys(n? \, 1) \rangle$

5.4 Weakening

The final action defined by the PROST-Objects method is to loosen the constraints on the after-state. This action results in the definition of two (or more) operations: one that loosens the constraints as appropriate for the test, and one (or more) that contains the residual part of the definition, the part of the operation not being tested.

$$C \mathrel{\widehat{=}} T_w \wedge T_r$$

Before weakening *ReadOneOpen*, it helps to 'simplify' it by exposing the hidden predicates in $\Xi FileSys$, because in this case, it is these predicates that are to be weakened. Since we are going to check that the read file remains open, but not that any other file does, we also split up the predicate $open = open'$ into three separate predicates.

```
┌─ ReadOneOpenExplicit ──────────────────────────────────────
│ ΔFileSys
│ n? : seq NAME
│ f! : seq FILE
├────────────────────────────────────────────────────────────
│ fsys = fsys'
│ open \ {n? 1} = open' \ {n? 1}
│ n? 1 ∈ open ⇒ n? 1 ∈ open'
│ n? 1 ∈ open' ⇒ n? 1 ∈ open
│
│ n? 1 ∈ open
│ ∧ #n? = 1
│     ⇒ f! = ⟨fsys(n? 1)⟩
└────────────────────────────────────────────────────────────
```

We simplify this, using $(A \Rightarrow B) \wedge (A \wedge C \Rightarrow D) \equiv A \Rightarrow B \wedge (C \Rightarrow D)$, to

```
┌─ ReadOneOpenExpSimp ───────────────────────────────────────
│ ΔFileSys
│ n? : seq NAME
│ f! : seq FILE
├────────────────────────────────────────────────────────────
│ fsys = fsys'
│ open \ {n? 1} = open' \ {n? 1}
│ n? 1 ∈ open' ⇒ n? 1 ∈ open
│
│ n? 1 ∈ open ⇒
│     n? 1 ∈ open'
│     ∧ (#n? = 1 ⇒ f! = ⟨fsys(n? 1)⟩)
└────────────────────────────────────────────────────────────
```

We are now in a position to weaken this definition, by choosing just the last predicate

```
┌─ ReadOneOpenTest ──────────────────────────────────────────
│ ΔFileSys
│ n? : seq NAME
│ f! : seq FILE
├────────────────────────────────────────────────────────────
│ n? 1 ∈ open ⇒
│     n? 1 ∈ open'
│     ∧ (#n? = 1 ⇒ f! = ⟨fsys(n? 1)⟩)
└────────────────────────────────────────────────────────────
```

We take the residual to be of the form $T_r = T_w \Rightarrow C$, which gives a residual orthogonal up to the declaration. After simplification, this is:

```
┌─ ReadOneOpenResSimp ─────────────────────────────────────
│ ΔFileSys
│ n? : seq NAME
│ f! : seq FILE
├──────────────────────────────────────────────────────────
│ ( n? 1 ∈ open ⇒
│     n? 1 ∈ open'
│     ∧ (#n? = 1 ⇒ f! = ⟨fsys(n? 1)⟩) )
│   ⇒
│
│     fsys = fsys'
│     ∧ open \ {n? 1} = open' \ {n? 1}
│     ∧ (n? 1 ∈ open' ⇒ n? 1 ∈ open)
└──────────────────────────────────────────────────────────
```

This choice is not the only pattern for weakening an operation. There are other ways of weakening that are useful, and that also satisfy the proof obligations. For example, the proof obligations are automatically satisfied if we choose any one of

1. $C = A \wedge B : T_w = A, T_r = A \Rightarrow B$

2. $C = A \wedge B : T_w = A \vee B, T_r = A \Leftrightarrow B$

3. $C = A \Leftrightarrow B : T_1 = A \Rightarrow B, T_2 = B \Rightarrow A$

4. $C = A \wedge B : T_w = A \vee B, T_1 = A \Rightarrow B, T_2 = B \Rightarrow A$

(Case 4 is a combination of cases 2 and 3. It is interesting because it allows the non-orthogonal tests A and B to be used, by the identities $A = (A \vee B) \wedge (B \Rightarrow A)$ and similarly for B.)

Our example falls into case 1.

6 Test Coverage

The residuals document what is not being tested. In the example above, let's assume that *ReadOneOpenTest* is the only test, and all the other leaves in the tree of specifications (see figure 1) are residuals. So the residuals at the first partitioning explicitly document that input sequences of length zero, and length greater than one, are not being tested. At the next partitioning we learn that (for sequences of length one) closed and non-existent files are not being tested. And at the weakening, we learn that three things are not being tested: that the file system is unchanged (from $fsys = fsys'$), that the subset of open files other than $n? 1$ is unchanged (from $open \setminus \{n? 1\} = open' \setminus \{n? 1\}$), and that for $n? 1$, that if it is open after then it was open before (from $n? 1 \in open' \Rightarrow n? 1 \in open$).

7 Reusing the Test Suite

The specification being used to generate the tests could itself be a refinement of some other pre-existing specification. For example, the specifications of a family of related

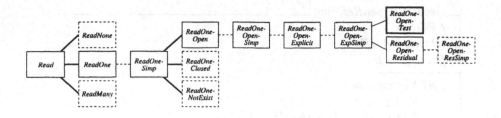

Fig. 1. The example tree of specifications. The specifications to the right of a link are abstractions of the specifications to the left. Partitions are shown as bold links, simplifications as dashed links, and weakenings as light links. Test leaves are shown as bold boxes, and residuals as dashed boxes.

products may be structured in such a way [Garlan & Delisle 1990], [Garlan & Notkin 1991]. Managed Object specifications form one such family.

If such a refinement relation exists, and if the other specification already has some tests defined, then those tests are valid tests for the new specification also (because abstraction is transitive).

This greatly reduces the burden on the test suite designer. New tests need be defined only for the *new* part of the specification, for example, extra operations, or extended parts of preconditions (in our interpretation, those before-states which were previously in the maximally non-deterministic 'unconstrained' region, but which have been refined into the 'interesting' region; see appendix A). All the common parts of the specification are already tested by the pre-existing tests.

Exploiting this reuse is a key feature of our structure for tests in PROST-Objects. The structure is difficult to capture formally using plain Z, but our ZEST specifications naturally fall in a refinement hierarchy, which is expressed as inheritance.

8 Testing ZEST class specifications

ZEST [Cusack & Rafsanjani 1992] is an object oriented extension to Z, particularly suited to the specification of Managed Objects [Wezeman & Judge 1994]. We are using it in PROST-Objects to specify both Managed Objects and their tests.

In ZEST, subclassing replaces the concept of refinement (the two concepts are not identical, but are close enough for our purposes) [Hall 1994]. So the abstract tests are *parent classes* of the tested class specification. We can use inheritance to capture, *formally within a specification*, the abstraction relation between tests described above. The inheritance tree also provides a framework for documenting the decisions made when building the tests for an object.

If the class specification under test is itself a child of another class, then its parent's tests are also valid tests for itself. Our test method enables us to include all parents' tests in the test suite. Not only can parents' tests themselves be reused, but the testing strategy—the choices of partitioning and weakening made at each step—can also be reused.

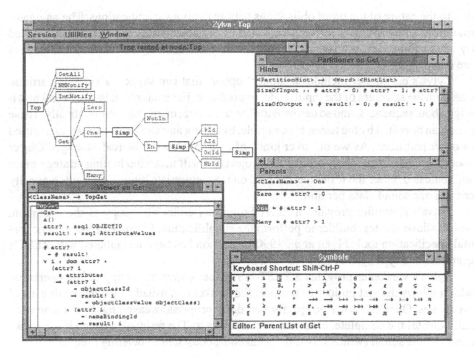

Fig. 2. A screen shot of Zylva, showing the Navigator (top left), the Partitioner (top right), a ZEST-Formaliser editor (bottom left), and a Symbol Palette (bottom right).

9 Tool Support

Many hundreds, or even thousands, of tests may well be needed for a real-world specification. Doing this systematically on paper (using any method, not just ours!) is time consuming, excruciatingly tedious, and error-prone; tool support is essential to make systematic test generation economically viable.

Under PROST-Objects, we are developing prototype tool support for our test method. This tool, called Zylva, provides support for the method actions of partitioning, simplifying, weakening, and for reusing parents' tests.

Zylva allows the tester to build a tree of specifications related by inheritance. (See figure 2.) The tool is built around a navigator, which allows movement around the existing test tree structure, and provides provides special subtools for partitioning, simplifying, and weakening, in order to grow and edit the tree. The tree structure is shown graphically; the kind of branch (Partition, Simplification, or Weakening) is indicated by colour; at each branch the status of the proof obligations (discharged or not) is indicated by the line style; the status of the leaves (unfinished, test purpose, residual, or parent tests) is indicated by box shape.

When a new set of parents is created by one of the subtools, some of the proof obligations are satisfied automatically by construction; the others need further work to be discharged. Also, editing a node in the middle of the tree causes potential inconsistencies, due to the nature of the proof obligations between various specifications. The navigator maintains a record of the current status, and of any potential inconsistencies introduced by such editing, and propagates them up the tree as necessary as the proof obligations are discharged.

Zylva's partitioner includes a 'hinter' option, that can suggest a suitable partition based on the structure of the input, state, or predicate. For example, it has built-in knowledge about sequences, and so can provide the 'zero, one, many' hint automatically. These hints can be edited by the tester, for example, by adding an extra $\#n? = 2$ case, and reused in other partitions. As we discover kinds of partitions useful for real Managed Object specifications during the course of the project, we will make the hinting strategy more sophisticated. Once the tester has decided on the appropriate hints, Zylva automatically creates one sound class per hint.

Zylva's simplifier provides a ZEST-Formaliser editor on a copy of the operation, which allows the test builder to perform the simplification. Formaliser is a generic formal specification tool [Flynn *et al.* 1990]; a version has been instantiated with ZEST's grammar and type system for PROST-Objects.

Zylva's weakener provides a ZEST-Formaliser editor on a copy of the operation, which allows the test builder to perform the weakening, usually by deleting unwanted predicates. Once the tester has performed the appropriate weakening, Zylva automatically creates the complete, orthogonal residual class. The residual is sound if the weakened class is sound (which it is if the weakening consisted of deleting predicates).

10 Conclusions

The insight that a test is an abstraction of an operation can be used to structure a hierarchy of tests.

The PROST-Objects test method

- provides rules for building and structuring this hierarchy

- suggests heuristics that minimise the proof obligations

- makes explicit the parts of the specification not being tested

- maximises reuse of tests between related specifications

The Zylva test tool provides direct support for the actions defined by the method. It also provides, by construction, partial automatic satisfaction of the proof obligations in the following cases:

	Partitioner	Weakener, with $T_r = T_w \Rightarrow C$
sound	yes	T_r sound if T_w sound
complete	if $\bigvee_i H_i$	yes
orthogonal	if $\neg (H_i \wedge H_j)$	yes

Such tool support relieves test builders of much of the low-level burden of defining tests, allowing them to concentrate on the skillful part: deciding just what are the 'good' tests.

11 Acknowledgements

PROST-Objects is a collaborative research project awarded by the United Kingdom's Department of Trade and Industry as part of its PROST programme (Programme of Research into Open System Testing). The partners are British Telecommunications plc, Logica UK Limited, and National Computer Centre Limited. The work described in this paper is the joint effort of all the team members on the project. Further work from the PROST-Objects project team is described elsewhere in this proceedings [Strulo 1995].

A An Aside on Interpretation

Consider a schema Op that defines a relation between before-states $State$ and after-states $State'$ (we ignore inputs and outputs for the moment; the argument easily generalises). The before-states $State$ can be partitioned into three regions:

'empty' region For these values of before-states the Op relation defines no after-state.

$$\{ State \mid \neg \, (\exists State' \bullet Op) \}$$

'unconstrained' region For these values of before-states the Op relation allows any valid after-state.

$$\{ State \mid (\forall State' \bullet Op) \}$$

'interesting' region For these values of before-states the Op relation allows some, but not all, valid after-states.

$$\{ State \mid (\exists State' \bullet Op) \wedge (\exists State' \bullet \neg \, Op) \}$$

The conventional interpretation of the schema Op as an operation is that in the 'empty' region, "anything can happen, including things outside the scope of specification". There is no formal distinction between the 'interesting' and 'unconstrained' regions, and the interpretation is that in these regions, "any one of the specified after-states occurs", which in the case of the 'unconstrained' region, means *any* state consistent with the state invariant.

In PROST-Objects, we choose to use a different interpretation. We interpret 'empty' to mean "the operation is forbidden"; this interpretation is sometimes called the 'firing condition' or 'enabling condition' interpretation [Josephs 1991]. We also make a distinction between the 'interesting' and 'unconstrained' regions. The tests defined in this paper are tests of the 'interesting' region only. We choose not to test the 'unconstrained' region; by definition any result is allowed, and so our test could not tell us anything useful. (Under our interpretation, we should also test the 'empty' region, to ensure the operation indeed cannot happen here. This aspect of 'negative' testing is not covered in this paper.)

We have chosen to use the firing condition interpretation in order to make some of our definitions simpler. Because of this choice, our operations tend to be written as implications, with the antecedents defining (part of) the 'interesting' region. This allows alternative partitions of the operation to be '∧-ed' together (as opposed to '∨-ed' together in the conventional interpretation), which fits in with the way we define inheritance when we move to an object oriented formalism.

This interpretation also allows us to specify and refine 'uncontrolled' operations that may occur spontaneously, but only when their firing condition is true [Strulo 1995]. Such operations are essential for our application area, Managed Objects, but they are not expressible under the conventional interpretation without resorting to 'tricks'.

References

[Bowen & Hall 1994]
Jonathan P. Bowen and J. Anthony Hall, editors. *Proceedings of the 8th Z User Meeting, Cambridge 1994*, Workshops in Computing. Springer Verlag, 1994.

[Cusack & Rafsanjani 1992]
Elspeth Cusack and G. H. B. Rafsanjani. ZEST. In Susan Stepney, Rosalind Barden, and David Cooper, editors, *Object Orientation in Z*, Workshops in Computing, chapter 10, pages 113–126. Springer Verlag, 1992.

[Flynn *et al.* 1990]
Mike Flynn, Tim Hoverd, and David Brazier. Formaliser—an interactive support tool for Z. In John E. Nicholls, editor, *Z User Workshop: Proceedings of the 4th Annual Z User Meeting, Oxford 1989*, Workshops in Computing, pages 128–141. Springer Verlag, 1990.

[Garlan & Delisle 1990]
David Garlan and Norman Delisle. Formal specifications as reusable frameworks. In Dines Bjørner, C. A. R. Hoare, and H. Langmaack, editors, *VDM'90: VDM and Z—Formal Methods in Software Development, Kiel*, volume 428 of *Lecture Notes in Computer Science*, pages 150–163. Springer Verlag, 1990.

[Garlan & Notkin 1991]
David Garlan and David Notkin. Formalizing design spaces: Implicit invocation mechanisms. In S. Prehn and W. J. Toetenel, editors, *VDM'91: Formal Software Development Methods, Noordwijkerhout, Volume 1: Conference Contributions*, volume 551 of *Lecture Notes in Computer Science*, pages 31–44. Springer Verlag, 1991.

[Hall 1994]
J. Anthony Hall. Specifying and interpreting class hierarchies in Z. In [Bowen & Hall 1994], pages 120–138.

[Josephs 1991]
Mark B. Josephs. Specifying reactive systems in Z. Technical Monograph PRG-19, Programming Research Group, Oxford University Computing Laboratory, 1991.

[Strulo 1995]
Ben Strulo. How firing conditions help inheritance. In Jonathan P. Bowen and Michael G. Hinchey, editors, *ZUM'95: 9th International Conference of Z Users, Limerick 1995*, Lecture Notes in Computer Science. Springer Verlag, 1995.

[Wezeman & Judge 1994]
Clazien Wezeman and Tony Judge. Z for Managed Objects. In [Bowen & Hall 1994], pages 108–119.

Improving Software Tests
using Z Specifications

Hans–Martin Hörcher

DST Deutsche System–Technik GmbH
Edisonstr. 3, D-24145 Kiel, Germany
Tel.: ++49-431-7109-478, Fax.: ++49-431-7109-503
e-mail hmh@informatik.uni-kiel.d400.de / hoercher@dst.dbp.de

Abstract. Formal Specifications become more and more important in the development of software, especially, but not only in the area of high integrity systems. Testing as a method to validate the functionality of a system against the specification will keep its justification also in a development process using formal specifications.

We demonstrate, where the problems lie when carrying out software integration tests using traditional testing techniques. It will then be demonstrated, how formal specifications can be used to achieve greater reliability and productivity during the software testing process by using extensive automatic tool support. This applies for the selection of test cases as well as the evaluation of test results, leading to a highly automated test process. First experiences from a case study will be given, in which we repeat the software integration test process for an application that has been developed by DST as part of the Cabin Intercommunication Data System (CIDS) for the new Airbus A330/340 family.

1 Introduction

Much has been said about the role of formal specifications during software development. The main advantage is, that it is possible to get away from the large amount of work during *a posteriori* error detection, replacing it by effort spent *a priori* during the construction of the system with an error free design. The use of formal specifications allows detection of errors and inconsistencies early during the software development process, optionally leading to a complete formal verification of each development step.

Nevertheless, with the current state of the art it is feasible that software test will still be needed, even when using formal specifications during software development. There are several reasons for this: Even if we have developed a system using complete formal verification of all development steps, it usually will be executed on a hardware that may fail, using an operating system that no one expects to be error free and being compiled using a compiler that has been developed traditionally. To make sure that these interfaces, which lie outside the scope of the specification, behave as assumed within the specification, testing will still be necessary.

Beyond that, today many models of application for formal methods do not follow a completely verified refinement development. At DST, for example, formal specifications very often are used during requirements analysis phase only. Further implementation path then follows traditional techniques for design and implementation, inspired by some guidelines for deriving design from the specification. This use of formal methods becomes even more attractive, when parts of the implementation process are done by a third party. Here the formal specification gains new importance as an interface to the implementing sub-contractor, who might be located in a different country or even continent. In this case, we usually do not have much control over the software production process, and the more important it becomes for us to have powerful and efficient mechanisms at hand to validate the products delivered by the sub-contractor.

In this paper we will give an overview over the recent developments that have been made at DST in the area of software testing against Z specifications. The ideas were born during development (and test) of the embedded control systems software development for the Airbus A330/340 Cabin Intercommunication Data System (CIDS) carried out at DST. Due to the criticality level of the system a lot of different test steps had to be carried out during development. These tests turned out to be much more expensive than originally expected and we tried to learn from this in order to find new ways to make this kind of software test more reliable and effective.

We will first give a summary of the traditional process of software testing and illustrate, which kind of problems we encountered with this. After this, we will summarize, how Z specifications may be used for systematically defining test cases for the implementation. Since test execution was done automatically, the greatest effort in this project turned out to be the task of evaluating the test results against the specification. Here we will demonstrate, how we will carry out this step automatically, using the Z specification as a reference. Finally, we will present first results from a case study, in which we apply these techniques to an existing project, the Cabin Illumination System (CIL) ([6]), which is part of the Airbus CIDS system.

A more detailed presentation of the ideas given in this paper can be found in [10]. Here, also other application areas rather than embedded systems software testing will be demonstrated. Although the ideas are presented for Z here, they might be easily applied to other specification languages.

2 The Testing Process

In this section we will outline, where the problems lie in the application of the classical software testing process. We will first give an overview over the steps applied here and point out work done in traditional testing. After that, we will focus on the use of test oracles, which play an important role within this process.

2.1 Steps during Traditional Software Test

Software test is usually carried out according to the following steps:

1. Test Data Selection

   Quality of the test case selection process has significant influence on the dependability of the whole process. If poor test cases are selected, the whole process may become meaningless. Much research has been made in the area of automating this process. Most of the approaches are based on the implementation's code, either using stochastical methods for generation or symbolic execution ([2, 5]). This seems quite natural, since in traditional software development process this usually is the only "specification" that has a formal semantics allowing detailed, automatic analysis. Using formal specifications these tasks can now be carried out along the specification. Here, different work has been done as well, either describing manual ([4, 1]) or automatic ([3]) test case selection.

We will show in Section 3.1 how test classes can be systematically derived from a Z specification. This procedure is similar to the one described in [3], for instance. The test classes derived make sure, that a complete coverage of the requirements will be achieved by them, which is one important completeness aspect. From these test classes, concrete test cases will then be derived afterwards.

Although the task of test data selection may become laborious, it mainly has to be carried out only once. Therefore, tool support would be helpful, of course, but, as pointed out later, we see much greater benefits from automating the task of test result evaluation.

2. Computation of Expected Results

In the classical theory of testing it is seen to be part of the test data selection process to formulate the expected outcome of the test cases, using a *test oracle*. As we will show in the next section, this is hardly possible in general, even with a formal specification available for automatic evaluation. Although it is not possible to supply concrete values for the expected results, the specification exactly describes the intended behavior of the system. In Section 3.2 we will show, how the specification can be used for automatic test result evaluation without serving as an oracle used to predict the expected outcome of the test in advance.

3. Test Execution

An automatic process for test execution is an essential prerequisite for carrying out tests on real live applications. Depending from the kind of test it will either be carried out within the final system or using special test environments. At DST, embedded systems software, for example, will be tested for logical errors before it is integrated into the target computer system. This is done using special test environments (see [8]) that allow detailed analysis of the system's behavior in a reproducible way, which is essential for an effective, automatic test execution. While in these environments parts of the target system are simulated by software, *hardware in the loop* test systems might be used to achieve the same functionality using the original target hardware. Other application areas require other kinds of tools, like, for example, capture and replay tools for GUI testing.

Since this topic is beyond the scope of this paper, the aspects of tool support for automatic test execution will not be handled further here.

4. Test Result Evaluation

As already outlined above, usually the existence of a *test oracle* is assumed to compute the expected outcome of a test case before it is executed. The task of test re-

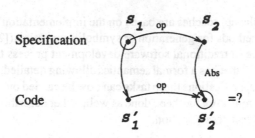

Fig. 1. Test result evaluation using a test oracle

sult evaluation then would simply consist of comparing these expected with the observed output. As we will show in the next section, this is not possible in general. We therefore will demonstrate, how we use the Z specification to automatically validate the observed output against the specification, without computing expected results in advance. From our experiences with the development of embedded systems software we expect to save much more efforts here, than this might be the case for an automatic test case selection tool. This may be illustrated by some figures: For one application from the CIDS system, the Cabin Interphone system (CIN), ≈ 200 test cases were derived, which print out on about 80 pages. The test protocol, which is produced by the automatic test tools, takes ≈ 2500 pages of output, which have to be analyzed.

5. Regression Testing

Regression test mainly consists of repeating the test execution and test result evaluation, while the test cases usually can expect to undergo at most minor modifications. Therefore, automating the test result evaluation process will help to save great efforts here as well.

2.2 Why Test Oracles fail

As pointed out in the previous section, part of the test case selection step is to determine the expected answer of the system when executing a specific test case. This is done, because

- the effort for this is expected to be lower, if done while thinking about the specific situation to find test cases,

- a human analyst might tend "to see, what he expects to see", if he still has to think about the observations made on the test output during evaluation,

- once determined, the expected output might be comparable to the observed behaviour automatically.

Thus, the following procedure would be followed, as it is sketched out in Figure 1:

1. Select a test case within the specification (S_1) and derive from it the corresponding test input values in the implementation's state space.

2. Computing the expected output (S_2) by using the specification of the operation as a test oracle and transforming the resulting value into the implementation's state space (S_2').

3. Compare expected and observed output after test execution.

We will now illustrate, which problems we encountered when following these guidelines during our projects.

Implicit Specifications Ideally, the requirements specification will be as abstract as possible in order to avoid implementation bias. This especially applies for Z specifications, which usually will be implicit, concentrating on properties rather than algorithms. Thus, an operation in Z may be seen as a relation between initial and final states that shall be legal under the operation, without giving a concrete algorithm. So it may be easy to check, if a given pair of states (S, S') is legal with respect to the specification, while in general it will not be possible to actively calculate the final state S' for a given initial state S efficiently.

Example: Sorting a sequence of integers

The specification of a sorting algorithm might be like this:

"The outcoming sequence shall contain the same elements as the incoming one, but the ith element shall be less or equal than the $(i + 1)$th."

For a given pair of sequences this requirement can be checked easily while it is not straightforward from this specification how to sort an incoming sequence.

Nondeterminism or Underspecification Sometimes specifications, or very often their refinement into an implementation will be nondeterministic. Underspecification during refinement gives the freedom to the implementor needed to achieve an efficient implementation. During test the problem arises here, that for a given expected after state (S_2) in the specification state space a large set of possible concrete values (S_2') in the implementation state space may be possible.

Example: Device Control

In the specification of the cabin illumination system (CIDS) for the Airbus the required reaction on switching the MAINon button is described as:

"All illumination units shall be switched to BRIGHT."

The implementation of the control software, however, will have to control each illumination unit individually. To implement the requirement stated above, the program will somehow iterate over all illumination units in order to switch them on individually. The order of iteration, however, is not determined by the specification, thus allowing simple iteration beginning in front of the cabin, in reverse order or even using some transformation tables (which indeed is the case), resulting in a large number of possible implementations with corresponding output.

"Generic" Test Data When testing transaction systems it very often will be necessary to define the test data – and therefore the expected outputs as well – using data depending on the actual system's state. We will call such test cases *generic* test cases, since they will be instantiated with concrete values during test execution.

Example: Stress Test

If a test case shall be executed several times in parallel in order to test the system's behavior under heavy load, it usually must contain components that will depend on the current system state (like a process sequence number) and therefore can only be instantiated during test execution. Here, the expected outputs must contain predicates as well.

Regression Testing Once the first test results have been collected and validated to be correct, it appears natural to compare new results resulting from regression tests against older ones. However, the problems discussed in 2.2 or 2.2 apply here as well. Our experience is, that these difficulties often make it impossible to directly compare the output against older results, leaving an enormous amount of manual work to be done here.

3 Testing against Z Specifications

In the previous section we have demonstrated, which problems we encountered during the test of various application areas. We will now show, how to overcome these problems using formal specifications as a reference. We use the Z notation ([14, 15]) for our purpose, but these ideas may be applied to other notations in a similar way. First we will show, how test cases may be derived from the specification. This process can be carried out partly supported by tools, giving evidence for completeness of the test cases. After that, we will present, how to automatically evaluate the test results against the specification. This will drastically increase efficiency and quality of this step.

3.1 Test Case Selection

The main aspect of Black Box Testing is to achieve complete coverage of the requirements described within the specification. With a Z specification as reference it becomes possible to systematically extract different situations for which the specification requires different behaviour. This is not only determined by the nature of the input variables of the system but also by the logical structure of the requirements. To make this structure explicit, each operation will be decomposed into a collection of subcases defining a set of equivalence classes (or a partition) on the input state space. For this we take the the operations predicate part and transform it into a Disjunctive Normal Form (DNF), consisting of a $\vee$ related collection of schemas, each of them containing predicates which contain $\wedge$ connectors only. Each schema describes an individual test class. [10] describes the procedure of test class derivation from a specification in more detail, in [3], a similar approach is described for VDM.

Let us consider, for example, the operation *AUTODIMop* from the CIL specification in [6]. This operation controls the automatic entry area dimming operation from the aircraft cabin:

$$\boxed{\begin{array}{l} AUTODIMop \\ \hline \Xi EAINDstate \\ \Xi SystemState \\ \Delta ILLstate \\ \hline cockDoor = open \wedge oilPres = high \Rightarrow \\ \quad ill' = ill \oplus \{\, a : domCAM\_EADIM \\ \qquad\qquad\qquad | \ CAM\_EADIM(a) < eaInd(fwd) \\ \qquad\qquad\qquad \bullet \ a \mapsto CAM\_EADIM(a)\} \\ cockDoor = closed \vee oilPres = low \Rightarrow \\ \quad ill' = ill \oplus \{\, a : domCAM\_EADIM \bullet a \mapsto eaInd(fwd)\} \end{array}}$$

The case analysis within the predicate part is complete and contains two alternatives. These will first result in two candidates for test classes, of which the first one does not require further transformation and therefore will directly result in the following test class:

$$\boxed{\begin{array}{l} AUTODIMop\_01 \\ \hline \Xi EAINDstate \\ \Xi SystemState \\ \Delta ILLstate \\ \hline cockDoor = open \\ oilPres = high \\ ill' = ill \oplus \{\, a : domCAM\_EADIM \\ \qquad\qquad\qquad | \ CAM\_EADIM(a) < eaInd(fwd) \bullet a \mapsto CAM\_EADIM(a)\} \end{array}}$$

Since the premisse of the second case contains a disjunction, it will be decomposed further, resulting in the following two test classes:

$$\boxed{\begin{array}{l} AUTODIMop\_02 \\ \hline \Xi EAINDstate \\ \Xi SystemState \\ \Delta ILLstate \\ \hline cockDoor \neq open \\ ill' = ill \oplus \{\, a : domCAM\_EADIM \bullet a \mapsto eaInd(fwd)\} \end{array}}$$

$$\boxed{\begin{array}{l} AUTODIMop\_03 \\ \hline \Xi EAINDstate \\ \Xi SystemState \\ \Delta ILLstate \\ \hline oilPress \neq high \\ ill' = ill \oplus \{\, a : domCAM\_EADIM \bullet a \mapsto eaInd(fwd)\} \end{array}}$$

Thus, we end up with the following representation of the operation $AUTODIMop$:

$$AUTODIMop \ \widehat{=}\ AUTODIMop\_01 \vee AUTODIMop\_02 \vee AUTODIMop\_03$$

Each of the schemas $AUTODIMop\_0$? describes an individual test class for the operation. For each of these classes, values have to be found to instantiate the required state. To determine these values, the same guidelines can be followed that are also used during traditional test data selection. These include:

- Select values for each input variable (including state components) that cover the range of legal values with respect to the constraints achieved for the current test class. These values include

  - random values from the legal range of value

  - upper and lower boundaries

  For many data structures, additional partitioning should be carried out in this step. Consider, for instance, an $x \in \mathbb{Z}$. Here, additional cases for $x < 0$, $x = 0$ and $x > 0$ should be selected.

  For sets and sequences, the notion of upper- and lower bound values must be extended. For sequences, at least an empty and a maximum lengthed sequence should be taken.

- For discrete values (i.e. free types) each value legal with respect to the constraints of the actual test class should be chosen at least once.

- White box test requirements might impose further test cases to be selected. These will result from decisions that were taken during refinement steps. Think, for example, of the implementation of a sequence as a dynamically allocated and extended array. Here, especially the case has to be tested, where the array is reallocated and extended.

Once achieved, the test cases will now be put together into test sequences in order not only to test single operation application, but also their interaction with each other.

Tool support exists within the DST Z-Tools in order to generate the test classes for an operation from its Z specification automatically. Test data will then be selected manually afterwards.

3.2 Test Result Evaluation

As already pointed out in 2.1, evaluation of test results turned out to be the most laborious task of the testing process, once automatic execution of tests has been achieved. Several reasons for this have been collected in 2.2. We will now show, how to overcome these problems by automatically evaluating the predicate parts of the Z-operations for validation of the test results.

How to overcome the Problems with Test Oracles Figure 2 shows how we propose to proceed for test result evaluation in order to overcome the problems discussed in 2.2:

1. Derive test cases (S_1) and corresponding test input data (S_1') as seen in Section 3.1.

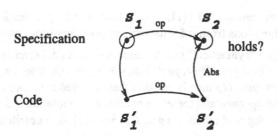

Fig. 2. Test result evaluation using formal specifications

2. Describe the expected output using predicates over the specifications state space instead of using concrete values.

3. Execute the test, achieving the the observed output (S_2') which will be transformed back into the specifications state space (S_2).

4. Check, if the resulting pair of states (S_1, S_2) fulfills the predicate defined for the expected output.

Operational schemas in Z directly describe, which after states (S_2) will be legal for a given input state (S_1) of the operation. Therefore, these schemas may serve directly as a description for the expected output of an operation which can be used as a basis for automatic test result evaluation.

Automatic Test Result Evaluation As motivated before, the Z specification of an operation may be regarded as the description of the expected output of an operations by means of predicates. It describes, which state transitions shall be legal for this operation without actually supplying an algorithm for computing such state changes. As seen before, this is one reason, why this specification usually will not be useful as test oracle. On the other hand, we have the implementation of this specification which we are going to test and which claims to be exactly the algorithm we are looking for, calculating the post state for a given pre state of the operation.

Our concept of test result evaluation follows exactly this idea. We will take the state transitions computed by the implementation for each test case, transform it back into the specification's state space and check it against the specifications by evaluating the predicates using the values observed. By this, the problems with the use of test oracles as described in 2.2 may be overcome.

For automatic test result evaluation against a Z specification proceed along the following steps:

Test Preparation

– Select test input data for the tests to be carried out (see 3.1).

– For each operation *op* supply predicates *pre_op* and *post_op*, evaluating the pre and post conditions for *op*. This can be done using the Z predicate compiler which has been developed at DST ([11, 12]). It takes the Z specification as an input and generates C functions from it for the pre and post conditions of each operation.

Note, that only a "syntactical" precondition is generated, i.e. an operation Op is split into the form $Op \mathrel{\widehat{=}} pre\_Op \wedge post\_Op$, where pre_Op does only contain undecorated variables. pre_Op may be weaker than the predicate 'pre Op' defined by the Z precondition operator and therefore precondition violations possibly might be detected only during evaluation of $post\_Op$. See [11] for a detailed discussion of this topic.

- Similar, provide code inv_state for each state invariant contained within state schemas.

- Supply a filter implementing the retrieve function between concrete data used in the implementation and the abstract data from the specification. At the moment, this will be done manually.

Test Execution

- Carry out the test as discussed in 2.1, achieving a collection of event/action sequences.

Test Result Evaluation

- Apply the retrieval filter to transform the event/action sequence into the specification's state space.

- At each operation activation that corresponds to some specific test case, check if the precondition of the test case actually holds. This is an important task to be carried out to actually validate the test data. Here we check, if the test case really covers the test class it has been designed for.

- Evaluate $pre\_op(S_1)$ and $inv\_state(S_1)$ using the pre state and the input parameters of the operation. If these evaluate $TRUE$, the operation has been called within its domain.

- Evaluate $post\_op(S_1, S_2)$ and $inv\_state(S_2)$. If one of them evaluates to $FALSE$, the test has been successful, since an error has been detected by the test.

There are two main reasons why this procedure will be more successful than the one usually underlying the test process: First, the observed output is checked in the specification's state space, eliminating the effects of possible nondeterminism within the data abstraction. Second, the specification is not used to actively calculate legal state transitions from pre and post conditions.

The second reason is important to be pointed out, since by this we overcome many of the problems usually related to executable specifications, mainly in connection with *implicit* ones, as this is the case for Z. This is important, since the main purpose of a formal specification still should be to improve the understanding of the problem rather than to execute it automatically. The application outlined within this paper shall be seen as an add on, to achieve even greater benefits during the later phases of the software development process. Therefore, it should not be necessary to adapt one's specification style in order to make the resulting specification executable. See [7] for a detailed discussion of this topic. In [9, 12] the executable subset of Z covered by the Z predicate compiler will be discussed further.

4 Case Study: The Airbus A330/340 CIL System

The *Cabin Intercommunication Data System (CIDS)* is an airborne fault-tolerant dual computer system to control the cabin communication functions implemented in the Airbus family A330/340. It comprises applications such as public address system, cabin interphone system, sign operation (e.g. fasten seat belts, no smoking) as well as monitoring and self test functionality. Although the system does not control flight relevant aircraft components, it still contains functionality which might indirectly impair human lives (e.g. emergency call functions, emergency evacuation signaling) and therefore had been classified as criticality level B according to DO178B (where "A" means highest and "D" means lowest criticality level, see [13]). During the requirements specifications phase, in which we used Structured Methods (Structured Analysis, Real-Time Analysis), some of the functions turned out to be surprisingly complex. Here we decided to apply formal specification techniques using the Z notation complementing the CASE specifications for critical applications.

Part of the development of the CIDS system for Airbus was an extensive testing process, during which we made many of the experiences described in Section 2. Due to this, the effort needed for testing had been under estimated at the beginning of the project. This, together with similar experiences during other projects, gave motivation for starting the developments described in Section 3. To find out, how these techniques apply in real life, we started a case study, where we repeated the integration testing process for the CIL application from the CIDS system.

The *Cabin Illumination System (CIL)* controls operation of the illumination units in the different cabin areas. [6] describes application and its specification in full detail. The size of the specification written for this application is about 35 pages of annotated Z specification (meaning ≈ 25 pages of pure Z), which result in ≈ 6000 (uncommented) lines of C-application code for 95 functions.

4.1 Testing Embedded Systems Software

We are now going to sketch out the scenario for tool supported integration test of embedded systems software against Z specifications, as we applied it during the case study. We will show, what tools currently are available and which further developments are planned at DST. We plan to demonstrate the current state of the system during the tool demonstrations at the ZUM'95.

Figure 3 gives an overview of the components of the test system and how they interact with each other. The rectangles stand for the different components of the system that are generated from the inputs placed above them.

The Integration Testbed is used to actually perform the tests. This is a tool, that allows execution of embedded systems software outside its target hardware within a simulated environment. The software under test will be "plugged" into a system which simulates

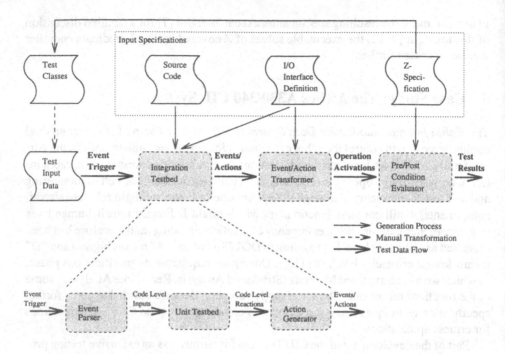

Fig. 3. A Test Environment for automatic Test of embedded Systems Software

the calling as well as underlying interface software layers in software, running on the development platform. Input to the system will be a textual representation of the events that shall trigger the system. These events will be analyzed by an Event Parser and transformed into actual values needed by the implementation. The software to be tested is embedded into a Unit Testbed, simulating all the components from the target system not available in the test environment. An Action Generator then transforms the programs reaction (like stub calls or variable changes) into sequences of actions as described in the I/O Interface Definition. This definition specifies, how each event from the requirements will be made available on implementation level. At the moment, only the Unit Testbed kernel is generated automatically from the sources, while the Event Parser and the Action Transformer will be written manually, based on the I/O Interface Definition.

The output from the integration testbed is passed into an Event/Action Transformer to be transformed into a sequence of operation activations and state variable changes from the Z specification. This transformer implements the retrieve function filter as described in 3.2. Written manually at the moment, it is planned to generate this component automatically from the I/O Interface Definition in the future.

Test cases will be derived from the Z specification by first generating the test classes automatically and then manually deriving the concrete input data (i.e. sequences of trigger events) from it. With these test cases, the test may now be executed automatically using the integration testbed.

For evaluation of the test results, the sequence of operation activations and state changes derived from the test output will be entered into the Pre/Post Condition Evaluator. This is mainly generated from the Z specification using the Schema Compiler as described in [11, 12]. It evaluates, if the state changes observed in the test output are legal with respect to the original specification. If the specification is violated, an error has been found within the implementation.

4.2 First Results

The scenario described in 4.1 describes the context of the case study. Two main questions have been imposed for this experiment:

- How do the test cases achieved by our techniques relate those originally defined for this application and how do the efforts compare?

- How does automatic test result evaluation apply to both – the original and the new test cases? Here, the efficiency aspect of the tools is of special interest.

We only have the final version of the application code delivered to the customer available for this experiment. Therefore we cannot expect to achieve realistic figures about the error detection rates for the different approaches as result of this case study.

Although we are still in an early stage of the case study, some initial results can already be given now. One general observation has be made quite early: The specification makes heavy use of sequential schema composition in order to model the external interface of the system on specification's top level. Expansion of these operations results in a lot of ∀ quantifiers that make the test classes generated by the system nearly useless. Apart from this, evaluation of these predicates during test result evaluation will probably lead to performance problems. Therefore, test classes will be generated for the individual operational schemas in order to exploit the underlying logic. The top level predicates do not imply further logical structures.

The original test plan for the CIL application contains 69 test cases. The test class generator produces about 50 test classes for the application, which we expect to result in ≈ 100 test cases when instantiated with test data. A detailed analysis must show, if all test cases originally defined for the application are covered by these. We already observed the fact, that the specification abstracts from some aspects that are explicitly checked as part of the original test cases. These were:

- Robustness aspects of the application (e.g. inconsistent CAM configuration data)

- additional events (e.g. Short Power Interrupt, CAM reconfiguration)

Apart from these aspects we expect the test cases derived from the specification to be a superset of those defined within the original test plan. The effort to achieve these test cases will be significantly lower than for the original ones, because their structure (and therefore the completeness argument) is automatically derived from the specification.

We do not yet have results from the test result evaluation. Although compilation of the specification works (i.e. the specification is executable with respect to the tool), we first need to finish the implementation of the Event/Action transformer before further activities can be started. The estimated effort for this will be ≈ one person week.

5 Final Remarks

We demonstrated, how – apart from the specification phase – further benefits can be drawn from a formal specification during validation of the implementation against the specification. We demonstrated, how the specification is used to systematically derive integration test cases that achieve complete coverage of the requirements. We also showed, how the Z specification may be used instead of a test oracle to validate the observed test results automatically. Finally we presented first experiences using these techniques within a real life application.

The techniques presented here may already be used during specification phase, early before any code is implemented, in order to validate the specification against the requirements. We realized this, when we started analyzing the test classes generated by the tools. Although they contain the same logical information as the original specification, the presentation in the disjunctive normal form gives completely new insight into the structure of the operations. Apart from that, the event/action traces described in 4.1 as input to the Event/Action Transformer do not need to result from test execution only. You might also ask the customer to specify, what reactions he expects to observe from the system in a specific situation, write this down in the same way and use the *Pre/Post Condition Evaluator* to validate this expectation against the specification. This approach to specification validation is described in [10] in further detail.

Acknowledgements Parts of the work presented here have been funded by the *Technologiestiftung Schleswig–Holstein* within the *CATI* project.

References

1. D. Carrington and P. Stocks. A Tale of two Paradigms: Formal Methods and Software Testing. In J. P. Bowen and J. A. Hall, editors, *Z User Workshop, Cambridge, June 1994*, Workshops in Computing, pages 51–68. BCS FACS, Springer Verlag, June 1994.
2. Richard Demillo and A. Jefferson Offutt. Experimental Results from an Automatic Test Case Generator. *ACM Transactions on Software Engineering*, 2(2):109–127, April 1993.
3. Jeremy Dick and Alain Faivre. Automating the Generation and Sequencing of Test Casees from Model–Based Specifications. In J.C.P. Woodcock and P.G. Larsen, editors, *FME'93: Industrial–Strength Formal Methods*, number 670 in Lecture Notes in Computer Science, pages 268–284. FME Europe, Springer Verlag, April 1993.
4. G.T.Sculard. Test Case Selection using VDM. In R.Bloomfield, L.Marshall, and R.Jones, editors, *VDM88: VDM – The Way Ahead*, number 328 in Lecture Notes in Computer Science, pages 178–186. VDM Europe, Springer Verlag, September 1988.
5. Walter Gutjahr. Automatische Testdatengenerierung zur Unterstützung des Softwaretests. *Informatik Forschung und Entwicklung*, 8(3):128–136, August 1993.
6. Ute Hamer and Jan Peleska. The Airbus A330/340 Cabin Communication System – A Z Application. In Michael G. Hinchey and Jonathan Bowen, editors, *Applications of Formal Methods*, Series in Computer Science. Prentice Hall International, to appear 1995.
7. Ian J. Hayes and Cliff B. Jones. Specifications are not (necessarily) executable. *Software Engineering Journal*, 4(6):330–338, November 1989.
8. Hans-Martin Hörcher. Das DST-Unittestbed zur automatisierten Durchführung von Unittests. *GI-Software Trends, Mitteilungen der Fachgruppe "Software–Engineering"*, 13(2):41–49, February 1993.

9. Hans-Martin Hörcher. Animation of implicit Specifications. Internal paper, DST Deutsche System–Technik GmbH, Kiel, Germany, 1994.
10. Hans-Martin Hörcher and Jan Peleska. The Role of Formal Specifications in Software Testing. In *Tutorial Notes for the FME'94 Symposium*. Formal Methods Europe, October 1994.
11. Erich Mikk. Automatic Compilation of Z Specifications into C for automatic Test Result Evaluation. Master's thesis, Christian Albrecht Universität Kiel, December 1993.
12. Erich Mikk. Compilation of Z Specifications into C for automatic Test Result Evaluation. In *Z User User Meeting - ZUM'95*, Workshops in Computing. BCS FACS, Springer Verlag, September 1995.
13. RCTA. DO-178-B, Software Consideratons in Airborne Systems and Equipment Certification. Technical Report RCTA Paper No. 548-92/SC167-177, RCTA, 1140 Connecticuit Avenue, Wasgington D.C., July 1992.
14. Mike Spivey. *The Z Notation – A Reference Manual*. International Series in Computer Science. Prentice Hall International, 2nd edition, 1992.
15. John B. Wordsworth. *Software Development with Z*. International Computer Science Series. Addison Wesley, 1992.

Compilation of Z Specifications into C for Automatic Test Result Evaluation

Erich Mikk

Christian-Albrechts-Universität zu Kiel
Institut für Informatik und Praktische Mathematik
Preußerstr. 1-9, D24105 Kiel, Germany
e-mail: erm@informatik.uni-kiel.d400.de

Abstract. If Z specifications are used as requirements specifications then test result evaluation leads to evaluation of schema predicates in states that are reached by the test. For automation of this approach Z operational schemas must be translated into programs that perform schema predicate evaluation. Predicate evaluation is straightforward; expressions are replaced by their values, logical connectives are evaluated using truth-tables, quantifiers and set constructions are evaluated using iteration. In order to exclude infinite iterations while evaluating quantifiers the schema compiler accepts besides finite quantifications only those which can be transformed into finite ones using term-rewriting techniques. These ideas are implemented in a Z predicate compiler.

1 Introduction

Formal specification methods provide a sound mathematical basis for system requirements descriptions. Having a formal specification of the system available suggest as a next step to use it during tests of the system. Software testing tries to verify the correctness of the system w.r.t. its specification in program states that are chosen for the test (states are functions assigning values to program variables or specification variables). Testing of software involves the choice of test cases, execution of tests and test result evaluation. We focus on the test result evaluation. A common way of test result evaluation is the formulation of expected results. Then test result evaluation is performed by comparison of results achieved by the test against expected results. Hörcher argues in [3] that formulation of expected results is hardly possible in general due to nondeterminism and underspecification in specifications. Our automatic test result evaluation mechanism that uses Z specifications as system's requirement specifications overcomes these problems.

The characteristic element of Z specifications is its schema. A schema can be used to describe constraints on system states (system invariants) or to describe state transitions. A test state is a legal state w.r.t. the specification, if the specification invariants hold. For a tested operation the initial and the final state must satisfy the predicate of the corresponding schema. Due to the implicit nature of Z specifications it is reasonable to substitute a test state in a schema predicate rather than to compute an expected outcome of a schema for a given initial state and compare it with an actual test result. The main characteristic of our approach is that the specification is not interpreted as an instruction for computing final states from initial ones but as a predicate indicating which pairs of initial and final states belong to the operation defined.

For automatic evaluation of test results a specification must be translated into code evaluating schema predicates in states reached by the test. This article describes a tool translating a subset of Z specifications into code for this purpose. The predicate compiler maps a schema to a C function that evaluates the schema predicate in states passed as parameter. Whereas the presentation will be built around Z specifications, the ideas can be applied to other model-oriented specification languages with an implicit specification style, s.a. VDM-SL. The target language of the compilation process is C due to performance reasons. However, our approach is independent of the target language of the compilation. Most of the problems addressed in this article are relevant even if functional or logic programming languages are considered as the target language.

A discussion about relating concrete test results to abstract specification states is beyond the scope of this article. Details are worked out in [5]. The predicate compiler is embedded into a testing framework which is presented in [3].

A comprehensive overview of specification based testing research is offered in [1]. Not all publications in this area consider automation of testing processes. Our contribution is an attempt to let the tedious and error-prone manual test result evaluation become automatic. This should help to increase the reliability of testing processes and decrease testing times. Industrial use of the schema predicate compiler for test result evaluation is reported in [3, 7].

This article is organized as follows. In Section 2 we present the semantics of predicates and discuss predicate evaluation issues. The intended automatic evaluation is achieved by restricting the set of input specifications to those which can be evaluated using at most finite iterations. This is the basis for the definition of executable predicates. Furthermore transformations for iteration elimination are defined. Section 3 considers the main transformation phases of the predicate compiler. Section 4 discusses the topic of code generation.

2 Evaluation of Schema Predicates

The key to test result evaluation is the evaluation of schema predicates in given states. So we first turn to the structure and meaning of predicates in Z. When we speak about execution of predicates, we mean their automatic evaluation. Z notation is a modified version of first order predicate logic, which is not decidable. Hence, our first task is to define a subset of executable predicates; "executable" corresponds to terminating computations.

Our evaluation of predicates is "brute force": expressions are replaced by their values, propositional calculus formulae are evaluated using truth tables, quantification and set construction (set comprehension) are evaluated using iteration. Hence, the executable subset of Z predicates are propositional calculus formulae, finite quantifications and finite set constructions (set comprehension). The syntactical definition of executable predicates provides a basis for their automatic recognition.

Such a subset of executable predicates is a severe restriction upon the input language. A further task in this Section is to extend the set of executable predicates by predicates which can be transformed into executable ones. A set of transformation rules for iteration elimination, whenever possible, is introduced for this purpose.

Z schema predicates are built up expressions which must be connected by relational symbols in order to integrate them into predicates (for predicate syntax see, e.g., [9]). Predefined relational symbols $=$ and $\in$ serve as a basis for the definition of further relations. After considering expression evaluation we discuss evaluation of predicates having form $expr_1\ REL\ expr_2$, where REL is a relational symbol.

2.1 Expression evaluation

Expressions in Z are built up using constants, like numbers, data structure building constructs and function applications. In [9] we find a discussion how functions are treated in Z:

> "In common with ordinary mathematical practice, Z regards functions as static relations between arguments and results; this contrasts with the view encouraged by some programming languages, where functions are methods for computing results from the arguments."

The notation of predicate logic is used for writing function definitions. We can distinguish between two kinds of function definitions: *explicit and implicit*. Explicit definitions are computing instructions, concerning how the result of the function can be calculated from arguments. Implicit definitions say nothing about "how to calculate", but about connections between arguments and results. From explicit definitions a program that implements them can easily be generated: this is a usual translation problem. Deriving algorithms from implicit specifications needs much more effort and is feasible only with human assistance.

Axiomatic or generic definitions in Z specifications can be used to introduce user-defined functions. Such functions are known to the whole specification. The schema compiler does not derive code for these function definitions since this involves issues of undecidability. In our approach the system user must provide code for functions itself or try to make use of the tools available. An implementation of our system uses the VDM Domain Compiler ([4]) that provides a library of functions for abstract data types. Consequently, the requirement that the user of the system must supply code for functions is not a severe restriction any more. Using code generated by the VDM Domain Compiler reduces possibility of errors in our testing system. This would be different, if a system user adds code to the system manually.

Function applications can be replaced by the corresponding results only if a function is applied in its definition domain. But how to deal with partially defined expressions?

The meaning of partially defined expressions is given in [8]. As argued there, *function application outside its definition domain is well-formed, but the value of function is undefined in this case.* This justifies the replacement of function applications outside their definition domain by a special value $\perp$ (bottom). Note that in order to construct this automatically the definition domain must be decidable. We assume that this is the case for input specifications of the predicate compiler.

2.2 Relations

Pre-defined relations: As mentioned in the beginning of this section, Z has two pre-defined relations: $=$ and $\in$. We refer to [8] for the definition of $=$:

> "$t_1 = t_2$ is **true** (assuming t_1 and t_2 have the same type) exactly when both t_1 and t_2 are defined and their values are equal, and is **false**, otherwise."

According to this definition, if one of the expressions t_1 or t_2 is not defined (replaced by $\perp$), then the value of the equality $t_1 = t_2$ is **false**; consequently, $\perp = \perp$ evaluates to **false**.

The definition of $\in$ can be given in a similar way ([8]):

> "$t_1 \in t_2$ is **true** (assuming t_1 has type a and t_2 has type $\mathbb{P}\,a$) exactly when both t_1 and t_2 are defined and the value t_1 is in set t_2, and is **false**, otherwise."

Note, that the value $\perp$ is always "trapped" by relations relating expressions (shown by induction over the structure of predicates), hence there is no need to give a definition of logical connectives for the case that one of their operands evaluates to $\perp$, because this cannot happen.

User-defined relations: Just like defining functions, the user can add new relations to specifications or use relations defined in libraries (e.g., the mathematical tool-kit in [9]). Similarly to the function case, we expect the user of the system to provide implementations for user-defined relations. The value $\perp$ that arises from partially defined expressions needs special consideration. Knowing the definition of basic relations ($=$ and $\in$), correct definitions and implementations for user-defined relations can be defined. This is not always intuitive as shown in the following example.

Example 1. Consider the definition of $\neq$:

$$\begin{array}{|l}
\hline
=[X]=\!\!=\!\!=\!\!=\!\!=\!\!=\!\!=\!\!=\!\!=\!\!=\!\!=\!\!=\!\!= \\
\quad \_\neq\_ : X \leftrightarrow X \\
\hline
\quad \forall x,y : X \bullet x \neq y \Leftrightarrow \neg\,(x = y) \\
\hline
\end{array}$$

According to the definition of $=$ the term $t_1 \neq t_2$ is **true**, if both t_1 and t_2 are defined and their values are not equal, or if at least one of them is not defined, and is **false**, otherwise. Note, that if at least one of t_1 or t_2 evaluates to $\perp$, then the result is **true**, rather than **false** or not defined.

2.3 Definition of Executable Predicates

A predicate is called executable if all quantification and set construction bounds appearing within the predicate are finite. This definition does not take into account the "dynamic" behavior of quantifiers or set construction, e.g. additional restrictions possibly allowing a restriction of the search space of the started iteration. For example, assume

N denoting the set of natural numbers, hence, $\exists x : N \bullet x < 10$ is not executable. In this special case the iteration could be interrupted when a suitable value is found. Of course, the additional information that is contained in the restriction $x < 10$ can be useful as well, but this is a matter of further transformations. The only intention of this restrictive definition is to capture a subset of executable predicates and to give syntactic criteria for their recognition. We are going to extend the set of executable predicates later in this Section by defining a set of transformations for iteration elimination whenever possible.

Representatives of executable predicates are propositional calculus formulae, quantifications over finite sets and set constructions requiring finite iteration.

2.4 Recognition of Non-executable Predicates

The constructs to determine iteration bounds are: quantifiers ($\forall, \exists, \exists_1$), λ- and μ-expressions, set comprehension and schema reference as expression. Once these constructs have been determined, the code generation module can check the iteration bounds while generating code for them: if an iteration bound is infinite, then the predicate is not executable. These cases involve:

- predefined sets $\mathbb{Z}, N, N_1$;

- the power-set constructor applied to an infinite set;

- user defined type constructors like sequences (seq), bags (bag), or (partial) functions.

Note that the list does not contain a reference to a given set. According to [9] we can not make any assumptions about the structure or cardinality of given sets unless they are explicitly made in a specification. Here, however, an additional assumption is made: it is assumed that a given set has a finite number of elements. This is natural within our testing framework.

2.5 Transformations that Extend the Set of Executable Predicates

The defined subset of executable predicates is a severe restriction on the input language: for most practical specifications a tool that accepts only executable predicates will not suffice. Our aim is to extend the set of executable predicates by those predicates which can be transformed into executable ones using a set of transformation rules. Obviously, the transformation rules must be correct in the sense that the application of the transformation does not change the meaning of a specification.

We use the notation from [6] to denote transformations by

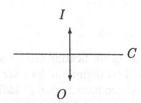

with I as input and O as output predicate, C as an applicability condition and $\updownarrow$ as equivalence rather than

$$\frac{C}{\vdash I \equiv O}$$

To shorten the formulation of transformation rules we make one *general assumption*:

- all predicates occurring in transformation rules are supposed to be syntactically valid and type-correct;

- all declaration parts of input predicates are normalized, i.e., the contribution made by a declaration part of some quantifier or set comprehension expression is made explicit to the property part of the construct.

The convincing argument for the correctness of the following rules is that they are derived from similar rules of predicate logic, taking into account that they deal with partially defined expressions rather than total expressions. A partially defined expression is defined on a certain domain for which all subexpressions of this given expression are defined. We call this domain the definition domain of this expression. Transformation rules leave definition domains intact, while manipulating expressions.

In our code-generation system these rules will be applied in a top-down fashion. Note that all rules represent equivalences: their application does not refine the specification (in sense of program refinement) and, therefore, there is no danger that they loose generality of a specification (the code is as abstract, i.e., nondeterministic, as the specification, not more concrete, as usually). In the following, a selection of transformation rules is given.

We begin with the quantifier elimination rule: if a predicate P does not contain x as free variable, denoted by the predicate $NotOccur(x, P)$, then the following transformations are valid ($\forall$- and $\exists$-elimination rules, resp.):

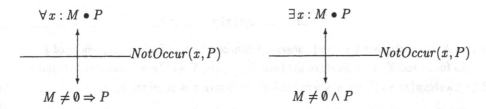

These rules are correct in a predicate calculus with total functions. Due to duality we have to show the correctness of only one of our transformations. Consider for the $\forall$-elimination rule the following two cases: M is not empty, M is empty (case "M is not defined" can not occur due to the general assumption that declaration parts in input predicates are normalized). The rule is obviously true if M is not empty. If M is empty, $M \neq \emptyset$ evaluates to false and the whole implication is true.

The next simplification rules can be shown to be correct using similar arguments. The $\exists$-*simplification* rule is:

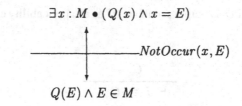

$$Q(E) \land E \in M$$

The importance of this rule is that the calculation of M is suspended: it is no longer needed for the evaluation of the quantifier (it can be useful while evaluating $E \in M$, but this term can be transformed with, e.g., the $\in$-simplification rule).

The $\in$-*simplification* rule is:

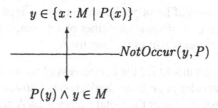

$$P(y) \land y \in M$$

As in the case with the $\exists$-simplification rule, the importance of this rule is that the calculation of M will be suspended.

Example 2. Assume seq N denotes the set of all sequences of natural numbers. When considering variable req : seq N the predicate $\exists l, r$: seq N $\bullet l \frown r = req$ might lead to an infinite iteration over the infinite set seq N during evaluation of the existential quantifier. Here, the following transformation helps to make the iteration finite:

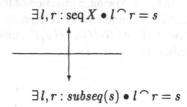

$$\exists l, r : subseq(s) \bullet l \frown r = s$$

where X is an arbitrary type and $subseq(s)$ denotes the set of all subsequences of s.

So for s : seq X, a predicate of the form $\exists l, r$: seq $X \bullet l \frown r = s$ can be rewritten as $\exists l, r : subtup(s) \bullet l \frown r = s$ where the infinite iteration is eliminated.

This is a small subset of the transformations that are currently implemented in the tool. An extension of our tool has a user interface allowing the user to add new transformations to the system.

3 Structure of the Predicate Compiler

The schema predicate compiler accepts a Z specification (possibly containing non-executable predicates) as input and generates C functions as code for it. For each schema a C function is generated: the declaration part of the schema defines parameters of the function and the predicate part of the schema is translated into code evaluating the predicates for states passed as parameters. The function returns a boolean value that indicates whether the schema predicate holds in the passed state or not.

3.1 Restrictions on the Input Language

The translation of Z operation schemas is defined for syntax-correct and type-correct specifications only. Executable predicates, and predicates that can be transformed into executable ones are accepted by the tool. Since the tool provides a feature for adding new transformations by the system user, the set of executable predicates is determined by the set of transformations which are used to transform non-executable predicates into executable ones.

Another restriction deals with generic schema references: actual parameters in a generic schema reference must be constant expressions. This restriction allows to generate all possible instances of generic schemas statically, i.e., without knowing the context of their applications.

The tool supports functions and relations of the mathematical tool-kit of [9]. New functions and relations can be added by the user, who can take the VDM Domain Compiler ([4]) to generate code for user-defined functions. Definition domains of partial functions are expected to be decidable.

A specification that does not fulfill requirements above mentioned is rejected with an error message.

3.2 Translation Schema

The specification undergoes a number of transformation steps before code for schemas is generated. These transformations simplify the structure of a specification and yield rather simple compiling specifications. Every transformation can be seen as a function mapping specifications to specifications.

1. **Textual expansion of abbreviations**
   Abbreviations do not add anything new to the expressive power of Z, but they just improve the readability of specifications. Abbreviations are expressed by axiomatic definitions.

2. **Conversion of free types into axiomatic definitions**
   Using free type definitions is a convenient way to define recursive structures such as trees. In [9] the meaning of free types is given by showing how to translate free type definitions into axiomatic definitions. This transformation is also performed by the compiler in order to simplify type inference.

3. **Generation of generic schema instances**
   Generic definitions define families of objects, which have the same behavior for different types. Since the set of all possible instances of generic schema definitions is statically determined by a specification (see restrictions in Section 3.1) one can generate generic instances during compile-time. So the specification is extended by instances of generic schemas.

4. **Textual expansion of hiding, pipes, sequential composition and pre-operation; transformation of schema inclusions**
   Since C lacks a schema notation and any rules of schema manipulation, one must implement these features in C. It is convenient to define the implementation for a restricted set of operations only. In order to manage other operations they must be expressed by "known" operations. Schema hiding, projection, piping, sequential composition and pre-operation are textually expanded to achieve elimination of these operations.

5. **Normalization of variable declarations**
   A variable definition in Z has two purposes: to bind a type to the variable and to define constraints restricting set of possible values of the variable. This means that different type expressions can denote the same type. A variable declaration is called to be in normal form if it uses only the maximum set of values as type expression. The maximum set of values has a unique denotation. Separation of type names (the maximum set of values of this type) and constraints that restrict the possible values of variables allows a considerable simplification of the type inference algorithm.

6. **Type inference**
   Determination of expression types is indispensable for a code generation step. It provides the basis for a optimization step, since optimizations are defined for type-correct terms only.

7. **Optimizing transformations**
   The transformations presented in Section 2.5 aim at eliminating infinite iterations while evaluating quantifiers, set comprehension, etc. These transformations can be applied to predicates even if an iteration needed for predicate evaluation is finite. Since these transformations lead to reduction of iterations during predicate evaluation, they are called *optimizations*.

8. **Type conversion**
   A particular Z type system allows construction of expressions, in which a function or a relation application assumes implicit type conversion of its arguments. For example, set union (∪) can be applied to relations as well: this application makes use of the fact that relations are sets of tuples. Such implicit type conversions are made explicit during this transformation step.

A detailed discussion about these transformations can be found in [5]. The application order of the following transformations is significant, because every transformation makes assumptions about the structure or about features of the input specification, and these features can be guaranteed only if previous transformations are finished. The type-inference algorithm used in the tool is very close to the type-checking algorithm presented in [10].

All transformations are independent of the target language C. Only the generation of generic schema instances is due to static typing in C.

After applying these transformations, code for schemas can be generated. We distinguish the generation of code for schema predicates and the generation of code for types. Since types and operations on them are implemented using the VDM Domain Compiler they are not compiled to C code but to VDM-SL domain equations. Code generation includes checking whether a predicate to be translated is executable or not.

4   Code Generation

Every schema is mapped to a C function; the schema name determines the name of the function, the schema signature determines formal parameters of the function and the schema property part determines the body of the function. The property part predicate will be translated into code evaluating the property. So the function generated returns a boolean value as result: it returns true, if the schema predicate holds in the state passed as actual parameter, and false, otherwise.

In order to execute the code it must be embedded into some environment defining global constants of the specification (for details see [3]).

4.1   Instantiation of Generic Relations and Functions

An *operation table* defines a mapping from Z function and relation names to corresponding function names in the VDM Domain Compiler (VDM DC). But mapping only Z function names to VDM function names is not sufficient, because Z functions and VDM functions are generic functions. There is a mechanism in the VDM DC to generate code for different instances of generic functions by determining actual generic parameters. This mechanism also determines the naming conventions for the functions generated in such a way. The code generated by the tool must use the same naming conventions: a function name is extended by the name of the abstract data type that is used as actual generic parameter instantiating types of the generic function.

4.2   Mapping Types to VDM-SL Types

Now, code generation for types is considered. The code generation module maps Z type names and type expressions built from type constructors and type names to the corresponding VDM-SL constructs. VDM-SL domain equations are generated during code generation.

We consider atomic type names first. Z has only one pre-defined type: $\mathbb{Z}$. This type is implemented using the *Intg* type from the VDM DC library. Another atomic type concerns given sets. Since an input specification does not contain sufficient information about how to implement given sets, no implementation is generated for them: given sets preserve their names in the code generated by the tool. It is expected that the implementation for given sets is defined by the environment in which the generated code is embedded.

Next, a mapping from Z type constructors to VDM-SL (refer to [2] for the VDM-SL description) type constructors is given. This mapping is defined by a *type table*. Here the simplified version of the actual type table is presented (relating Z type constructors to their implementations in VDM-SL):

- $\mathbb{P}\,T$ is implemented by T-**set**: the set of finite sets of elements of T.

- $T_1 \times T_2 \times \ldots \times T_n$ is implemented by product type $T_1 \times T_2 \times \ldots \times T_n$: the set of all ordered tuples mk-$(t_1, t_2, \ldots, t_n)$ with $t_i \in T_i$.

- Binding $\langle\!\langle p_1 : T_1; \ldots; p_n : T_n \rangle\!\rangle$ of schema type S is implemented as a set of VDM-SL composite values (records) mk-$S(e_1, \ldots, e_n)$ with $e_i \in T_i$.

- Z free type definition

$$T ::= d_1 \langle\!\langle E_1 \rangle\!\rangle \mid d_2 \langle\!\langle E_2 \rangle\!\rangle \mid \ldots \mid d_n \langle\!\langle E_n \rangle\!\rangle$$

is implemented by the union type

$$T = d_1 : E_1 \mid d_2 : E_2 \mid \ldots \mid d_n : E_n,$$

where E_i can depend on the union type T recursively.

- seq T is implemented by T^*: the set of finite sequences of elements of T.

- bag T is implemented by T-**bag**: the set of finite bags of elements of T.

- $T_1 \to T_2$ is implemented by $T_1 \xrightarrow{m} T_2$: the set of all finite maps from T_1 to T_2.

The final code for a specification may contain two different code pieces for the same schema. First, every schema defines a function evaluating its predicate in states passed as parameter. Second, for a schema reference as expression a record type and corresponding field selector functions are generated.

4.3 Partially Defined Expressions

The code generated must consider partially defined expressions. In this Section we describe how this is done using *exception handling*. We use VDM-SL notation to discuss it ([2]).

According to the Z syntax application of relations involves the step from partially defined expressions to predicates. The general idea of evaluating partially defined expressions and relations is presented in Section 2: function application outside its definition domain is replaced by a special value $\bot$; evaluation of relations ($=$, $\in$ and user defined relations) considers this special value $\bot$ also (for relation definitions see Section 2.2).

Consider a partially defined function

$$f : D_1 \to\!\!\!\to [D_2]$$

(the parentheses [ ] express that the value of the function is either a value from D_2 or nil, i.e. f is partially defined on D_1). Assume that def_f is the definition domain of f:

$$def_f \subseteq D_1 \land fun : def_f \to D_2$$

(f is a total function over domain def_f). The function body specification for f has the following form

$$f(arg) \overset{\text{def}}{=} ($$
$$\text{(if } (arg \in def_f)$$
$$\text{then } (\dots \text{return}(value);)$$
$$\text{else exit EXCEPTION};$$
$$))$$

This definition has the following meaning: if arg is a member of def_f then the function value is calculated and returned as result of the function, otherwise no value is returned and the function is terminated with raising an exception (the exit EXCEPTION clause). Here we expect that the context calling f handles this irregular termination of the function. Now our usage of exception handling is described. Consider the relation

$$expr_1 \text{ REL } expr_2.$$

Assume that some subexpression of $expr_1$ or $expr_2$ contains a call of a partially defined function f. Further assume that the VDM-SL function is_rel implements the relation REL and that the value of the evaluation of this relation must be assigned to a variable $current$. So the code without exception handling looks like

$$current := is\_rel([\![expr_1]\!], [\![expr_2]\!]),$$

where $[\![.]\!]$ denotes the compiling function of Z expressions. For the case $[\![expr_1]\!]$, $[\![expr_2]\!]$ or that their subexpressions are partially defined (i.e. some function call can finish with exit EXCEPTION) we add the following clause for exception handling

$$\text{trap EXCEPTION with } current := FALSE;$$
$$\text{in } \quad current := is\_rel([\![expr_1]\!], [\![expr_2]\!]).$$

These concepts are implemented using the *longjump* feature of C.

5 Concluding Remarks

This article describes a tool that translates Z schemas into executable code for test result evaluation. For every schema a function evaluating its predicate is generated. The function returns a boolean value as result which indicates whether the state passed as a parameter satisfies the schema predicate or not. The tool accepts an executable subset of predicates only. The notion of executable predicates guarantees that the code generated by the tool always terminates for any state passed as a parameter.

In our testing context it is not necessary to formulate explicit results by listing the explicit values. Instead any Z specification may be regarded as a general form of expected results. In contrast to the use of executable specifications, our approach has the advantage of allowing exactly the level of abstraction appropriate for the problem to be specified. This becomes possible because of interpreting implicit Z specifications as predicates that describe connections between initial and final states.

First industrial experiences of using the predicate compiler exist. In [3] the schema predicate compiler was used to evaluate test results of avionic embedded system software. In [7] the predicate compiler was used to validate real-time constraints of tests for a tramway crossing control system. Here the external behavior of the system was captured in a sequence of pairs of events and their occurrence times. A Z specification described real-time constraints of the system w.r.t. such sequences. Code generated by the predicate compiler checked the validity of these constraints on the system execution sequences.

The case studies have shown that requirement specifications could be used directly without modifications for automatic test result evaluation using our approach. But extra work had to be done to handle infinite iterations and to increase the performance of the final code. For every new specification class we had to add new transformations eliminating infinite iterations before we succeeded to generate code for it. In [7] we saw that the final code spent too much time constructing new sequences from an input sequence. Here we added transformations that replaced quantification over sequences by quantification over sequence indices.

Acknowledgments I would like to thank Hans-Martin Hörcher for his constant support and advice while both writing this article and implementing the schema predicate compiler. I also thank W.-P. de Roever for reading the draft version of this article and making valuable remarks and corrections.

Parts of the work presented here have been funded by the *Technologiestiftung Schleswig–Holstein* within the *CATI* project.

References

1. D. Carrington and P. Stocks. A tale of two paradigms: Formal methods and software testing. Workshops in Computing, pages 51–68. Springer-Verlag, 1994.
2. John Dawes. *The VDM-SL Reference Guide*. Pitman Publishing, 1st edition, 1991.
3. Hans-Martin Hörcher. Improving Software Tests using Z Specifications. In *Z User Workshop*, Lecture Notes in Computer Science. Z User Group, Springer-Verlag, 1995.
4. Hans-Martin Hörcher and Uwe Schmidt. Programming with VDM Domains. In D. Bjørner, H. Langmaack, and C.A.R. Hoare, editors, *VDM'90 – VDM and Z*, number 428 in Lecture Notes in Computer Science, pages 122–134. VDM Europe, Springer Verlag, April 1990.
5. Erich Mikk. Automatic Compilation of Z Specifications into C for Automatic Test Result Evaluation. Master's thesis, Christian Albrecht Universität Kiel, December 1993.
6. Helmut A. Partsch. *Specification and Transformation of Programs*. Springer, 1990.
7. Jan Peleska. Formal specification & test procedures for the ELPRO LET tramway crossing system. Technical report 9505, JP Consulting, Kiel, Germany, 1995.

8. J. M. Spivey. *Understanding Z: A Specification Language and its Formal Semantics*, volume 3 of *Cambridge Tracts in Theoretical Computer Science*. Cambridge University Press, January 1988.

9. J. M. Spivey. *The Z Notation: A Reference Manual*. Prentice Hall International Series in Computer Science, 2nd edition, 1992.

10. J. M. Spivey and B. A. Sufrin. Type inference in Z. volume 428 of *Lecture Notes in Computer Science*, pages 426–438. VDM-Europe, Springer-Verlag, 1990.

Language

Equal Rights for Schemas in Z

Samuel H. Valentine
University of Brighton
Department of Computing
Moulsecoomb, Brighton BN2 4GJ

Abstract

Schemas are a part of the Z notation, but not well integrated into the mathematical language. The paper shows how to make Z more unified and more flexible by removing the distinction between expressions and schema-expression. The Cartesian product can then be taken as a special case of a schema.

As example, it is shown how to embed operation schemas into the mathematical language, and have a neat calculus expresses relational schemas algebra.

Introduction

The schema is often regarded as the distinctive feature of Z, and as one which gives Z an edge over competing notations like VDM. The use of state schemas and operation schemas at any level paragraphs is practically fundamental to Z specification work (see Hayes 87 say). On the other hand the way in which schemas can be used inside the mathematical language, and in particular their use as types, are seldom exploited to the full. Potter, Sinclair & Till 91 for example, write "... schema types, which we shall have no occasion to use Z" and many other texts never show an awareness of the possibility.

The aim in this paper is to show the possibility and some of the benefits of using schemas at all levels. It explains the meaning of the schema calculus, and it advocates some modest changes to the definition of Z to allow schema calculus operations to be used throughout the mathematical language.

We take Spivey 92 as our definitive document of Z as far as the subject-matter of this paper is concerned. It differs from Spivey 89 only in the addition of schema enumeration of schema pipeline.

In general there is no suggestion are to some extent independent of each other, but rather each other even when together. They are in

Equal Rights for Schemas in Z

Samuel H Valentine,
University of Brighton
Department of Computing
Moulsecoomb, Brighton BN2 4GJ

Abstract

Schemas are a part of the Z notation, but not well integrated into its mathematical language. The paper shows how to make Z more unified and more flexible by removing the distinction between expression and schema-expression. The Cartesian product can also be taken as a special case of a schema.

As examples, it is shown how to embed operation schemas into the mathematical language, and how schema calculus expresses relational database algebra.

Introduction

The schema is often regarded as the distinctive feature of Z, and as one which gives it an edge over competing notations like VDM. The use of state schemas and operation schemas as top-level paragraphs is practically ubiquitous in Z specification work (see Hayes 87 etc.). On the other hand the ways in which schemas can be used inside the mathematical language, and in particular their use as types, are seldom exploited to the full. Potter, Sinclair & Till 91, for example, write " ... schema types, which we shall have no occasion to use ..." and many other authors never even show an awareness of the possibility.

The aim of this paper is to show the possibility and some of the benefits of using schemas at all levels. It explains the meaning of the schema calculus, and it advocates some modest changes to the definition of Z to allow schema calculus operations to be used throughout the mathematical language.

We take Spivey 92 as our defining document for Z; as far as the subject-matter of this paper is concerned, this differs from Spivey 89 only in the addition of schema renaming and of schema piping.

The generalisations to be suggested are to some extent independent of each other, but reinforce each other when taken together. They are in brief:

a) identifying a schema with the set of bindings which satisfy it, and hence merging schema-expressions with expressions;
b) allowing the operations of renaming, selection, hiding and decoration to be used on bindings, on schemas, on sets of schemas and so on;
c) legitimising schemas of null signature;
d) treating Cartesian products as special cases of schemas.

Scope and Type

The type language of Z is very simple, being built up from basic types, the powerset constructor and the two sorts of product constructor. Types are global in scope and have a finite description, which is unique provided that in the case of schema types we list the components in some canonical order such as alphabetical order of the component names.

Z is a statically block-structured notation, and statically typed. The word "static" describes judgements which can be made by immediate inspection of the text of a specification, without consideration of the values of variables or the truth of predicates. A Z type-checking program can decide unambiguously for every occurrence of every variable where its declaration is and what type it has. If this cannot be done, the specification is ill-typed and has no meaning.

A declaration is composed of basic-declarations, which are written separated by semicolons or newlines. Each basic-declaration is either explicit or is a schema use. A declaration is assembled from its constituent basic-declarations by including all the declared variables and their types. The same variable name may occur more than once, provided that the types are identical, and in that case all the corresponding constraints still apply. If the same variable is declared in the same declaration with two or more different types, we have a type error.

An explicit basic-declaration declares one or more variables as a member of some set, as for example

    x, y, z: s

where the three variables x, y, z, are declared as members of the set s. The type of x, y and z is determined by the type of s, and these three names are then in scope for the construction in which this declaration appears (that is, a schema, a comprehension, a quantification or a λ- or μ-expression). The set on which the declaration is made, "s" in this example, can be written as any expression. This expression is not itself within the scope of the declaration, so as an extreme example

    s: $\mathbb{P}$ s

is a valid declaration of an inner "s" as a subset of an outer "s", provided the latter is already declared. Although this example may be regarded as obscure or in bad style, the only effective way to ensure that this issue does not arise indirectly would be completely to ban all re-use of names, making it difficult to write large specifications at all.

A schema type has any number of components, identified by their distinct names, each of which has its type. Two schema types are compatible if where they have the same component names, the corresponding types are the same. They are identical if they are compatible and do have all the same component names. The value of a schema is a set of bindings, where each binding consists of all the pairs of the name and the value of each component.. Since a schema is a set, it can be used as the operand of an explicit declaration as described above.

A schema can also be used as a declaration in itself. When it is used in this way, its type is known. All the components of the schema are effectively declared, each with its corresponding type, as if they had been written at the point where the schema appeared as the declaration. The values of these declared components are then constrained to satisfy the property of the schema.

We recall that type and scope are interpreted statically, so variables occurring in the definition of the schema are taken as within the scope corresponding to where they are written, without regard to where the schema is used. To take a rather artifical example, if we have

$$a == \{ 1, 2, 3, 4 \}$$
$$S \triangleq [\, x\text{: } a \,]$$

then the predicate

$$\exists \, a\text{: } \{ 4, 5, 6 \} \bullet \{ S \mid x = a \} \neq \emptyset$$

is well-typed and true. The inclusion of the schema S causes the effective declaration

$$x\text{: } \mathbf{Z}$$

with the property

$$x \in a_{outer}$$

The comprehension is then further constrained with the predicate

$$x = a_{inner}$$

Using Schemas

A schema can be taken as a term. It has a type, the set-of-schema type, and a value, the set of bindings which satisfy the property of the schema. The type and the set together incorporate all the meaning of the schema; there's nothing else.

There are three ways we can use schemas in the mathematical language of Z:

a) We can use a schema as a term or expression denoting a set. So we can use a schema as the operand of an explicit declaration, and we can form expressions like $\mathbb{P}$ S, or S $\cap$ T, where S and T are schemas of identical type, and predicates like x $\in$ S, where x is a binding of the schema's type. We can also form "schema-expressions" using the "schema-calculus". We show below how this can be regarded as manipulations of schemas as sets.

b) We can use a schema as a predicate. The variables of the schema must already exist in the current context, with the right type, and the predicate is true if the values of those variables satisfy the property of the schema.

c) We can use a schema as a declaration, as we described above. This is equivalent to a declaration of the schema's variables as members of their types, together with a use of the schema as predicate to constrain their values.

The usual convention of defining a state and a family of operations using schemas as described in Hayes 87 etc. does not constitute an extra way of using schemas in this classification. That convention leaves formally unstated how the schemas are to be used, and filling in the formal details can only be done using the schemas in the ways given above. We give examples of how this is done in the section "Operation Schemas and the Mathematical Language" below.

Identifying a Schema with the Set of Bindings which Satisfy it

The idea of fully identifying a schema with the set of bindings which satisfy it, even at the level of concrete syntax, was proposed in Arthan 92 and arises naturally from the observation that the whole meaning of a schema is implicit in that set.

In Spivey 92 the use of set operations on schemas is allowed, but the result is a set of elements of schema type which is not usable directly in the other ways permitted to schemas. Separately from that, schema-calculus can be used to create schema-expressions, which are in every way like schemas but restricted to appearance as the operand in schema-definition paragraphs.

The proposal here is to relax the restrictions in both these cases; if a schema is identified with the set of its bindings, any expression whose value is such a set can be recognised as a schema, and the schema-calculus can also be considered as a family of operations on such sets.

First we need to clarify the rule that the names of schemas are drawn from the same vocabulary as other variable names. Spivey 92 gives a syntax with terminal symbols

"Word"

and

"Schema-Name - Same as Word but used to name a schema".

We must interpret this as meaning that the two are to be distinguished only by the type-checking process rather than by any more explicit means, so that there is no syntactic bar on using any schema-valued expression as a schema.

Renaming, Hiding and Selection

Renaming, hiding and selection are described in Spivey 92 with each being allowed in different contexts. A renaming of a schema takes the form S [new1/old1, new2/old2, ...] where S is some schema reference with components old1, old2, ..., and new1, new2, ... are the component names which are to replace them in an otherwise similar schema. Hiding is done by writing expressions of the form S \ (x1, x2, ...) where S is any schema expression and x1, x2, ... are the names of those of its components which are to be hidden. Selection takes the form b.x where b is any binding and x is the name of the component which is to be selected.

We propose to generalise these as follows:
a) to allow all of renaming, hiding and selection to be done on any expression of appropriate type;
b) to allow all of renaming, hiding and selection to be used on bindings, on schemas (sets of bindings), on sets of schemas and so on;
c) to introduce a new variant of selection to be written in the form S.(x1, x2, ...), as an alternative to hiding all the components of S not listed.

The effect of this generalisation is that where selection using the form S.(x1, x2, ...), renaming or hiding are applied to a binding the result is a binding with modified components, where they are applied to a schema the result is a schema, and similarly for sets of schemas and so on. If selection using the form S.x is applied to a binding, the result is the component value; if it is applied to a schema the result is a set of values, and similarly for sets of schemas and so on.

Decoration

The "decorations" are the characters ' ? ! and possibly others which may be added to the ends of words. Decorated names are distinct from those with different decorations or none, so for example x, x', x? and x! are all separate variables which may be unrelated.

According to Spivey 92 the names of ordinary variables may be decorated, but the names of schemas may not; if a decoration is applied to the name of a schema it is taken as being the operation of decoration, whereby the decoration is applied to the names of all the components. Thus if S is a schema with components x, y and z, S' is a similar schema with components x', y' and z'.

Our proposal is to unify schema-names with other names, and also to allow schema-valued expressions wherever we allow schemas. Accordingly we extend the operation of decoration to apply to any schema-valued expression delimited by any sort of brackets. Thus (S)' where S stands for any schema-valued expression, will mean the result of S with all its component names dashed. We also extend the operation of decoration to apply to bindings, as well as to sets of bindings (schemas) and similarly to sets of sets of bindings and so on indefinitely.

We now allow the names of schemas to be decorated, but in order to preserve compatibility with the existing convention, we say that where S is a simple name, and if S' is not otherwise declared in the current context, we take it that S' is an abbreviation for (S)', and similarly for all names and decorations.

The θ Operator

The θ operator is used to denote a particular binding. If S is a schema, θS is the binding whose component names are taken from S, with types and values taken from the types and values of those names in the context of use.

The type of a θ-expression is newly created, and is not necessarily the same as any other type (though most useful when it is). We recall that types are always global, so the scope of the names and variables used to establish the type and value of the θ-expression are irrelevant to the type of the θ-expression itself.

We propose here to generalise θ to allow as its operand any schema-valued expression, bracketed if necessary. The names which this expression would declare can immediately be discovered, and the type and value are then established using the types and values of those names in the context of use. We shall make use of this generalisation of θ below, writing terms like θ [S; T] and θ (∃ S • T).

For decorated terms we preserve the existing rule, as described in Spivey 92, that terms of the form θS', where S is a simple unbracketed name, create the θ-term using the value of the decorated variables but with the undecorated type. We extend this rule to apply in the same way to terms of the form θ(S)', where S is any schema-valued expression and we have a decoration written outside the brackets.

The Schema-Calculus

The proposal here is to allow the free use within the mathematical language of all constituents of what Spivey calls "schema-expressions". Thus we allow free use of a schema in "horizontal form", where we write

[Schema-text]

With our generalised operand to θ, we can define this as meaning the same as the comprehension

{ Schema-text • θ [Schema-text] }

where θ [Schema-text] is just a way of denoting the binding of all the components of "Schema-text". Writing it like this makes it clear that a schema in this form can be used wherever a comprehension can be used.

We also allow the use anywhere within the mathematical language of the operations of "schema-calculus" namely:
a) the schema quantifiers, $\exists$ $\forall$ $\exists_1$
b) the logical schema operators, $\neg$ $\wedge$ $\Rightarrow$ $\vee$ $\Leftrightarrow$ $\upharpoonright$
c) the precondition and composition operators, pre ; $\gg$.

We explain below what all these operations mean, in order to legitimise the proposed generalisation whereby we can use them on any expressions of schema type.

a) The three sorts of schema quantification are shown by expressions of the form $\exists$ S • T, $\forall$ S • T, and $\exists_1$ S • T, where S is some schema text and T is a schema expression. In all cases the schemas must be compatible, and the result is a schema with those signature variables present in T but absent in S. The variables of T which are not preserved are described as "hidden". In the formal description below we shall express our results as comprehensions, and we can make use of our generalisation of the operand of θ by writing the term of the comprehension as θ ($\exists$ S • T). That is, the expression θ ($\exists$ S • T) is to be understood as the binding of all the variables included in the signature of T which are not in the signature of S.

We propose to drop the rule given in Spivey 92 to the effect that all variables given in S must also appear in T.

Dropping that rule sharpens the need for a clear understanding of the scope rules here, which Spivey does not state explicitly. With ordinary predicate quantifications of the form $\exists\ S\ \bullet\ p$, and so on, where p is some predicate, it is well established that the whole quantification is within the scope of the variables introduced by the schema-text S. It therefore seems consistent to say, where we have a schema quantification like $\exists\ S\ \bullet\ T$, that T is within the scope of the variables introduced by S.

This scope rule applies only to the variables textually within T, of course, so if we have

$$x:\ \mathbb{N}$$
$$T \triangleq [\ y,\ z:\ \mathbb{N}\ |\ z > y > x\ ]$$

and consider the schema

$$\exists\ x,\ y:\ \mathbb{N}\ |\ x > y\ \bullet\ T$$

the variable x in T retains its outer reference, while y is identified with that in the quantification. The resultant schema has only z in its signature. If on the other hand we wrote directly

$$\exists\ x,\ y:\ \mathbb{N}\ |\ x > y\ \bullet\ [\ y,\ z:\ \mathbb{N}\ |\ z > y > x\ ]$$

the x of the schema would be within the scope of the quantification, so the meaning of the whole expression would be quite different, though still having just z in its signature.

i) We can now define schema existential quantification with the rewriting rule, whereby

$$\exists\ S\ \bullet\ T$$

means

$$\bigcup\ \{\ S\ \bullet\ \{\ \{\ \theta(S)\ \};\ T\ \bullet\ \theta\ (\exists\ S\ \bullet\ T)\ \}\ \}$$

The term $\{\ \theta(S)\ \}$ is the singleton set of the binding of the variables of S. It is a schema-valued expression and so can be used as a declaration. Its effect is to redeclare the variables in the signature of S identically in the inner comprehension, where the declarations of T are then conjoined with them. The rule as a whole says that the result is the set of all bindings of the unhidden variables which are consistent with S and T, where T is evaluated in the context of S.

ii) Similarly for schema universal quantification we say that

$$\forall\ S\ \bullet\ T$$

means

$$\bigcap\ \{\ S\ \bullet\ \{\ \{\ \theta(S)\ \};\ T\ \bullet\ \theta\ (\forall\ S\ \bullet\ T)\ \}\ \}$$

where as before the term $\theta\ (\forall\ S\ \bullet\ T)$ is just a way of specifying the binding which has the components present in T and absent from S. The result is the set of all bindings of the unhidden variables which are consistent with S and T for all possible values of the variables of S.

The above definitions use generalised union, $\bigcup$, and generalised intersection, $\bigcap$, which are themselves defined in terms of $\exists$ and $\forall$ respectively. Thus the schema versions of $\exists$ and $\forall$ are being defined in terms of the predicate versions, as one might expect..

iii) To deal with the other quantifier, $\exists_1$, we define a new function, which we call uniqueInRange

$$\underline{\quad[\ X,\ Y\ ]\quad}$$
$$\text{uniqueInRange: } (X \leftrightarrow Y) \to \mathbb{P}\ Y$$

$$\forall\ R: X \leftrightarrow Y\ \bullet\ \text{uniqueInRange } R = \{\ y: Y\ |\ \exists_1\ x: X\ \bullet\ x \mapsto y \in R\ \}$$

where uniqueInRange gives the elements which occur exactly once in the range of the relation to which it is applied.

Then we can define schema unique existential quantification with the rewriting rule to the effect that
$$\exists_1\ S\ \bullet\ T$$
means
$$\text{uniqueInRange } (\bigcup\ \{\ S\ \bullet\ \{\ \{\ \theta(S)\ \};\ T\ \bullet\ \theta\ S \mapsto \theta\ (\exists\ S\ \bullet\ T)\ \}\ \})$$
which says that the result is the set of all bindings of the unhidden variables which are consistent with S and T for exactly one configuration of the variables of S.

b) The logical schema operators, $\neg\ \wedge\ \Rightarrow\ \vee\ \Leftrightarrow\ |$, form new schemas from existing schemas by simple logical combinations of their types and properties.

i) Schema negation is written as $\neg\ S$, where S is any schema. The result is to be a schema with the same type as S, but with a property true for exactly the values where the property of S is false. We can define this formally as a generic function:

$$\underline{\quad[\ X\ ]\quad}$$
$$\neg: \mathbb{P}\ X \to \mathbb{P}\ X$$

$$\forall\ S: \mathbb{P}\ X\ \bullet\ \neg\ S = \{\ x: X\ |\ x \notin S\ \}$$

This definition fully and correctly captures the description given for schema negation, but applies to all sets, whether of schema type or not. It is the definition of set complement, a function not currently given in the mathematical tool-kit for Z because it is considered usually better style to gain its effect by other means. Since the formal definition for negation is giving us set complement as well, however, we can leave it like that.

The stylistic danger with set complementation, and equally with schema negation, is the use of the generic parameter, which is implicitly instantiated to the whole type. If we have some schema

$$S \triangleq [\ x: N \ | \ x < 42 \ ]$$

the value of the schema $\neg \ S$ is effectively

$$notS \triangleq [\ x: Z \ | \ x \notin N \lor x \geq 42 \ ]$$

whereas the natural assumption might have been that we were working within a "local universe" of the set N. If the negated schema is used only in contexts where the constraint $x \in N$ is automatically reimposed, this can still count as good style, but it is an abuse of the type system to rely on the whole extent of a type whose value has been established by implicit instantiation. Schema negation should never be used in ways which commit such "type abuse".

ii) Schema conjunction is the formation of expressions of the form $S \land T$, where S and T are schemas whose types are compatible. The result of this expression is a schema whose type is obtained by merging the types of S and T, and whose property is the conjunction of the two separate properties. This can be immediately defined as equivalent to the horizontal schema $[\ S; \ T \ ]$, which in turn is equivalent to the comprehension $\{ \ S; \ T \ \bullet \ \theta \ [\ S; \ T \ ] \ \}$.

iii) The next three logical schema operators can now be defined with generic equations for any schema expressions of compatible type:

$$(S \Rightarrow T) = \neg \ (S \land \neg \ T)$$
$$(S \lor T) = (\neg \ S \Rightarrow T)$$
$$(S \Leftrightarrow T) = ((S \Rightarrow T) \land (T \Rightarrow S))$$

We saw above that schema negation is just a special case of set complement. There are similar correspondences of other logical operators with set operators, namely

$$(S \land T) = S \cap T$$
$$(S \land \neg T) = S \setminus T$$
$$(\neg \ (S \Rightarrow T)) = S \setminus T$$
$$(S \lor T) = S \cup T$$

and $S \Leftrightarrow \neg \ T$ is equivalent to the "symmetric difference", which is not given in the mathematical tool-kit of Spivey 92, but is commonly described in the set theory literature.

In these equivalences, however, the set operations are defined for sets of any type, with the types identical for the two operands, while the schema operations are defined only for schema types, which can be different from each other so long as they are compatible.

iv) The schema projection operator, "↾", is a variant of schema conjunction where the components of the result are only those of the second operand. It can be defined with the generic equation

$$S ↾ T = \{ S; T \bullet \theta(T) \}$$

where S and T must be compatible schemas.

c) The remaining operators are those intended for use with operation schemas under the usual conventions of the meanings of decorations. The operations are pre ; ≫.

i) "pre" is a specialised version of hiding which removes all variables decorated with ' or !. If we write θ (pre S) for the θ-term of the variables of S without those decorated with ' or !, we can give the equation

$$\text{pre } S = \{ S \bullet \theta \text{ (pre S) } \}$$

ii) The schema composition S ; T composes the two schemas by identifying the dashed components of S with the undashed components of T, then hiding them. Let us define "State" as the schema which has property "true", and signature containing all the names which occur dashed in S and undashed in T, and if we have that State' is compatible with S and State is compatible with T, we can give the equation

$$S ; T =$$
$$[(\exists \text{ State}' \bullet S); (\exists \text{ State} \bullet T) \mid$$
$$\{ \text{ State}' \mid S \bullet \theta\text{State}' \} \cap \{ \text{ State} \mid T \bullet \theta\text{State} \} \neq \varnothing]$$

In this definition we first declare S with State' hidden, and T with State hidden, which gives us the right signature for the answer. Then we add the constraint that for each configuration of those variables, there shall be at least one tuple common to the comprehensions { State' | S • θState' } and { State | T • θState }.

iii) Similarly the schema piping S ≫ T pipes the two schemas by identifying the output components of S with the input components of T, then hiding them. Let us define "Pipe" as the schema which has property "true", and signature containing all the names which occur in S decorated with ! and in T decorated with ?, and if we have that Pipe! is compatible with S and Pipe? is compatible with T, we can give the equation

$$S ≫ T =$$
$$[(\exists \text{ Pipe!} \bullet S); (\exists \text{ Pipe?} \bullet T) \mid$$
$$\{ \text{ Pipe!} \mid S \bullet \theta\text{Pipe!} \} \cap \{ \text{ Pipe?} \mid T \bullet \theta\text{Pipe?} \} \neq \varnothing]$$

The Syntax of Schema Expressions

The proposal to allow schema-expressions to be used as expressions causes no difficulties with regard to type or meaning, but does create a syntactic problem. In Spivey 92 schema-expressions are segregated from the rest of the notation, and join it only by appearance as the operand of the schema assignment operator, "≙". This means that symbols used in schema-expressions can be the same as those used with a different meaning in ordinary expressions. Under the proposed unification of the two, the symbol "≙" merges with the symbol "==", but we may be introducing ambiguity for the symbols

$$[\ ] \ \exists \ \forall \ \exists_1 \ \neg \ \wedge \ \Rightarrow \ \vee \ \Leftrightarrow \ \upharpoonright \ \setminus \ ;$$

King, Sørensen & Woodcock 88, although like Spivey keeping schema-expression separate from expression, give different symbols for the schema operations, requiring different keystrokes for entry at the terminal, and somewhat similar in printed form to the others but visibly larger.

If the current proposal is to work it seems inevitable that we must provide distinct symbols in the case of the symbols

$$[\ ] \ \upharpoonright \ \setminus \ ;$$

because the uses of the symbols are just different in the two contexts. We propose therefore that this should be done on the lines given in King, Sørensen & Woodcock 88.

In the case of the square brackets, [,], there are three uses, namely
i) for horizontal schema description, where the new symbol is needed,
ii) to contain renaming lists,
iii) to contain generic parameter lists.
These last two uses can share the same symbol, as they can always be distinguished by the context and by whether the immediately preceding name is generic.

In the case of the symbols

$$\exists \ \forall \ \exists_1 \ \neg \ \wedge \ \Rightarrow \ \vee \ \Leftrightarrow$$

there is a systematic analogy between the way the corresponding operations treat schemas and the way they treat predicates, and it is feasible to treat the symbols as denoting a single operator which treats its operand appropriately in the two cases. This means that for parsing purposes the analyses of predicates and of expressions must be merged into a single hierarchy of operator binding precedences.

Logical Operations and Logical Schema Operations

If we have a schema expression such as S ∨ T in a context expecting a predicate, we can interpret it as a schema-disjunction which is then taken as a predicate, or alternatively as two schemas each separately taken as a predicate, then put together with a simple disjunction. The meaning of schema-disjunction is such that the end result is the same either way. If we have this schema expression in a context expecting a set or a declaration, however, we must interpret it as a schema-disjunction. The natural choice, therefore, is to stipulate that we interpret disjunction as schema-disjunction when both its operands are schemas.

What we have said about disjunction applies equally to the other logical operations in our list, namely

$$\exists \; \forall \; \exists_1 \; \neg \; \wedge \; \Rightarrow \; \vee \; \Leftrightarrow$$

For all of them, it makes no difference whether we treat a supplied schema argument as a predicate, then interpret the operation as an operation on predicates, or interpret the operation as one on schemas, then treat the result as a predicate. As we saw with disjunction above, if we do not want the result to be a predicate we must choose the schema-operation interpretation, so it seems natural to take that interpretation in all cases.

We may note that our clarification of the exact nature of the schema-quantifications, above, was designed to ensure that it behaves as we have said here. That is, in predicate position, the text ∀ S • T has the same meaning whether we are interpreting it as a schema-quantification or as an ordinary quantification. This can be proved fairly directly from the definitions given above.

Where the argument (for ¬), or the second argument (for the quantifications) or either argument (for ∧ ⇒ ∨ ⇔) is a predicate, we have the simple predicate logical operation as would be expected. If we wish to be sure that some schema-expression S is taken as a predicate we can write "S ⇔ true", for example.

Schemas of Null Signature

There is a consistent analogy between predicates and schemas of null signature. Spivey 92 allows schemas with no components to arise indirectly, but dismisses them as "not very useful". His concrete syntax does not allow an explicit declaration to be empty. We propose that declarations should be allowed to be empty in all contexts.

So if p is any predicate, we can form a schema [| p]. This is an expression in Z, which can be used wherever an expression can be used, and can be equated to a name. In particular we can write

    True ≙ [| true]

and

    False ≙ [| false]

where True is the set containing only the null binding, and False is the empty set of the same type.

If we use these schemas in contexts expecting predicates, "True" will have value "true" and "False" will have value "false". These are ordinary Z terms, so we can set up their type:

    Bool == { True, False }

and make declarations using that type such as

    x: Bool

If we want to equate our "Bool" x with a predicate p, we can write either

    x = [| p]

or

    x ⇔ p

The ability to attach names to predicates in this way can be used in the presentation of proofs. For example we can write

    Induction ≙

      [| (∀ s: $\mathbb{P}$ $\mathbb{N}$ |

        0 ∈ s ∧ (∀ n: s • n + 1 ∈ s) • s = $\mathbb{N}$)]

then use the name "Induction" whenever we want to refer to the contained predicate.

It can also be useful to turn a predicate into a schema so that it can become an operand in schema-calculus. For example if S is any schema and we write

    True ∨ S

the result is the schema with the same signature as S, but with property always true. This makes it easy to discuss schema normalisation precisely. It has also proved useful in work with executable Z to be able to separate establishing a signature from the assertion of the associated property, so that the latter could be done later than the former (Valentine 94, 95).

It would be possible to go further and identify predicates totally with schemas of null signature. Although this offers an attractive simplification of Z in some ways, the identification cannot be seen as wholly natural, and it causes the primitive notion of predicate to cease to be primitive.

Schema Types and Cartesian Products

In Z as currently defined, there are two distinct but analogous product types. There is the Cartesian product, of two or more components identifed by their positions, with individual tuples which are members of such products. There are also schemas, of zero or more components identified by their names, with individual bindings which are members of such schemas.

It seems natural to bring these together by regarding a Cartesian product as a special case of a schema, where the components are identified by their positions in the list. In some contexts, using 1, 2, 3, and so on as component names could cause confusion with the representation of numbers, so it is suggested that component names #1, #2, #3, and so on should be used.

The proposal is that all the existing concrete syntax and properties of Cartesian products should be retained, but that they should be identified with schemas whose component names are #1, #2, etc. Similarly, tuples are now identified with bindings in the same way.

Relations, as subsets of Cartesian products, also become schemas, and this allows some rather neat definitions in the mathematical tool-kit. For example relational inverse can be expressed as schema renaming:
$$\forall R: X \leftrightarrow Y \bullet R^\sim = R [\#2/\#1, \#1/\#2]$$
domain and range can be defined using selection, saying
$$\forall R: X \leftrightarrow Y \bullet \text{dom } R = R . \#1 \wedge \text{ran } R = R . \#2$$
domain restriction can be defined with
$$\forall s: \mathbb{P} X; R: X \leftrightarrow Y \bullet s \lhd R = \{ p: R \mid p . \#1 \in s \}$$
and we can give similarly abbreviated definitions of all the tool-kit functions on relations.

For selection of a component from a tuple we are writing t.#1, t.#2, and so on, where t is any tuple. Brien & Nicholls 92 suggest the form t.1, t.2, for tuple selection, but the form using #1, #2, is preferable if we want to be consistent with a more widely applicable notation.

This identification allows the conversion between a Cartesian product and a schema with ordinary names to be accomplished by mere renaming. It also clarifies the questions about Cartesian products with a single component, or with none, both of which it is now possible to represent.

Within tools (for example in the Z-- interpreter, Valentine 95) this identification means that properties which previously had to be supported analogously for both Cartesian products and schemas now need to be implemented only once, and this helps to ensure both the proper coverage of the facilities, and their correct implementation. Input and output routines still need to cater for all the existing concrete syntax, and to use the Cartesian product forms where appropriate.

Operation Schemas and the Mathematical Language

The convention for writing Z specifications is to give a schema representing a state, and a family of schemas representing operations on that state. The operations are specified in terms of undashed and dashed versions of the state variables, corresponding to the "before" and "after" states respectively. They are often written using the Δ convention, where we make definitions of the form

ΔState ≙ [State; (State)']

Inputs and outputs using the conventions of decoration with ? and ! respectively are a complication which we may wish to bypass. One satisfactory way of getting rid of explicit inputs and outputs is to represent them both as sequences of values which form part of the state, where an operation can then detach the head of an input sequence or add to the tail of an output sequence. We shall assume that this has been done where appropriate, and therefore we can ignore the decorations ? and ! in what follows.

In large specifications we need to modularise, often by forming some sort of hierarchy. One way of doing this, called "promotion", is to embed an operation directly into a higher-level operation, with predicates to relate the two, then hide the local variables. Thus we write something like

LocalOp ≙ [ΔLocalState | ...]
FrameOp ≙ [ΔLocalState; ΔGlobalState |
          θLocalState = ... ∧
          θLocalState' = ...]
GlobalOp ≙ ∃ ΔLocalState • FrameOp ∧ LocalOp

For examples of use see Barden, Stepney & Cooper 94.

The limitation of promotion is that, as is immediately apparent, it can only be used where a use of the global operation implies at most one use of the local operation. For a fully general approach, we need to be able to use operation schemas with the same freedom with which we manipulate functions and relations, and one way of ensuring this is simply to convert the operation schema into a corresponding relation. If we have some

Op: ℙ ΔState

for example, and we want to define the equivalent relation

Rel: State ↔ State

we can do so by writing

Rel = { Op • θState ↦ θState' }

We can often work directly with operation schemas, however. To take a simple but useful example, if we have some state "State", and some initial schema

Init: ℙ State

and operations

Op1, Op2, Op3: ℙ ΔState

and some sequence of operations is applied in order

OpHistory: seq (ℙ ΔState)

and we wish to specify the corresponding sequence of states

StateHistory: seq State

we can do so by saying

StateHistory 1 ∈ Init ∧

∀ i: dom OpHistory •

∃ OpHistory i •

θState = StateHistory i ∧

θState' = StateHistory (i + 1)

This could be expressed without the relaxations of notation we have suggested in this paper, but slightly less directly.

Another nice example is schema minimisation, (see Gravell 90). Suppose we have some nondeterministic operation schema

Op ≙ [ΔState; cost: **Z** | ...]

where "cost" is some measure of the cost of the operation, which we want to minimise. We can express the optimum choice as

BestOp ≙ [Op | cost = min { { θState }; Op • cost }]

Schema Calculus and Relational Algebra

A perhaps unexpected consequence of allowing free use of schema-calculus on sets is that we can now easily and directly express relational algebra (see Codd 72, Date 86, Ullman 82; for use with Z see also Sufrin & Hughes 85, Semmens & Allen 91).

In relational database theory a "relation" is a finite set of "records" subject to the constraints of a "relation scheme". We can represent such a relation scheme with a Z schema. The "columns" are the components of the schema. The "attributes" are the component names, and the "domains" are the component types.

A particular "relation" is then a finite set of bindings subject to the constraints of the relation scheme. The "rows" or "tuples" or "records" correspond to the bindings. If we have a relation scheme "S", we can declare a particular relation "R" as perhaps

R: F S

Usually some of the attributes are identified as forming the "key". To express this we can divide the definition of the relation scheme

Key ≙ [.....]

Scheme ≙ [Key; ....]

then to declare the set of relations where "Key" is indeed a key we could say

Relations ≙

{ R: F Scheme |

{ R • θKey ↦ θScheme } ∈ Key ↠ Scheme }

that is, for each valid relation there is a function from the key attributes to the rest of the attributes.

Although the purest approach to relational algebra identifies attributes only by their names, some treatments and some implementations allow them to be identified by their position. If so, we can reflect this in Z by renaming the attributes as #1, #2, #3 and so on.

Relational algebra, as established in Codd 72, has eight operations. We describe briefly how each appears in schema-calculus.

i) "Union" is exactly as in Z.

ii) "Intersection" is exactly as in Z.

iii) "Set difference" is exactly as in Z.

iv) "Selection" is the formation of a subset of the relation, most simply expressed in Z by using a comprehension.

v) "Projection" is the restriction to particular attributes, most directly expressed in Z by using the proposed generalisation of selection, writing terms of the form S . (x, y, z ...). Alternatively we can use schema projection or schema hiding, possibly accompanied by renaming.

vi) The "(Cartesian) product" of two relations is the relation which has extended tuples containing in all combinations the content of the tuples in each relation. It is represented in Z by first renaming enough to guarantee that all component names are distinct between the two relations, then forming the schema conjunction.

vii) The "natural join" of two relations is the set of all records composed by putting together records of each of the relations where the overlapping attributes have equal values. It corresponds exactly in Z with schema conjunction, where we are assuming that if component names in the two relations are the same, they refer to the same thing and are of the same type.

viii) The "relational division" or "quotient" is defined as follows. If we have two relations R, S, where the attributes of S all occur also in R, and we form the relation "R Divide S" we first identify the quotient attributes as those of R with those of S removed. A tuple is present in the quotient if every way of extending its attributes using values of those attributes taken from a tuple in S produces a tuple found in R. This corresponds exactly in Z to the schema universal quantification ∀ S • R.

Summary of Suggested Modifications

We summarise for convenience all the modifications suggested above to Z as defined in Spivey 92.
- allow any schema-valued expression to appear as a declaration,
- allow any schema-expression to appear as an expression,
- clarify that "Words" and "Schema-Names" are drawn from the same vocabulary,
- allow renaming, hiding, selection and decoration on any expression whose type is that of binding, set of binding, set of set of bindings, and so on, with appropriate components,
- introduce selection in the form S . (x, y, z ...) to hide all components except those listed,
- allow schema names to be decorated, but treat S' as an abbreviation of (S)' where necessary,
- allow as operand to θ any schema-valued expression,
- clarify that the term of a quantification is within the scope of its declaration in all cases,
- allow variables in the declaration part of schema quantifications which do not appear in the signature of the term,
- widen schema-negation to cover all cases of set complement,
- treat "≙" and "==" as identical,
- provide distinct symbols for schema-calculus uses of [ ] ↾ \ ;,
- interpret ∃ ∀ ∃₁ ¬ ∧ ⇒ ∨ ⇔ as schema operators wherever possible,
- allow all declarations to be empty,
- identify Cartesian products with schemas with components #1, #2, ...

Summary

We have explained schema-calculus and shown some examples of its use. We have suggested some modest enhancements to Z all of which are upwards compatible, in the sense that existing legal Z specifications remain legal and with the same meaning, except for the proposal to introduce some new symbols for schema-calculus work. This separation of the symbols is not a new suggestion and is already required by some tools.

It is finally suggested that tool-builders should experiment with implementations of the enhancements suggested here.

Acknowledgements

The ideas in this paper have been greatly stimulated by discussion within the BSI Standards Panel IST/5/-/52: Z Notation (alias ISO panel SC22 WG19). In particular, some of the suggestions were made by Rob Arthan and by Stephen Brien. Rob Arthan also read earler drafts of this paper and pointed out some obscurities and a mistake. Views expressed here are not necessarily shared by other members of the panel, however, and the proposals of this paper may not all be incorporated in the forthcoming standard.

My thanks also to Dan Simpson, University of Brighton, for fruitful discussion and support. The work was undertaken with him as part of the project "Models Algebra and Mechanical Support in Z", under SERC grant number GR/J 52020.

References and Bibliography

- Arthan R D. Issues for Z Concrete Syntax, ICL Secure Systems, FST Project, Ref: DS/FMU/IED/WRK036, February 1992.
- Arthan R D. Proposal to allow General Use of Expressions as Schemas in Z, Z Standards Committee Document D-105, June 1992.
- Barden R, Stepney S & Cooper D. Z in Practice, Prentice Hall 1994.
- Brien S & Nicholls J E (editors). Z Base Standard Version 1.0, Oxford University Computer Laboratory 1992.
- Codd E F. Relational Completeness of Data Base Sublanguages, in Rustin R (editor), Data Base Systems, Prentice Hall 1972.
- Date C J. An Introduction to Database Systems, Volume 1, (fourth edition), Addison Wesley 1986.
- Gravell A M. Minimisation in Formal Specification and Design, in Nicholls J E (editor), Proceedings of Fourth Annual Z User Meeting, Springer 1990.
- Hayes I (editor). Specification Case Studies, Prentice Hall 1987.
- King S, Sørensen I H & Woodcock J C P. Z: Grammar and Concrete and Abstract Syntaxes, Oxford University Computing Laboratory, Technical Monograph PRG-68, July 1988.
- Potter B, Sinclair S & Till D. An Introduction to Formal Specification and Z, Prentice Hall 1991.
- Semmens L & Allen P. Using Yourdon and Z, in Nicholls J E (editor), Proceedings of Fifth Z User Meeting, Springer 1991
- Spivey J M. Understanding Z, Cambridge University Press 1988.
- Spivey J M. The Z Notation — A Reference Manual (first edition), Prentice Hall 1989.
- Spivey J M. The Z Notation — A Reference Manual (second edition), Prentice Hall 1992.
- Sufrin B & Hughes J. A Tutorial Introduction to Relational Algebra, Programming Research Group, Oxford, July 1985.
- Ullman J D. Principles of Database Systems (second edition), Computer Science Press 1982.
- Valentine S H. Operation Schemas in Z–, University of Brighton Department of Computing, June 1994.
- Valentine S H. The Programming Language Z–, Information and Software Technology, Vol 37, No 5, May 1995.

Structuring Z Specifications: Some Choices

Anthony MacDonald and David Carrington

Software Verification Research Centre
Department of Computer Science
The University of Queensland
Queensland 4072, Australia

email: {anti, davec}@cs.uq.oz.au

Abstract. This paper investigates the issue of structuring Z specifications. It uses examples from a large specification to examine both conventions for using Z and notational extensions including Object-Z. Because of the importance of good structure within a specification, specifiers need to be aware of a range of techniques and where each is applicable.

1 Introduction

Application of formal methods and, in particular, Z[1, 15] is increasing. Z is a notation rather than a specific method and the notation provides many ways to specify systems. This raises a question: is structure important? To answer this question, we must consider what Z specifications aim to achieve and whether satisfying these aims is assisted by appropriate choices of structure.

Spivey[14] argues that formal specifications provide a precise description of a system without unduly constraining how the system achieves it. Specifications allow questions about the system to be answered without having to "disentangle the information from a mass of detailed program code, or to speculate about the meaning of phrases in an imprecisely worded prose description". Spivey also states that a formal specification can act as the single reference point between customer, designer and programmer. Specifications are supposed to capture system requirements clearly and concisely.

Can structure help achieve these aims? Normal Z practice is to structure a specification as formal schemas embedded in informal explanatory text. However with Z's increasing popularity, the size of specifications is also increasing. The problems faced when writing/reading large specifications often mirror those associated with large pieces of programming code. Just as techniques were developed and used to structure code at a higher level than procedures, techniques must be used to structure large specifications at a higher level than schemas. Approaches applicable to large software systems may be applicable to large specifications. An *'in the large'* structuring technique should enable large specifications to satisfy the aims of formal specifications. Any additional structuring technique should complement using schemas for lower-level structure. The components should be independently understandable and should compose easily to form an understandable whole without reference to irrelevant detail. Unfortunately there is no single correct technique: different techniques suit different situations. The paper demonstrates five different structuring techniques to highlight the choices available to specifiers.

Kernighan and Plauger[6] assert that "Good programming cannot be taught by teaching generalities. The way to learn to program well is by seeing, over and over, how real programs can be improved by the application of a few principles of good practice and a little common sense". We feel this applies equally to specifications and in this paper we attempt to highlight the suitability of each technique for particular situations.

The paper is divided into the following sections. Section 2 introduces the techniques using examples from a specification of the Production Cell Case Study[8]. Section 3 compares the suitability of each technique as applied to both small and large specifications. The paper closes with conclusions from the comparison.

2 Z Specification Styles

This section demonstrates five techniques for structuring Z specifications, starting with the simplest, or flat approach. A logical progression from this is the "Oxford" style which partitions the specification into components and which can be seen in Hayes[4, Chapter 2]. The third style parameterises similar components to avoid duplication. Hayes[4, Chapter 5] provides an example. The final two techniques use extensions to Z to impose structure: the first uses a library/module extension of Z developed by Wildman and Hayes[4, 5] and the second uses Object-Z[2, 12]. Combining object-orientation with Z has been investigated by many people, including Schuman and Pitt[13] and Hall[3]. Two collections[16, 7] capture recent work on extending Z with object-oriented features. This section first discusses the production cell case study and then specifies part of it using the five techniques.

2.1 The Production Cell Case Study

The original production cell can be found in a metal processing plant in Karlsruhe, Germany. Forschungszentrum Informatik (FZI), Karlsruhe, used the production cell as the basis of a study on the use of formal methods for critical software systems[8]. The first step of the study generated a requirements document[9] which attempts to capture, in plain language, the specification of a software controller for the production cell. To simplify the model, the system specified is a simulation of the production cell and the simulator (see figure 1) runs cyclically. The production cell simulation (the system) takes a metal blank as input. The metal blank is transported by a conveyer belt (the feed belt) to an elevating rotary table. The elevating rotary table rotates and rises vertically to present the metal blank for the robot to pick up. The robot takes the metal blank to a press. The press presses the metal blank and the robot retrieves the metal blank. The robot transports the metal blank to a second conveyer belt (the deposit belt). The deposit belt transports the metal blank to a crane. The

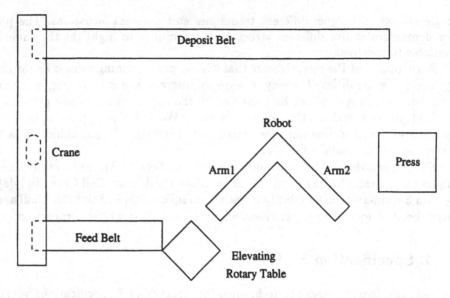

Fig. 1. The production cell

crane picks up the metal blank and transports the metal blank to the feed belt and completes the cycle.

The system actions are further complicated by a desire for speed and efficiency. To accommodate these aims, the robot is fitted with two arms. The arms are placed perpendicularly to each other and must rotate together. The arms can extend/retract and load/unload independently. A second consequence of the aim for speed and efficiency is the system should be able to handle multiple metal blanks in different parts of the system concurrently.

2.2 Production Cell Specification Overview

Our Z specification[10, 11][1] of the system has been built in a bottom-up manner. The first part of the specification is the independent specification of each component. The system is subdivided into the following components: feed belt, elevating rotary table, robot, press, deposit belt and crane. The independent specification of each component captures the local state information and the allowable operations on the state. The independent specifications do not take into account relationships between different components. For example, the robot cannot pick up a metal blank from the elevating rotary table if no metal blank is there; however this information is not relevant to the independent specifications of either the robot or the elevating rotary table. The independent specifications, each with their own local state, are combined to give the overall state of the system. Operations from the independent specifications are promoted to apply to the total state and are adjusted to take into consideration the relationships

---

[1] [10] is a complete production cell specification; this paper is extracted from [11].

between components. The system is modelled by the overall state and the corresponding operations, but the operations are partial and are only guaranteed to succeed (i.e., have a specified behaviour) when their pre-conditions are satisfied. The final section of the specification extends the partial operations to total operations. This is achieved by specifying possible error cases for each partial operation as an error schema. Joining the partial operation and the error schema via the schema calculus generates the final operation.

Five structuring techniques are explained using the robot section of the specification. The robot has several components and is introduced using the flat style. The other techniques are introduced relative to this initial specification.

2.3 Flat

The robot has two arms, called arm1 and arm2, and four discrete orientations.

- The first orientation, called *load_arm1*, aligns arm1 to the elevating rotary table while arm2 is not aligned with any other production cell component.
- The second orientation, *load_arm2*, aligns arm2 to the press and arm1 is not aligned.
- The third orientation, *unload_arm2*, aligns arm2 over the deposit belt for unloading and arm1 is not aligned.
- The fourth orientation, *unload_arm1*, aligns arm1 to the press and arm2 is not aligned.

Robot_Orientation ::= *load_arm1* | *load_arm2* | *unload_arm2* | *unload_arm1*

The robot's arms each have two attributes: the first, *Component_Loaded*, records whether the robot's arm is carrying a blank (*loaded*) or not (*unloaded*) and the second, *Arm_Extent*, records whether the arm is *retracted* or *extended*.

Component_Loaded ::= *loaded* | *unloaded*

Arm_Extent ::= *retracted* | *extended*

Initially, both arm1 and arm2 are *unloaded* and *retracted* and the robot's orientation is *load_arm1*.

```
__Robot _____      __Init_Robot _____
 robot_orientation : Robot_Orientation  Robot
 arm1_load : Component_Loaded          _____
 arm2_load : Component_Loaded           robot_orientation = load_arm1
 arm1_extent : Arm_Extent               arm1_load = unloaded
 arm2_extent : Arm_Extent               arm2_load = unloaded
_____        arm1_extent = retracted
                                        arm2_extent = retracted
                                       _____
```

For safety reasons, the robot must not rotate if either arm is extended. The robot's rotation does not change either the arm loaded status or the arm extension status, but only accesses the relevant variables.

```
┌─ Rotate_Robot_0 ─────────────────────────────────────────────
│ Δ Robot
│ new_pos? : Robot_Orientation
├───────────────────────────────────────────────────────────────
│ arm1_extent = retracted ∧ arm2_extent = retracted
│ robot_orientation' = new_pos?
│ arm1_load' = arm1_load ∧ arm2_load' = arm2_load
│ arm1_extent' = arm1_extent ∧ arm2_extent' = arm2_extent
└───────────────────────────────────────────────────────────────
```

Operations associated with arm1 do not change the robot's orientation or attributes of arm2.

```
┌─ Δ Arm1 ──────────────────────────────────────────────────────
│ Δ Robot
├───────────────────────────────────────────────────────────────
│ robot_orientation' = robot_orientation
│ arm2_load' = arm2_load
│ arm2_extent' = arm2_extent
└───────────────────────────────────────────────────────────────
```

To extend arm1, the robot must be at one of two orientations: *load_arm1* or *unload_arm1*. At this level of abstraction, the interactions between the arm and either the elevating rotary table or the press are ignored. The load state of the arm is not changed. The result of an *Extend_Arm1_0* operation is the arm extended. A successful retract operation leaves arm1 retracted without changing the load state of the arm or moving the robot.

```
┌─ Extend_Arm1_0 ──────────────┐   ┌─ Retract_Arm1_0 ─────────────┐
│ Δ Arm1                       │   │ Δ Arm1                       │
├──────────────────────────────┤   ├──────────────────────────────┤
│ robot_orientation ∈          │   │ arm1_load' = arm1_load       │
│     {load_arm1, unload_arm1} │   │ arm1_extent' = retracted     │
│ arm1_load' = arm1_load       │   │                              │
│ arm1_extent' = extended      │   │                              │
└──────────────────────────────┘   └──────────────────────────────┘
```

Loading arm1 only occurs when the robot is at orientation *load_arm1* and the arm is extended and unloaded. A load arm1 operation causes the arm to be loaded with no other change. Unloading arm1 has as its pre-condition that the orientation is *unload_arm1* and the arm loaded, as its post-condition that the arm is unloaded.

```
┌─ Load_Arm1_0 ────────────────┐   ┌─ Unload_Arm1_0 ──────────────┐
│ Δ Arm1                       │   │ Δ Arm1                       │
├──────────────────────────────┤   ├──────────────────────────────┤
│ robot_orientation = load_arm1│   │ robot_orientation = unload_arm1│
│ arm1_load = unloaded         │   │ arm1_load = loaded           │
│ arm1_extent = extended       │   │ arm1_extent = extended       │
│ arm1_load' = loaded          │   │ arm1_load' = unloaded        │
│ arm1_extent' = arm1_extent   │   │ arm1_extent' = arm1_extent   │
└──────────────────────────────┘   └──────────────────────────────┘
```

Operations for arm2 are almost identical except that they occur at orientations *load_arm2* and *unload_arm2* instead of orientations *load_arm1* and *unload_arm1*. They are not shown since they do not contribute directly to the discussion. However, the existence of multiple arms is important to the discussion.

2.4 Partitioned

The second technique partitions the robot into components. Each arm is specified and then included in the robot.

```
┌─ Arm1 ─────────────────────────────────────────
│ arm1_load : Component_Loaded
│ arm1_extent : Arm_Extent
```

The given type *Robot_Orientation* must be declared before the arms as the arm operations, except for retraction, are orientation-dependent. As access to the robot attribute *robot_orientation* is not possible, the operations require an input parameter. The input parameter, *current_pos?*, contains the arm's current orientation and is used to ensure that extending, loading and unloading of an arm occurs at the correct orientations.

```
┌─ Extend_Arm1 ──────────────────      ┌─ Load_Arm1 ──────────────────
│ ΔArm1                                │ ΔArm1
│ current_pos? : Robot_Orientation     │ current_pos? : Robot_Orientation
├──────────────────────────────       ├──────────────────────────────
│ current_pos? ∈                       │ current_pos? = load_arm1
│   {load_arm1, unload_arm1}           │ arm1_load = unloaded
│ arm1_load' = arm1_load               │ arm1_extent = extended
│ arm1_extent' = extended              │ arm1_load' = loaded
│                                      │ arm1_extent' = arm1_extent
```

Arm2 is defined similarly and is omitted for brevity.

The robot's state schema is composed of three component state schemas. Defining the schema *Robot_0* is unnecessary here but simplifies later promotion operations. An extra operation, *Generate_Orientation*, outputs the robot's current orientation for input to the arm operations.

$$Robot\_0 \cong [robot\_orientation : Robot\_Orientation]$$

```
┌─ Robot ────────────────────       ┌─ Generate_Orientation ──────────
│ Robot_0                           │ ΞRobot
│ Arm1                              │ current_pos! : Robot_Orientation
│ Arm2                              ├──────────────────────────────
│                                   │ current_pos! = robot_orientation
```

Promoting operations to the robot state is straight-forward and involves a simple framing schema and the schema calculus. The framing schema for arm1 constrains the state variables an arm1 operation can change to only those from the *Arm*1 state.

$$Arm1\_Ops \,\,\hat{=}\,\, \Delta Robot \wedge \Xi Robot\_0 \wedge \Xi Arm2$$

$$Load\_Arm1\_0 \,\,\hat{=}\,\, Generate\_Orientation \gg (Arm1\_Ops \wedge Load\_Arm1)$$

Without the input parameter to the arm1 operations, separate framing schemas are required for each operation, making the specification larger and more complicated. The robot's arms are identical in function and differ only in the orientations that an arm can be extended, loaded or unloaded. The next three structuring styles specify a generic arm which is instantiated twice.

2.5 Parameterised

The parameterised specification defines a generic arm which is instantiated twice in the robot. A generic arm is specified as an independent entity parameterised on *Orientation*. An arm does not contain information to restrict its actions to specific orientations but captures the existence of such orientations. The arm state schema contains two new state variables, *load_pos* and *unload_pos* which restrict the orientations at which loading and unloading of an arm can occur. The variables *load_pos* and *unload_pos* are never changed (after initialisation) and ΔArm is redefined to equate these variables (replacing the conventional definition).

```
┌─ Arm[Orientation] ─────────────
│ load : Component_Loaded
│ extent : Arm_Extent
│ load_pos : ℙ Orientation
│ unload_pos : ℙ Orientation
└────────────────────────────────
```

```
┌─ ΔArm[Orientation] ─────────────
│ Arm[Orientation]
│ Arm'[Orientation]
├─────────────────────────────────
│ load_pos' = load_pos
│ unload_pos' = unload_pos
└─────────────────────────────────
```

The arm operations differ slightly from those specified in the partitioned version. Apart from the obvious addition of a generic parameter, an extend operation can always occur at this level of abstraction, i.e. *Extend* has no pre-conditions and the post-condition is an extended arm. The load and unload operations compare the input *current_pos?* with the variable *load_pos/unload_pos* to ensure the operations proceed at valid orientations.

```
┌─ Extend[Orientation] ───────────
│ ΔArm[Orientation]
├─────────────────────────────────
│ extent' = extended
│ load' = load
└─────────────────────────────────
```

```
┌─ Load[Orientation] ─────────────
│ ΔArm[Orientation]
│ current_pos? : Orientation
├─────────────────────────────────
│ current_pos? ∈ load_pos
│ extent = extended
│ load = unloaded
│ extent' = extent
│ load' = loaded
└─────────────────────────────────
```

Compared to the flat specification, the robot's state schema changes in two ways: the number of state variables decreases by hiding arm details within the generic arm schema and predicates constraining where each arm can be extended and loaded/unloaded are added. This is an example of a common trade-off to be made when writing Z specifications: either putting constraints as part of a state invariant or putting them on every operation that might otherwise violate them. Extra constraints on the arms are needed because the arm has been specified without knowledge of the robot. For example, at the level of an arm it is logical to allow extension unconditionally, but for safety reasons the robot allows arm extension at certain orientations only.

$$
\begin{array}{|l}
\hline \_Robot_____ \\
\quad robot\_orientation : Robot\_Orientation \\
\quad arm1, arm2 : Arm[Robot\_Orientation] \\
\hline
\quad arm1.extent = extended \Leftrightarrow robot\_orientation \in \{load\_arm1, unload\_arm1\} \\
\quad arm1.load\_pos = \{load\_arm1\} \\
\quad arm1.unload\_pos = \{unload\_arm1\} \\
\quad arm2.extent = extended \Leftrightarrow robot\_orientation \in \{load\_arm2, unload\_arm2\} \\
\quad arm2.load\_pos = \{load\_arm2\} \\
\quad arm2.unload\_pos = \{unload\_arm2\} \\
\hline
\end{array}
$$

The robot's operations change in several ways. Framing schemas[2] are used to promote the arm operations to the robot state. These restrict an operation to apply to either $arm1$ or $arm2$. The framing schema for arm1, $Arm1\_Ops$, equates both $robot\_orientation$ and $arm2$ and binds $arm1$ and $arm1'$ to Arm and Arm'.

$$
\begin{array}{|l}
\hline \_Arm1\_Ops_____ \\
\quad \Delta Robot \\
\quad \Delta Arm[Robot\_Orientation] \\
\hline
\quad arm2' = arm2 \land robot\_orientation' = robot\_orientation \\
\quad arm1 = \theta Arm \land arm1' = \theta Arm' \\
\hline
\end{array}
$$

Binding $arm1$ to Arm is necessary as the expression $arm1.extent$ is valid, while $arm1.Load$ is not. By including both $\Delta Robot$ and $\Delta Arm[Robot\_Orientation]$ and binding $arm1$ to Arm, the framing schema allows the $Load$ operation on an Arm to be promoted and used at the robot level on $arm1$.

$$Load\_Arm1\_0 \;\widehat{=}\; Generate\_Orientation \gg$$
$$((Arm1\_Ops \land Load[Robot\_Orientation]) \setminus (\Delta Arm[Robot\_Orientation]))$$

Using the $Generate\_Orientation$ schema defined in section 2.4, the current orientation is piped to the load arm operation. The arm operation is restricted

---

[2] This technique requires an additional framing schema to be used at initialisation.

to changing only *arm*1's state variables with all $\Delta Arm[Robot\_Orientation]$ occurrences hidden ($\Delta Arm[Robot\_Orientation]$ is present in both $Arm1\_Ops$ and $Load[Robot\_Orientation]$).

The preceding techniques use standard Z. The next two techniques use extensions to Z to structure the specification.

2.6 Library

Specifying the robot using libraries is a logical extension of the parameterised specification. An arm library (see figure 2) is specified and used by the robot. The library specified is parameterised by the type *Orientation*. Parameterisation is applied to the library as a whole rather than the individual schemas and simplifies the use of genericity. The library is also textually enclosing allowing the *load_pos* and *unload_pos* state variables of the parameterised version to be constants within the library. A special definition of ΔArm is not required. Larger changes occur at the robot level when the arm is used. Prior to use, a library must be instantiated. As the specification requires two arms, the instantiated library must be decorated. The values of *load_pos* and *unload_pos* can be set here.

> **instantiate** $Arm1 :: Arm[Robot\_Orientation]$
>
> $Arm1 :: load\_pos = \{load\_arm1\}$
>
> $Arm1 :: unload\_pos = \{unload\_arm1\}$

The robot's state schema consists of three component state schemas.

> $Robot\_0 \;\hat{=}\; [robot\_orientation : Robot\_Orientation]$

─── Robot ──────────────────────────────
$Robot\_0$
$arm1\text{-}Arm1 :: State$
$arm2\text{-}Arm2 :: State$
──────────────────────────────
$arm1\text{-}extent = extended \Leftrightarrow robot\_orientation \in \{load\_arm1, unload\_arm1\}$
$arm2\text{-}extent = extended \Leftrightarrow robot\_orientation \in \{load\_arm2, unload\_arm2\}$
──────────────────────────────

The qualification *arm*1- ensures all variables of $Arm1 :: State$ are prefixed with *arm*1-. Otherwise, the variables in the two arm schemas are combined as happens normally in Z when schemas are combined. The decoration, *Arm*1, that occurs in the instantiation, only applies to the schemas in the library. Wildman and Hayes[5] choose this approach "because it separates the concerns of qualification of names and decoration of schemas".

The framing schema for arm1 is simplified to

> $Arm1\_Ops \;\hat{=}\; \Delta Robot \wedge \Xi Robot\_0 \wedge arm2\text{-}Arm2 :: \Xi State$

which allows changes to arm1's state variables only. Loading arm1 becomes

> $Load\_Arm1\_0 \;\hat{=}\; Generate\_Orientation \gg (arm1\text{-}Arm1 :: Load \wedge Arm1\_Ops)$

---

library *Arm[Orientation]*

An arm has two state variables. The first, *load*, records whether an arm is *loaded* or *unloaded* and an arm is initially *unloaded*. The second, *extent*, records whether the arm is *retracted* or *extended* and is initialised to *retracted*. Associated with an arm are two constants, *load_pos* and *unload_pos*, which are the orientations where an arm can be loaded or unloaded. The values of the constants are set at instantiation time.

$Arm\_Extent ::= retracted \mid extended$

| $load\_pos : \mathbb{P}\ Orientation$ | | $unload\_pos : \mathbb{P}\ Orientation$ |

_State_____
$load : Component\_Loaded$
$extent : Arm\_Extent$

_Init_____
$State$
___
$load = unloaded$
$extent = retracted$

Arm extension and retraction always succeed (at this level of abstraction) and change the value of *extent* while leaving *load* unchanged.

_Extend_____
$\Delta State$
___
$extent' = extended$
$load' = load$

_Retract_____
$\Delta State$
___
$extent' = retracted$
$load' = load$

An arm can only load and unload at certain orientations and to perform a *Load* or *Unload* operation, the arm must compare the current arm orientation to the set of allowable load or unload orientations. These operations change only the *load* variable of the arm.

_Load_____
$\Delta State$
$current\_pos? : Orientation$
___
$current\_pos? \in load\_pos$
$extent = extended$
$load = unloaded$
$extent' = extent$
$load' = loaded$

_Unload_____
$\Delta State$
$current\_pos? : Orientation$
___
$current\_pos? \in unload\_pos$
$extent = extended$
$load = loaded$
$extent' = extent$
$load' = unloaded$

endlib

Fig. 2. The Arm Library

2.7 Object-Z

Specifying an arm (see figure 3) in Object-Z is similar to the library specification. The specification is simpler because promoting operations in Object-Z is easier. This means that we do not bother to parameterise the arm class. The operations differ because Object-Z operations have a delta (Δ) list that defines which state variables potentially change. This simplifies the operations by removing the equating of variables that do not change. For example, *Extend* becomes a simple two line schema. Operations without a delta list inspect the state without causing any change.

Component_Loaded ::= *loaded* | *unloaded*
Arm_Extent ::= *retracted* | *extended*

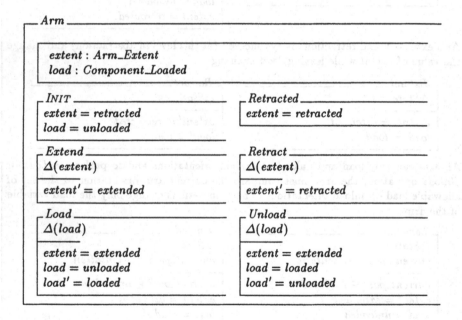

Fig. 3. The Arm Object

The major difference occurs where operations are promoted to the robot level (see figures 4 and 5). An intermediate level of structure is introduced via the *RobotBase* class which encapsulates all the information about the robot's orientation. It serves to simplify the *Robot* class specification. No framing schemas are necessary and the load operation for arm1 is

$$Load\_Arm1 \; \hat{=} \; arm1.Load \wedge base.LoadArm1Orientation$$

$Robot\_Orientation ::= load\_arm1 \mid unload\_arm1 \mid load\_arm2 \mid unload\_arm2$

__RobotBase_____

$\boxed{orientation : Robot\_Orientation}$ __INIT_____
$\boxed{orientation = load\_arm1}$

__Rotate_____
$\Delta(orientation)$
$new\_pos? : Robot\_Orientation$
$orientation' = new\_pos?$

$Arm1Orientation \; \widehat{=} \; [orientation \in \{load\_arm1, unload\_arm1\}]$
$Arm2Orientation \; \widehat{=} \; [orientation \in \{load\_arm2, unload\_arm2\}]$
$LoadArm1Orientation \; \widehat{=} \; [orientation = load\_arm1]$
$LoadArm2Orientation \; \widehat{=} \; [orientation = load\_arm2]$
$UnloadArm1Orientation \; \widehat{=} \; [orientation = unload\_arm1]$
$UnloadArm2Orientation \; \widehat{=} \; [orientation = unload\_arm2]$

Fig. 4. The RobotBase Object

__Robot_____

$base : RobotBase$
$arm1, arm2 : Arm$

$arm1 \neq arm2$
$arm1.extent = extended \Leftrightarrow base.Arm1Orientation$
$arm2.extent = extended \Leftrightarrow base.Arm2Orientation$

$Init \; \widehat{=} \; arm1.Init \wedge arm2.Init \wedge base.Init$
$Rotate \; \widehat{=} \; arm1.Retracted \wedge arm2.Retracted \wedge base.Rotate$
$Extend\_Arm1 \; \widehat{=} \; arm1.Extend \wedge base.Arm1Orientation$
$Extend\_Arm2 \; \widehat{=} \; arm2.Extend \wedge base.Arm2Orientation$
$Retract\_Arm1 \; \widehat{=} \; arm1.Retract$
$Retract\_Arm2 \; \widehat{=} \; arm2.Retract$
$Load\_Arm1 \; \widehat{=} \; arm1.Load \wedge base.LoadArm1Orientation$
$Load\_Arm2 \; \widehat{=} \; arm2.Load \wedge base.LoadArm2Orientation$
$Unload\_Arm1 \; \widehat{=} \; arm1.Unload \wedge base.UnloadArm1Orientation$
$Unload\_Arm2 \; \widehat{=} \; arm2.Unload \wedge base.UnloadArm2Orientation$

Fig. 5. The Robot Object

Each of the specification techniques achieves the same end via different methods. The following section discusses the positive and negative aspects of each technique.

3 Comparison

Initially, the discussion of each technique is constrained to its suitability for specifying the robot and its arms. However, the robot and its arms have a simple relationship and the components operate independently of each other at the robot level. The second half of the discussion deals with each technique when applied to a more complex state with many components that interact to give the final operations.

3.1 Structuring in the Small

The size[3] of the robot specification differs little between the techniques. The Object-Z version is the shortest at a page and a half while the parameterised version is the longest at two pages. An interesting point is that factoring components does not necessarily decrease specification size.

It can be noted that the partitioned style, which specifies each arm before inclusion in the robot state, does not work successfully for components that are dependent on some aspect of the higher-level state. This problem is avoided in the other component-based styles by using generic parameters. A component-based specification requires the components to be independently understandable.

The parameterised specification is the least intuitive to understand of all the specifications with both the framing schemas and the use of hiding making the specification complex, while the Object-Z specification is probably the simplest to understand.

Looking at specific details of the specifications, there are no interesting differences between the specifications of the arms other than the simplification and lack of duplication gained by reusing the arm rather than specifying two similar arms. However, there are three kinds of difference at the robot level: differences relating to the robot's state schema and initialisation, differences relating to the *Rotate_Robot* operation, and differences in the arm operations at the robot level. The differences in the arm operations at the robot level center on the different promotion techniques used by the structuring styles.

The main difference between the state schemas is the addition of predicates in the parameterised, library and Object-Z versions. These predicates make the constraints on the robot state explicit rather than implicit via the operations as in the simpler flat and partitioned versions. The initialisations are almost identical except for the initialisation of the instantiated generic arms in the parameterised style which requires a framing schema to nominate which arm

---

[3] This is the size of the complete specifications, not the fragments included in this paper.

is being initialised. It can be inferred that parameterisation is not suited to structuring where multiple instantiations of the same type are needed, but is more suited where each instance is based on a different type.

Across the five different specification techniques, there are three distinct specifications styles for *Rotate_Robot_0*. The first occurs in both the flat and parameterised versions and requires the post-state of every variable of the robot to be described explicitly.

Rotate_Robot_0

$\Delta Robot$
$new\_pos? : Robot\_Orientation$

$arm1.extent = retracted \wedge arm2.extent = retracted$
$robot\_orientation' = new\_pos?$
$arm1'.load = arm1.load \wedge arm2'.load = arm2.load$
$arm1'.extent = arm1.extent \wedge arm2'.extent = arm2.extent$

The second occurs in both the partitioned and library versions and uses the structure of the specification to equate whole components rather than individual state variables.

Rotate_Robot_0

$\Delta Robot$
$new\_pos? : Robot\_Orientation$

$arm1\text{-}extent = retracted \wedge arm2\text{-}extent = retracted$
$robot\_orientation' = new\_pos?$
$\theta arm1\text{-}Arm1 :: State' = \theta arm1\text{-}Arm1 :: State$
$\theta arm2\text{-}Arm2 :: State' = \theta arm2\text{-}Arm2 :: State$

The third occurs only in the Object-Z version. This version clearly identifies the components involved. Expanding this definition gives

---
Rotate

$\Delta(orientation)$
$new\_pos? : Robot\_Orientation$

$arm1.extent = retracted \wedge arm2.extent = retracted$
$base.orientation' = new\_pos?$

The delta list enumerates the state variables that may be changed by this operation. This removes the need to specify explicitly that other state variables do not change. The last two specifications of the rotate operation are more focussed as they abstract away from irrelevant detail.

The arm operations at the robot level differ for each technique. The flat style, while being relatively easy to follow, has considerable duplication, no information hiding and does not encourage reuse. The partitioning style provides information hiding, but has the duplication problems associated with the flat technique. As already stated, parameterised structuring is not suited to multiple instantiations. Framing schemas are needed to equate the unused state components and to restrict the operation to the relevant arm. The parameterised style also uses hiding. While neither framing schemas nor hiding are difficult concepts, their use is counter-intuitive, especially when compared to the final two techniques. The library mechanism is considerably easier to understand with a framing schema needed only to equate the unused state components. Object-Z is simpler again with no need for framing schemas.

For a simple system where the components do not interact with each other (i.e. an arm1 operation at the robot level does not affect arm2 at all), the library mechanism or Object-Z seem to be the easiest to use and understand. The decision between them would be based on whether the specifier wishes to use an object-oriented approach or not.

3.2   Structuring in the Large

The second half of the production cell specification involves combining all the components and promoting and combining the component operations to operations allowed on the production cell. Promotion is complicated by relationships between components at the production cell level. These relationships either impose additional constraints on operations defined for a single component or require defining system-level operations as combinations of component operations. Defining these relationships requires read access to state variables of multiple components, but changing these variables is mostly not required.

The flat style is not appropriate as components are being considered. The arguments for not using the flat style at the robot level also apply at the production cell level. These include lack of information hiding and duplication which leads to a cumbersome specification (thirteen state variables with no more than five accessed by any operation). Components can be included in the production cell in two ways: as instances (declaration) or by direct inclusion (schema inclusion). The discussion of component inclusion is in two parts, the first considering alternatives in Z and the second considering Object-Z.

Z Instantiating a schema promotes the state variables but not the operations. As was shown earlier using the parameterisation style, such operations are normally promoted using framing schemas. For the production cell, only direct inclusion of components is discussed.

$$Cell \;\widehat{=}\; Feed\_Belt \land ERT \land Robot \land Press \land Deposit\_Belt \land Crane$$

Generating the production cell operations from the component operations using promotion and combination uncovers issues not present in a smaller, less complex specification. Firstly, extra constraints on operations at the production cell level are needed. For example a robot arm, as long as the robot is at a correct orientation, can always be extended within the robot component. However for safety reasons, the robot's arm extensions are constrained at the production cell level where the robot must interact with either the elevating rotary table or the press. Another issue is how to specify which state variables are unchanged, especially when most of the production cell operations access state variables from only one or two components. The discussion focuses on the robot, but now considers its relationships with the rest of the cell.

Unchanged State Variables Consider the operation *Load_Arm1_1* which combines the robot operation *Load_Arm1_0* and the elevating rotary table operation *Unload_ERT_0*. The operation changes the state variables associated with both the robot and elevating rotary table leaving the state variables associated with all other components unchanged. Equating unchanged components can be approached in several ways. The simplest method is to equate each of the unused states independently.

$$
\begin{array}{|l}
\hline \text{\_\_} Load\_Arm1\_1 \text{_____} \\
\Delta Cell \\
Load\_Arm1\_0 \\
Unload\_ERT\_0 \\
\hline
\theta Feed\_Belt' = \theta Feed\_Belt \\
\theta Press' = \theta Press \\
\theta Deposit\_Belt' = \theta Deposit\_Belt \\
\theta Crane' = \theta Crane \\
\hline
\end{array}
$$

However, with a large collection of components this produces long lists and is a tedious process. It would be clearer if it were possible to equate the whole state except the components changed by the operation as shown below.

$$
\begin{array}{|l}
\hline \text{\_\_} Load\_Arm1\_1 \text{_____} \\
\Delta Cell \\
Load\_Arm1\_0 \\
Unload\_ERT\_0 \\
\hline
\theta Cell' \setminus (Robot', ERT') = \theta Cell \setminus (Robot, ERT) \\
\hline
\end{array}
$$

This is not valid Z as hiding is not allowed to be used with theta (θ). There are two equivalent methods which both involve declaring the sub-state before use. The first and less intuitive method uses existential quantification.

$$
Cell\_0 \;\widehat{=}\; \exists\, ERT \bullet (\exists\, Robot \bullet Cell)
$$

The second uses hiding and is intuitively closer to what is required. The definition for *Cell_0* should contain the state variables to be hidden in parentheses rather than the state schemas to which the variables belong. However, we believe this minor extension captures clearly what is required.

$$
Cell\_0 \;\widehat{=}\; Cell \setminus (Robot, ERT)
$$

Either method can be used in the following schema which is valid Z.

```
┌─ Load_Arm1_1 ─────────────────────────────────
│ ΔCell
│ Load_Arm1_0
│ Unload_ERT_0
├───────────────────────────────────────────────
│ θCell_0' = θCell_0
└───────────────────────────────────────────────
```

The production cell has eleven different sub-states for seventeen operations and having to pre-declare each sub-state and include a predicate in each operation is cumbersome. Moving the statement of what remains unchanged from the predicate section to the declaration section of a schema and avoiding naming each substate is preferred. A new naming convention is proposed enabling the specifier to specify that the whole state remains unchanged except for a list of state variables. The list of state variables can also contain a schema reference where the schema reference is a state schema and it is equivalent to listing all the variables from the state schema. The new naming convention has the form $\Xi A \nabla(B)$ and is equivalent to $\Xi A \setminus (\Delta B) \wedge \Delta B$. An example of the new naming convention is demonstrated by the interactions between the robot and the elevating rotary table at the production cell level.

$$Load\_Arm1\_1 \,\hat{=}\, \Xi Cell\nabla(Robot, ERT) \wedge Load\_Arm1\_0 \wedge Unload\_ERT\_0$$

where
$$\Xi Cell\nabla(Robot, ERT) \,\hat{=}\, \Xi Cell \setminus (\Delta Robot, \Delta ERT) \wedge \Delta Robot \wedge \Delta ERT$$

Extra Constraints To capture these constraints, extra predicates are added to operations at the production cell level. Adding extra predicates is quite straightforward and is discussed for later comparison with Object-Z. An example is the schema *Extend_Arm1_1* which constrains the arm to extend only if a collision can not occur between the loaded arm and a blank on either the elevating rotary table or in the press.

```
┌─ Extend_Arm1_1 ───────────────────────────────
│ Ξ Cell∇(Robot)
│ Extend_Arm1_0
├───────────────────────────────────────────────
│ (robot_orientation = load_arm1 ∧ table_position = ready_to_unload
│       ∧ arm1_load = loaded ⇒ table_loaded = unloaded)
│ (robot_orientation = unload_arm1 ∧ press_position = open_for_arm1
│       ∧ arm1_load = loaded ⇒ press_loaded = unloaded)
└───────────────────────────────────────────────
```

Error schemas Error schemas should be mentioned before discussing Object-Z. The operations defined previously for the production cell are partial operations. An operation is partial when its behaviour is defined for only some states. Error schemas are specified and then combined with partial operations to produce total operations which are always defined and these either output an *ok* report or a report specifying the error that occurred. This means the behaviour of the operation is specified in *all* circumstances. The error schema and the total operation for loading arm1 are shown as an example.

$Report ::= ok \mid wrong\_robot\_orientation \mid avoid\_collision\_between\_blanks \mid \cdots$

$Success \mathrel{\widehat{=}} [r! : Report \mid r! = ok]$

```
┌─ Extend_Arm1_Error ─────────────────────────────────────────
│ Ξ Cell
│ r! : Report
├─────────────────────────────────────────────────────────────
│ robot_orientation ∉ {load_arm1, unload_arm1} ⇒
│         r! = wrong_robot_orientation
│ robot_orientation = load_arm1 ⇒
│         (table_position = ready_to_unload ∧ table_loaded = loaded
│         ∧ arm1.load = loaded ⇒ r! = avoid_collision_between_blanks)
│ robot_orientation = unload_arm1 ⇒ ⋯
└─────────────────────────────────────────────────────────────
```

$Extend\_Arm1 \mathrel{\widehat{=}} (Extend\_Arm1\_1 \wedge Success) \vee Extend\_Arm1\_Error$

Object-Z Object-Z can be structured using either inheritance or instantiation; for this case study, instantiation is used. Instantiation models the production cell more naturally as a collection of structured components.

Unused State Variables Promoting the component operations to the production cell and combining them to form production cell operations is simple in Object-Z, as shown in the robot example, with none of the Z problems of having to equate unused state variables.

Extra Constraints Adding extra constraints to an operation is not as simple and can be attempted in several ways. The first adds an extra local operation containing the extra predicates. An example of an extra constraint and its use is shown for the extend arm operation which restricts the arm's extension to avoid collisions between blanks.

```
┌─ Extra_Extend_Arm1 ─────────────────────────────────────────
│ (robot.base.orientation = load_arm1 ∧ ert.position = ready_to_unload
│         ∧ robot.arm1.load = loaded ⇒ ert.load = unloaded)
│ (robot.base.orientation = unload_arm1 ∧ press.position = open_for_arm1
│         ∧ robot.arm1.load = loaded ⇒ press.load = unloaded)
└─────────────────────────────────────────────────────────────
```

$Extend\_Arm1 \mathrel{\widehat{=}} robot.Extend\_Arm1 \wedge Extra\_Extend\_Arm1$

In some cases the extra constraints could be operations exported from one of the component classes. This was demonstrated in the *Robot* class definition where the operation *Arm1Orientation* from the *RobotBase* constrains the *Extend_Arm*1 operation. This method has some drawbacks: the extra constraints may not be able to be placed in a single class. The constraint *Extra_Extend_Arm*1 is an example because of inter-object dependencies. A possible solution is for the component classes to provide an operation equivalent to each possible combination of state variables and every possible value of the state variable and combine these essentially boolean operations (predicates) at the top-level. For the arm class, the additional operations would include the following.

┌ *Arm_Retracted*
extent = *retracted*

┌ *Arm_Extended*
extent = *extended*

┌ *Arm_Loaded*
load = *loaded*

┌ *Arm_Unloaded*
load = *unloaded*

Combining simple operations to capture the extra constraints has disadvantages. The combination of simple operations generates a schema of approximately the same size and arguably greater complexity than the extra operation. Placing the simple operations in the classes is conceptually sound, but increases the complexity of the class without an equivalent lessening of the complexity at the higher level. Placing a complete extra operation within a component class also causes problems by placing application-specific information within a low-level, reusable class. This can be overcome using inheritance and placing the extra operations in an intermediate class. The intermediate class is a wrapper around the reusable class making it usable in the application.

The extra constraints could become class invariants. This may require rewriting the predicates to ensure the system is not over-constrained. The extra constraint *Extra_Extend_Arm*1 is suitable as part of a class invariant. Some constraints only need hold for the duration of an operation and not at any other time. For example, moving the press requires the robot's arms be retracted and yet loading the press requires a robot arm to be extended. In our specification, it is not possible to capture as a class invariant that the robot arms must be retracted if the press is moving, yet are allowed to be extended at other times. Adding extra predicates to the class invariant makes them apply to all operations in the class. However, the constraints may be required only for a subset of the class operations and this can make the specification harder to understand.

Error Schemas Error schemas in Object-Z suffer the same problems and have the same solutions as extra constraints. Adding error schemas to the production cell highlights a stylistic problem. Unlike the familiar Z style which embeds schemas in informal text, the Object-Z class syntax does not encourage specifiers to include informal explanations within the class. Instead, class definitions are embedded in informal text. This means that each class needs to be kept as small as possible. This can be achieved by using inheritance and instantiation to build complex classes in stages.

4 Conclusions

The paper has investigated five different structuring techniques for Z specifications. The five techniques were introduced and compared using part of the production cell specification. The specification of the robot is small enough to understand and yet complex enough to enable the investigation and comparison of the structuring techniques. The same structuring techniques were also applied to a larger context, the complete production cell, to investigate if the approaches that are appropriate in the small are also valid in the large.

Each structuring technique has benefits and is suited to different situations. The flat technique is suitable for small specifications where the use of components is unnecessary. Each component of the production cell except the robot was built in this fashion except in the Object-Z version. The partitioned technique provides a suitable style when each component is able to be independently specified. Structuring using parameterisation is useful if the instances have distinct parameters, but quickly becomes complex for multiple instances with the same parameter. The library structuring method embodies the advantages of the previous two techniques. The library mechanism overcomes the lack of reuse in the partitioned style and the instantiation problems of the parameterised style. With a larger specification, the focus moved from structuring to promotion difficulties and the promotion technique used by the partitioned and library techniques was found to be suitable. An extension to Z was suggested and involved the extension of the xi (Ξ) naming convention with the symbol ∇. The new naming convention enables the specifier to specify that the whole state remains unchanged except for the list of state variables following the symbol, ∇. Object-Z provides a concise structuring and promotion method for small systems, but in larger systems some problems were encountered dealing with extra constraints on top-level operations and with error schemas.

Specifiers need to be aware of a range of structuring techniques and understand when each is applicable. The paper has provided some realistic examples of techniques currently in use. As Z is applied to larger specification tasks, well-structured specifications are even more important.

Acknowledgements

We wish to thank Claus Lewerentz and Thomas Lindner for developing the production cell case study and assisting us with it, Luke Wildman and Ian Hayes for their assistance on using the library extension to Z, and Roger Duke, Jin Song Dong and Gordon Rose for discussions on Object-Z.

References

1. S. M. Brien and J. E. Nicholls. Z base standard version 1.0. Technical Monograph PRG-107, Programming Research Group, Oxford University Computing Laboratory, November 1992.

2. R. Duke, G. Rose, and G. Smith. Object-Z: a Specification Language Advocated for the Description of Standards. Technical Report 94-45, Software Verification Research Centre, Dept. of Computer Science, University of Queensland, 1994.

3. A. Hall. Using Z as a specification calculus for object-oriented systems. In D. Bjørner, C.A.R. Hoare, and H. Langmaack, editors, *VDM'90: VDM and Z!*, volume 428 of *Lect. Notes in Comput. Sci.*, pages 290–318. Springer-Verlag, 1990.

4. Ian Hayes, editor. *Specification Case Studies*. International Series in Computer Science. Prentice Hall, London, UK, 2nd edition, 1993.

5. Ian Hayes and Luke Wildman. Towards libraries for Z. In J. P. Bowen and J. E. Nicholls, editors, *Z User Workshop: Proceedings of the Seventh Annual Z User Meeting, London, December 1992*, Workshops in Computing. Springer-Verlag, 1993.

6. Brian W. Kernighan and P. J. Plauger. *The Elements of Programming Style*. McGraw-Hill, second edition, 1978.

7. Kevin Lano and Howard Houghton, editors. *Object-oriented specification case studies*. Prentice Hall, 1994.

8. C. Lewerentz and T. Lindner. *Case Study Production Cell: A Comparative Study in Formal Software Development*. Lecture Notes in Computer Science. Springer-Verlag, 1994. in press.

9. T. Lindner. *Task Description*. In [8].

10. Anthony MacDonald and David Carrington. Z specification of the Production Cell. Technical Report TR94-46, Software Verification Research Centre, Univ. of Queensland, Australia, November 1994.

11. Anthony MacDonald and David Carrington. Structuring Z Specifications: Some Choices. Technical Report TR95-19, Software Verification Research Centre, Univ. of Queensland, Australia, January 1995.

12. G. Rose. Object-Z. In S. Stepney, R. Barden, and D. Cooper, editors, *Object Orientation in Z*, Workshops in Computing, pages 59–77. Springer-Verlag, 1992.

13. S. Schuman and D. Pitt. Object-oriented subsystem specification. In L. Meertens, editor, *Program Specification and Transformation*, pages 313–341. North-Holland, 1987.

14. J. M. Spivey. An introduction to Z and formal specifications. *Software Engineering Journal*, 4(1):40–50, January 1989.

15. J. M. Spivey. *The Z Notation: A Reference Manual*. Prentice Hall, second edition, 1992.

16. S. Stepney, R. Barden, and D. Cooper, editors. *Object Orientation in Z*. Workshops in Computing. Springer-Verlag, 1992.

Experiments with the Z Interchange Format and SGML *

Daniel M. Germán and D.D. Cowan

Dept. of Computer Science
University of Waterloo
Waterloo Ont. N2L 3G1
dmg@csg.uwaterloo.ca and dcowan@csg.uwaterloo.ca

Abstract. Standards, if widely accepted, encourage the development of tools and techniques to process objects conforming to that standard. This paper describes a number of experiments using available tools to process text containing Z specifications adhering to the existing Z Interchange Format. The experiments resulted in tools that could be used in specific programming environments where Z was used to describe software systems.

1 Introduction

The benefits of a standard language for representing specifications and programs, and a standard method of electronically interchanging documents containing these representations are clear. If we all use the same language, communication of concepts, ideas and descriptions is greatly simplified. Further, if the same electronic format is used and there are enough users of a language, then organizations are willing to make the investment required to create tools for processing and transforming these documents. Standard representations produce another benefit, they also allow tools to be linked to produce more powerful toolkits. The specification language Z is no exception to these observations.

Z is a language that uses a wide range of mathematical symbols. Unfortunately not all those symbols are available in ASCII or any other commonly accessible character set. Although this lack of availability of symbols has not been an impediment to the creation of Z documents, their interchange between products and platforms has been seriously hindered. For instance, someone can create Z documents under Microsoft Windows[2] using the public domain and shareware fonts available for Z symbols. However, if these documents are processed using UNIX-based tools they will appear unintelligible, as there is no transparent way to translate the Z files from one system to the other.

LaTeX[1] has become the *de-facto* standard for the interchange of Z documents, and many Z tools can process LaTeX documents using zed.sty[2], fuzz.sty[3], oz.sty[4]. Although LaTeX is a sufficient solution, it is not an optimal one. For example, a LaTeX Z document might include formatting instructions that are not relevant to the Z specification. Inclusion of this extra material makes parsing of LaTeX documents more complex

---

* The research described in this paper was supported by IBM Canada, the Information Technology Research Centre of Ontario and Inforium Technologies Inc.
[2] Microsoft Windows is a trademark of Microsoft Corporation

than necessary. In order to import those documents a parser has to understand the complete syntax of LaTeX and zed.sty, and might still fail with some complicated structures that involve complex macros and formatting operations. In addition, LaTeX is not well known outside academia.

An argument can be made that the use of LaTeX provides an adequate solution to the problem of a standard interchange format. However, a tagging scheme such as the one supported by the *Z Interchange Format* [5] and the Standard Generalized Markup Language (SGML[3]), represents a simpler and more elegant solution.

The standardization committee decided to create an interchange format that would facilitate the sharing of Z documents between tools and platforms. This format should not have any character set dependencies, and should be easy to parse and generate, so that tools to manipulate the format can be easily constructed. These criteria lead to the choice of SGML as the underlying technology for the interchange format. SGML is an ISO standard for the definition of languages that describes the structure of text [6]. SGML relies on the inclusion of tags in the text of the document to define its structure. In addition, the use of SGML is spreading rapidly, and there are many commercial and public-domain tools available to author and process SGML documents.

2 The Z Interchange format

Appendix D of the draft of the Z Standard [5] includes the description of the Z Interchange Format. The definition states that:

> **The Z Interchange Format** defines a portable representation of Z, allowing Z documents to be transmitted between different machines.[4]

We believe that the definition should be extended to include sharing of Z documents between different products.

In order to try and make the paper self-contained, the remainder of this section contains a very brief description of SGML and the Z Interchange Format.

2.1 SGML

SGML is the abbreviation for "Standard Generalized Markup Language", and is a metalanguage to define markup languages that specify the structure of a document. The markup language does not specify how the document is processed. For example, tags in a markup language could specify the author or the title of an article, but not the type of font or its size when it is printed or displayed. When a document is marked up using an SGML tag set, the text is tagged with information about its structure.

Each element in a tagging language defined using SGML and the relationships among the tags determines a class of documents. A class is known as a DTD (Document Type Definition) in SGML jargon. Any SGML document should have a corresponding DTD.

---

[3] See 2.1 for a description of SGML

[4] Taken from [5], page 171.

Normally, SGML documents use "<tag>" to specify the start of an element, and "</tag> " to specify its end. *Elements* are the logical units of SGML documents. *Entities* are similar to macros: they define names that will eventually be replaced[5]. Normally non-common special symbols (such as many matematical symbols used in Z) are unavailable in typical character sets. SGML entities allow the inclusion of those symbols without commiting to any particular character set.

Figure 1 is an example of a small SGML document, namely a tagged memo:

```
<memo>
<from>John Doe</from>
<to>Bob Smith</to>
<body>
<paragraph>
The party is today!
</paragraph>
</body>
</memo>
```

Fig. 1. A Memo encoded using SGML

From Figure 1 we can infer that a memo is composed of *from*, *to* and *body* elements, and *body* is composed of sequence of *paragraph* elements (in this case a single one).

2.2 The SGML representation of Z

The function of the Z Interchange Format is to facilitate the interchange of Z documents and not necessarily represent their full structure. The DTD for the Z Interchange Format is similar but not identical to the grammar of Z. In particular, at the lowest level of the DTD for the interchange format there are sequences of identifiers and operators, while in the grammar of Z there are specific rules on how those identifiers and operators can be joined together. The declaration of entities is implementation dependent and therefore is not specified in the standard.

2.3 An example of a declaration of an element

Figure 2 shows the declaration of the element *axdef* (axiomatic definition). The semantics of this declaration are similar to those of a BNF production: *axdef* generates a *decpart* (declaration part) followed by an optional *axpart* (axiom part).

---

[5] This definition of entities is not complete, but is sufficient in the context of the Z Interchange Format.

```
<!ELEMENT axdef  - -  (decpart, axpart?) >
```

Fig. 2. The element declaration for *axdef*

2.4 The final result

Figure 3 shows the SGML encoding of a complete schema, and Figure 4[6] shows its normal representation. Notice the use of "&*id*;" to represent the special symbols of Z in SGML. They are called *entities* and avoid the use of special character symbols in the SGML file.

```
<schemadef>Update
<decpart>
<declaration>&Delta;CheckSys</declaration>
<declaration>a? : ADDR</declaration>
<declaration>p? : PAGE</declaration></decpart><axpart>
<predicate >working' = working &oplus; &lcub; a? &map; p?
&rcub;</predicate></axpart></schemadef>
```

Fig. 3. The SGML representation of an schema

The Z Interchange Format is simple enough that a person can enter Z in SGML directly. However, the representation is not very readable.

$$
\begin{array}{|l}
\hline
\textit{Update} \\\hline
\Delta\,\textit{CheckSys} \\
a?:ADDR \\
p?:PAGE \\\hline
\textit{working}' = \textit{working} \oplus \{a? \mapsto p?\} \\\hline
\end{array}
$$

Fig. 4. A Z schema in its normal representation

---

[6] This example was taken from [7]

3 Some Applications of the Z Interchange Format

The Z Interchange Format has several advantages including:

- conceptual simplicity,

- reliance on the ASCII character set,

- easy to parse,

- and use of a standard (SGML).

We would expect these features to be present in an interchange format, but the inclusion of SGML has produced some interesting side effects. Our research group is interested in the application of document systems incorporating SGML to various aspects of software engineering. When we became aware of the Z Interchange Format, we decided to conduct some experiments to assess its usefulness. Our experiments were oriented in two directions: building an environment for the editing of Z documents; and including Z specifications in literate programming. In each case we tried to use available tools that supported SGML.

3.1 Building an environment to author Z documents using SGML tools

Preparing documents containing specifications written in the Z Interchange Format requires an environment that can display all the mathematical symbols used in Z. LATEX is an example of such an environment. Since the release of shareware and public domain fonts for Apple MacIntosh, and Microsoft Windows it is possible to write Z using word processors such as WordPerfect [7] or Microsoft Word [8].

SGML editors are available for several different operating environments, and we believed that one of these editors in conjunction with the Z Interchange Format could be used as the foundation for an editor for Z documents. This editor called *ZEdit*, was produced for Microsoft Windows using Rita [8] a structured editor for documents, the *Live*PAGE[9] SGML Document Browser[9], and Visual Basic. The two tools were connected through the file system and used Direct Data Exchange (DDE) for communication.

Rita is a structured editor for documents [8] that is capable of reading and saving SGML documents in ASCII format. Rita is grammar directed, and understands the structure of a given DTD. The user is restricted by the DTD and can only insert valid SGML constructs into a document; hence, the user can only create documents that comply with a given document grammar. In addition, Rita uses a menu to prompt the user as to which structures can be inserted in a document at an any given point. In simple terms we can view Rita as a word processor that only allows the user to produce "correct" SGML documents, that is documents that conform to a given DTD.

---

[7] WordPerfect is a trademark of WordPerfect Corp.

[8] Word is a trademark of Microsoft Corp.

[9] *Live*PAGE is a trademark of Inforium Technologies Inc.

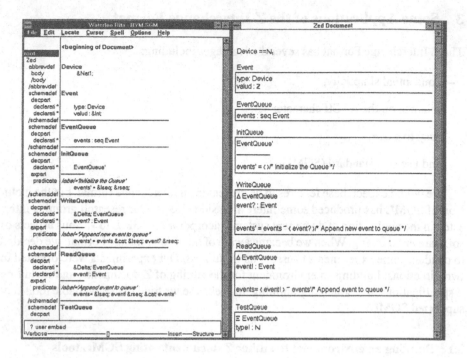

Fig. 5. A session of *ZEdit*

The *Live*PAGE SGML Document Browser is one of a series of tools developed by Inforium Technologies Inc. to manipulate documents stored in SGML format. The Browser takes as input a DTD, an SGML file and a style-sheet to specify how the SGML file is to be typeset. The output of the *Live*PAGE Browser is a typeset document on the screen. A printed typeset document can also be produced.

Figure 5 illustrates an editing session using Rita and the *Live*PAGE Browser; Rita is on the left in the Figure. Since we are using a regular keyboard we enter the entities for the special symbols rather than the symbols themselves. Rita is almost WYSIWYG, but it only has one font for display purposes, and is not capable of showing the special Z character fonts. Since the *Live*PAGE Browser has a flexible style sheet we use it to display the document. Once a modification is saved, the Browser on the right in the Figure automatically displays the changes using the complete Z symbols.

Using these combined tools the user does not need to know that *ZEdit* is tagging the Z specifications with SGML. In fact, users never need to examine the source files. We also developed a converter from the Z Interchange Format to the zed.sty LaTeX style. This converter provides quite strong evidence that the Z Interchange Format is easy to process: the converter is a simple lex program.[10]

To create *ZEdit* it was only necessary to create a style-sheet that will typeset the Z In-

---

[10] This program is available at ftp://csg.uwaterloo.ca/pub/dmg/zif2tex.tar

terchange Format in a common representation. Since the amount of work was minimal, it certainly demonstrates some of the benefits of using SGML.

3.2 Literate Programming and Z

Knuth coined the term *Literate Programming* [10] to describe the technique of combining compilable source code with descriptive prose in a single master document, called a *literate program*. He developed WEB as an implementation of his ideas. WEB was a tagging scheme; the user had to incorporate cryptic tags in the document to specify how to process text and code fragments. The tags used in WEB allowed a programmer to produce documentation using the TEX document formatting system and to process source code using a Pascal compiler. Two different programs called *Weave* and *Tangle* are used to process this tagged master document. *Weave* produces the program documentation that can be processed by TEX; *Tangle* produces Pascal source code capable of being processed by a compiler. Unfortunately there were no tools other than standard editors to assist the programmer in creating literate programs. Thus, the programmer had to remember numerous tags, and their correct placement to obtain a syntactically correct WEBdocument. Of course the resulting document was unreadable without further processing by TEX.

Ryman proposed an extension to Knuth's ideas by incorporating formal methods into the literate programming environment[11]. Mixing formal specifications, source code, and plain text in one document has numerous advantages. Code can be described both in natural language and using formal specifications, making the code easier to understand, and yet reducing potential ambiguities. Since all information about the software is contained in one document, changes in formal specifications are more likely to incorporated in the code, and vice versa. Storing the three components in the same environment, is likely to reduce inconsistency and redundancy.

We decided to create a literate programming environment using an SGML-based tagging scheme that will incorporate formal specifications, source code and documentation including multimedia entities such as pictures, diagrams, sound and video clips. The environment was built by "gluing" together available tools.

WEB uses tags to markup the components that constitute the structure of the literate program. *Tangle* and *Weave* convert the WEB file into a Pascal program and a TEX file respectively. We replaced the original markup tags with SGML-based tags, and produced a DTD for literate programming. Although the new environment is not conceptually better, it has several advantages.

Available SGML tools support literate programming in a more interactive environment. For example, we could use the *ZEdit* environment described in the previous section to isolate authors from the details of SGML tags and the production of syntactically correct documents[11]. This approach is certainly better than forcing the programmer to edit literate programs containing raw WEB code with embedded cryptic markup. Another

---

[11] Notice that this process creates documents compliant with the DTD for the Z Interchange Format; such documents not necesarily conform to the concrete Z, since the former does not mimic the latter

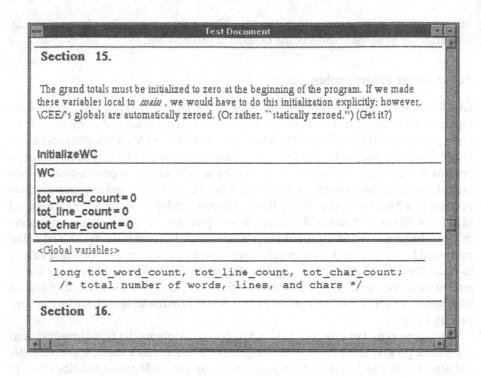

Fig. 6. A literate program that includes Z specifications

benefit of SGML markup is the ability to include many different types of digital information. Pictures, images and sounds can be easily inserted and played with SGML viewers. Even Z formal specifications could be explained using sound clips embedded in the document. For a complete description of the tools and the issues involved see [12].

Constructing a literate programming environment including formal specifications is straightforward. We just add the Z Interchange Format to our DTD for literate programming, and merge the style-sheet from *ZEdit* with one for literate programming. This modified environment then easily supports the insertion of Z specifications into our literate programs. Furthermore, the Z specifications can be easily extracted from the literate programming document and converted to another format to be further processed by a type-checker. Figure 6 shows a partial literate program generated with the prototype.

Since formal specifications play an important role in software development, we believe they must be easily integrated with other software documents. This experiment with literate programming, SGML, and the Z Interchange Format demonstrates the ease with such integration can be achieved.

4 Word Processors, SGML, Publishing, the World-Wide Web and Z

There is growing interest in and support for SGML, because of the pressure to standardize the representation of documents. A number of different groups have recognized that standards are beneficial and are promoting the use of SGML. For example, some editors recommend that their writers use SGML to create their manuscripts, and some organizations require that the documentation of their contracted systems be delivered in SGML. Because of this growing support, there are some word-processors such as FrameMaker [12] that already implement some support for SGML, while other companies have promised that new versions of their products will support SGML. Specifically, Microsoft Word and WordPerfect have recently announced they will provide SGML suport on their products.

Using these tools with the Z Interchange Format will make it easier to author Z specifications. Furthermore, the specifications will be easily portable to Z tools to have them verified.

The World-Wide Web[13] and its clients (Netscape, Mosaic, etc.) present another example of SGML-compliant tools that might be used in conjunction with the Z Interchange Format. The World-Wide Web is distributed repository of hypertext documents located in the Internet. Documents on the World-Wide Web are tagged with HTML. HTML is a markup language defined with a DTD (similarly to the way the Z Interchange Format is defined). Version 2 of HTML does not include most of the entities required in the Z Interchange Format, but it might be possible that in the future is is augmented for that purpose. Thus, documents containing Z specifications could be stored and browsed on the World-Wide Web[13].

5 Conclusions

Our limited experiments with available tools strongly supports the thesis that the Z Interchange Format fulfills its objectives: it is an effective way to interchange Z documents between different tools and platforms.

Furthermore, the results of these experiments indicate that the use of the Z Interchange Format can be used to create editors for Z using standard SGML tools, and that the Z specifications can be easily included in other SGML documents.

The Z Interchange Format will definitely make it easier to share specifications, as well as make them accessible through inclusion in books, documentation and the World Wide Web. Furthermore, by choosing SGML for the Z Interchange Format, the processing of Z documents can increasingly benefit from current and future SGML technology.

Acknowledgments

Many thanks to our anonymous referees for their constructive comments.

---

[12] FrameMaker is a trademark of Frame Technologies

[13] Visit http://www.comlab.ox.ac.uk/archive/z/html-z.html on the World-Wide Web) for more information

References

1. L. Lamport, $\LaTeX$ User's Guide & Reference Manual. Reading, Massachusetts, USA: Addison-Wesley Publishing Company, 1986.
2. J. M. Spivey, "A guide to the zed style option." Oxford University Computing Laboratory, December 1990.
3. J. M. Spivey, The fUZZ Manual. Computing Science Consultancy, 2 Willow Close, Garsington, Oxford OX9 9AN, UK, 2nd ed., 1992.
4. P. King, "Printing Z and Object-Z $\LaTeX$ documents." Department of Computer Science, University of Queensland, May 1990.
5. S. M. Brien and J. E. Nicholls, "Z base standard," Technical Monograph PRG-107, Oxford University Computing Laboratory, 11 Keble Road, Oxford, UK, Nov. 1992. Accepted for standardization under ISO/IEC JTC1/SC22.
6. C. Goldfarb, SGML Handbook. Oxford University Press, 1990. (0-19-853737-9).
7. J. M. Spivey, The Z Notation: A Reference Manual. Prentice Hall International Series in Computer Science, 2nd ed., 1992.
8. D. Cowan, E. Mackie, G. Pianosi, and G. d. V. Smit, "Rita – An Editor and User Interface for Manipulating Structured Documents," Electronic Publishing, Origination, Dissemination and Design, vol. 4, pp. 125–150, September 1991.
9. Inforium Inc., Waterloo Ont. Canada, LivePage Browser Tutorial and Reference Manual, Version. 2.0, 1994.
10. D. E. Knuth, "Literate programming," The Computer Journal, vol. 27, pp. 97–111, May 1984.
11. A. Ryman, "Formal Methods and Literate Programming," in Proceedings of the Third IBM Software Engineering ITL, June 1993.
12. D. Morales-Germán, "An SGML based Literate Programming Environment," in Proceedings of the 1994 CAS Conference, pp. 42–49, November 1994.
13. T. Berners-Lee, R. Cilliau, A. Loutonen, H. F. Nielsen, and A. Secret, "The World Wide Web," Communications of the ACM, vol. 37, pp. 76–82, August 1994.

Panel Session

The Future of Industrial Formal Methods

The Future of Formal Methods in Industry

Anthony Hall, Praxis, UK

I have worked on several projects where we have used formal methods to specify and design systems. By far the largest part of our work has been *specification* of the sequential behaviour of the system mostly using the state and operations model that has become known as the "Established Strategy" for Z. We have also tried to carry out refinements to move from a specification to a design. In critical areas we have carried out proofs of the design, sometimes by hand and sometimes using model-checking tools.

I believe the current state of the art can be summed up as follows:

1. The specification problem is solved.

   The specification of the sequential behaviour of systems using model-based specification languages like Z is well understood. In my division of Praxis it is almost routine. We have applied it to large, operational systems with great success, and I believe we can teach the necessary skills so there is no barrier, in principle, to its wide adoption.

2. Design is hard.

   Formal methods do not solve the design problem. The conventional wisdom in the formal methods community is that design is refinement: this is not true for large systems. Design is really all about system structure — its architecture — and architecture has almost nothing to do with function. One challenge for formal methods is to formalise this notion of architecture and give a formal account of how architecture organises a collection of functional components in such a way that they collectively satisfy the system specification.

   It is also at the design level that issues of concurrency become dominant, and connections must be found between state-based sequential notations and process-based concurrency notations.

3. Proof is very hard.

   I doubt whether manual proof will ever play a significant role in the development of any but the most critical software. What I think is happening is that automatic verification — decision procedures and model checkers — are becoming vastly more powerful and will be increasingly used to verify specifications. This may lead to a lowering of the level of abstraction of specifications, so that checking is tractable; more optimistically, we may find ways of "refining" abstract specifications into finite-state, checkable forms.

David Parnas, McMaster University, Canada

The position that I will take at this panel is that there are two distinct issues that determine whether or not one can apply mathematical methods routinely in software development. The first of these is the question of document information content. It is important that each formal description contain exactly the information needed, i.e. that a specification be a specification and not a model. The second requirement is that the notation recognise what Dijkstra has termed "the smallness of the human skull" — our inability to work productively with long and complex formulae. In this part of the discussion, I will refer to the work of N.G. de Bruijn and his colleagues in Eindhoven who have discussed and exploited the differences between the notation used by logicians and that used by "working mathematicians". The vernacular of mathematics is different from the formal notation of the logicians and the difference is, in my opinion, essential to the practical use of mathematics in industry.

Nico Plat, Cap Volmac, The Netherlands

In the past seven years I have been working with and on formal methods, first in an academic setting and for the last two and a half years in an industrial setting. During that period I have seen several attempts of industrial application of formal methods — with different degrees of success — however usually only with an experimental character. With a few exceptions, formal methods are not routinely used by software development companies.

Apart from the technical advances that have been made in the past decade, I consider the better understanding of the way formal methods can be applied as major progress: this will help in avoiding disillusions caused by over-optimistic expectations of what formal methods can do. Nevertheless, if we want to get into a situation in which formal methods are more routinely used, I think that formal methods will have to be sold better than is done now, and a service-oriented attitude of formal method providers (of tools, consultancy, training) will be a decisive factor. This includes:

– a focus on providing solutions to the problems that clients have. This is reflected upon formal methods in two ways:

1. Formal methods can easily be *part of* a solution, but they seldomly *form* a solution in their entirety. They merely have a number of confirmed advantages over conventional techniques, such as the provision of abstraction and proof facilities. A more direct link to the client's problems is needed, and a solution is then usually only possible in terms of an integrated service.

2. It must be made more plausible that formal methods indeed provide the claimed advantages, preferably not in terms of qualitative measures but in terms of quantitative ones, for example to make the efforts needed to construct a formal specification visible. Many formal method case studies have been carried out, but only few come up with hard figures. An easy-access public database (e.g. on a World-Wide Web server) making such case studies available to a wider audience will lower the barrier to using formal methods.

– it is unrealistic to expect that clients change their established software processes drastically, yet promoters of formal methods have a tendency to work in this direction. It is therefore essential that integration between formal methods and conventional techniques is sought and that support for these "combined" formal/informal methods/techniques is provided.

In the *Afrodite* project (ESPRIT-III project nr. 6500) an object oriented extension of VDM-SL, called VDM^{++}, has been tested on a number of industrial case studies. The project has resulted in an integrated service offering, called *Venus*, consisting of tool support for combined usage of Rumbaugh's OMT and VDM^{++}, education, training, and consultancy.

John Rushby, SRI International, USA

I will restrict my remarks to what I know of the industrial use of formal methods in the USA. I will ignore the exercises conducted for regulatory purposes; these are adequately covered in the survey by Craigen, Gerhart, and Ralston; also, in my opinion, they reflect accidents of history, rather than reliable guides to future practice.

To my knowledge, there are no cases in the USA where truly formal methods are used in a routine manner on the main development path of significant industrial products; so far, all industrial uses of formal methods have been of an exploratory or experimental nature, undertaken either as tech-transfer or evaluation exercises. However, some of these exercises have been of a substantial scale. In most cases, the formal methods were developed and applied by a specialist organization or university, whose costs were met by some external (though not necessarily government) funding source; in some cases the "host" organization was also partially funded by the external agency.

I will divide the projects considered into two groups, labeled "documentation" and "analysis".

Formal Methods for Documentation. There are two fairly large projects that I would like to draw your attention here. The first is the TCAS II requirements specification developed by Nancy Leveson and her students. TCAS is a safety-critical airborne system used to alert pilots of airplanes that are on a potential collision course and to advise them of appropriate evading action (the advisories have regulatory force). A modified statechart-like notation was developed and used for this exercise, which was actually a respecification to supplement (it actually supplanted) a pseudocode specification.

The second example is application of CoRE (Consortium Requirements Engineering method, where "Consortium" is the Software Productivity Consortium — a research center funded by a collection of aerospace companies) to the avionics of the Lockheed C-130J. CoRE is derived from the A-7 or SCR (Software Cost Reduction) requirements specification methods developed at the Naval Research Laboratory.

There are two features of these exercises that are noteworthy (apart from their substantial size, and apparent success). First, in both cases, the formal methods researchers attempted to develop a notation and methodology that would be comfortable to the industrial practitioners involved, and actually evolved their notation in response to early trials. Second, both examples are concerned with requirements specification and address issues of systems, rather than software, engineering.

Formal Methods for Analysis. One of the distinctive features of industrial use of formal methods in the USA, compared with Europe, is the stress on automated analysis (theorem proving, model-checking, and related technologies such as language inclusion and reachability analysis), and the focus on debugging and exploration, rather than "proving correctness."

Compared with software, hardware (especially processor) designs tend to be fairly similar to each other, and rather regular. This has made it worthwhile to invest considerable resources in developing highly automated procedures that can be applied to many designs.

The problems of interest in processors tend to be those associated with complex data dependencies — the handling of pipeline stalls and bypass registers, and their interaction with branch prediction, external interrupts, cache misses, and instruction prefetching and so on.

I will describe three application areas where formal methods have been used for significant analyses.

1. Hardware

The Collins AAMP5 is a microprocessor for avionics applications. It has the same (large and complex) instruction set as the earlier AAMP2 (there are thirty of these on board each 747-400), but its implementation is optimized for performance and uses 500,000 transistors to provide semiautonomous instruction prefetch, on-chip instruction and data caches, and a pipelined, microcoded data processing unit. The AAMP is a stack architecture: many of its 209 instructions take zero addresses and implicitly apply to the top few data values on the stack, and leave their results on the stack. For efficiency, the top registers of the stack are cached in internal registers and there is significant design complexity in managing the necessary stack adjustments as the cache fills or empties.

Under NASA sponsorship, SRI and Collins Commercial Avionics jointly developed formal specifications in PVS for the full microarchitecture of the data processing unit of the AAMP5 (its interactions with the caches and prefetch unit were abstracted, though later examined with state exploration and model-checking techniques), and for most of the macroarchitecture (enough to fully specify over 100 of the most important instructions). The PVS specifications were subjected to formal inspections by Collins engineers, who found little difficulty in reading them and detected several errors. Conversely, two minor errors in the microcode were identified through the process of formal specification (which forced consideration of some unusual cases). A core set of instructions (representing different classes) were selected for formal microcode verification using the PVS theorem prover. Once the infrastructure for highly automated proof of AAMP5 microcode had been developed, it was possible to verify instructions in a known instruction class at the rate of about one a day; new instruction classes required about a week. Because the AAMP series is very mature, Collins did not expect to find any bugs in the AAMP5 microcode, so two errors were (with the agreement of SRI management) covertly seeded in the microcode supplied to SRI. One of these was consciously designed to reflect the kind of bug that had proved hard to find by conventional techniques; the other was a "natural" bug that had escaped detection and made it into an early fabrication of the AAMP5, where it was found in tests of application code. Both bugs were systematically detected during the process of formal verification.

The success of the AAMP5 project has led to a further project in which formal specification and verification are being applied to a simpler processor (of the kind used for ultra-critical functions such as autoland): unlike AAMP5, formal methods are being used in the main design loop (PVS specifications are the primary design documents), the intention is to fully verify the microarchitecture and microcode, and much more of the formal specification and verification is being done by Collins engineers, rather then SRI researchers. As an indication of the economies available through reuse, the full AAMP-FV microarchitecture was fully specified by Collins engineers in 50 hours, compared with over 1000 hours required (mostly by SRI) for the AAMP5.

Other chip companies are now experimenting with the formal methods technology developed for formal verification of the AAMP processors.

2. Protocols

Most of the other applications of formal methods to industrial hardware designs have involved model-checking, or the related technology of language containment.

Cache-coherence protocols for shared-store multiprocessors have seen a lot of activity. David Dill (Stanford) and engineers at Sun have used the Murphi state exploration system in the design loop as a debugging aid. Ken McMillan (Cadence) and engineers at Silicon Graphics used the symbolic model checker CTL for the same purpose. SMV was also used to detect errors in the cache coherence protocol of the Futurebus standard. Kurshan (AT&T) has used the COSPAN (based on language inclusion and omega-automata) to find a "showstopper" bug in an AT&T switch protocol that had been under simulation for 13 months.

3. Mode Tables

Hoover and colleagues at ORA Corp, working with engineers at Honeywell, developed a specialized BDD-based analyzer to check the mode-transition tables used in avionics for limited forms of consistency and applied it to part of the MD-11 autopilot. Honeywell are reportedly delighted with the product (now called "Tablewise").

Summary. There is more industrial use of formal methods in the US than you might think. The successful applications use formal methods to solve specific problems, and the methods are often tailored to the problems concerned. These problems are mostly in requirements and systems engineering rather than the routine aspects of software engineering. The successful tools make heavy use of powerful automation: even the theorem provers use BDDs and decision procedures.

Chris Sennett, DRA, UK

The industrial use of formal methods has been held up by the fact that the formal methods community and the software development community are almost disjoint. Formal methods people are academics, motivated to produce elegant mathematics, measuring their output in published papers and with a network built on academic contacts. Software developers are engineers, motivated to produce working systems quickly and cheaply and communicating via an entirely different network of contacts. As a result, formal methods seem largely irrelevant to workers at the coal face. For this to change it is essential for there to be more interaction between the communities and I would have thought that the onus for change ought to lie with the formal methods community as it is the provider of the techniques which the developers should use. In a word, formal methods must get real.

There seem to be two basic motivations for wanting to use formal methods: one is to produce software quickly and at low cost and the other is to demonstrate integrity (safety, security, reliability, etc.). In practice, these two motivations have conflicted. Integrity has been seen to be a question of formal verification, which has always been bedevilled by a tendency to undervalue any proof which is less than complete. This has resulted in the development of large scale formal proofs which have been extremely expensive. This has given formal methods the reputation of being far too expensive for normal use. Perhaps even more worrying is the fact that these large scale formal proofs have been of little use to integrity, as important safety and security issues get lost in a forest of detail.

As a consequence I would like to see formal methods develop in the following areas.

1. To be capable of being deployed selectively, where they are most effective, and for realistic problems. In my experience, formal methods are effective when algorithms are complicated and when representations are not obvious. Use of formal methods in these situations pays real dividends in terms of the time to get a working program provided the dividend is not wasted in specifying the whole program. To be realistic and helpful, formal methods should also be capable of being applied to realistic languages, C for example, and to the sort of problem which actually concerns programmers such as arithmetic exceptions and storage management.

2. To serve the purposes of integrity, formal methods should be used understandably and deal with the concepts and domains familiar to the system engineers. This will usually involve many different models and consequently a combination of methods, some of which will be informal. A particularly important feature is that it should be possible to reason about and analyse a high level design. When am I going to be able to analyse a system built as a set of components at the conceptual level of X-Windows?

Formal methods have a potential to revolutionise software development and system design, but until they are capable of addressing large scale and messy real-world issues, they are simply irrelevant.

Object-Orientation

Specifications and Their Use in Defining Subtypes

Barbara Liskov[1] and Jeannette M. Wing[2]

[1] Massachusetts Institute of Technology
Laboratory for Computer Science
545 Technology Square
Cambridge, MA 02139, USA

[2] Computer Science Department
Carnegie Mellon University
5000 Forbes Avenue
Pittsburgh, PA 15213, USA

Abstract. Specifications are useful because they allow reasoning about objects without concern for their implementations. Type hierarchies are useful because they allow types that share common properties to be designed as a family. This paper is concerned with the interaction between specifications and type hierarchies. We present a way of specifying types, and show how some extra information, in addition to specifications of the objects' methods, is needed to support reasoning. We also provide a new way of showing that one type is a subtype of another. Our technique makes use of information in the types' specifications and works even in a very general computational environment in which possibly concurrent users share mutable objects.

1 Introduction

Object-oriented programming languages support a programming methodology based on data abstraction in which programs are composed of modules, each implementing an abstract data type. The type is abstract because it is possible to interact with its objects only by calling their operations or methods. The type's implementation (e.g., a class) defines a representation for the type's objects (e.g., a set of instance variables) and provides implementations of the methods based on this representation. The representation details are encapsulated: they are accessible only to the class, and hidden from users.

Encapsulation is useful because it allows us to reimplement a type with another class without affecting its users, assuming the new class has the same behavior as the old one. However, to make sense out of this "behavior" requirement we need a way of defining what the required behavior is. The old implementation is not a sufficient definition since it includes details that may or may not be important. For example, if a collection object has a method that returns an element that matches some predicate, the implementation of the method will make a choice if there are several elements that match. Is this choice part of the required behavior? Or, can a different implementation make a different choice?

The way to capture behavior is to define it separately from implementations in a *specification*. Specifications have long been a cornerstone of the data abstraction methodology, but have received less attention in work on object-oriented programming. Furthermore, specification techniques for data abstractions aren't quite right for object types because they assume methods belong to the type, not the object.

This paper provides a specification technique tailored to the needs of object types. Our approach allows a type to have multiple implementations and makes it convenient to define the subtyping relation. Our specifications are based on the Larch formal specification technique, which means that they have a precise mathematical meaning that serves as a firm foundation for reasoning, e.g., that a class implements a type correctly. We also discuss informal specifications based on our approach.

Specifications are used to reason about object behavior in abstract terms. Two kinds of properties are of interest: *invariant* properties, which are properties true of all states, and *history* properties, which are properties true of all sequences of states. For example, an invariant property of an integer counter might be that its value is always greater than or equal to zero; a history property might be that its value always increases. Both invariant and history properties are examples of *safety* properties ("nothing bad happens"). We might also want to prove *liveness* properties ("something good eventually happens"), e.g., the value of the integer counter eventually reaches 100, but our focus here will be just on safety properties.

Our type specifications define the behavior of methods of objects of the type (i.e., the "instance methods") but not any of the additional methods, usually called "class methods", that do not belong to particular objects. These additional methods are used to create new objects from scratch (we will refer to such methods as *creators*) and may also be used for additional purposes, e.g., to maintain statistics about various properties of the objects. Leaving these methods out is desirable because this is a place where implementations might differ (e.g., different classes that implement the same type might have different creators); in addition, different creators may be needed for subtypes.

In the absence of the creators, however, there is no way to prove invariant properties. The problem is that such properties are proved by *data type induction*, in which the creators are used for the basis step, and the methods for the induction step. If there are no creators defined in the specification, there is no basis step. To compensate for this loss, we add explicitly stated *invariants* to our specifications.

Specifications are also useful for defining type hierarchy. In strongly-typed languages such as Simula 67, Modula-3, and Trellis/Owl, subtypes are used to broaden the assignment statement. An assignment

$$x: T := E$$

is considered to be legal provided the type of expression E is a subtype of the declared type T of variable x. Once the assignment has occurred, x will be used according to its "apparent" type T, with the expectation that if the program performs correctly when the actual type of x's object is T, it will also work correctly if the actual type of the object denoted by x is a subtype of T. Intuitively, the subtype's objects must behave "the same" as the supertype's as far as anyone using the supertype's objects can tell. This paper gives a definition of the subtype relation that ensures the subtypes' objects behave properly.

Our definition ensures that all history and invariant properties that can be proved about supertype objects also hold for subtype objects. In particular, just as we add invariants to specifications to state a type's invariant properties, we add *constraints* to state its history properties explicitly. Proofs that a subtype ensures constraints of a supertype are done in terms of type specifications directly.

Thus, the paper makes two contributions:

1. It provides a way of specifying object types that allows a type to have multiple implementations and makes it convenient to define the subtyping relation. The technique requires creators to be specified separately from types, but still supports the invariant properties needed for reasoning about the type's objects.

2. It provides a new definition of the subtype relation. Our technique requires that additional information be included in the type's specification. It works even in a very general environment in which possibly concurrent users share mutable objects. Our technique is also constructive: One can prove whether a subtype relation holds by proving a small number of simple lemmas.

Others have worked on both of the problems attacked in this paper. For example, many have proposed Z as the basis of specifications of object types[7, 10, 5]; Goguen and Meseguer use FOOPS[13]; Leavens and his colleagues use Larch[16, 18, 9]. Though many of these researchers separate the specification of an object's creators from its other methods, no one has identified the problem posed by the missing creators, and thus no one has provided an explicit solution to this problem.

Others such as America[1], Cusack[6], and Dhara and Leavens[9] have proposed rules for determining whether one type is a subtype of another. Many of these approaches are not constructive, i.e., they tell you what to look for, but not how to prove that you got it. Other work[27, 17] is couched in formalisms that we believe are not very easy for programmers to deal with. In contrast, our subtype definition is constructive and takes the form of a simple checklist of rules, which programmers can use in a formal or informal way. Furthermore, only we have a technique that works in a general environment in which objects can be shared among possibly concurrent users. The rule for proving the subtype relation given in this paper is simpler than the one given in our own earlier work[22], but it requires more information in specifications, which may be a disadvantage.

The remainder of the paper is organized as follows. In Section 2 we describe our model of computation. Section 3 describes our specification technique and the extra information needed to make up for the loss of data type induction. Section 4 describes our subtype relation and the extra information needed for it to work in a very general computational environment. We close with a summary of what we have accomplished.

2 Model of Computation

We assume a set of all potentially existing objects, *Obj*, partitioned into disjoint typed sets. Each object has a unique identity. A *type* defines a set of *values* for an object and a set of *methods* that provide the only means to manipulate that object.

Objects can be created and manipulated in the course of program execution. A *state* defines a value for each existing object. It is a pair of two mappings, an *environment* and a *store*. An environment maps program variables to objects; a store maps objects to values.

> *State* = *Env* × *Store*
> *Env* = *Var* → *Obj*
> *Store* = *Obj* → *Val*

Given an object, x, and a state ρ with an environment, e, and store, s, we use the notation x_ρ to denote the value of x in state ρ; i.e., $x_\rho = \rho.s(\rho.e(x))$. When we refer to the domain of a state, $dom(\rho)$, we mean more precisely the domain of the store in that state.

We model a type as a triple, $\langle O, V, M \rangle$, where $O \subseteq Obj$ is a set of objects, $V \subseteq Val$ is a set of legal values, and M is a set of methods. Each method for an object is a *constructor*, an *observer*, or a *mutator*. Constructors of an object of type τ return new objects of type τ; observers return results of other types; mutators modify the values of objects of type τ. A type is *mutable* if any of its methods is a mutator. We allow "mixed methods" where a constructor or an observer can also be a mutator. We also allow methods to signal exceptions; we assume termination exceptions, i.e., each method call either terminates normally or in one of a number of named exception conditions. To be consistent with object-oriented language notation, we write $x.m(a)$ to denote the call of method m on object x with sequence of arguments a.

Objects come into existence and get their initial values through *creators*. Unlike other kinds of methods, creators do not belong to particular objects, but rather are independent operations. They are the "class methods"; the other methods are the "instance methods." (We are ignoring other kinds of class methods in this paper.)

A *computation*, i.e., program execution, is a sequence of alternating states and statements starting in some initial state, ρ_0:

$$\rho_0 \; S_1 \; \rho_1 \; \cdots \; \rho_{n-1} \; S_n \; \rho_n$$

Each statement, S_i, of a computation sequence is a partial function on states. A *history* is the subsequence of states of a computation. A state can change over time in only three ways[3]: the environment can change through assignment; the store can change through the invocation of a mutator; the domain can change through the invocation of a creator or constructor. We assume the execution of each statement is atomic. Objects are never destroyed:

$$\forall \, 1 \leq i \leq n \, . \, dom(\rho_{i-1}) \subseteq dom(\rho_i).$$

3 Specifications of Types and Creators

3.1 Type Specifications

While we do not wish to endorse one formal specification method over another, we do assume a type specification includes the following information:

---

[3] This model is based on CLU semantics[20].

– The type's name;

– A description of the type's value space;

– For each of the type's methods:

  • Its name;

  • Its signature (including signaled exceptions);

  • Its behavior in terms of pre-conditions and post-conditions.

In our work we use formal specifications in the two-tiered style of Larch[14]. In the first tier, Larch *traits*, written in the style of algebraic specifications, are used to define a vocabulary of *sort* and *function* symbols. These symbols define a term language, where each term denotes a value (of a particular sort). For example, the term "{}" might be used to denote the empty set value and the term "$s \cup \{i\}$" might be used to denote the set value equal to the union of the set s and the singleton set $\{i\}$. Axioms and inductive rules of inference are used to determine when two terms, and hence two values, are equal. For example, such axioms would let us prove the commutativity of $\cup$ so that we could show that the two terms "$s \cup \{i\}$" and "$\{i\} \cup s$" denote the same set value. In our examples, we stick to standard notation for functions on sets, sequences, and tuples with their usual mathematical properties; in this paper we omit the Larch traits that would be used to specify such details. Many other specification languages like Z[26], OBJ3[12], ACT-ONE[11] could just as easily be adopted to describe a type's value space.

In the second tier, Larch *interfaces* are used to describe the behaviors of an object's methods. For example, Figure 1 gives a specification for a set type whose objects have methods *insert*, *delete*, *select*, *elements*, and *equal*. The **uses** clause defines the value space for the type by identifying a sort. For example, the **uses** clauses in the figure indicates that all values of objects of type set are denotable by terms of sort S, as introduced in the MSet trait. The values of this particular sort are mathematical sets. The **methods** clause provides a specification for each method. Since a method's specifications need to refer to the method's object, we introduce a name for that object in the **for all** line. For each method, pre- and post-conditions are written in terms of **requires**, **modifies**, and **ensures** clauses. The pre-condition is the predicate of the **requires** clause; if this clause is missing, the pre-condition is trivially "true." For example, *select's* precondition checks to see if the set has any elements. The post-condition is the conjunction of the **modifies** and **ensures** clauses. *Insert's* post-condition says that the set's value may change because of the addition of its integer argument.[4] The **ensures** clause of the *elements* method makes use of the *same_els* function; we assume this is defined in the MSet trait (it returns true iff the set and the sequence contain exactly the same elements and furthermore the sequence has no duplicates).

A **modifies** $x_1, \ldots, x_n$ clause is shorthand for the following predicate

$$\forall\, x \in (dom(pre) - \{x_1, \ldots, x_n\}) \,.\, x_{pre} = x_{post}$$

---

[4] Notice that we rely on the meaning of set union to guarantee that if i is already in the set, then inserting i again will not change the set's value.

set = **type**
    uses MSet (set **for** S)
    for all *s*: set
    methods

        insert = **proc** (*i*: int)
           modifies *s*
           ensures $s_{post} = s_{pre} \cup \{i\}$

        delete = **proc** (*i*: int) **signals** (not_in)
           modifies *s*
           ensures if $i \in s_{pre}$ **then** $s_{post} = s_{pre} - \{i\}$
                 else signal not_in

        select = **proc** ( ) **returns** (int)
           requires $s_{pre} \neq \{\}$
           ensures $result \in s_{pre}$

        elements = **proc** ( ) **returns** (sequence[int])
           ensures $same\_els(s_{pre}, result)$

        equal = **proc** (*t*: set) **returns** (bool)
           ensures $result = (s = t)$

Fig. 1. A Type Specification for Sets

which says only objects listed may change in value. A **modifies** clause is a strong statement about all objects not explicitly listed, i.e., their values may not change. If there is no **modifies** clause then nothing may change.

In the **requires** and **ensures** clauses *x* stands for an object, x_{pre} stands for its value in the initial state, and x_{post} stands for the object's value in the final state.[5] Distinguishing between initial and final values is necessary only for mutable types, so we suppress the subscripts for parameters of immutable types (like integers). We need to distinguish between an object, *x*, and its value, x_{pre} or x_{post}, because we sometimes need to refer to the object itself, e.g., in the *equal* method, which determines whether two (mutable) sets are identical. *Result* is a way to name a method's result parameter.

Methods may terminate either normally or exceptionally. Specifiers can introduce values for exceptions that the method may signal through a **signals** clause in the method's header. *Delete* signals the exception not_in if the element to be deleted is not in the set.

In general, for a given method specification

    m = **proc** (args) **returns** (result) **signals** $(e_1, \ldots, e_n)$
        requires ReqPred

---

[5] Referring to an object's final value is meaningless in pre-conditions, of course.

> **modifies** $x_1, \ldots, x_n$
> **ensures** EnsPred

m.pre is ReqPred, *m.post* is ModPred $\land$ EnsPred, where ModPred is the predicate defined earlier for the meaning of a **modifies** clause, and *m.pred*, the method's associated first-order predicate, is:

> ReqPred $\Rightarrow$ (ModPred $\land$ EnsPred)

Thus, a method's behavior is defined only when the pre-condition is satisfied; it is undefined otherwise.

The form of a Larch interface specification can easily be adapted for informal specifications. The **uses** clause is simply a description of the values, possibly using well-understood mathematical concepts. The specifications of the methods have **requires**, **modifies**, and **ensures** clauses, but the predicates are given informally (in English) in terms of the concepts introduced in the **uses** clause. We have used specifications like this with considerable success in our own work, and they are described in Liskov and Guttag[21].

3.2 Specifying Creators

Objects are created and initialized through creators. We do not include creators in type specifications so that different implementations of a type can have different creators and also so that subtypes can have different creators from their supertypes.

Figure 2 shows specifications for three different creators for sets. The first creator creates a new empty set, the second creates a singleton set, and the third creator allows the implementor to represent the set as a hash table, where the given integer argument determines the hash table's size. These examples show that different implementations may require different information when new objects are created, and therefore it is a good idea to allow the creators to be defined differently for them.

> create = **proc** ( ) **returns** (set)
>   **ensures** $result_{post} = \{\} \land$ **new**($result$)

> create_single = **proc** (i: int) **returns** (set)
>   **ensures** $result_{post} = \{i\} \land$ **new**($result$)

> create_hash = **proc** (n: int) **returns** (set)
>   **requires** $n > 0$
>   **ensures** $result_{post} = \{\} \land$ **new**($result$)

Fig. 2. Three Creators for Sets

As an argument to the special predicate **new**, *result* stands for the object returned, not its value. The assertion **new**(x) stands for the predicate:

$$x \in dom(post) - dom(pre)$$

Recall that objects are never destroyed so that $dom(pre) \subseteq dom(post)$.

3.3 Type Specifications Need Explicit Invariants

By not including creators we lose a powerful reasoning tool: data type induction. Data type induction is used to prove type invariants. The base case of the rule requires that each creator of the type establish the invariant; the inductive case requires that each non-creator preserve the invariant. Without the creators, we have no base case, and therefore we cannot prove type invariants!

To compensate for the lack of data type induction, we state the invariant explicitly in the type specification by means of an **invariant** clause. In the case that the invariant is trivial (i.e., identical to "true"), as it is for sets, the **invariant** clause can be omitted. Figure 3 gives a specification of a bounded_set type for which the invariant is not trivial. The sort in this case is a pair $\langle bound, els \rangle$, where $bound$ is a natural number, and els is a mathematical set of integers. The invariant is that a bounded_set's size never exceeds its bound. In general, the predicate $\phi(x_\rho)$ appearing in an **invariant** clause for type τ stands for the predicate $\forall x : \tau, \rho : State . \phi(x_\rho)$. We assume that components of the set's value, e.g., bound, remain unchanged if not otherwise stated explicitly in the post-conditions of Figure 3.

Stating invariants explicitly in a type specification has three consequences. First, all creators for a type τ must establish τ's invariant, I_τ:

– For each creator for type τ, show $I_\tau(result_{post})$.

Second, each method of the type must *preserve the invariant*. To prove it, we assume each method is called on an object of type τ with a legal value (one that satisfies the invariant, I_τ and show that any value of an object it produces or modifies is legal:

– For each method m of τ, assume $I_\tau(x_{pre})$ and show $I_\tau(x_{post})$.

For example, we would need to show *insert*, *delete*, *select*, *elements*, and *equal* each preserves the invariant for bounded_set. Informally the invariant holds because *insert's* pre-condition checks that there is enough room in the set for another element; *delete* either decreases the size of the set or leaves it the same; *select*, *elements*, and *equal* do not change the set at all. The proof ensures that methods deal only with legal values of an object's type.

Third, the absence of data type induction limits the kinds of invariant properties we can prove about objects. All invariant properties must follow from the conjunction of the type's invariant and invariants that hold for the entire value space. For example, we could prove that insertion on bounded_sets is commutative by appealing to the commutativity of $\cup$, which, as stated earlier, holds for all mathematical set values. Since the explicit invariant limits what invariant properties can be proved, the specifier needs to be careful when defining it.

However, it is important that the invariant be just strong enough and no stronger. A general rule is that specifications should be as weak as possible: they should state only

bounded_set = **type**
    uses BSet (bounded_set **for** S)
    for all s: bounded_set
    invariant $| s_\rho.els | \leq s_\rho.bound$
    methods

        insert = **proc** (i: int)
            requires $| s_{pre}.els | < s_{pre}.bound$
            modifies s
            ensures $s_{post}.els = s_{pre}.els \cup \{i\}$

        delete = **proc** (i: int) **signals** (not_in)
            modifies s
            ensures if $i \in s_{pre}.els$ **then** $s_{post}.els = s_{pre}.els - \{i\}$
                    else signal not_in

        select = **proc** ( ) **returns** (int)
            requires $s_{pre}.els \neq \{\}$
            ensures $result \in s_{pre}.els$

        elements = **proc** ( ) **returns** (sequence[int])
            ensures $same\_els(s_{pre}.els, result)$

        equal = **proc** (t: bounded_set) **returns** (bool)
            ensures $result = (s = t)$

Fig. 3. A Type Specification for Bounded Sets

what a user needs to depend on and nothing more. Weak specifications are desirable because they give more freedom to the implementor, and also, as we shall see, because they accommodate more subtypes. This rule affects the specifications of the methods as well as the invariant. For example, the specification of *select* is non-deterministic, leaving the implementor free to choose whatever element is most convenient in that particular implementation; it also permits us to define the fifo_set subtype discussed in the next section.

In summary, the invariant plays a crucial role in our specifications. It captures what normally would be proved through data type induction, allowing us to reason about properties of legal values of a type without having specifications of creators.

4 Subtypes and the Subtype Relation

4.1 Specifying Subtypes

To state that a type is a subtype of some other type, we simply append a **subtype** clause to its specification. We allow multiple supertypes; there would be a separate **subtype** clause for each. Figure 4 gives an example. What distinguishes the fifo_set type from its set supertype is the more constrained behavior of its *select* method. In addition, it has an extra method, *remove*.

A subtype's value space may be different from its supertype's. Here, we use sequences (as defined in the Sequence trait) to represent values for fifo_set. The invariant indicates that only sequence values with no duplicate elements are legal values of fifo_set objects. (Again, we assume the *no-duplicates* function is defined appropriately in the Sequence trait.) Under the **subtype** clause we define an *abstraction function*, A, that relates these sequence values to set values in the obvious manner. The **subtype** clause also lets specifiers rename methods of the supertype, e.g., *oldest* for *select;* all other methods of the supertype are "inherited" without renaming, e.g., *insert*, *delete*, *elements*, and *equal*. Any supertype method may be further constrained. In particular, *oldest* ensures that the returned element is the one that was first inserted longest ago. The additional *remove* method removes the oldest element from the set.

4.2 Subtype Relation Defined in Terms of Specifications

Various authors have defined subtyping by relating two type specifications[1, 15, 6, 3, 22]. The commonality among these subtype definitions is that they all capture the following two properties, stated informally:

- Values of the subtype relate to values of the supertype.

- Behaviors of the subtype methods relate to behaviors of corresponding supertype methods.

As introduced in the previous section, we use *abstraction functions* to relate value spaces. These are similar to America's transfer functions[1], Leavens's simulation relations[15], and Bruce and Wegner's coercion functions[3]. Since we include explicit invariants in type specifications we must make sure that the subtype definition requires that abstraction functions *respect the invariant*: the abstraction function must map legal values of the subtype to legal values of the supertype.

For relating methods, we require that the predicate of the subtype method implies that of the supertype. (Recall that the predicate for method m is $m.pre \Rightarrow m.post$.) This requirement guarantees that a method called on a subtype object will still exhibit acceptable behavior according to the corresponding supertype method's specification. This semantic requirement on the methods' pre- and post-conditions is analogous to the syntactic requirement on the methods' signatures[2, 24, 4] (see our contra/covariance rules below). Except for minor differences, our way of relating methods is similar to America's[1], Leavens's[15], and Cusack's[6].

More formally, given two types, σ and τ, each of whose specifications respectively preserves its invariant, I_σ and I_τ, we define the subtype relation, $<$, in Figure 5.

Using this definition, it is a straightforward exercise to show that fifo_set is a subtype of set. Preserving the invariant is trivial because the invariant for sets is trivial. The reason that the *oldest* method simulates the behavior of *select* is that the specification of *select* does not constrain which element of the set is returned; the specification of *oldest*

fifo_set = **type**
    uses Sequence (fifo_set **for** Seq)
    for all s: fifo_set
    invariant *no-duplicates*(s_ρ)
    methods

        insert = **proc** (i: int)
            modifies s
            ensures if $i \in s_{pre}$ **then** $s_{pre} = s_{post}$
                    else $s_{post} = s_{pre} \parallel [i]$

        delete = **proc** (i: int) **signals** (not_in)
            modifies s
            ensures if $(i \in s_{pre})$ **then** **signal** not_in
                    else $s_{post} = rem(s_{pre}, i)$

        oldest = **proc** ( ) **returns** (int)
            requires $s_{pre} \neq [\ ]$
            ensures $result = first(s_{pre})$

        elements = **proc** ( ) **returns** (sequence[int])
            ensures *same_els*$(s_{pre}, result)$

        equal = **proc** (t: set) **returns** (bool)
            ensures $result = (s = t)$

        remove = **proc** ( )
            modifies s
            ensures if $s_{pre} \neq [\ ]$ **then** $s_{post} = rest(s_{pre})$
                    else $s_{pre} = s_{post}$

    subtype of set (oldest **for** select)

    $A : Seq \to S$
    $\forall q : Seq.$
        $A([\ ]) = \{\}$
        $A(q \parallel [i]) = A(q) \cup \{i\}$

Fig. 4. A Type Specification for Fifo_set

DEFINITION OF THE SUBTYPE RELATION, ¦: $\sigma = \langle O_\sigma, S, M \rangle$ is a *subtype* of $\tau = \langle O_\tau, T, N \rangle$ if there exists an abstraction function, $A : S \to T$, and a renaming map, $R : M \to N$, such that:

1. The abstraction function respects invariants:

   – *Invariant Rule.* $\forall s : S \, . \, I_\sigma(s) \Rightarrow I_\tau(A(s))$

   This rule implies that A must be defined for all values of S that satisfy $I_\sigma(s)$; A can be partial since it need not be defined for values of S that do not satisfy $I_\sigma(s)$.

2. Subtype methods preserve the supertype methods' behavior. Let m_τ of τ be the corresponding renamed method m_σ of σ (i.e., $R(m_\sigma) = m_\tau$). R may be partial, since it need not be defined on extra methods of σ, and must be onto and one-to-one. The following rules must hold:

   – *Signature rule.*

     • *Contravariance of arguments.* m_τ and m_σ have the same number of arguments. If the list of argument types of m_τ is α_i and that of m_σ is β_i, then $\forall i \, . \, \alpha_i < \beta_i$.

     • *Covariance of result.* Either both m_τ and m_σ have a result or neither has. If there is a result, let m_τ's result type be γ and m_σ's be δ. Then $\delta < \gamma$.

     • *Exception rule.* The exceptions signaled by m_σ are contained in the set of exceptions signaled by m_τ.

   – *Methods rule.* For all $x : \sigma$:

     • *Pre-condition rule.* $m_\tau.pre[A(x_{pre})/x_{pre}] \Rightarrow m_\sigma.pre$.

     • *Predicate rule.* $m_\sigma.pred \Rightarrow m_\tau.pred[A(x_{pre})/x_{pre}, A(x_{post})/x_{post}]$

     where $P[a/b]$ stands for predicate P with every occurrence of b replaced by a.

Fig. 5. Definition of the Subtype Relation (Without Constraints)

simply resolves this non-determinism in favor of the element that has been in the set the longest. The additional method, *remove*, causes no problem because the rule imposes no constraints on extra methods.

Our definition would not allow bounded_set to be a subtype of set. The pre-condition for set's *insert* is "true" since it is always legal to insert more elements in a set. Using our definition, we could not show that the pre-condition rule holds for *insert*. Intuitively, this restriction makes sense since a user of a set would be surprised if inserting did not cause the element to appear in the (unbounded) set. On the other hand, set could be a subtype of bounded_set since even though a user of a bounded_set might expect to reach the bound eventually, it is not surprising that any particular call has not reached the bound. However, the specification of bounded_set is a bit odd, because it is not possible for a user to observe the bound. If we were to change bounded_set's *insert* to signal an exception instead of assuming a non-trivial pre-condition, then by the exception rule we could show that set is not a subtype of bounded_set.

4.3 Adding Constraints to Type Specifications

The subtype rule given in the preceding section is perfectly adequate when procedures are considered in isolation and when there is no aliasing. For example, consider a procedure

get_max = **proc** (*s*: set) **returns** (int)
        ensures $\forall x : Int \,.\, x \in s_{pre} \Rightarrow result \geq x$

This procedure could be implemented by using the *elements* method to obtain the sequence of elements and then iterating over the sequence. In the execution of a call *get_-max(f)*, where *f* is a fifo_set, *get_max* actually calls subtype methods, but since each of these simulates the behavior of the corresponding supertype method, the procedure will execute correctly.

The definition is not sufficient, however, in the presence of aliasing, and also in a general computational environment that allows sharing of mutable objects by multiple users. Consider first the case of aliasing. The problem in this case is that within a single procedure a single object is accessible by more than one name, so that modifications using one of the names are visible when the object is accessed using the other name. For example, suppose S is a subtype of T and that variables

x: T
y: S

both denote the same object (which must, of course, belong to S or one of its subtypes). When the object is accessed through *x*, only T methods can be called. However, when it is used through *y*, S methods can be called and the effects of these methods are visible later when the object is accessed via *x*. To reason about the use of variable *x* using the specification of its type T, we need to impose additional constraints on the subtype relation.

Now consider the case of an environment of shared mutable objects, such as is provided by object-oriented databases (e.g., Thor[19] and Gemstone[23]). (In fact, it was our interest in Thor that motivated us to study the meaning of the subtype relation in the first place.) In such systems, there is a universe containing shared, mutable objects and a way of naming those objects. In general, lifetimes of objects may be longer than the programs that create and access them (in fact, objects might be persistent) and users (or programs) may access objects concurrently and/or aperiodically for varying lengths of time. Of course there is a need for some form of concurrency control in such an environment. We assume such a mechanism is in place, and consider a computation to be made up out of atomic units (i.e., transactions) that exclude one another. The transactions of different computations can be interleaved and thus one computation is able to observe the modifications made by another.

Now let's consider the impact of having subtyping in such an environment. As an example, suppose one user installs an object that maps string names to sets. Later, a second user enters a fifo_set into that object mapped under some string name; a binding like this is analogous to assigning a subtype object to a variable of the supertype. After this, both users occasionally access that fifo_set object. The second user knows it is a fifo_set and accesses it using fifo_set methods. The question is: What does the first user need to know in order for his or her programs to make sense? We think it ought to be sufficient for a user to know only about the "apparent type" of the object. Thus the first user ought to be able to reason about his or her use of the fifo_set object using the invariant and history properties of set.

History properties are of course especially of interest for mutable types. In general we can formulate history properties as predicates over state pairs: For any computation, c,

$$\forall x : \tau, \rho, \psi : State \, . \, [\, \rho < \psi \wedge x \in dom(\rho) \,] \Rightarrow \phi(x_\rho, x_\psi)$$

where $\rho < \psi$ means that state ρ precedes state ψ in c. Notice that we implicitly quantify over all computations, c, and we do not require that ψ is the immediate successor of ρ. We can prove a history property by showing that it holds after the invocation of each method. Actually we only need to do the proof for each mutator:

– *History Rule*: For each mutator m of τ, show $m.pred \Rightarrow \phi[x_{pre}/x_\rho, x_{post}/x_\psi]$.

where ϕ is a history property on objects of type τ.

Our subtype rule lets us show that invariant properties of the supertype are preserved by its subtype. But what about history properties? Let us look at an example before we answer this question. Suppose we have a fat_set type that has all the methods of set except *delete*; fat_sets only grow while sets grow and shrink. The subtype relation discussed in Section 4.2 would allow us to prove that set is a subtype of fat_set. However, someone using a fat_set can deduce that once an element is inserted in the fat_set, it remains there forever. This history property does not hold for sets, and therefore, if the object in question is actually a set, and some other user is using it as a set, the user who views it as a fat_set will be able to observe surprising behavior (namely, the set shrinks). Therefore, set should *not* be a subtype of fat_set, and to disallow such subtype relations we need to extend our subtype definition.

To obtain a subtype relation that preserves history properties, we first add some information to specifications, a *constraint* clause that describes the history properties of the type explicitly [6]. In particular, we add to the specification for fat_set:

constraint $\forall x : int \, . \, x \in s_\rho \Rightarrow x \in s_\psi$

Similarly, we could add:

constraint $s_\rho.bound = s_\psi.bound$

---

[6] The use of the term "constraint" is borrowed from the Ina Jo specification language[25], which also includes constraints in specifications.

to the specification of bounded_set to declare that a bounded_set's bound never changes. The predicate appearing in a **constraint** clause is an abbreviation for a history property. For example, fat_set's constraint expands to the following: For any computation, c,

$$\forall s : fat\_set, \rho, \psi : State \; . \; [\; \rho < \psi \wedge s \in dom(\rho) \;] \Rightarrow$$
$$[\forall x : int \; . \; x \in s_\rho \Rightarrow x \in s_\psi]$$

Just as we had to prove that methods preserved the invariant, we must show that they *satisfy the constraint* by proving it for each mutator using the history rule. The constraint replaces the history rule as far as users are concerned: users can make deductions based on the constraint but they cannot reason using the history rule directly.

Next we extend our definition of the subtype relation to require that subtype constraints ensure supertype constraints. The full definition of our subtype relation is summarized in Figure 6, where C_σ and C_τ are the constraints given in the specifications of types σ and τ, respectively. We assume each type specification preserves invariants and satisfies constraints.

Returning to our set and fat_set example, we see that the constraint rule is not satisfied since set has the trivial constraint "true," which does not imply fat_set's constraint. On the other hand, fifo_set is a subtype of set because it also has the trivial constraint.

As another example, consider a varying_set type that allows a bound to increase but does not provide any method to make such a change. It has the constraint:

constraint $s_\rho.bound \leq s_\psi.bound$

Bounded_set is a subtype of this type, and so is a dynamic_set type with all the methods of varying_set plus *change_bound*, which increases the bound:

change_bound = **proc** (*n*: int)
    requires $n \geq s_{pre}.bound$
    modifies s
    ensures $s_{post}.bound = n$

Intuitively, the non-determinism in the supertype is resolved in two different ways in the two subtypes: varying_set allows the bound to increase but does not require it; bounded_set does not take advantage of this opportunity, but dynamic_set does. This is another example of how subtypes can tighten the non-determinism present in the supertype.

4.4 Discussion

Our first subtype definition (Figure 5) is similar to others[1, 6, 15], but it does not go far enough. It fails to rule out subtype relations that would permit surprising behavior in the presence of shared mutable objects. Besides us, only Dhara and Leavens[9, 8] address this case. Rather than add explicit constraints in specifications, they place restrictions on the kinds of aliasing allowed in their programs. Their solution is limited to the special case of single-user, single-program environments.

We have also worked out another general approach that requires an *extension map* instead of explicit constraints[22]. This extension map is defined for all extra mutators introduced by the subtype and requires "explaining" the behavior of each extra mutator

DEFINITION OF THE SUBTYPE RELATION, i: $\sigma = \langle O_\sigma, S, M \rangle$ is a *subtype* of $\tau = \langle O_\tau, T, N \rangle$ if there exists an abstraction function, $A : S \to T$, and a renaming map, $R : M \to N$, such that:

1. The abstraction function respects invariants:

   – *Invariant Rule.* $\forall s : S \cdot I_\sigma(s) \Rightarrow I_\tau(A(s))$

2. Subtype methods preserve the supertype methods' behavior. If m_τ of τ is the corresponding renamed method m_σ of σ, the following rules must hold:

   – *Signature rule.*

     • *Contravariance of arguments.* m_τ and m_σ have the same number of arguments. If the list of argument types of m_τ is α_i and that of m_σ is β_i, then $\forall i \cdot \alpha_i < \beta_i$.

     • *Covariance of result.* Either both m_τ and m_σ have a result or neither has. If there is a result, let m_τ's result type be γ and m_σ's be δ. Then $\delta < \gamma$.

     • *Exception rule.* The exceptions signaled by m_σ are contained in the set of exceptions signaled by m_τ.

   – *Methods rule.* For all $x : \sigma$:

     • *Pre-condition rule.* $m_\tau.pre[A(x_{pre})/x_{pre}] \Rightarrow m_\sigma.pre$.

     • *Predicate rule.* $m_\sigma.pred \Rightarrow$
        $m_\tau.pred[A(x_{pre})/x_{pre}, A(x_{post})/x_{post}]$

3. Subtype constraints ensure supertype constraints.

   – *Constraint Rule.* For all $x : \sigma \cdot C_\sigma(x) \Rightarrow$
      $C_\tau[A(x_\rho)/x_\rho, A(x_\psi)/x_\psi]$

Fig. 6. Definition of the Subtype Relation

as a program expressed in terms of non-extra methods. For example, the *remove* method of fifo_set would be explained by the program:

   $i := s.oldest(); \; s.delete(i)$

To show that history properties are preserved for non-extra mutators, we use the methods rule.[7] However, because the properties are not stated explicitly, they cannot be proved for the extra methods. Instead extra methods must satisfy any possible property, which is surely guaranteed if the extra methods can be explained in terms of the non-extra meth-

---

[7] A more constraining methods rule that requires subtype methods to have identical pre-conditions to those of the corresponding supertype methods is needed. Thanks to Ian Maung (private communication) for pointing this out.

ods. Showing that the subtype constraint is stronger than the supertype's takes care of all the methods, not just the extra ones.

The approach using explicit constraints is appealing because it is so simple. In addition, it allows us to rule out unintended properties that happen to be true because of an error in a method specification. Having both the constraint and the method specifications is a form of useful redundancy: If the two are not consistent, this indicates an error in the specification. The error can then be removed (either by changing the constraint or some method specification). Therefore, including constraints in specifications makes for a more robust methodology.

On the minus side is the loss of the history rule. Users are not permitted to use the history rule because if they did, they might be able to prove history properties that a subtype did not ensure. Therefore the specifier must be careful to define a strong enough constraint. In our experience the desired constraint is usually obvious, but here is an example of an inadequate constraint: Suppose the definer of fat_set mistakenly gave the following constraint:

$$\textbf{constraint } |s_\rho| \leq |s_\psi|$$

Then users would be unable to deduce that once an element is added to a fat_set it will always be there (since they are not allowed to use the history rule).

5 Conclusions

This paper makes two contributions. First it provides a specification technique for object types that allows creators to be specified separately from types. Separating the creators is important because it allows different implementations of a type to have different creators, and also because it allows subtypes to have different creators from supertypes. However, leaving out creators leads to the loss of data type induction, which is needed to prove type invariants. We make up for this loss by including explicit invariants in our specification. Our specifications also contain explicit constraints; these identify a minimal set of history properties that methods of the type and all its subtypes must preserve.

Our specifications are based on Larch; we believe that this is a particularly readable and easy-to-use approach for programmers. In addition, it is easy to give informal specifications that have the same general form as our formal ones. In our experience, informal specifications that have a prescribed format and information content are very useful in program development.

Our second contribution is a new definition of the subtype relation. We argue that all properties provable about objects of a supertype should also hold for objects of its subtypes. This very strong definition ensures that if one user reasons about a shared object using properties that hold for its apparent type, that reasoning will be valid even if the object actually belongs to a subtype and is manipulated by other users using the subtype methods. Our definition of subtyping highlights the importance of history properties by making constraints explicit in specifications and by requiring that subtypes guarantee supertype constraints. Explicit constraints allow us to give a simple and straightforward definition of subtyping that works even in a very general environment in which possibly concurrent users share mutable objects.

This paper showed how these two contributions interact with each other. In the presence of subtyping, type specifications have to change to accommodate the separation of creators from an object's other methods. At the same time, type specifications must contain sufficient information for users to prove that one type is a subtype of another.

Acknowledgments

Special thanks to John Reynolds who provided perspective and insight that led us to explore alternative definitions of subtyping and their effect on our specifications. We thank Gary Leavens for helpful verbal and e-mail discussions on subtyping and pointers to related work. In addition, Gary, John Guttag, Mark Day, Sanjay Ghemawat, and Deborah Hwang, Greg Morrisett, Eliot Moss, Bill Weihl, Amy Moormann Zaremski, and the referees gave useful comments on earlier versions of this paper.

This paper originally appeared under the same title in the ACM Conference Proceedings of OOPSLA '93, *SIGPLAN Notices*, Vol. 28, No. 10, October 1993, pp. 16-28. The paper, "A Behavioral Notion of Subtyping," by Liskov and Wing, published in *ACM TOPLAS*, Vol. 16, No. 6, November 1994, is a long version of this material combined with that in [22].

This research was supported for Liskov in part by the Advanced Research Projects Agency of the Department of Defense, monitored by the Office of Naval Research under contract N00014-91-J-4136 and in part by the National Science Foundation under Grant CCR-8822158; for Wing, by the Avionics Lab, Wright Research and Development Center, Aeronautical Systems Division (AFSC), U. S. Air Force, Wright-Patterson AFB, OH 45433-6543 under Contract F33615-90-C-1465, Arpa Order No. 7597.

References

1. Pierre America. Designing an object-oriented programming language with behavioural subtyping. In J. W. de Bakker, W. P. de Roever, and G. Rozenberg, editors, *Foundations of Object-Oriented Languages, REX School/Workshop, Noordwijkerhout, The Netherlands, May/June 1990*, volume 489 of *LNCS*, pages 60–90. Springer-Verlag, NY, 1991.
2. Andrew P. Black, Norman Hutchinson, Eric Jul, Henry M. Levy, and Larry Carter. Distribution and abstract types in Emerald. *IEEE TSE*, 13(1):65–76, January 1987.
3. K.B. Bruce and P. Wegner. An algebraic model of subtype and inheritance. In Francois Bancilhon and Peter Buneman, editors, *Advances in Database Programming Language*, pages 75–96. Addison-Wesley, Reading, MA, 1990.
4. Luca Cardelli. A semantics of multiple inheritance. *Information and Computation*, 76:138–164, 1988.
5. D. Carrington, D. Duke, R. Duke, P. King, G. Rose, , and P. Smith. Object-Z: An object oriented extension to Z. In *FORTE89, International Conference on Formal Description Techniques*, December 1989.
6. Elspeth Cusack. Inheritance in object oriented Z. In *Proceedings of ECOOP '91*. Springer-Verlag, 1991.
7. Elspeth Cusack and Michael Lai. Object-oriented specification in LOTOS and Z, or my cat really is object-oriented! In J. W. de Bakker, W. P de Roever, and G. Rozenberg, editors, *Foundations of Object Oriented Languages*, pages 179–202. Springer Verlag, June 1991. LNCS 489.

8. Krishna Kishore Dhara. Subtyping among mutable types in object-oriented programming languages. Master's thesis, Iowa State University, Ames, Iowa, 1992. Master's Thesis.
9. Krishna Kishore Dhara and Gary T. Leavens. Subtyping for mutable types in object-oriented programming languages. Technical Report 92-36, Department of Computer Science, Iowa State University, Ames, Iowa, November 1992.
10. David Duke and Roger Duke. A history model for classes in object-Z. In *Proceedings of VDM '90: VDM and Z*. Springer-Verlag, 1990.
11. H. Ehrig and B. Mahr. *Fundamentals of Algebraic Specification 1*. Springer-Verlag, 1985.
12. J.A. Goguen, C. Kirchner, H. Kirchner, A. Megrelis, J. Meseguer, and T. Winkler. An introduction to OBJ3. In J.-P. Jouannaud and S. Kaplan, editors, *Proceedings, Conference on Conditional Term Rewriting*, pages 258–263. Springer-Verlag, 1988. LNCS 308.
13. Joseph A. Goguen and Jose Meseguer. Unifying functional, object-oriented and relational programming with logical semantics. In Bruce Shriver and Peter Wegner, editors, *Research Directions in Object Oriented Programming*. MIT Press, 1987.
14. John V. Guttag, James J. Horning, and Jeannette M. Wing. The Larch family of specification languages. *IEEE Software*, 2(5):24–36, September 1985.
15. Gary Leavens. Verifying object-oriented program that use subtypes. Technical Report 439, MIT Laboratory for Computer Science, February 1989. Ph.D. thesis.
16. Gary T. Leavens. Modular specification and verification of object-oriented programs. *IEEE Software*, 8(4):72–80, July 1991.
17. Gary T. Leavens and Krishna Kishore Dhara. A foundation for the model theory of abstract data types with mutation and aliasing (preliminary version). Technical Report 92-35, Department of Computer Science, Iowa State University, Ames, Iowa, November 1992.
18. Gary T. Leavens and William E. Weihl. Reasoning about object-oriented programs that use subtypes. In *ECOOP/OOPSLA '90 Proceedings*, 1990.
19. B. Liskov. Preliminary design of the Thor object-oriented database system. In *Proc. of the Software Technology Conference*. DARPA, April 1992. Also Programming Methodology Group Memo 74, MIT Laboratory for Computer Science, Cambridge, MA, March 1992.
20. B. Liskov, R. Atkinson, T. Bloom, E. Moss, J.C. Schaffert, R. Scheifler, and A. Snyder. *CLU Reference Manual*. Springer-Verlag, 1981.
21. B. Liskov and J. Guttag. *Abstraction and Specification in Program Design*. MIT Press, 1985.
22. Barbara Liskov and Jeannette M. Wing. A new definition of the subtype relation. In *Proceedings of ECOOP '93*, pages 118–141, Kaiserslautern, Germany, July 1993. Springer-Verlag. LNCS 707.
23. David Maier and Jacob Stein. Development and implementation of an object-oriented DBMS. In S.B. Zdonik and D. Maier, editors, *Readings in Object-Oriented Database Systems*, pages 167–185. Morgan Kaufmann, 1990.
24. Craig Schaffert, Topher Cooper, Bruce Bullis, Mike Kilian, and Carrie Wilpolt. An introduction to Trellis/Owl. In *Proceedings of OOPSLA '86*, pages 9–16, September 1986.
25. John Scheid and Steven Holtsberg. Ina Jo specification language reference manual. Technical Report TM-6021/001/06, Paramax Systems Corporation, A Unisys Company, June 1992.
26. J.M. Spivey. *The Z Notation: A Reference Manual*. Prentice-Hall, 1989.
27. Mark Utting. *An Object-Oriented Refinement Calculus with Modular Reasoning*. PhD thesis, University of New South Wales, Australia, 1992.

How Firing Conditions Help Inheritance

Ben Strulo

BT Laboratories
Martlesham Heath, Ipswich IP5 7RE, United Kingdom

Abstract. We tackle an under-explored question in Z semantics; what does an operation schema mean? We explain why this is important by describing our requirement to specify objects with active and passive behaviours. We show that one of the interpretations we uncover in order to represent active behaviour can also considerably simplify object-oriented inheritance.

1 Introduction

How do we understand a formal specification? The specification represents a mathematical model which is then linked informally to the real world. In Z, it is well understood how the core of the language defines a model made up of sets. But between this model and the real world there are conventions and interpretation which are seldom made explicit. In particular, this paper focuses on the "state and operations" conventions whose interpretation is revealed to be important in our application area, network management.

Within BT we use Z to specify the behaviour of objects within telecommunications networks. We are particularly interested in a network manager's view of resources within a network. Typically the manager views such resources as active objects whose behaviour towards the manager is influenced by invisible factors which she does not wish to model. When specifying such objects in Z we have found the need to capture properties not usually described in Z specifications. Formalizing these properties has forced us to look more closely at the meaning of an operation in Z. We have identified several different interpretations of operations that all allow us to represent both these active behaviours and the usual passive behaviours.

The network management objects that we are specifying are typically defined using object-oriented inheritance. This design technique allows useful general objects to be standardized and then reused in a natural way to define more specialized objects for particular contexts. We wish to reflect this construction process within our Z specifications. Representing inheritance within Z turns out to be particularly straightforward in one of the interpretations we have identified.

In this paper we discuss the reasons for classifying behaviours as active or passive. We describe how these two sorts of behaviour have led us to study the possible interpretations of certain common Z conventions. Finally we show how, by choosing a specific and non-standard interpretation for all Z operations, we can simplify and clarify the specification of the object-oriented inheritance process.

2   Active versus Passive Behaviour

We concern ourselves here with the common Z convention of describing an object as a state schema and a collection of operation schemas. It is usually understood that these operations make up the only interface to that object; if the object changes state it does so because one of these operations was performed. It may be that these operations are all controlled by the user of the object and so the object sits passively waiting to be told to act. This is the case for abstract data types and many Z specifications are of this form.

But some objects are not naturally viewed in this way. When describing a system informally it is common to say that under some circumstances some part of a system acts "by itself". Clocks and alarms are obvious examples but in fact, such an active description is appropriate whenever the real cause of the behaviour is not visible at the current level of abstraction.

As an example consider a router in a network. The manager may wish to be notified when the router becomes congested. But this is caused by a complicated sequence of events at interfaces which the manager does not control and in fact does not even wish to model. As far as the manager is concerned this event happens spontaneously and outside of her control, though she does know when it happens; she receives the notification. This is very different from the operations that *are* controlled by the manager, such as disabling the router. It would be most undesirable if this operation happened on its own!

We have been using Z and also ZEST, an object-oriented extension of Z, to specify network management objects [12, 2]. These objects are typically active, just like the router above. So when we specify them we have used a convention which identifies each operation as either controlled by the manager or spontaneous. But this identification is not enough. When we tried to use notions of refinement and inheritance it became clear that we also had to clarify our understanding of how operation schemas are interpreted as real operations.

3   The State and Operations Conventions

The conventions used to specify sequential systems in Z are so common and straightforward that they have been incorporated into the Z language and described in standard textbooks like [8, 11]. A schema is used to represent the possible states of a system. Then operation schemas are written to describe the ways this state can change. The signatures of the operation schemas include a primed and an unprimed version of the signature of the state schema. In this way the operation schemas can be understood as a relation between states, a transition relation from an unprimed "before" state to primed "after" states. There is an additional convention identifying inputs to the operation with a ? and outputs with a !.

Spivey considers the systems described by these specifications as abstract data types but the same ideas are commonly used to describe any object that is passively controlled at one interface. This way of specifying sequential systems is generally presented as the most common use of Z. We will describe the implied interpretation as the "established" interpretation.

3.1 The Established Interpretation

We have a clear understanding of how the state schema represents the states of the object and how the operation schemas define a transition relation for each operation. So what is left to understand?

Operations are represented by a relation over states. Thus an operation has some before states for which it defines after states and others about which it is silent. The set of before states for which an after state is defined (the domain of the relation) is called the pre-condition. In the established interpretation, the pre-condition is interpreted as the states in which the operation is defined. Outside these states the operation is considered to be undefined. Thus the precondition defines the "domain of definition" of the operation.

"Undefined" is usually interpreted to mean that the operation can do anything at all including fail to terminate. This general notion is often called divergence and is a useful modelling concept. It is convenient to the specifier because it allows him to concentrate on specifying the important cases, leaving the unimportant cases undefined for later consideration. Sometimes, however, it has a more important and particular use. This arises particularly when the underlying model of our system is valid only under certain assumptions. If these are violated then our system behaves in a way not correctly represented by any behaviour of our model. We can represent these exceptions as divergence.

For example imagine an object which performs integer division. We model an operation which is sent two numbers and returns a result.

$$
\begin{array}{|l}
\hline
\_\, Divide \,\underline{\hspace{6cm}} \\
\; x?, y? : \mathbb{Z} \\
\; z! : \mathbb{Z} \\
\hline
\; y? \neq 0 \\
\; z! = x?\, \mathrm{div}\, y? \\
\hline
\end{array}
$$

In the real world the object may raise some sort of exception if the divisor is zero. Then entirely different behaviour may result; perhaps the system is halted. By modelling this as undefined we are not saying that the object returns some unknown result. Instead we are saying that this exception behaviour may not be adequately modelled by *any* operation which returns a number. Because we will not be sending zero divisors we prefer a simpler behaviour model which may not adequately describe this exception.

Thus in the established interpretation, the operation schema describes what will happen when the operation is invoked. It also describes when this behaviour is guaranteed (within the pre-condition).

3.2 The Firing Condition Interpretation

For active behaviours there is actually little need to specify a domain of definition, unlike the passive behaviours suited to the established interpretation. Instead of saying that an active behaviour is undefined from some states we can identify the states in which it can happen at all. We call these states the "firing condition" of the operation. Specifiers of objects with active behaviours sometimes use this alternative interpretation of the pre-condition of such operations. In other words, they interpret the operation to be defined everywhere but to be possible only within its pre-condition.

The behaviours for which the established interpretation is used are usually passive. The pre-condition and the post-condition are like two halves of a contract — the object guarantees to perform the operation to get the appropriate post-condition as long as the environment meets its obligation to only request this when the pre-condition is satisfied.

The behaviours for which the firing condition interpretation is used are usually active and so the object does them even though the environment may not want it to. Here the pre-condition and the post-condition are both constraints on the object — the pre-condition states the only times the object is allowed to act, and the post-condition what it can do. If the object has any flexibility about what it can do then that must be explicitly present as non-determinism within the predicate.

3.3 The Hybrid Approach

It is natural to describe certain objects as having both active and passive behaviours. This interpretation for "reactive systems" is described in Josephs [4] and has also been used by others. Furthermore we have identified that the established interpretation is a natural way of describing passive behaviours and the firing condition interpretation is a natural way of describing active behaviours. Specifiers, like Josephs, have used them by mixing the two as needed, rather than sticking to one or the other. Since we have already decided to distinguish active and passive behaviour anyway this approach seems satisfactory at a specification level.

However, when we try to deal formally with such specifications we find this approach less natural. For example, when refining the specification of an object we may wish to define an operation where it was previously divergent. This means we need to know where the operation is defined. Thus in this hybrid approach, the definition of refinement needs to take account of whether each operation is active or passive.

3.4 Object-Oriented Specification

Object-oriented approaches to specification are likely to be useful and are being integrated with Z [10]. In BT we wish to adopt an object-oriented approach to specification that is compatible with the flavour of object-orientation practised in the network management standards and described in [13].

Inheritance is a way of building specialized children from the specification of more general parents. It provides a tool for reusing general specifications in a structured way. However we are interested in a specific form of inheritance which constructs children that are refinements of their parents, sometimes called strong or strict inheritance. In this way we know that the children do not contradict their parents' specification; indeed they are implementations of them. By producing children that are refinements of their parents we obtain useful substitutability properties. In particular, if we know how to manage a parent object we can manage any child object in the same way. Thus inheritance turns out to be a convenient way to model the management view of resources within a network.

To explain how to use inheritance in Z we need to show how to create a child object from the specification of its parent. We want this construction technique to be expressive and simple to understand but also to give a child object that is[1] a refinement of its parent. Obviously the way we do this depends on our definition of refinement. But it turns out to be much easier in some interpretations than in others.

Hall, in [3] (and earlier papers), has presented a detailed and clear description of how object oriented designs can be modelled within Z. But, as he points out, to allow natural specialization within Z we must sometimes expand the non-determinism of the parent operation further than we would expect under the established interpretation. This turns out to be exactly the natural way of writing the parent operation in the firing condition interpretation.

We present this in more detail using a simple example.

4 An Example

We present a simple example to describe using the various interpretations. The example is a simplified version of the manager's view of our network router. It has two modes;

$$Mode ::= enable \mid disable$$

Its state includes the mode and the current error number.

State _____
$mode : Mode$
$error : \mathbb{Z}$
_____

It has three operations. It can be enabled (as long as it is currently disabled) by the _Enable_ operation. It can be disabled (as long as it is currently enabled) by the _Disable_ operation. When enabled it may become congested at which time it emits a congestion alarm (operation _Alarm_) which includes the current error number. Of course it doesn't become congested when it is disabled.

5 How do the interpretations work?

We can now take the three interpretations described above and see how the example looks in each. In each case we also present simple conditions for refinement. Refinement is an important topic which has been extensively studied and we will not attempt to cover it completely. We will simply look briefly at the special case where the state space of the concrete specification is the same as that of the abstract specification (Spivey's "operation refinement"). Though this is not adequate in general it is enough to differentiate the interpretations we present.

---
[1] perhaps subject to a consistency proof obligation

5.1 The Established Interpretation

In this interpretation the pre-condition of every operation represents the operation's domain of definition — outside this set the operation is divergent. Thus all operations are always possible. This is very convenient for passive behaviours but not so good for active ones.

To describe active operations we first need to fill in their behaviour outside their firing condition. We can say that they cause no state-change there. Because a Z operation gives outputs of the same type everywhere we may need to make the output outside the firing condition take some special "not present" value. We will still need to mark operations as active or passive so that the reader can understand that the active behaviours only "really" happen when they have some effect, and that when they really happen they can't be prevented by the environment.

So our example operations become

```
__ Enable _____
  ΔState
 _____
  mode = disable ∧ mode' = enable ∧ error' = error
```

```
__ Disable _____
  ΔState
 _____
  mode = enable ∧ mode' = disable ∧ error' = error
```

```
__ Alarm _____
  ΔState
  error! : ℙℤ
 _____
  mode = mode'
  mode = enable ∧ error' = error + 1 ∧ error! = {error} ∨
  mode = disable ∧ error' = error ∧ error! = ∅
```

Note that the alarm operation is slightly clumsy. We have had to represent the output as a set to show that sometimes it doesn't happen. At least the description does represent explicitly the positive assertion that the alarm cannot happen when the *mode* is *disable*.

Operation refinement in this interpretation has been presented many times and is quite straightforward. Firstly the concrete specification can be less non-deterministic. If the abstract operation allows either of, say, two possible outcomes then the concrete one can choose to allow only one.

Secondly the refined operation must be defined on as many before states but can be defined on more. Thus the refined pre-condition must be as large as larger. Of course since our active operations are defined everywhere then their pre-condition can't get any larger.

More formally, if *Aop* is the abstract operation and *Cop* the concrete then we require two conditions:

$$\vdash \text{pre } Aop \land Cop \Rightarrow Aop$$

This makes the concrete operation stronger than the abstract within the abstract domain of definition.

$$\vdash \text{pre } Aop \Rightarrow \text{pre } Cop$$

This makes the concrete operation's pre-condition weaker.

5.2 The Firing Condition Interpretation

In this interpretation the pre-condition of every operation represents its firing condition — outside this set the operation is impossible. Thus to model our passive operations we need to find a new way to represent their divergence.

For some before states an operation may impose no constraint on its after states. This occurs in particular when the explicit predicate is identically true. These are the states where, for operation *Aop*, with state *State*, we have[2]

$$\forall State' \bullet Aop$$

But in these states the operation is, more or less, completely undefined. It certainly does not define any restriction on the after state though it is not *explicitly* divergent. We can interpret this lack of constraint as divergence rather than the usual interpretation as a weakly but well-defined operation. Thus we can now identify three different regions for an operation rather than the usual two, inside and outside the pre-condition. Consider an operation *Aop*, with state *State*, in this interpretation. We can define:

1. The "Unconstrained" region:

$$Unconstrained \cong \forall State' \bullet Aop$$

   These states are where the operation is very loose about its after state and allows any valid state. These are the states from which the operation is divergent.

2. The "Empty" region:

$$Empty \cong State \land \neg \exists State' \bullet Aop$$

   These states are outside the (usual) pre-condition and so the operation defines no after state. The operation is outside its firing condition and so we consider it to be defined to be impossible here.

3. The "Interesting" region:

$$Interesting \cong (\exists State' \bullet Aop) \land (\exists State' \bullet \neg Aop)$$

   These states are all the rest, where some but not all after-states are allowed. The operation is thus both well-defined and possible.

---

[2] For simplicity we do not consider outputs, which need quantification just like the after-state.

The domain of definition of the operation is then made up of regions 2 and 3 where the operation is explicitly defined to be either possible or impossible for each state.

So our example operations now become

$$\begin{array}{|l}\hline \_FiringEnable _____ \\ \Delta State \\ \hline mode = disable \Rightarrow mode' = enable \wedge error' = error \\ \hline \end{array}$$

Note how the operation is explicitly loosely defined for $mode = enable$.

$$\begin{array}{|l}\hline \_FiringDisable _____ \\ \Delta State \\ \hline mode = enable \Rightarrow mode' = disable \wedge error' = error \\ \hline \end{array}$$

$$\begin{array}{|l}\hline \_FiringAlarm _____ \\ \Delta State \\ error! : \mathbb{Z} \\ \hline mode = enable \\ error' = error + 1 \wedge error! = error \\ mode' = mode \\ \hline \end{array}$$

Now the alarm operation is simpler to read, and in particular the output is more clearly represented. The pre-condition of $mode = enable$ here represents the positive assertion that the Alarm is not possible when the router is disabled. Since the operation is defined throughout we also know that it will always be possible when the router is enabled.

In this interpretation we can no longer use the standard notion of refinement. It makes no sense to weaken the pre-condition of *FiringAlarm* and suddenly allow alarms where no alarms were allowed before. But by paying attention to the domain of definition we can translate the notion of operation refinement from the established interpretation in a fairly straightforward way. Because the divergent part of the abstract operation is already maximally weak any refinement must be a strengthening. So we require

$$\vdash Cop \Rightarrow Aop$$

This requirement also prevents an operation becoming possible where it was impossible. However we also require that the concrete operation doesn't become impossible where it was defined and possible.

$$\vdash (\exists State' \bullet Aop) \wedge (\exists State' \bullet \neg Aop)$$
$$\Rightarrow$$
$$(\exists State' \bullet Cop) \wedge (\exists State' \bullet \neg Cop)$$

These refinement rules impose the same requirements as the usual ones but they also constrain us not to change the defined part of the pre-condition.

We can clarify the difference between this and the established interpretation by considering our three regions under both interpretations. They are:

- The *Unconstrained* region where the operation is divergent in the firing condition interpretation but well-defined in the established interpretation.

- The *Empty* region where the operation defines no after state. These are the states where the operation is divergent in the established interpretation while defined but impossible in the firing condition interpretation.

- The *Interesting* region where both interpretations agree that something is being defined.

Using these regions we can tabulate how the two interpretations describe the sorts of behaviour we have identified.

	Established	Firing Condition
Divergent	Empty region $\neg$ pre Aop	Unconstrained region $\neg$ pre $(State' \wedge \neg Aop)$
Defined	Interesting and Unconstrained pre Aop	Interesting and Empty $State \wedge$ pre $(State' \wedge \neg Aop)$
Firing Condition	All regions $State$	Interesting region pre $Aop \wedge$ pre $(State' \wedge \neg Aop)$
Cannot express	After states $\equiv \emptyset$	After states $\equiv State'$
Convenient for	Passive behaviour	Active Behaviour

5.3 The Hybrid Approach

We have defined refinement for both interpretations. We can simply combine these definitions for the hybrid approach where both interpretations are used. So our object is described by the three operations *Enable*, *Disable* and *FiringAlarm*. The refinement rules we should choose are simply the established ones for passive behaviours and the firing condition ones for active behaviour. Unfortunately this means the conditions for refinement depend on the identification of active and passive behaviours.

6 Object-oriented Inheritance

For the purpose of object oriented specification we need to consider how we can construct a specialized version of an abstract operation. We require this new concrete operation to be a refined version of the abstract operation. This means that it is an implementation of the abstract operation but that it is more constrained and hence more specialized. We present first the sort of inheritance mechanism that is useful under the established interpretation.

In this interpretation, given an abstract operation Aop, we wish to be able to generally strengthen Aop but only within its pre-condition. And we separately wish to be able to weaken its pre-condition. This can be done by constructing the concrete operation, *Cop*, in terms of two new schemas, *Strengthen* and *Weaken*, as

$$Cop \mathrel{\hat=} (\text{pre } Aop \lor Weaken) \land$$
$$(\text{pre } Aop \Rightarrow Aop) \land$$
$$(Weaken \Rightarrow Strengthen)$$

We can see the two new schemas as asserting the *Strengthen* post-condition within an explicit domain of definition given by *Weaken*.

Thus we might wish to make a more concrete *Enable* operation that works in both *enable* and *disable* states. We could write

```
┌─ ConcreteEnable ──────────────────────────────
│ ΔState
├────────────────────────────────────────────────
│ pre Enable ∨ mode = enable
│ pre Enable ⇒ Enable
│ mode = enable ⇒ mode' = enable
└────────────────────────────────────────────────
```

This is a refinement as we would hope.

This construction process is satisfactory, but it is also rather complex to implement in Z. It requires the calculation of the pre-condition as well as schema negation and disjunction. But we need such a complex mechanism to provide the flexibility to expand the pre-condition of the abstract operation while still asserting its post-condition within its pre-condition. In the firing condition interpretation these difficulties go away.

In this interpretation, because our first refinement rule requires us to create a strengthening, we can use simple conjunction:

$$Cop \mathrel{\hat=} Aop \land Strengthen$$

This construction process is obviously very simple but in particular it is well-supported by Z using mechanisms like schema inclusion.

So now we can build the refined version of the *FiringEnable* operation.

```
┌─ ConcreteFiringEnable ────────────────────────
│ FiringEnable
├────────────────────────────────────────────────
│ mode = enable ⇒ mode' = enable
└────────────────────────────────────────────────
```

What could be simpler!

7 Related Work

Our work has defined a notion of refinement to ensure that any refinement is "compatible" or "substitutable". In an object-oriented context such a relationship is often called sub-typing and, just as there are many definitions of refinement, there are many of sub-typing.

When objects are entirely under the control of a single user then a reasonable consensus has been reached as to how refinement works. However we are working with a model in which some operations on an object are under the sole control of the user, while others are used in the most inconvenient way possible, perhaps by "daemons". In Liskov and Wing [5] a model in which all operations are shared with other users is considered. They use pre-conditions to mean firing conditions and either require pre-conditions to remain unchanged (similar to our requirement for active behaviours) or they require the refined behaviour to preserve a specification of the histories of the abstract behaviour. Since all of Liskov and Wing's behaviours are active in our sense they need not differentiate between undefined and impossible.

As is usual in Z, we represent our operations by a single schema which is a relation between states. Other approaches like the Refinement Calculus [6, 7] or the B language [1] add a pre-condition to this relation which allows them to express "miraculous" behaviour which occurs in states inside the pre-condition but with no post-condition. Such behaviour cannot be implemented but its presence simplifies refinement. This distinction between divergent and miraculous behaviour is similar to our distinction between divergent and impossible behaviour. However miraculous is always more refined than any possible behaviour while our impossible behaviour is only more refined than divergence. Thus our refinement rules need to add a constraint that non-divergent behaviour must not be refined into nothing. We gain the ability to specify that behaviour must be possible at the expense of losing some of the mathematical elegance of these other theories of refinement.

8 Conclusions

The state and operations conventions are additional to the core of Z and have a less clear formal semantics. Applications involving active behaviour of objects provide a test for the usefulness of the usual interpretations of these conventions. We have shown how the established interpretation means that active behaviours must be represented clumsily while the usual alternative, the hybrid approach, requires us to predicate refinement on an understanding of the type of an operation.

To resolve this situation we have settled upon the firing condition interpretation for our network management objects. This means we must use an unconventional description of passive operations. However the interpretation has proved particularly convenient because it provides such a simple characterization of object-oriented inheritance. We can model general inheritance by schema inclusion which is a particularly straightforward operation in Z.

9 Acknowledgements

These conclusions have arisen from work on the PROST-Objects project undertaken for the United Kingdom's Department of Trade and Industry. The project consortium is studying how formal methods can be used to model and then to test objects in general and, in particular, the managed objects defined by international network management standards. The partners in the project are Logica UK Limited, British Telecommunications plc, and National Computer Centre Limited. The work described here is the joint effort of all the team members on the project. Further work from the PROST-Objects project team is described elsewhere in this proceedings [9]. I must thank my colleagues at BT, particularly Elspeth Cusack, and also the anonymous referees for comments which have considerably improved the style and content of this paper.

References

1. Jean-Raymond Abrial. *Assigning Meanings to Programs - the B Book*. Cambridge University Press, 1995 (to appear).
2. Elspeth Cusack and Clazien Wezeman. Deriving Tests for Objects Specified in Z. In J.P. Bowen & J.E. Nicholls, editors, *Z User Workshop, London 1992*. Workshops in Computing, Springer-Verlag, 1992.
3. Anthony Hall. Specifying and Interpreting Class Hierarchies in Z. In J.P. Bowen & J.A. Hall, editors, *Z User Workshop, Cambridge 1994*. Workshops in Computing, Springer-Verlag, 1994.
4. Mark B. Josephs. *Specifying Reactive Systems in Z*. PRG Technical Report, PRG-TR-19-91, Oxford University Computing Laboratory, 1991.
5. Barbara H. Liskov and Jeanette M. Wing. A Behavioral Notion of Subtyping. *ACM Transactions on Programming Languages and Systems*, November 1994.
6. Carroll Morgan. *Programming from Specifications*. Prentice Hall International, second edition, 1994.
7. Carroll Morgan and Trevor Vickers. *On the Refinement Calculus*. Springer-Verlag, 1992.
8. J. M. Spivey. *The Z Notation – a Reference Manual*. Prentice Hall International, second edition, 1992.
9. Susan Stepney. *Testing as Abstraction*. In J.P. Bowen & M.G. Hinchey, editors, *ZUM'95: 9th International Conference of Z Users* (this proceedings). Lecture Notes in Computer Science, Springer-Verlag, 1995.
10. Susan Stepney, Rosalind Barden and David Cooper (Eds.) *Object Orientation in Z*. Springer Verlag, 1992.
11. J. B. Wordsworth. *Software Development using Z*. Addison Wesley, 1992.
12. Clazien Wezeman and Tony Judge. Z for Managed Objects. In J.P. Bowen & J.A. Hall, editors, *Z User Workshop, Cambridge 1994*. Workshops in Computing, Springer-Verlag, 1994.
13. ISO/IEC IS 10165 (Parts 1–4) *Management Information Services, Structure of Management Information*. 1990.

Extending $\mathcal{W}$ for Object-Z

Graeme Smith

Software Verification Research Centre
Department of Computer Science
University of Queensland
Australia

Abstract. This paper presents a logic for Object-Z which extends $\mathcal{W}$, the logic for Z adopted as the basis of the deductive system in the Z Base Standard. The logic provides a basis on which tool support for reasoning about Object-Z specifications can be developed. It also formalises the intended meaning of Object-Z constructs and hence provides an abstract, axiomatic semantics of the language.

1 Introduction

Object-Z [Ros92, RD93, DRS94] is an extension of Z in which the existing syntax and semantics of Z are retained and new constructs are introduced to facilitate specification in an object-oriented style. The enhanced structuring improves the readability of large specifications. It also enables the possibility of modular verification and refinement. This is dependent, however, on the development of a formal semantics, a proof system and, ultimately, tools for Object-Z.

Since Object-Z is a conservative extension of Z, in the sense that the existing syntax and semantics of Z are retained, a natural approach to developing a semantics has been to extend already existing semantics of Z. A denotational semantics of an early version of Object-Z [DKRS91] which extended the denotational semantics of Z of Spivey [Spi88] was described in [DD90]. More recently, a denotational semantics of Object-Z extending that of the Z Base Standard [BN92] has been proposed [GR94].

Similarly, an axiomatic semantics or logic for reasoning about Object-Z specifications could be developed as an extension to an existing logic for Z. Approaches for reasoning in Z have been proposed by Spivey [Spi88], Woodcock and Loomes [WL88], Diller [Dil90] and Wordsworth [Wor92] but none of these attempt to give a complete set of proof rules for Z. Progress towards such a set of rules, however, has been made in $\mathcal{W}$ [WB92]. These rules have been adopted by the Z Base Standard and (largely) proved sound with respect to its denotational semantics. This proof of soundness together with the fact that $\mathcal{W}$ has been successfully encoded on several theorem provers (e.g. see [Mar93]) makes it an ideal choice as the basis for a logic for Object-Z.

The objective of this paper is to extend the rules in $\mathcal{W}$ to cover the additional constructs in Object-Z. This is facilitated by extensions to $\mathcal{W}$ itself. Section 2 introduces Object-Z and Section 3 provides an overview of $\mathcal{W}$ and details the extensions adopted to accommodate Object-Z. Sections 4 to 7 present the proof rules for reasoning about Object-Z specifications.

Throughout the paper, the following naming conventions are adopted.

d	declarations
p, q	predicates
Φ, Ψ	sets of predicates
s, t, u, v	expressions
x, y, z	variables
a	object references
b	bindings
S	schemas
A, B	classes
OP	operations
Op	operation names
T	types
X	generic types
D	type or axiomatic definitions
λ	rename lists

2 Object-Z

The major extension in Object-Z is the *class schema* which captures the object-oriented notion of a class by encapsulating a single state schema with all the operation schemas which may affect its variables. The class schema is not simply a syntactic extension but also defines a type whose instances are *object references*, i.e. identifiers which reference objects of the class[1]. By declaring variables of class types within a class, objects which refer to and use other objects may be specified.

An Object-Z class schema, often referred to simply as a class, is represented syntactically as a named box possibly with generic parameters. In this box there may be local type and constant definitions, at most one state schema and associated initial state schema, and zero or more operations. In addition to these explicit definitions, a class may *inherit* the definition of one or more other classes. It may also have a visibility list restricting the way objects of the class may be used, and a history invariant for capturing liveness properties; these, however, are not discussed in this paper.

For this paper, the basic structure of a class is as follows.

```
_ ClassName[generic parameters]_____
  inherited class designators
  local type and constant definitions
  state schema
  initial state schema
  operations
_____
```

---

[1] In an early version of Object-Z [DKRS91], instances of a class schema were objects of the class. However, work on the language, aimed at implicit support for object identity and object sharing, has motivated a revised semantics based on object references.

A class designator consists of a class name, an instantiation of that class' generic parameters and possibly a rename list. The rename list allows constants, state variables and operations to be renamed. For example, given a generic class $A[X_1, X_2]$ with state variables x and y and operations Op_1 and Op_2, the following is a class designator which instantiates the generic types X_1 and X_2 with actual types t_1 and t_2 and renames y to z and Op_2 to Op_3.

$$A[t_1, t_2][z/y, Op_3/Op_2]$$

When a class is inherited, its local types and constants are implicitly available in the inheriting class. Any types or constants with the same name occurring in both the inherited and the inheriting class are semantically identified and hence must have compatible definitions.

The inherited class' state schema and initial state schema are implicitly conjoined with those declared explicitly in the inheriting class. The inherited class' operations are implicitly available except where the name of an operation in the inheriting class is the same as that of an operation of the inherited class. In this case, the inherited operation is implicitly conjoined with that in the inheriting class.

The local type and constant definitions of a class have the same syntax as global type and constant definitions in Z. Their scope, however, is limited to the class in which they are declared. A constant is associated with a fixed value which cannot be changed by any operations of the class. However, the value of constants may differ for different objects of the class.

The state schema is nameless and has its declarations partitioned by a Δ into *primary* and *secondary* variables. Secondary variables are implicitly available for change in any operation and are usually defined in terms of the primary variables. For example, the state schema of a 'square' class may have *side* as a primary variable and *area* and *perimeter* as secondary variables.

$side : \mathbb{N}$
Δ
$area : \mathbb{N}$
$perimeter : \mathbb{N}$

$area = side * side$
$perimeter = 4 * side$

Both constants and state variables can be object references. They are declared using a class designator either as a reference to a member of a particular class, e.g. $a : A[t_1, t_2]$, or to a member of a collection of classes belonging to a particular inheritance hierarchy[2], e.g. $a : \downarrow A[t_1, t_2]$ is a reference to an object of class $A[t_1, t_2]$ or any class derived from $A[t_1, t_2]$ by inheritance. The constants and variables of an object can be accessed using

---

[2] Object references to a member of a class in an arbitrary collection of classes [DD93] or to objects "contained" in a class [DD94] can also be declared. These notions will not be discussed in this paper.

the dot notation, e.g. $a.x$ denotes the constant or variable x of the object referenced by a. The dot notation is also used to specify that an object is in its initial state, e.g. $a.INIT$, and to apply an operation to an object, e.g. $a.Op_1$.

The initial state schema is named with the keyword $INIT$ and has only a predicate part. It implicitly includes both the declarations and the predicates of the state schema.

The operations are defined either as *operation schemas* or *operation expressions*. They are interpreted in the class' local environment (i.e. the global environment of the specification enriched with the local type and constant definitions) enriched with the declarations and predicates of the state schema in both primed and unprimed form. Therefore, if $a : A[t_1, t_2]$ is declared in the state schema of a class, $a.Op_1$ is a valid operation definition.

An operation schema extends the notion of a schema in Z by adding to it a *delta-list*. The delta-list is a list of the primary variables which the operation may change when it is applied to an object of the class; all other primary variables remain unchanged[3]. When two or more operations are combined to define a new operation, however, their delta-lists are united so that the new operation may change any variable which any of its constituent operations could have changed. In this way, delta-lists enable a more flexible calculus for combining operations than would be allowed by Z schemas alone.

For example, given the operations Inc_x and Inc_y which increment the variables x and y respectively,

```
┌─ Inc_x ──────────────        ┌─ Inc_y ──────────────
│ Δ(x)                          │ Δ(y)
├──────────────────────        ├──────────────────────
│ x' = x + 1                    │ y' = y + 1
└──────────────────────        └──────────────────────
```

the operation expression $Inc\_x \wedge Inc\_y$ is equivalent to the following operation schema.

```
┌─ Inc_x ∧ Inc_y ──────────────────
│ Δ(x, y)
├──────────────────────────────────
│ x' = x + 1
│ y' = y + 1
└──────────────────────────────────
```

Apart from conjunction $\wedge$, other operation operators are the *parallel operator* $\|$, the *choice operator* $[]$, the *enrichment operator* $\bullet$, hiding and renaming[4].

The parallel operator $\|$ is a binary operator introduced into Object-Z to allow specification of inter-object communication. The operator identifies and equates input variables in either operation with output variables in the other operation having the same basename, i.e. apart from the ? or !. The identified input variables are hidden in the resulting operation; the output variables are not hidden and so may be equated with other

---

[3] When no delta-list appears in an operation, this is equivalent to having an empty delta-list, i.e. no primary variables can change.

[4] Object-Z also has a sequential composition operator which will not be discussed in this paper.

input variables in subsequent parallel compositions.

The choice operator [] is a binary operator which allows the specification of nondeterministic choice between operations. The meaning of Op_1 [] Op_2 is that either operation Op_1 occurs or operation Op_2 occurs but not both. For example, Inc_x [] Inc_y is equivalent to the following operation schema.

$$
\begin{array}{|l}
\underline{Inc\_x \;[]\; Inc\_y} \\
\Delta(x, y) \\
\hline
(x' = x + 1 \land y' = y \\
\lor \\
y' = y + 1 \land x' = x)
\end{array}
$$

The enrichment operator • allows operations to be interpreted within the class' local environment enriched with the declarations and predicates of another operation or schema. It is particularly useful when specifying operations occurring on particular objects in a set of objects: references to the selected objects appear in the declarations of the first operation and can be used in the definition of the second. For example, given the declaration $s : \mathbb{P}\, A[t_1, t_2]$ in the state schema of a class, the operation which involves the parallel composition of two distinct objects referenced from s performing Op_1 is as follows.

$$
[\, a_1, a_2 : s \mid a_1 \neq a_2 \,] \bullet a_1.Op_1 \;\|\; a_2.Op_1
$$

Hiding and renaming are defined as in Z. Since the primary and secondary variables (in both primed and unprimed form) are only available to an operation definition (and not part of it), they cannot be hidden or renamed.

3 $\mathcal{W}$

$\mathcal{W}$ is a Gentzen-style sequent calculus. Axioms and theorems are expressed using *sequents* and further theorems are derived by the application of *inference rules*. The expression of these rules is facilitated by the use of *meta-functions* defined outside the logic.

3.1 Sequents

Sequents are of the form

$$
d \,|\, \Psi \vdash \Phi
$$

The *antecedent* of the sequent consists of a list of declarations d and a set of predicates Ψ and the *consequent* is a set of predicates Φ. A sequent is *valid* if at least one of the predicates in Φ is true in the global environment of the specification enriched by d and satisfying all predicates in Ψ. Both the antecedent and the consequent can be empty. An empty antecedent is equivalent to the empty declaration and the single predicate *true*.

An empty consequent is equivalent to the single predicate *false*. (In effect, predicates in Ψ are conjoined whereas predicates in Φ are disjoined.)

For example, given a Z specification which includes the schema

```
_S_____
  x, y : N
  _____
  x < y
_____
```

the following sequents are valid. (S is a declaration in these sequents, i.e. it abbreviates $x, y : N \mid x < y$ as in Z.)

$$S \vdash x \in N$$
$$S \vdash y \in N$$
$$S \vdash x < y$$

An Object-Z specification comprises not only a global environment but also a local environment for each specified class. To reason about properties within these local environments, we extend $\mathcal{W}$ so that sequents can be interpreted within a particular class context as follows.

$$A :: \quad d \mid \Psi \vdash \Phi$$

The validity of the sequent is determined in the global environment of the specification enriched with the local environment of class A. For example, if the specification contains the following class

```
_A_____
  | x : N
  |_____
  | x < 10
  ⋮
_____
```

then the following are valid sequents.

$$A :: \quad \vdash x \in N$$
$$A :: \quad \vdash x < 10$$

3.2 Inference Rules

Inference rules for manipulating sequents are of the form

$$\frac{premisses}{conclusion} \quad (\,name\,)\,[\,proviso\,]$$

The *premisses* are zero or more sequents and the *conclusion* is a single sequent. The *name* identifies the rule and the *proviso* is a predicate which must be true in the global environment of the specification for the rule to be applicable. An inference rule is *sound* if, whenever the proviso is satisfied and the premisses are valid, the conclusion is also valid.

As an example, consider the following rule for logical disjunction.

$$\frac{p_1 \vdash \quad p_2 \vdash}{p_1 \vee p_2 \vdash} \quad (\vee \vdash)$$

Such a rule (with empty consequents) is intended to be used in combination with $\mathcal{W}$'s "rule-lifting" meta-theorem. This meta-theorem allows a common declaration to be added to every antecedent in a rule or a common predicate to be added to either every antecedent or every consequent. Therefore, given that the above rule is sound, so is the following rule which is derived by adding the predicate q to each consequent.

$$\frac{p_1 \vdash q \quad p_2 \vdash q}{p_1 \vee p_2 \vdash q}$$

This rule states that if we can deduce the predicate q from the predicates p_1 and p_2 then we can deduce q from the predicate $p_1 \vee p_2$.

We extend the notation for representing provisos so that they can be interpreted in a class context. For example, the proviso $A :: p$ states that p must be true in the global environment of the specification enriched with the local environment of the class A.

So that $\mathcal{W}$'s inference rules can be used in the context of a class, the following meta-theorem is introduced.

If the rule $\qquad \dfrac{d_1 \mid \Psi_1 \vdash \Phi_1}{d_2 \mid \Psi_2 \vdash \Phi_2} \quad [\, p \,] \qquad$ is sound,

then the rule $\qquad \dfrac{A :: \quad d_1 \mid \Psi_1 \vdash \Phi_1}{A :: \quad d_2 \mid \Psi_2 \vdash \Phi_2} \quad [\, A :: p \,] \qquad$ is also sound.

This meta-theorem also allows us to define rules involving a single class without explicitly stating a class context.

For each existing meta-theorem of $\mathcal{W}$ (e.g. "rule-lifting"), we also introduce a similar meta-theorem to hold within a given class context.

3.3 Meta-Functions and Operators

$\mathcal{W}$ makes use of meta-functions to facilitate the expression of rules. The meta-function α returns the names of the variables declared in a declaration or schema expression. The meta-function ϕ returns the names of free variables in an expression, predicate, declaration or schema expression.

$\mathcal{W}$ also has a meta-substitution operator $\odot$[5] which allows a binding to be substituted into an expression, predicate or declaration. When a binding is substituted into an expression, the free variables in the expression which are in the domain of the binding are replaced by the corresponding expressions in the range of the binding. For example, if the domain of the binding b contains x and y then:

$b\odot(x \in y)$ \quad\quad evaluates to \quad\quad $b.x \in b.y$

When substituting into a declaration the substitution is made into the types only, i.e. for b as above:

$b\odot(x : y)$ \quad\quad evaluates to \quad\quad $x : b.y$

The meta-substitution operator is used to define renaming in $\mathcal{W}$. For example, to rename x to y in the predicate p the following notation is adopted.

$(\!| \; x \rightsquigarrow y \; |\!) \odot p$

To simplify notation, it is assumed that a rename list is of the form $x_1 \rightsquigarrow s_1, \ldots, x_n \rightsquigarrow s_n$. Thus, the predicate of $S[\lambda]$ where S is defined as in Section 3.1 is denoted

$(\!| \; \lambda \; |\!) \odot (x < y)$

The meta-function α is extended for Object-Z to also return the names of the types[6] and attributes (i.e. constants and primary and secondary variables) of a class. This set includes both the types and attributes explicitly declared in the class and those implicitly included via inherited classes. To define it, we therefore need first to define a meta-function ι which returns the set of inherited classes of a class. This meta-function needs to account for the fact that the actual parameters of an inherited class may depend on the generic parameters of the class into which it is inherited. For example, if the class $B[X_1]$ inherits the class $A[X_2]$ with its generic parameter X_2 instantiated with X_1 then the set of classes inherited by the class $B[t]$ will include $A[t]$.

In general, given a class $B[X_1, \ldots, X_n]$ with inherited classes $A_1[u_1, \ldots, u_j], \ldots, A_m[v_1, \ldots, v_k]$, we have

$\iota(B[t_1, \ldots, t_n]) = \{A_1[b\odot u_1, \ldots, b\odot u_j], \ldots, A_m[b\odot v_1, \ldots, b\odot v_k]\}$
\quad where $b = (\!| \; X_1 \rightsquigarrow t_1, \ldots, X_n \rightsquigarrow t_n \; |\!)$

α is then defined for classes as follows where α_d is an additional meta-function which returns the names of the explicitly declared (i.e. non-inherited) types and attributes of a class.

$\alpha(A[\lambda]) = (\!| \; \lambda \; |\!) \odot \alpha(A)$
$\alpha(B) = \alpha_d(B) \cup \alpha(A_1) \cup \ldots \cup \alpha(A_n)$ \quad where $\iota(B) = \{A_1, \ldots, A_n\}$

---

[5] We adopt the symbol '$\odot$' used in the Z Base Standard rather than the symbol '.' used in $\mathcal{W}$.

[6] In the case of a free type, the names of the constants in its axiomatic representation (see Spivey [Spi92] for details) are also returned by α.

In a similar fashion, the meta-functions α_1 and α_2 are defined to return the names of the primary and secondary variables of a class respectively.

We also extend the meta-function ϕ and the meta-substitution operator $\odot$ of $\mathcal{W}$ to treat references to the state schema, initial state schema and operations of a class as variables rather than schema references. This enables us to rename class operations and is also necessary to define inference rules where we want to use definitions from one class in the context of another. In such cases, the definitions must not include (as free variables) references to the class' schemas as these will, in general, have a different meaning in the other class context. Examples of these uses of ϕ and $\odot$ appear in the following sections.

Further meta-functions and operators required for the rules in this paper will be defined as they are needed.

4 Classes

Given the definition of a generic class $A[X_1, \ldots, X_n]$, we have a set of inference rules from which properties of the class can be derived. These rules define the class' local environment, state schema, initial state schema and operations.

4.1 Local Definitions

The meta-function η is introduced to return a predicate representing the properties of the local definitions of a class. These properties are inherited with a class, whereas other properties (e.g. those derived by induction) are not. η is defined in terms of a meta-function η_d which returns the properties of the explicitly declared local definitions. It is defined, in turn, by a meta-function η_D which returns the properties of individual type or axiomatic definitions as follows ($\equiv$ is meta-equivalence).

$$\eta_D[T_1, \ldots, T_n] \equiv true$$
$$\eta_D(T == s) \equiv T = s$$
$$\eta_D(d \mid p) \equiv [d] \wedge p$$
$$\eta_D(T ::= x_1 \mid \ldots \mid x_m \mid y_1 \langle\!\langle s_1[T] \rangle\!\rangle \mid \ldots \mid y_k \langle\!\langle s_k[T] \rangle\!\rangle) \equiv$$
$$\quad x_1 \in T, \ldots, x_m \in T$$
$$\quad y_1 \in s_1[T] \rightarrowtail T \wedge \ldots \wedge y_k \in s_k[T] \rightarrowtail T$$
$$\quad \text{disjoint}\langle \{x_1\}, \ldots, \{x_m\}, \operatorname{ran} y_1, \ldots, \operatorname{ran} y_k \rangle$$
$$\quad \forall z : \mathbb{P}\, T \bullet \{x_1, \ldots, x_m\} \cup y_1 (\!\mid s_1[z] \mid\!) \cup \ldots \cup y_k (\!\mid s_k[z] \mid\!) \subseteq z \Rightarrow T \subseteq z$$

In defining the properties of an axiomatic definition we place schema brackets around its declaration to form a schema which can then be used as a predicate. The properties of a free type are derived from its axiomatic definition as given in Spivey [Spi92].

Given that $A[X_1, \ldots, X_n]$ has local definitions $D_1, \ldots, D_m$,

$$\eta_d(A[t_1, \ldots, t_n]) \equiv b \odot (\eta_D(D_1) \wedge \ldots \wedge \eta_D(D_m))$$
$$\text{where } b = (\!\mid X_1 \rightsquigarrow t_1, \ldots, X_n \rightsquigarrow t_n \mid\!)$$

To account for inherited classes (possibly with renaming) we have

$$\eta(A[\lambda]) \equiv \langle\!|\,\lambda\,|\!\rangle \odot \eta(A)$$
$$\eta(B) \equiv \eta_d(B) \wedge \eta(A_1) \wedge \ldots \wedge \eta(A_n) \quad \text{where } \iota(B) = \{A_1, \ldots, A_n\}$$

When we are reasoning within the local environment of class $A[X_1, \ldots, X_n]$ instantiated with actual parameters $t_1, \ldots, t_n$, we can use the axioms describing the local types and axiomatic definitions. This is captured by the following rule.

$$\frac{A[t_1, \ldots, t_n] :: \quad \eta(A[t_1, \ldots, t_n]) \vdash}{A[t_1, \ldots, t_n] :: \quad \vdash} \quad (\textit{LocalDefs})$$

For example, assuming class A of Section 3.1 has no other local definitions,

$$\eta(A) \equiv [x : N] \wedge x < 10$$
$$\equiv x \in N \wedge x < 10$$

To prove the sequent $A :: \quad \vdash x \in N$ is valid, we use the *LocalDefs* rule in combination with the "rule-lifting" meta-theorem to obtain the following rule.

$$\frac{A :: \quad x \in N \wedge x < 10 \vdash x \in N}{A :: \quad \vdash x \in N}$$

Since the premiss is obviously valid, the conclusion must also be valid.

4.2 State Schema

The keyword $STATE$ is introduced to identify the schema corresponding to the (mutable) state of the class. This schema consists of the inherited state schema extended with the primary and secondary variable declarations and predicate of the explicit state schema.

To refer to the inherited state schema the name $\uparrow STATE$ is used. Similarly, the name $\uparrow INIT$ refers to the inherited part of the initial state schema and $\uparrow Op$ to the inherited part of an operation Op.

If $A[X_1, \ldots, X_n]$ has the state schema

d_1
Δ
d_2
p

we have

$$\frac{A[t_1, \ldots, t_n] :: \quad STATE = }{[\uparrow STATE;\ b \odot d_1;\ b \odot d_2 \mid b \odot p] \vdash}{A[t_1, \ldots, t_n] :: \quad \vdash} \quad (\textit{StateDef})\,[\,q\,]$$

where $q \equiv b = (\!|\, X_1 \rightsquigarrow t_1, \ldots, X_n \rightsquigarrow t_n \,|\!)$

The antecedent of the premiss is a predicate equating the two schema expressions. This rule is intended to be used in combination with "rule-lifting" in a similar fashion to *LocalDefs*.

4.3 Initial State Schema

The schema corresponding to the initial state of a class consists of the inherited initial state schema conjoined with the state schema of the class extended with the predicate of the explicit initial state schema.

If $A[X_1, \ldots, X_n]$ has the initial state schema

```
__INIT_____
  p
|_____
```

we have

$$\frac{A[t_1, \ldots, t_n] :: \quad INIT = \uparrow INIT \wedge [\, STATE \mid b \odot p\,] \vdash}{A[t_1, \ldots, t_n] :: \quad \vdash} \quad (\, InitDef\,)\,[\,q\,]$$

where $q \equiv b = (\!|\, X_1 \rightsquigarrow t_1, \ldots, X_n \rightsquigarrow t_n \,|\!)$

4.4 Operations

For notational convenience, whenever an operation appears as an expression, predicate, declaration or schema expression in the logic, we interpret it as the schema part of the operation only (i.e. without the delta-list).

The schema part of an operation consists of the inherited part of the operation conjoined with the schema part of the explicit operation definition interpreted in the local environment of the class enriched with $\Delta STATE$ (i.e. $STATE \wedge STATE'$).

For each operation definition

$$Op \cong OP$$

in $A[X_1, \ldots, X_n]$ we have

$$\frac{A[t_1, \ldots, t_n] :: \quad Op = \Delta STATE \bullet (\uparrow Op \wedge b \odot OP) \vdash}{A[t_1, \ldots, t_n] :: \quad \vdash} \quad (\, OpDef\,)\,[\,p\,]$$

where $p \equiv b = (\!|\, X_1 \rightsquigarrow t_1, \ldots, X_n \rightsquigarrow t_n \,|\!)$

The meta-function Ω is introduced to return the set of names of operations of a class. It is defined in terms of a function Ω_d which returns the explicitly defined operations of the class.

$$\Omega(A[\lambda]) = \langle\!\!\langle\, \lambda\, \rangle\!\!\rangle \odot \Omega(A)$$
$$\Omega(B) = \Omega_d(B) \cup \Omega(A_1) \cup \ldots \cup \Omega(A_n) \quad \text{where } \iota(B) = \{A_1, \ldots, A_n\}$$

If a particular operation name Op occurs in one or more inherited classes but not in the inheriting class then there is an implicit operation named Op in the inheriting class. The schema part of Op consists of the schema part of the inherited operation interpreted in the local environment of the class enriched with $\Delta STATE$.

$$\frac{A[t_1, \ldots, t_n] :: \quad Op = \Delta STATE \bullet \uparrow Op \vdash}{A[t_1, \ldots, t_n] :: \quad \vdash} \quad (\,ImplicitOpDef\,)\,[\,p\,]$$

where $p \equiv \iota(A[t_1, \ldots, t_n]) = \{A_1, \ldots, A_m\}$
$\qquad Op \in \Omega(A_1) \cup \ldots \cup \Omega(A_m)$
$\qquad Op \notin \Omega_d(A[t_1, \ldots, t_n])$

5 Inheritance

The inherited part of a schema is defined as the conjunction of the same-named schemas of each inherited class. Before the conjunction takes place, each schema must be fully expanded so that it does not include references to other schemas of its class.

5.1 Inherited State Schema

The inherited state schema of a class is the conjunction of each of the state schemas of the inherited classes.

$$\frac{A_1 :: \ \vdash STATE = S_1 \quad \ldots \quad A_n :: \ \vdash STATE = S_n}{B :: \ \vdash \uparrow STATE = S_1 \wedge \ldots \wedge S_n} \quad (\,\uparrow STATE\,)\,[\,p\,]$$

where $p \equiv \iota(B) = \{A_1, \ldots, A_n\}$
$\qquad \phi(S_1) \subseteq \alpha(A_1)$
$\qquad \ldots$
$\qquad \phi(S_n) \subseteq \alpha(A_n)$

The provisos of the form $\phi(S_i) \subseteq \alpha(A_i)$ ensure that the schemas S_i are expressed in fully expanded form (i.e. in terms of $\alpha(A_i)$ only and not in terms of references to the state, initial state or operations of A_i).

5.2 Inherited Initial State Schema

The inherited initial state schema of a class is the conjunction of each of the initial state schemas of the inherited classes.

$$\frac{A_1 :: \ \vdash INIT = S_1 \quad \dots \quad A_n :: \ \vdash INIT = S_n}{B :: \ \vdash \uparrow INIT = S_1 \wedge \dots \wedge S_n} \quad (\uparrow INIT)\,[\,p\,]$$

where $p \equiv \iota(B) = \{A_1, \dots, A_n\}$
$\phi(S_1) \subseteq \alpha(A_1)$
$\dots$
$\phi(S_n) \subseteq \alpha(A_n)$

5.3 Inherited Operations

The schema of the inherited part of an operation is the conjunction of each of the schemas of the operations of the same name occurring in the inherited classes. Not all of the inherited classes need have such an operation.

$$\frac{A_1 :: \ \vdash Op = S_1 \quad \dots \quad A_k :: \ \vdash Op = S_k}{B :: \ \vdash \uparrow Op = S_1 \wedge \dots \wedge S_k} \quad (\uparrow Op)\,[\,p\,]$$

where $p \equiv \iota(B) = \{A_1, \dots, A_n\}$
$Op \in \Omega(A_1) \cap \dots \cap \Omega(A_k)$
$Op \notin \Omega(A_{k+1}) \cup \dots \cup \Omega(A_n)$
$\phi(S_1) \subseteq \alpha(A_1)$
$\dots$
$\phi(S_k) \subseteq \alpha(A_k)$

5.4 Renaming

An inherited class may have attributes and operations renamed. The properties of the resulting class are obtained by applying the renaming to the properties of the original class.

$$\frac{A :: \ \vdash p}{A[\lambda] :: \ \vdash (\!|\,\lambda\,|\!)\odot p} \quad (\,ClassRenaming\,)$$

6 Operation Expressions

The properties of the schema parts of operation expressions can be expressed in terms of Z schema expressions. These expressions can then be reasoned about using the rules of $\mathcal{W}$.

To avoid ambiguity, the schema part of an operation is denoted explicitly, where necessary, by the notation $[\,OP\,]$ (i.e. the schema which includes (the schema of) OP as a declaration). For example, $[\,OP_1\,] \wedge [\,OP_2\,]$ is a schema expression whereas $OP_1 \wedge OP_2$ is an operation expression.

6.1 Conjunction

The schema of the operation $OP_1 \wedge OP_2$ is the conjunction of the schemas of OP_1 and OP_2.

$$\frac{OP_1 \wedge OP_2 = [OP_1] \wedge [OP_2] \vdash}{\vdash} \quad (\text{ OpConjunction })$$

6.2 Parallel

To define the parallel operator, the meta-functions $\beta_?$ and $\beta_!$ are introduced to return the basenames of the inputs and outputs of an operation respectively.

The schema of the operation $OP_1 \parallel OP_2$ can be formed by

- noting any input variables of the schema of each operation which have the same basename as an output variable of the schema of the other operation,

- renaming each such input variable to the name of the output variable,

- and conjoining the resulting schemas.

$$\frac{\begin{array}{c} OP_1 \parallel OP_2 = [OP_1][y_1!/y_1?, \ldots, y_m!/y_m?] \\ \wedge \\ [OP_2][x_1!/x_1?, \ldots, x_n!/x_n?] \vdash \end{array}}{\vdash} \quad (\text{ Parallel }) \, [\, p\,]$$

where $p \equiv \beta_!(OP_1) \cap \beta_?(OP_2) = \{x_1, \ldots, x_n\}$
$\beta_?(OP_1) \cap \beta_!(OP_2) = \{y_1, \ldots, y_m\}$

6.3 Choice

To define the choice operator, the meta-function δ_{OP} is introduced to return the names of the variables in the delta-list of an operation.

$\delta_{OP}[\Delta(x_1, \ldots, x_n) \; d \,|\, p] = \{x_1, \ldots, x_n\}$
$\delta_{OP}[d \,|\, p] = \varnothing$
$\delta_{OP}(a.Op) = \varnothing$
$\delta_{OP}(OP_1 \wedge OP_2) = \delta_{OP}(OP_1) \cup \delta_{OP}(OP_2)$
$\delta_{OP}(OP_1 \parallel OP_2) = \delta_{OP}(OP_1) \cup \delta_{OP}(OP_2)$
$\delta_{OP}(OP_1 \; [] \; OP_2) = \delta_{OP}(OP_1) \cup \delta_{OP}(OP_2)$
$\delta_{OP}(OP_1 \bullet OP_2) = \delta_{OP}(OP_1) \cup \delta_{OP}(OP_2)$
$\delta_{OP}(OP[\lambda]) = \delta_{OP}(OP)$
$\delta_{OP}(OP \setminus (x_1, \ldots, x_n)) = \delta_{OP}(OP)$

Note that the delta-list of $a.Op$ is empty because this operation doesn't change any of the variables of the class in which it occurs. Although the object referenced by a is changed, the reference a itself is not. Also, the delta-list of an operation is not changed by renaming or hiding since these do not affect the class' variables.

The schema of the operation $OP_1 \; [] \; OP_2$ can be formed by

- adding to the schema of OP_1 the predicate which states that all variables in OP_2's delta-list which are not in OP_1's delta-list are unchanged,

- adding to the schema of OP_2 the predicate which states that all variables in OP_1's delta-list which are not in OP_2's delta-list are unchanged,

- and disjoining the resulting schemas.

$$\frac{OP_1 \,[]\, OP_2 = \begin{array}{l} [\, OP_1 \,|\, x_1' = x_1 \wedge \ldots \wedge x_n' = x_n \,] \\ \quad\quad \vee \\ [\, OP_2 \,|\, y_1' = y_1 \wedge \ldots \wedge y_m' = y_m \,] \vdash \end{array}}{\vdash} \quad (\textit{Choice}) \,[\,p\,]$$

where $p \equiv \delta_{OP}(OP_2) \backslash \delta_{OP}(OP_1) = \{x_1, \ldots, x_n\}$
$\delta_{OP}(OP_1) \backslash \delta_{OP}(OP_2) = \{y_1, \ldots, y_m\}$

6.4 Enrichment

If in the environment enriched with the schema of OP_1 we can show that the schema of OP_2 is equal to $[\,d \mid p\,]$, where the free variables of d do not include any variables declared in OP_1, then the schema of $OP_1 \bullet OP_2$ is equal to $[\,OP_1;\, d \mid p\,]$.

$$\frac{OP_1 \vdash OP_2 = [\,d \mid p\,]}{\vdash OP_1 \bullet OP_2 = [\,OP_1;\, d \mid p\,]} \quad (\textit{Enrichment}) \, [\,\phi(d) \cap \alpha(OP_1) = \varnothing\,]$$

For example, consider evaluating the operation expression

$$[\,s : \mathbb{P}\mathbb{N}\,] \bullet [\,x : s\,]$$

Since the following sequent is valid

$$[\,s : \mathbb{P}\mathbb{N}\,] \vdash [\,x : s\,] = [\,x : \mathbb{N} \mid x \in s\,]$$

and $\phi(x : \mathbb{N}) = \varnothing$, using *Enrichment* we can deduce

$$[\,s : \mathbb{P}\mathbb{N}\,] \bullet [\,x : s\,] = [\,s : \mathbb{P}\mathbb{N};\, x : \mathbb{N} \mid x \in s\,]$$

6.5 Renaming

The schema of the operation $OP[\lambda]$ is the schema of OP renamed by λ.

$$\frac{OP[\lambda] = [\,OP\,][\lambda] \vdash}{\vdash} \quad (\textit{OpRenaming})$$

6.6 Hiding

The schema of the operation $OP \setminus (x_1, \ldots, x_n)$ is the schema of OP with $x_1, \ldots, x_n$ hidden.

$$\frac{OP \setminus (x_1, \ldots, x_n) = [OP] \setminus (x_1, \ldots, x_n) \vdash}{\vdash} \quad (\text{OpHiding})$$

7 Objects

To reason about an object of a class, we employ the dot notation to refer to its types, constants, pre-state (unprimed) variables and post-state (primed) variables. In addition to the notation $a.x$ of Object-Z, therefore, we introduce the notation $a.T$ to denote the local type T of the class of the object referenced by a and $a.x'$ to denote the post-value of the primary or secondary variable x of the object referenced by a.

7.1 Local Types

A local type will be the same for all objects of the class.

$$\frac{\vdash a_1 \in A \quad \vdash a_2 \in A}{\vdash a_1.T = a_2.T} \quad (\text{LocalTypes}) \, [\, T \in \alpha(A) \,]$$

7.2 Class Membership

A meta-substitution operator $\circledast$ is introduced to enable the substitution of the types, attributes and primed variables of an object into a predicate. It is defined in terms of the meta-substitution operator $\odot$ as follows.

$a \circledast p = b \odot p$
where $a \in A$
$\qquad \{x_1, \ldots, x_n\} = \alpha(A)$
$\qquad \{x_k, \ldots, x_n\} = \alpha_1(A) \cup \alpha_2(A)$
$\qquad b = (\!| \, x_1 \rightsquigarrow a.x_1, \ldots, x_n \rightsquigarrow a.x_n, x'_k \rightsquigarrow a.x'_k, \ldots, x'_n \rightsquigarrow a.x'_n \, |\!)$

Similarly, $\circledast$ is defined for substitution of the types, attributes and primed variables of an object into expressions and declarations.

If an object is a member of a class A then its types, attributes and primed variables satisfy the properties of A's local environment and the predicate of A's state schema (in both primed and unprimed form).

$$\frac{\vdash a \in A \quad A :: \vdash STATE = S}{\vdash a \circledast (\eta(A) \wedge \Delta S)} \quad (\text{ClassMembership}) \, [\, \phi(S) \subseteq \alpha(A) \,]$$

7.3 Initialising Objects

The predicate $a.INIT$ is true if and only if the attributes of a satisfy the predicate of A's initial state schema.

$$\frac{\vdash a \in A \quad A :: \; \vdash INIT = S}{\vdash a.INIT \Leftrightarrow a \circledast S} \qquad (\textit{Initialisation}) \; [\, \phi(S) \subseteq \alpha(A) \,]$$

7.4 Applying Operations to Objects

The meta-function δ is introduced to return the set of names of variables in the delta-list of an operation of a class. It is defined in terms of a meta-function δ_d which returns the explicitly declared delta-list of the operation.

Given a class $A[X_1, \ldots, X_n]$ with operation definition $Op \; \hat{=} \; OP$,

$$\delta_d(A[t_1, \ldots, t_n], Op) = \delta_{OP}(OP)$$

To account for the delta-lists of the inherited parts of operations we have

$$\delta(A[\lambda], Op) = \langle\!| \; \lambda \; |\!\rangle \odot \delta(A, Op)$$
$$\delta(B, Op) = \delta_d(B, Op) \cup \delta(A_1, Op) \cup \ldots \cup \delta(A_k, Op) \quad \text{if } Op \in \Omega_d(B)$$
$$= \delta(A_1, Op) \cup \ldots \cup \delta(A_k, Op) \quad \text{if } Op \notin \Omega_d(B)$$
$$\text{where } \iota(B) = \{A_1, \ldots, A_n\}$$
$$Op \in \Omega(A_1) \cap \ldots \cap \Omega(A_k)$$
$$Op \notin \Omega(A_{k+1}) \cup \ldots \cup \Omega(A_n)$$

The schema of the operation $a.Op$ is formed by

- taking the schema of Op of a's class,

- equating all primary variables not in Op's delta list with their primed counterparts,

- substituting the types, attributes and primed variables of a into the declaration and the predicate of the resulting schema

- and then hiding the primed and unprimed variables of a's class (leaving only the variables declared explicitly in Op).

$$\frac{\vdash a \in A \qquad A :: \; \vdash Op = [\,d \,|\, p\,]}{\vdash a.Op =} \qquad (\textit{OpApplication}) \; [\, p \,]$$
$$[\, a \circledast d \,|\, a \circledast (p \wedge x_1' = x_1 \wedge \ldots \wedge x_n' = x_n) \,]$$
$$\backslash (y_1, \ldots, y_m, y_1', \ldots, y_m')$$

$$\text{where } p \equiv Op \in \Omega(A)$$
$$\alpha_1(A) \backslash \delta(A, Op) = \{x_1, \ldots, x_n\}$$
$$\alpha_1(A) \cup \alpha_2(A) = \{y_1, \ldots, y_m\}$$
$$\phi[\,d \,|\, p\,] \subseteq \alpha(A)$$

For example, given the following class

$$
\begin{array}{|l}
\hline
\_A \underline{\hspace{8cm}} \\
\quad
\begin{array}{|l}
\hline
x, y : N \\
\hline
\end{array} \\
\\
\quad
\begin{array}{|l}
\hline
\_Op \underline{\hspace{6cm}} \\
\Delta(x) \\
in? : N \\
\hline
x' = x + in? \\
\hline
\end{array} \\
\\
\vdots \\
\hline
\end{array}
$$

using *StateDef*, *OpDef* and *Enrichment* we can deduce

$$A :: \ \vdash Op = [\, x, x', y, y' : N; \ in? : N \,|\, x' = x + in? \,]$$

Therefore, given $a : A$, since $\alpha_1(A) = \{x, y\}$, $\alpha_2(A) = \varnothing$ and $\delta(A, Op) = \{x\}$, using *OpApplication* we can deduce

$$
\begin{aligned}
a.Op &= ([\, a \circledast (x, x', y, y' : N; \ in? : N) \,|\, a \circledast (x' = x + in? \wedge y' = y) \,]) \\
&\quad \backslash (x, x', y, y') \\
&= ([\, x, x', y, y' : N; \ in? : N \,|\, a.x' = a.x + in? \wedge a.y' = a.y \,]) \\
&\quad \backslash (x, x', y, y') \\
&= [\, in? : N \,|\, a.x' = a.x + in? \wedge a.y' = a.y \,]
\end{aligned}
$$

7.5 Polymorphism

If a is a member of $\downarrow A$ then a's class is either A or a class derived from A by inheritance.

$$\frac{\vdash a \in \downarrow A}{\vdash a \in A_1 \vee \ldots \vee a \in A_n} \quad (\text{Polymorphism}) \, [\, p \,]$$

where $p \equiv (\iota^{-1})^*(\!|\, \{A\} \,|\!) = \{A_1, \ldots, A_n\}$

8 Conclusion

This paper has presented a logic for Object-Z which extends $\mathcal{W}$, the logic for Z adopted as the basis of the deductive system in the Z Base Standard. The sequent notation of $\mathcal{W}$ was extended to allow reasoning within the local environments of Object-Z classes and inference rules to cover the additional constructs and operators of Object-Z were presented.

The extended logic forms a basis on which tool support for reasoning about Object-Z specifications can be developed. Preliminary encoding of the rules has begun on the

Ergo theorem prover [UW94]. Its encoding on theorem provers which already support $\mathcal{W}$ (e.g. [Mar93]) should also be feasible.

As well as a basis for reasoning, the logic also provides an abstract, axiomatic semantics of Object-Z formalising many constructs in the language which have only been described informally previously. The logic is yet to be proved sound; however, a denotational semantics of Object-Z is currently being developed and a proof of soundness with respect to this will be carried out when it is completed.

While the logic provides a basis for reasoning about Object-Z specifications, it doesn't explain how such reasoning should proceed. Even the simplest proofs require a sequence of inference rules to be applied which may not always be obvious. Therefore, a practical proof system for Object-Z requires, in addition to the logic, tactics which perform sequences of proof steps corresponding to commonly performed proofs. In particular, it is foreseen that tactics should be available which utilise the modularity inherent in Object-Z specifications.

Acknowledgements

The author would like to thank Roger Duke, Alena Griffiths, Wendy Johnston and Gordon Rose for many helpful discussions. This work is supported by the Australian Research Council (ARC).

References

[BN92]   S.M. Brien and J.E. Nicholls. Z Base Standard: Version 1. Technical Report PRG-107, Oxford University Computing Laboratory, 1992.

[DD90]   D. Duke and R. Duke. Towards a semantics for Object-Z. In D. Bjørner, C.A.R. Hoare, and H. Langmaack, editors, *VDM'90: VDM and Z!*, volume 428 of *Lecture Notes in Computer Science*, pages 242–262. Springer-Verlag, 1990.

[DD93]   J. Dong and R. Duke. Class union and polymorphism. In C. Mingins, W. Haebich, J. Potter, and B. Meyer, editors, *Technology of Object-Oriented Languages and Systems (TOOLS 12 & 9)*, pages 181–190. Prentice-Hall International, 1993.

[DD94]   J. Dong and R. Duke. The geometry of object containment. Technical Report 94-17, Software Verification Research Centre, Department of Computer Science, University of Queensland, Australia, 1994. To appear in Object-Oriented Systems (OOS).

[Dil90]   A. Diller. *Z: An Introduction to Formal Methods*. John Wiley and Sons, 1990.

[DKRS91]   R. Duke, P. King, G. Rose, and G. Smith. The Object-Z specification language: Version 1. Technical Report 91-1, Software Verification Research Centre, Department of Computer Science, University of Queensland, Australia, 1991.

[DRS94]   R. Duke, G. Rose, and G. Smith. Object-Z: a specification language advocated for the description of standards. Technical Report 94-45, Software Verification Research Centre, Department of Computer Science, University of Queensland, Australia, 1994.

[GR94]   A. Griffiths and G. Rose. A semantic foundation for object identity in formal specification. Technical Report 94-21, Software Verification Research Centre, Department of Computer Science, University of Queensland, Australia, 1994.

[Mar93]   A. Martin. Encoding $\mathcal{W}$: A logic for Z in 2OBJ. In J.C.P. Woodcock and P.G. Larsen, editors, *FME'93: Industrial-Strength Formal Methods*, volume 670 of *Lecture Notes in Computer Science*, pages 462–481. Springer-Verlag, 1993.

[RD93]   G. Rose and R. Duke.   An Object-Z specification of a mobile phone system.   In
           K. Lano and H. Haughton, editors, *Object-Oriented Specification Case Studies*, pages
           110–129. Prentice-Hall International, 1993.

[Ros92]   G. Rose. Object-Z. In S. Stepney, R. Barden, and D. Cooper, editors, *Object Orien-
           tation in Z*, Workshops in Computing, pages 59–77. Springer-Verlag, 1992.

[Spi88]   J.M. Spivey. *Understanding Z: A specification language and its formal semantics*, vol-
           ume 3 of *Cambridge Tracts in Theoretical Computer Science*. Cambridge University
           Press, 1988.

[Spi92]   J.M. Spivey. *The Z Notation: A Reference Manual (2nd Ed.)*. Series in Computer
           Science. Prentice-Hall International, 1992.

[UW94]   M. Utting and K. Whitwell. Ergo user manual. Technical Report 93-19, Software
           Verification Research Centre, Department of Computer Science, University of Queens-
           land, Australia, 1994.

[WB92]   J.C.P. Woodcock and S.M. Brien. $\mathcal{W}$: A logic for Z. In J.E. Nicholls, editor, *Z User
           Workshop*, Workshops in Computing, pages 77–98. Springer-Verlag, 1992.

[WL88]   J.C.P. Woodcock and M. Loomes. *Software Engineering Mathematics*. Pitman, 1988.

[Wor92]   J.B. Wordsworth.   *Software Development with Z: A Practical Approach to Formal
           Methods in Software Engineering*. International Computer Science Series. Addison-
           Wesley, 1992.

Applications II

A Formal Semantics for a Language with Type Extension

Peter Bancroft

Queensland University of
Technology
Brisbane 4001, Australia
e-mail: bancroft@fit.qut.edu.au

Ian Hayes

University of Queensland
Brisbane 4072, Australia
e-mail: ianh@cs.uq.oz.au

Abstract. The purpose of this paper is to give a formal semantics for a language which includes type extension. Used in association with pointer variables, this forms the basis of object-orientation in the languages Oberon and Oberon-2 which have evolved from Modula-2. The focus is on the meaning of assignment because this is the important difference between such languages and the strongly-typed Pascal family. An abstract syntax is defined using the Z notation and the static and dynamic semantics are given in a denotational style.

1 Introduction

In languages such as Pascal and Modula-2, program variables are declared to be of a certain type and may then be assigned only values that belong strictly to that type. In the latest of this family of languages, Oberon [RW92, Wir88b] and Oberon-2 [MW91], this restriction has been relaxed for record and pointer types to achieve inheritance. The same approach to the development of an object-oriented Ada9X is suggested in [Sei91].

Type extension [Wir88a] is the creation of new record types from existing record types by the addition of extra fields. A record extension is considered to be compatible with the original record type in that the values of the extension may be assigned to variables of that type. Of course the extra fields cannot be stored in such a variable and are "dropped off". This is known as the projection of the extended value onto the original type. Being able to extend record types in this way allows the creation of a record hierarchy, one of the common features of object-oriented languages.

Pointer types may also be extended. A pointer variable is declared to refer to a particular base record type; however it may be assigned a reference to a record extension of its base type. The difference between this pointer assignment and the corresponding record assignment is that the extra fields of the record extension remain accessible through the pointer. Such access is subject to a type guard (which is essentially an assertion that the dynamic type of the pointer is an extension of the type named in the guard). For example, in Oberon we write,

```
vr(tid)^.fld
```

meaning that variable *vr* has been extended so that it has dynamic type *tid*, a pointer to a record with a field name *fld*. Such an expression could be used in assignment as an r-value to read the content of *fld* or as an l-value to update the field's value.

When a pointer variable references an extended value, it is also possible to take advantage of the dynamic type to assign the extended value to any other compatible variable. Again this requires the use of a type guard and is known as guarded assignment. This is written,

```
vr1 := vr(tid)
```

Here, *vr* is asserted to have dynamic type *tid* which must be the same type or an extension of *vr1*. Its value may therefore be assigned to *vr1*. The guard is necessary only when the static type of *vr* is not an extension of the static type of *vr1*.

Type guarded expressions have the potential to abort. To avoid this, Oberon includes the assertion

```
vr IS tid
```

which may be used in the guards of *if* and *while* statements when relying on the dynamic type of an extended variable.

Type extension is not the only feature of Oberon that characterises it as an object-oriented language but it does facilitate single inheritance. It is the purpose of this paper to formalise type extension. While most of the commands of the type-strict languages are unaffected, the impact on the meaning of assignment is significant.

In Section 2, an abstract syntax for a language with type extension is given. It includes type and variable declaration and a definition for the expressions of the language. We do not include all the usual constructs but prefer to concentrate on those features that are directly related to assignment. Section 3 presents a static semantics for the language by defining a static environment and giving the well-formedness conditions for types, variables and expressions. These are essentially type-checking constraints. In Section 4, a dynamic environment is defined. The dynamic semantics for variables and expressions are given in Sections 4.2 and 4.3 respectively. The projection function is defined in Section 4.4 and is an important feature of the ensuing definition of assignment in Section 5. We use the Z notation [Hay93, Spi92] to present a formal semantics for a small Oberon-like language in the style of [Hay94].

2 Abstract Syntax

For our purposes it is sufficient to give the abstract syntax of type declarations (Section 2.1), variable references (Section 2.2) and expressions (Section 2.3).

2.1 Type declarations

We introduce a basic set of names (identifiers) for the declaration of types and variables in the language.

[*Id*]

The possible types of the language are given by a disjoint union which represents the most general view of type declaration.

$$Type ::= nat \mid bool \mid record\langle\langle Id \twoheadrightarrow Type\rangle\rangle \mid ptr\langle\langle Type\rangle\rangle \mid$$
$$extend\langle\langle Id \times (Id \twoheadrightarrow Type)\rangle\rangle \mid typeid\langle\langle Id\rangle\rangle$$

Included in this free type definition are record, pointer, extension and named types. The function *record* takes a partial function from field names (*Id*) to their types and constructs a record type. The function *ptr* is used to construct a pointer type to a particular base type. An extend type consists of an *Id* (which we later constrain to be the name of a record or another extend type) together with some extra fields. The function *typeid* allows us to give new names to existing types.

2.2 Variable references

Variable references are elements in the following disjoint union.

$$Varref ::= vref\langle\langle Id\rangle\rangle \mid fieldref\langle\langle Varref \times Id\rangle\rangle \mid deref\langle\langle Varref\rangle\rangle \mid$$
$$typeref\langle\langle Varref \times Id\rangle\rangle$$

The function *vref* takes a name $id \in Id$ and produces a variable *vref(id)*. The function *fieldref* is used to select a field from a record or extend variable, while *deref* is used to dereference a pointer variable. The function *typeref* is used to create type-guarded variable references (to deal with extended pointer variables).

Example 1. Consider the following variable references, in Oberon-like notation:

n, r (a record), p (a pointer), r.f1, r.f2, p^ and p(tid).

In the abstract syntax defined above, we write

$n, r, f1, f2, p, tid : Id$

$vref(n), vref(r), vref(p) \in Varref$

$fieldref(vref(r), f1), fieldref(vref(r), f2) \in Varref$

$deref(vref(p)) \in Varref$

$typeref(vref(p), tid) \in Varref$

2.3 Expressions

Expressions are Boolean or natural constants, variable references or type assertions. We omit the usual unary and binary Boolean or natural operators as they are not needed in defining type extension.

$$Expr ::= boolean\langle\langle \mathbb{B}\rangle\rangle \mid natural\langle\langle \mathbb{N}\rangle\rangle \mid exvar\langle\langle Varref\rangle\rangle \mid is\langle\langle Varref \times Id\rangle\rangle$$

As there is no Boolean type declared in the Z notation, we define $\mathbb{B} ::= True \mid False$. N is the set of natural numbers. The function *exvar* takes a variable reference and returns an expression, which is of particular interest to us as the right side of an assignment statement (Section 5.1). The function *is* asserts the dynamic type of a variable (Section 3.3).

Example 2. Given the Oberon-like expressions, n, b and r is tid we write,

$exvar(vref(n))$

$exvar(vref(b))$

$is(vref(r), tid)$

3 Static Semantics

A static environment (Section 3.1) classifies identifiers as either variables or types and in both cases gives the corresponding type associated with the identifier. The environment is used to express constraints in the construction of types in Section 3.2 and to perform type checking on variables in Section 3.3. In Section 3.4 expressions are type checked. Much of the detail from this section is omitted as it is routine. The full paper is available from *ftp.fit.qut.edu.au*. We retain only the novel features and the definitions required in Section 4. For easier reading, we also omit stating the obvious universal quantification of variables in axiomatic definitions, throughout.

3.1 Static Environment

Before we define a static environment we unfold *typeid* types to identify the base types which they alias. Unfolding the aliased types at this point makes it easier later to identify the extensions of record and pointer types. The *alias* function determines the *id* at the end of a sequence of *typeid* declarations. For types other than *typeid* the *id* is mapped to itself.

$$\mid \quad alias : (Id \nrightarrow Type) \nrightarrow (Id \rightarrow Id)$$

A static environment comprises three functions. Given a name ($\in Id$), a static environment defines it to be either a type or a variable declaration. The *al* function associates all type names with their aliases as defined above.

$\_\_$*StatEnv*$_____$

$types : Id \nrightarrow Type$
$vars : Id \nrightarrow Type$
$al : Id \nrightarrow Id$

$\rule{6cm}{0.4pt}$

$\mathrm{dom}\, types \cap \mathrm{dom}\, vars = \{\,\}$
$types \in \mathrm{dom}\, alias$
$al = \{id : \mathrm{dom}\, types \bullet id \mapsto alias(types)(id)\}$

Example 3 (A static environment). The following is a typical declaration block in an Oberon-like language:

```
var n:nat; b:bool;
type rec1 = record f1:nat; f2:bool end;
     recid1 = rec1;
var r:recid1;
type ptr1 = pointer to rec1;
     ptrid1 = ptr1;
var p:ptrid1;
type ext1 = extend rec1 with f3:nat end;
var e:ext1;
```

In our notation the static environment is given by,

$$\rho.\mathit{types} = \{\mathit{rec}1 \mapsto \mathit{record}\{f1 \mapsto \mathit{nat}, f2 \mapsto \mathit{bool}\}, \mathit{recid}1 \mapsto \mathit{typeid}(\mathit{rec}1),$$
$$\mathit{ptr}1 \mapsto \mathit{ptr}(\mathit{typeid}(\mathit{rec}1)), \mathit{ptrid}1 \mapsto \mathit{typeid}(\mathit{ptr}1),$$
$$\mathit{ext}1 \mapsto \mathit{extend}(\mathit{rec}1, \{f3 \mapsto \mathit{nat}\})\}$$
$$\rho.\mathit{vars} = \{n \mapsto \mathit{nat}, b \mapsto \mathit{bool}, r \mapsto \mathit{typeid}(\mathit{recid}1),$$
$$p \mapsto \mathit{typeid}(\mathit{ptrid}1), e \mapsto \mathit{typeid}(\mathit{ext}1)\}$$
$$\rho.\mathit{al} = \{\mathit{rec}1 \mapsto \mathit{rec}1, \mathit{recid}1 \mapsto \mathit{rec}1, \mathit{ptr}1 \mapsto \mathit{ptr}1$$
$$\mathit{ptrid}1 \mapsto \mathit{ptr}1, \mathit{ext}1 \mapsto \mathit{ext}1\}$$

In Oberon, a variable may be declared as an unnamed record, extend or pointer type. e.g. `var rec:record f1:T1; f2:T2 end;` The functions *rectype*, *exttype* and *ptrtype* combine the appropriate named and unnamed types in a given environment.

$$\mathit{exttype} : \mathit{StatEnv} \rightarrow \mathbb{P}\ \mathit{Type}$$
$$\mathit{rectype} : \mathit{StatEnv} \rightarrow \mathbb{P}\ \mathit{Type}$$
$$\mathit{ptrtype} : \mathit{StatEnv} \rightarrow \mathbb{P}\ \mathit{Type}$$

We use these definitions, to declare two functions that will enable us to determine the base types from which named and unnamed extend and pointer types are constructed.

$$\mathit{extbasetype} : \mathit{StatEnv} \rightarrow (\mathit{Type} \nrightarrow \mathit{Type})$$
$$\mathit{ptrbasetype} : \mathit{StatEnv} \rightarrow (\mathit{Type} \nrightarrow \mathit{Type})$$

It is necessary to identify equivalent types in a given static environment — in particular those aliased using the *typeid* function. We define *eqtype* to be the (name) equivalence class of types, for a given environment.

$$\mathit{eqtype} : \mathit{StatEnv} \rightarrow (\mathit{Type} \leftrightarrow \mathit{Type})$$

$$\mathit{eqtype}(\rho) = \{id1, id2 : Id \mid \mathit{alias}(\rho.\mathit{types})(id1) = \mathit{alias}(\rho.\mathit{types})(id2) \bullet$$
$$(\mathit{typeid}(id1), \mathit{typeid}(id2))\} \cup \mathrm{id}\ \mathit{Type}$$

Pointer, record and extend types are central to the object-oriented nature of the language we are defining. The type hierarchy created by the extension of record types facilitates single inheritance. With the preceding definitions we are in a position to describe the relationship between pointer and record types and their extensions.

$$dirext : StatEnv \rightarrow (Type \leftrightarrow Type)$$
$$recext : StatEnv \rightarrow (Type \leftrightarrow Type)$$
$$ptrext : StatEnv \rightarrow (Type \leftrightarrow Type)$$
$$ext : StatEnv \rightarrow (Type \leftrightarrow Type)$$

$$dirext(\rho) = \bigcup\{t1, t2 : Type \mid t2 \mapsto t1 \in extbasetype(\rho) \bullet$$
$$(eqtype\ \rho)(\mid \{t1\} \mid) \times (eqtype\ \rho)(\mid \{t2\} \mid)\}$$
$$recext(\rho) = (dirext(\rho))^+$$
$$ptrext(\rho) = \{p1, p2 : ptrtype(\rho) \mid$$
$$ptrbasetype(\rho)(p1) \mapsto ptrbasetype(\rho)(p2) \in recext(\rho)\}$$
$$ext(\rho) = recext(\rho) \cup ptrext(\rho) \cup$$
$$\{t : (rectype(\rho) \cup exttype(\rho) \cup ptrtype(\rho)) \bullet t \mapsto t\}$$

Any (named or unnamed) extend type (and its equivalent types) is a direct extension of the type identified by its first argument (and its equivalent types) in the type declaration, for a given environment. The transitive closure of this relation is $recext(\rho)$, giving all record extensions. A pointer type $p2$ is an extension of another pointer type $p1$ if the type referred to by $p2$ is a record extension of the type referred to by $p1$. The expression, $t1 \mapsto t2 \in ext(\rho)$ may be interpreted as "$t1$ is extended by $t2$", where $ext(\rho)$ is the union of $recext(\rho)$, $ptrext(\rho)$ and the identity relation on $rectype$, $exttype$ and $ptrtype$. We include the latter in $ext(\rho)$ as we consider any $rectype$, $exttype$ or $ptrtype$ to be an extension of itself.

In the declaration of *extend* types, the *extbasetype* must be a *named record* or *extend* type but it is necessary to avoid circular definitions. For example, it makes no sense to declare *extend* type $e1$ with a first argument of *exttype* $e2$ and then to declare $e2$ as an *extend* type with first argument $e1$. To avoid this we define a well-formed static environment.

$$WFStatEnv : \mathbb{P}\ StatEnv$$

3.2 Type Constraints

The following auxiliary function is used to simplify the treatment of record and extend types. All *rectype* and *exttype* types may be "flattened" to become a partial function from *Id* to *Type*, essentially representing the fields of the type. The definition, as it is given, allows a field name to be redeclared (as indicated by the relation from *Id* to *Type*). In the definition of well-formed types below we place constraints on *rectype* and *exttype* types to effectively make the relation a total function.

$$fields : WFStatEnv \rightarrow (Type \nrightarrow (Id \leftrightarrow Type))$$

The well-formedness constraints for types are given by the function *WFType*.

$$WFType : Type \rightarrow \mathbb{P}\ WFStatEnv$$

The types *nat* and *bool* are well-formed in any environment. A *record* type is well-formed when the type of each of its fields is well-formed. A *pointer* type can only refer to well-formed *record* or *extend* type. The first argument of an *extend* type must be the name of a well-formed *rectype* or *exttype*, avoiding circular definitions (since it is defined only over a well-formed static environment); each of the field types (in the second argument) must be well-formed; and the names of the added fields must be different from the names of the fields of the base *rectype* or *exttype*. A *typeid* type is well-formed if the actual type to which the name refers is well-formed.

3.3 Type Checking Variables

There are also constraints on variable references to consider. The function *WFVarref* takes any *Varref* and returns the set of environments in which that variable is well-formed. The type of any *Varref* is obtained using the auxiliary function *TPVarref*.

$$WFVarref : Varref \rightarrow \mathbb{P}\ WFStatEnv$$
$$TPVarref : Varref \rightarrow (WFStatEnv \nrightarrow Type)$$

A *vref* is well-formed in any environment in which its type is well-formed. A *fieldref* is well-formed in any environment in which its first argument is a *Varref* of some well-formed *rectype* or *exttype* and its second argument is the name of one of the fields of that *rectype* or *exttype*. A *deref* is well-formed in any environment in which its argument is a *Varref* of a well-formed *ptrtype*. A *typeref* is well-formed if its first argument is a well-formed *Varref* and its second argument is the *Id* of a *pointer* type which is an extension of the type of the first argument (which must therefore be a pointer type).

The type of a *vref* is the type of the variable to which it refers in the given environment. The type of a *fieldref* is the type of the selected field. The type of a *deref* is the *ptrbasetype* of its argument. The type of a *typeref* is the type to which its second argument refers (giving in effect, the *dynamic type* of its first argument).

Example 4 (Refer to example 3).

$$TPVarref(vref(n))(\rho) = nat$$
$$TPVarref(vref(r))(\rho) = typeid(recid1) \qquad \text{(named record type)}$$
$$TPVarref(fieldref(vref(r), f1))(\rho) = nat$$
$$TPVarref(fieldref(vref(e), f3))(\rho) = nat$$
$$TPVarref(deref(vref(p))(\rho) = typeid(rec1) \qquad \text{(named record type)}$$

3.4 Type Checking Expressions

We define functions for establishing the environments in which each of our allowed expressions is well-formed and to determine the type of any legal expression.

$WFExpr : Expr \rightarrow \mathbb{P}\ WFStatEnv$

$TPExpr : Expr \rightarrow (WFStatEnv \nrightarrow Type)$

Boolean and natural expressions are well-formed in any well-formed static environment and are of the expected types. The type of an *exvar* expression is the type of the corresponding *Varref* and is well-formed whenever that *Varref* is well-formed. It is worth noting that there is a type guard expression given by *exvar(typeref(vr, id))*, the type of which is equal to the type denoted by *id*. This is needed for guarded assignment in Section 5.4. The *is* expression has a well-formed *typeref* as argument and is of type *bool*.

4 Dynamic Semantics of Expressions

A dynamic environment and a store are defined so that we can assign meaning to the well-formed variables and expressions of the language. We firstly consider the values that may be assigned to each of the declared types.

4.1 Value of Types

We assume a set of locations in a store.

[*Loc*]

The values that may be attributed to variables and expressions of our allowed types are given by,

$Value ::= natval\langle\langle\mathbb{N}\rangle\rangle \mid boolval\langle\langle\mathbb{B}\rangle\rangle \mid recval\langle\langle Id \nrightarrow Value\rangle\rangle \mid refval\langle\langle Loc\rangle\rangle$

The functions *natval* and *boolval* take a natural number and a boolean respectively, and return a *Value*. Similarly, the functions *recval* and *refval* are used to construct appropriate values.

Val is a function that returns the set of values for each of the declared types. In other words, we can use *Val* to associate certain values with each of our allowed types.

$Val : WFStatEnv \rightarrow (Type \rightarrow \mathbb{P}\ Value)$

$(\forall id : \mathrm{dom}\,\rho.types \cup \mathrm{dom}\,\rho.vars \bullet$
$\quad (Val\ \rho)(nat) = natval(\!| \ \mathbb{N}\ |\!) \land$
$\quad (Val\ \rho)(bool) = boolval(\!| \ \mathbb{B}\ |\!) \land$
$\quad (Val\ \rho)(record(flds)) = recval(\!| \ \{valflds : (\mathrm{dom}\,flds) \rightarrow Value \mid$
$\qquad\qquad (\forall id : \mathrm{dom}\,flds \bullet valflds(id) \in (Val\ \rho)(flds(id)))\} \ |\!) \land$
$\quad (Val\ \rho)(ptr(t)) = refval(\!| \ Loc\ |\!) \land$
$\quad (Val\ \rho)(extend(id, flds)) =$
$\qquad (Val\ \rho)(record((fields\ \rho)(extend(id, flds)))) \land$
$\quad (Val\ \rho)(typeid(id)) = (Val\ \rho)(\rho.types(\rho.al(id))))$

The possible values for the type *nat* (*bool*) are those produced by the *natval* (*boolval*) function. The values of a record type are the *recval* images of the allowed (*Id*, *Value*) pairs for each field of the record. As *extend* may be applied only to *rectype* and *exttype*, the values of *extend* types are record values. *Pointer* values are locations in a store. The values of a *typeid* are the allowed values of its base type, determined by the *al* function in the static environment.

4.2 Value of Variables

Having established what values may be given to each of the types we can now set up a store and a dynamic environment to account for the value of a variable. A stored value consists of a type and a value. The constructor function *sval* takes a type and a value and returns a *StoredVal*.

$$
\begin{array}{|l}
\_StoredVal_____ \\
dyntype : Type \\
dynval : Value \\
\hline
\end{array}
$$

$$sval == (\lambda t : Type;\ v : Value \bullet (\mu\, StoredVal \mid dyntype = t \wedge dynval = v))$$

A store is a partial function from locations to stored values.

$$Store == Loc \nrightarrow StoredVal$$

Example 5 (A typical store). Given $l1, l2, l3, l4 \in Loc$

$$
\begin{aligned}
\sigma == \{ & l1 \mapsto sval(nat, natval(5)), \\
& l2 \mapsto sval(typeid(rec1), recval\{f1 \mapsto natval(8), f2 \mapsto boolval(True)\}), \\
& l3 \mapsto sval(ptr(typeid(rec1)), refval(l4)) \\
& l4 \mapsto sval(typeid(ext1), \\
& \qquad recval\{f1 \mapsto natval(8), f2 \mapsto boolval(True), f3 \mapsto natval(3)\}) \}
\end{aligned}
$$

In location *l2* we have a *recval* consisting of two fields *f1* and *f2* with *natval* and *boolval* values respectively. We would expect that *l4* has been allocated dynamically as it is referred to by the *refval* in location *l3*. This is an example of the dynamic extension of a pointer. The *dyntype* of location *l3* is *ptr(rec1)* but when we dereference the pointer, we obtain location *l4* which has a *dyntype* of *typeid(ext1)*, an extension of the record type *rec1*.

A dynamic environment is required to define the location of each declared variable.

$$DynEnv == Id \nrightarrow Loc$$

Example 6 (Refer to examples 3 and 5).

$$\rho = \{n \mapsto l1, r \mapsto l2, p \mapsto l3\}$$

Note again that the *dynval* value for *p* (*l3*) in example 5 is *refval(l4)* which gives the location of an extension of the static type (*rec1*) to which this *ptr* type refers. The *dyntype* for the dereferenced pointer stored in *l4* is *ext1*. The pointer *varref p* has been extended.

We shall need to be able to extract the value in a location in a store for field selectors of *rectype* and *exttype* variables. To do this we define the following auxiliary functions.

$$select : Id \rightarrow (Value \nrightarrow Value)$$

$$select(id) = \{flds : Id \nrightarrow Value \mid id \in \operatorname{dom} flds \bullet recval(flds) \mapsto flds(id)\}$$

For nested records, a sequence of field selectors is required.

$$selectfield : \operatorname{seq} Id \rightarrow (Value \nrightarrow Value)$$

$$selectfield(s) = {}^{\circ}_{9}/(s\,{}^{\circ}_{9}\,select)$$

The *selectfield* function allows us to successively apply *select* to each *id* in the field selector sequence until the required field is reached.

We are now in a position to determine the meaning of any *Varref* in a given dynamic environment for a given store. We recall that a *Varref* may be any one of four disjoint types (Section 2.1). We firstly define a schema for the location associated with a variable, consisting of a location and a field selector sequence, together with a value constructor function, *dval*.

```
┌─ DynLoc ────────────────────────────────
  dynloc : Loc
  dynflds : seq Id
└─────────────────────────────────────────
```

$$dval == (\lambda\, loc : Loc;\ s : \operatorname{seq} Id \bullet (\mu\, DynLoc \mid dynloc = loc \wedge dynflds = s))$$

The dynamic type of a pointer variable is (a pointer to) the *dyntype* identified as the first argument in the location of the dereferenced variable in a store. We include this in the following definition as it is used to give the meaning of a *typeref*.

$$dtype : Varref \rightarrow (WFStatEnv \nrightarrow DynEnv \nrightarrow (Store \nrightarrow Type))$$
$$MVarref : Varref \rightarrow (WFStatEnv \rightarrow DynEnv \nrightarrow (Store \nrightarrow DynLoc))$$

$$dtype(vr) = (\lambda \rho_S : WFStatEnv \mid TPVarref(vr)(\rho_S) \in ptrtype(\rho_S) \bullet$$
$$(\lambda \rho_D : dom(MVarref(vr)(\rho_S)) \bullet$$
$$(\lambda \sigma : dom(MVarref(vr)(\rho_S)(\rho_D)) \bullet$$
$$ptr((\sigma(MVarref(deref(vr))(\rho_S)(\rho_D)(\sigma)).dynloc).dyntype))))$$

$$MVarref(vref(id)) =$$
$$(\lambda \rho_S : WFStatEnv \bullet (\lambda \rho_D : DynEnv \mid id \in dom\,\rho_D \bullet$$
$$(\lambda \sigma : Store \bullet dval(\rho_D(id), \langle \rangle))))$$

$$MVarref(fieldref(vr, id)) =$$
$$(\lambda \rho_S : WFStatEnv \bullet (\lambda \rho_D : dom(MVarref(vr)(\rho_S)) \bullet$$
$$(\lambda \sigma : dom(MVarref(vr)(\rho_S)(\rho_D)) \bullet$$
$$(\textbf{let}\, ln == (MVarref(vr)(\rho_S)(\rho_D)(\sigma)).dynloc \bullet$$
$$(\textbf{let}\, fs == (MVarref(vr)(\rho_S)(\rho_D)(\sigma)).dynflds \bullet$$
$$dval(ln, fs \frown \langle id \rangle)))))))$$

$$MVarref(deref(vr)) =$$
$$(\lambda \rho_S : WFStatEnv \bullet (\lambda \rho_D : dom(MVarref(vr)(\rho_S)) \bullet$$
$$(\lambda \sigma : dom(MVarref(vr)(\rho_S)(\rho_D)) \bullet$$
$$(\textbf{let}\, ln == (MVarref(vr)(\rho_S)(\rho_D)(\sigma)).dynloc \bullet$$
$$(\textbf{let}\, fs == (MVarref(vr)(\rho_S)(\rho_D)(\sigma)).dynflds \bullet$$
$$dval(refval^{-1}(selectfield(fs)((\sigma(ln)).dynval)), \langle \rangle)))))))$$

$$MVarref(typeref(vr, id)) =$$
$$(\lambda \rho_S : WFStatEnv \bullet (\lambda \rho_D : dom(MVarref(vr)(\rho_S)) \bullet$$
$$(\lambda \sigma : dom(MVarref(vr)(\rho_S)(\rho_D)) \mid$$
$$(\rho_S.types(id), dtype(vr)(\rho_S)(\rho_D)(\sigma)) \in ext(\rho_S) \bullet$$
$$MVarref(vr)(\rho_S)(\rho_D)(\sigma))))$$

The meaning of a *vref* is the location to which the given variable reference refers in the given store, coupled with an empty sequence. A *fieldref* refers to the location of the *Varref* that provides its first argument but adds its second argument (another field) to the existing field selector sequence. The value of a *deref* is a location with an empty sequence; the location is obtained by determining the field defined by the field selector sequence of its argument, using *selectfield*. The meaning of a *typeref* is the meaning of its first argument but is constrained with respect to the types of its arguments. The type of the value in the store to which the pointer variable *vr* refers must be an extension of the type denoted by *id*.

In a type-strict language, the dynamic type is always the same as the static type — pointers can only reference values of their base type. In a language with type extension the dynamic type can be any extension of the static type. (If the *ptrbasetype* is an unnamed type this is always a value of the static type, as only named types can be extended.) The definition of a well-formed store expresses this constraint.

$WFStore : WFStatEnv \rightarrow (DynEnv \nrightarrow \mathbb{P}\,Store)$

$WFStore(\rho_S)(\rho_D) =$
$\quad \{\sigma : Store \mid (\forall\, vr : Varref \mid TPVarref(vr)(\rho_S) \in ptrtype(\rho_S) \bullet$
$\quad (TPVarref(vr)(\rho_S) \mapsto dtype(vr)(\rho_S)(\rho_D)(\sigma)) \in ext(\rho_S))\}$

4.3 Value of Expressions

We provide a dynamic semantics for the allowable expressions of the language. Our main interest is in the value of *exvar*. *MExpr(exvar (vr))* is used in the semantics of assignment statements to determine an r-value. The meaning of an expression is defined to be a value in a given well-formed store.

$MExpr : Expr \rightarrow (WFStatEnv \rightarrow DynEnv \nrightarrow (Store \nrightarrow Value))$

$MExpr(natural(n1)) = (\lambda \rho_S : WFStatEnv \bullet (\lambda \rho_D : DynEnv \bullet$
$\quad (\lambda \sigma : Store \mid \sigma \in WFStore(\rho_S)(\rho_D) \bullet natval(n1))))$
$MExpr(boolean(b1)) = (\lambda \rho_S : WFStatEnv \bullet (\lambda \rho_D : DynEnv \bullet$
$\quad (\lambda \sigma : Store \mid \sigma \in WFStore(\rho_S)(\rho_D) \bullet boolval(b1))))$
$MExpr(exvar(vr)) =$
$\quad (\lambda \rho_S : WFStatEnv \bullet (\lambda \rho_D : dom(MVarref(vr)(\rho_S)) \bullet$
$\quad (\lambda \sigma : Store \mid \sigma \in WFStore(\rho_S)(\rho_D) \cap$
$\qquad\qquad\qquad dom(MVarref(vr)(\rho_S)(\rho_D)) \bullet$
$\quad (\text{let } ls == MVarref(vr)(\rho_S)(\rho_D)(\sigma) \bullet$
$\quad selectfield(ls.dynflds)((\sigma(ls.dynloc)).dynval)))))$
$MExpr(is(vr,id)) =$
$\quad (\lambda \rho_S : WFStatEnv \bullet (\lambda \rho_D : dom(MVarref(vr)(\rho_S)) \bullet$
$\quad (\lambda \sigma : Store \mid \sigma \in WFStore(\rho_S)(\rho_D) \cap$
$\qquad\qquad\qquad dom(MVarref(vr)(\rho_S)(\rho_D)) \bullet$
$\quad \text{if } (typeid(id), dtype(vr)(\rho_S)(\rho_D)(\sigma)) \in ext(\rho_S)$
$\quad \text{then } boolval(True)$
$\quad \text{else } boolval(False))))$

MExpr(exvar(vr)) has the value in the location associated with *MVarref(vr)* for the given environments and store. An *is* expression is an assertion that the dynamic type of *vr* is an extension of the type identified by *id* and is associated with the appropriate *boolval*.

4.4 Projection

In defining a semantics for assignment we need the following definition to account for the intended difference between assigning extended values to records (where the extra fields are dropped) and pointers (where the extra fields are still accessible). The following function will be used.

$$\begin{array}{|l} \hline proj : WFStatEnv \rightarrow Type \nrightarrow (Value \rightarrow Value) \\ \hline \mathrm{dom}(proj\ \rho) = \mathrm{ran}\,\rho.types \\ (\forall\, t : rectype(\rho) \cup exttype(\rho);\ flds : Id \nrightarrow Value \bullet \\ \quad proj(\rho)(t)(recval(flds)) = recval(\mathrm{dom}(fields\ \rho(t)) \lhd flds)) \\ (\forall\, t : Type;\ v : Value \mid t \notin (rectype(\rho) \cup exttype(\rho)) \bullet proj(\rho)(t)(v) = v) \end{array}$$

Projection is defined for all types in a given dynamic environment. Any value may be projected onto any type. Most importantly, the projection of a record value, *recval(flds)*, into a *rectype* or *exttype* has *flds* domain restricted to the fields of that type; other fields are dropped off. For projection type *t* other than *rectype* or *exttype*, the projected value remains unchanged. When a *refval* is projected onto a pointer type of which it is an extended value the whole *refval* is retained, giving extended pointer values access to any extra fields. The definition allows us to apply the *proj* function uniformly to the r-value of an assignment (Section 5.1) which is projected onto the type of the left side *Varref*. Section 5.2 ensures assignment compatibility so that we will only use *proj* to project a value onto a type to which it belongs or of which it is an extended value.

5 Assignment with Type Extension

5.1 Abstract Syntax of Commands

The assignment command is fundamental to our language. It comprises a left side variable reference (which may be *vref*, *fieldref*, *deref* or *typeref*) and a right side expression.

$$\begin{array}{|l} \hline \_Assign\_ \\ \hline lhs : Varref \\ rhs : Expr \\ \hline \end{array}$$

The commands of the language are defined by a disjoint union. At this point we are only interested in assignment, though we assume that commands such as *if*, *while*, etcetera are included with their usual meanings.

$$Com ::= assign\langle\!\langle Assign \rangle\!\rangle \mid \dots$$

5.2 Static Semantics for Commands

Type compatibility of named types is defined in terms of the *ext* relation in Section 3.1. The identity on types is also included.

$$\begin{array}{|l} \hline TypeCompat : WFStatEnv \rightarrow (Type \leftrightarrow Type) \\ \hline TypeCompat(\rho) = ext(\rho) \cup \mathrm{id}\ Type \end{array}$$

$(t1, t2) \in \mathit{TypeCompat}(\rho)$ is read as, "t2 is assignable to t1". With this definition, we can express the static type consistency constraints for commands as given by the *WFCom* function.

$$\mathit{WFCom} : \mathit{Com} \rightarrow \mathbb{P}\ \mathit{WFStatEnv}$$

$$\mathit{WFCom}(assign(asn)) = \{\rho : \mathit{WFVarref}(asn.lhs) \cap \mathit{WFExpr}(asn.rhs) \mid$$
$$(\mathit{TPVarref}(asn.lhs)(\rho), \mathit{TPExpr}(asn.rhs)(\rho)) \in \mathit{TypeCompat}(\rho)\}$$

The requirement for assignment is that the type of the right side expression is compatible with the type of the left side variable reference.

5.3 Dynamic Semantics for Commands

Any command may change the values of variables in the store. To give meaning to commands we need to model these possible changes. We define a transition function on a well-formed store.

$$\mathit{Tr} == \mathit{WFStore} \rightarrow\!\!\!\!\rightarrow \mathit{WFStore}$$

Updating a value (in particular, for a *fieldref*) in a store requires the use of the following auxiliary function. The first argument is the field selector sequence (empty except for a *fieldref*). The second argument is a given new (field) value. The result is a partial function from the old (record) value to the updated (record) value.

$$\mathit{update} : \mathrm{seq}\, \mathit{Id} \rightarrow \mathit{Value} \rightarrow (\mathit{Value} \rightarrow\!\!\!\!\rightarrow \mathit{Value})$$

$$\mathrm{dom}\, \mathit{update}(fs)(nv) = \mathrm{dom}\, \mathit{selectfield}(fs)$$
$$\mathit{update}(\langle\rangle)(nv)(v) = nv$$
$$(\forall \mathit{flds} : \mathit{Id} \rightarrow\!\!\!\!\rightarrow \mathit{Value} \bullet (\forall id : \mathrm{dom}\, \mathit{flds} \bullet$$
$$\quad \mathit{update}(\langle id\rangle \frown fs)(nv)(\mathit{recval}(\mathit{flds})) =$$
$$\quad \mathit{recval}(\mathit{flds} \oplus \{id \mapsto \mathit{update}(fs)(nv)(\mathit{flds}(id))\})))$$

When the value to be updated is a *fieldref* the update function unwinds the recursion as it steps through the field selector sequence, until the appropriate field is reached.

The meaning of a command is given by the function *MCom*.

$$MCom : Com \to (WFStatEnv \to (DynEnv \nrightarrow Tr))$$

$$MCom(assign(asn)) =$$
$$(\lambda\rho_S : WFStatEnv \bullet$$
$$(\lambda\rho_S : \mathrm{dom}\, MVarref(asn.lhs)(\rho_S) \cap \mathrm{dom}\, MExpr(asn.rhs)(\rho_S) \bullet$$
$$(\lambda\sigma : \mathrm{dom}\, MVarref(asn.lhs)(\rho_S)(\rho_D) \cap$$
$$\mathrm{dom}\, MExpr(asn.rhs)(\rho_S)(\rho_D) \bullet$$
$$(\textbf{let}\, lhsloc == MVarref(asn.lhs)(\rho_S)(\rho_D)(\sigma) \bullet$$
$$(\textbf{let}\, rhsval ==$$
$$proj(\rho_S)(TPVarref(asn.lhs)(\rho_S))(MExpr(asn.rhs)(\rho_S)(\rho_D)(\sigma)) \bullet$$
$$(\textbf{let}\, ty == \sigma(lhsloc.dynloc).dyntype;$$
$$val == \sigma(lhsloc.dynloc).dynval \bullet$$
$$\sigma \oplus \{lhsloc.dynloc \mapsto$$
$$(ty, update(val, lhsloc.dynflds)(rhsval))\})))))))$$

The effect of assignment is that the location in the store of the left side variable reference is updated with the projection of the value of the right side expression into the type of the right side variable reference. (The *proj* function may only change the r-value if it is a record or extend type, otherwise $rhsval = MExpr(asn.rhs)(\rho_S)(\rho_D)(\sigma)$.)

5.4 Guarded Assignment

When a pointer to a named type has been assigned a value that is a strict extension of its static type, it is then possible to use it as a variable expression on the right side of an assignment using its dynamic type. The *typeref* function from Section 2.1 together with *exvar* from Section 2.3 assert that the dynamic type of a variable reference expression is an extension of a given named type. An assignment command such as this is called *guarded assignment* and is of the form,

$$assign(vr1, exvar(typeref(vr2, tid)))$$

This corresponds to the Oberon syntax,

```
vr1 := vr2(tid);
```

The intention is that *vr1* is assigned the value of *vr2* under the assertion that the dynamic type of *vr2* is an extension of the type represented by *tid*. The static semantics for *typeref* ensures that the type represented by *tid* is an extension of the static type of *vr2* and the static semantics of assignment guarantees that *tid*)) is an extension of the static type of *vr1*. It is also possible to use guards on the left side of an assignment as in the case of assigning new values to the fields of an extended pointer variable. In our notation this is achieved in the following way.

$$assign(deref(typeref(vr2, tid)), exp)$$

In Oberon, we write,

```
vr2(tid)^ := exp;
```

The intention is that *vr2* has been extended so that its dynamic type is an extension of *tid*. This means that the dereferenced variable can be assigned the *recval* given by *exp*. This situation can also be applied in the case of *fieldref* to access the extended fields.

5.5 Type Assertion

Our expression syntax and semantics allows us to write the following important boolean expression referring to the dynamic type of a variable reference. (It is only applicable to pointers to named types as they are the only variables allowed to dynamically assume extended values in our discussion.)

$$is(vr, tid)$$

This expression asserts that the variable reference *vr* has a dynamic type which is an extension of the type named *tid* (Section 4.3). In Oberon, we write,

```
vr IS tid
```

6 Conclusion

We have given a formal semantics for the novel features of a language which incorporates type extension. The emphasis has been on assignment because this is the most important aspect of such a language. In Section 5.3 we have given a dynamic semantics for assignment. This single definition incorporates record, pointer and guarded assignment. Guarded assignment allows a pointer whose dynamic type is an extension of its static type to be assigned to a *vref* with a static type which is an extension of the pointer's static type. This is sometimes referred as "reverse" assignment.

References

[Hay93] I. J. Hayes(editor). *Specification Case Studies*. Prentice Hall International, second edition, 1993.

[Hay94] I. J. Hayes. A small language definition in Z. Technical Report UQ-SVRC-TR94-50. Software Verification Research Centre. University of Queensland. December, 1994.

[MW91] H. Mössenböck and N. Wirth. The programming language Oberon-2. *Structured Programming*, 12(4). 1991.

[RW92] M. Reiser and N. Wirth. *Programming in Oberon: Steps beyond Pascal and Modula*. Addison-Wesley, 1992.

[Sei91] E. Seidwitz. Object-oriented programming through type extension in Ada9X. *Ada Letters*, 11(2): 86-97. March/April 1991.

[Spi92] J. M. Spivey. *The Z Notation: A Reference Manual*. Prentice Hall International, second edition, 1992.

[Wir88a] N. Wirth. Type extensions. *ACM Transactions on Programming Languages and Systems*, 10(2):204-214. April 1988.

[Wir88b] N. Wirth. The programming language Oberon. *Software Practice and Experience*. July 1988.

From Z to Code:
A Graphical User Interface for a
Radiation Therapy Machine

Jonathan Jacky[*] and Jonathan Unger

Radiation Oncology Department RC-08
University of Washington
Seattle, WA 98195

Abstract. We wrote a formal specification in Z for the graphical user interface of
a radiation therapy machine. We implemented our specification in a Pascal dialect
on a workstation that uses the X window system to manage the keyboard and display. We initially model the user interface as a collection of separate Z operation
schemas based on the informal prose requirements. From these we derive a state
machine model, represented as a state transition table whose entries are schema
names from the Z specification. Our state transition table format compactly represents nested states that are modelled in Z by schema inclusion. We implement
each table entry as a Pascal function or procedure. We also implement a dispatcher
that selects the proper state transition whenever any X event occurs; our dispatcher
is the X event loop. Our dispatcher is a table-driven interpreter that can handle
any state transition system expressed in the format we defined. We model the dispatcher in Z and formally derive some of its code. Experiences implementing and
testing the program are described.

1 Introduction

The Clinical Neutron Therapy System at the University of Washington is a cyclotron
and treatment facility that provides particle beams for cancer treatments with fast neu-
trons, production of medical isotopes, and physics experiments. This paper concerns the
control program for the operator's console that therapists use to set up and deliver treatments to patients. The programmable part of the console is a workstation that runs a
commercial real-time operating system [2] and uses the X window system to manage
the keyboard and display [11, 12]. This console is just one component of a large control system that includes several computers and many non-programmable elements. The
delegation of functions among the software and hardware components is described elsewhere [7, 8].

2 Why use a formal notation?

We already have a thorough description of our system in prose and pictures that the users
consider to be complete. Why go to all the effort of writing a second description of the
same behaviors in a formal notation?

[*] email jon@radonc.washington.edu, telephone (206)-548-4117

Safety issues motivate much of our formalization effort. We want to be able show that our machine meets generic requirements for safety and completeness such as those proposed by Jaffe, et al [10]. Our initial efforts concentrated on describing system behaviors [4, 5, 6].

In the present effort we apply the formal notation to a different purpose: describing the internal structure and workings of our implementation. In this paper we use Z as a detailed design notation and show how we derive program code from Z texts.

What are the advantages of formal notation, compared to carefully written technical prose? We find that the advantages have little to do with precision and freedom from ambiguity. When you are describing the behavior of a device that is designed for some specialized purpose, it is not difficult to write prose that is clear, precise and free of ambiguity: "The radiation beam cannot turn on unless sensors indicate that the therapy room door is closed." The real advantages of formal notations arise from their brevity and their ability to support systematic progression from specification through design to code.

Formal descriptions can be shorter than prose because we can abbreviate recurring patterns. We observe that our prose description of the user interface is repetitious; many operations work almost the same way. We see an opportunity to make the program small by factoring out common features and similar behaviors. This factoring out is only suggested by the prose description; in our formal specification we make it explicit. The Z texts are no mere paraphrase of the prose requirements. They are a different expression of the same behaviors, in a form that is more concise and better organized to serve as a guide for programming.

Formal notations support systematic development. They provide the only way we have to really show that a specification, design and code all describe the same thing. This paper presents an example of such a development using Z.

We formalize only those aspects of the the user interface which are pertinent to this design task. We do not attempt to treat "look and feel" aspects such as the appearance of the display. They are already described in sufficient detail in the informal requirements [8].

3 Informal requirements

The purpose of the treatment console program is to help ensure that patients are treated correctly, as directed by their prescriptions. The program reads a database of prescriptions for many patients. Each patient's prescription usually includes several different beam configurations called fields. Each field is defined by many machine settings that must be set properly to deliver each prescribed treatment. The console program enables the operator to choose patients and fields from the prescription database, and ensures that the radiation beam can only turn on (through a separate nonprogrammable subsystem) when the prescribed settings for the chosen field have been achieved.

The informal requirements for the console program, including but not limited to its user interface, describe the activities associated with about a dozen different screens. For example, one screen shows the fields available for the currently selected patient. Operators select these screens by pressing dedicated function keys, and can also position a cursor over particular items on each screen and select them (for example to choose a field in order to load its prescribed settings). Operators can also enter or modify values for some items after they select them, by typing at the workstation keyboard.

4 Formal model

We wrote a formal description [9] of every operation described in the prose requirements [8], using the Z notation as described in [14]. The following summary is simplified for brevity.

Each operation from the prose is modelled by one or more Z operation schemas on the *Console* state. The *display* state variable indicates which of the screen designs is currently visible. The *edit* state variable indicates what kind of user interaction is currently in progress: it is *idle* when no interaction is in progress and the console is waiting for input, *editing* when the user is entering or modifying a value and the console is waiting for the user to type a character, etc. The *item* state variable indicates the name (not the value) of the item which is being modified, while *buffer* models the (possibly incomplete) string that the user edits. Most operations are only *Available* when editing is not already in progress; otherwise, the console is *Engaged*.

$$EDIT ::= idle \mid editing \mid \ldots$$

```
┌─ Console ──────────────────────────────────
│ display : DISPLAY
│ edit : EDIT
│ item : NAME
│ buffer : STRING
│
│ ...
└────────────────────────────────────────────
```

$$Available \; \hat{=} \; [\; Console \mid edit = idle \;]$$
$$Engaged \; \hat{=} \; [\; Console \mid edit \in \{ \; editing, \ldots \; \} \;]$$

Operations occur whenever the user provides input at the workstation by typing or pressing a function key (we do not use the workstation "mouse"). Every input event is modelled by the *Event* operation schema. Each value of *INPUT* represents a different X event, but we can ignore many of them.

```
┌─ Event ────────────────────────────────────
│ ΔConsole
│ input? : INPUT
└────────────────────────────────────────────
```

$$Ignore \; \hat{=} \; Event \wedge \Xi Console$$

When the console is *Available*, the user may select a new *display* by pressing a function key. We need a function *name* to associate the *input?* (the function key that the

user presses) with the name of the *DISPLAY* that the user wants to see. The newly selected *display'* appears and console remains *Available*. Here we do not need to model
the details of updating the display contents.

```
┌─ SelectDisplay ─────────────────────────────────────────
│ Event
├─────────────────────────────────────────────────────────
│ Available
│ Available'
│ name input? ∈ DISPLAY
│ display' = name input?
└─────────────────────────────────────────────────────────
```

The *SelectDisplay* operation schema corresponds to a single operation in the informal
prose description. Other operations from the informal description must be modelled as
several Z operations. Each editing operation is modelled as at least three Z operations:
one to begin *Edit*ing (when the user presses an appropriate function key), another to
Get each keystroke and modify the buffer (usually just by adding the new character to
the end), and a third to *Accept* the new value when editing is finished (signalled when
the user types a terminator character, such as the RETURN key).

```
┌─ Edit ──────────────────────────────────────────────────
│ Event
├─────────────────────────────────────────────────────────
│ Available
│ Engaged'
│ buffer' = empty
│ display' = display
└─────────────────────────────────────────────────────────
```

```
┌─ Get ───────────────────────────────────────────────────
│ Event
├─────────────────────────────────────────────────────────
│ Engaged
│ Engaged'
│ buffer' = modify buffer input?
│ display' = display
│ item' = item
└─────────────────────────────────────────────────────────
```

```
┌─ Accept ────────────────────────────────────────────────
│ Event
├─────────────────────────────────────────────────────────
│ Engaged
│ Available'
│ input? ∈ terminator
│ buffer' = buffer
│ display' = display
└─────────────────────────────────────────────────────────
```

Edit and *Accept* are building blocks; we specialize them to describe particular editing operations. The simplest example occurs when the operator types a message to annotate an event log (which is mostly written automatically). The user presses the WRITE LOG MESSAGE function key to invoke the *EditMessage* operation, a specialization of *Edit*. This results in the *EditingMessage* state:

```
┌─ EditMessage ──────────────────────────────────
│ Edit
│ ──────────────────────────────────────────────
│ input? = log_message
│ item' = name log_message
└─────────────────────────────────────────────────
```

$$EditingMessage \;\widehat{=}\; [\; Engaged \mid item = name\; log\_message \;]$$

As the user types, *Get* collects characters in *buffer*. When the user types a terminator character, the *WriteMessage* operation, a specialization of *Accept*, writes the completed *message!* to the event log.

```
┌─ WriteMessage ─────────────────────────────────
│ Accept
│ message! : STRING
│ ──────────────────────────────────────────────
│ EditingMessage
│ message! = buffer
└─────────────────────────────────────────────────
```

Most editing operations work in a similar fashion, but the values that they collect change the underlying machine state instead of merely appearing in a file.

5 Combining the operations

The Z texts in Section 4 do not describe any explicit control structure that invokes operations when they are requested. We must design and implement this control structure. We begin by combining all top-level operations into a single *ConsoleOp* schema (not including building blocks such as *Event*, *Edit* and *Accept* which are only used to define other operations).

$$ConsoleOp \;\widehat{=}\; SelectDisplay \vee EditMessage \vee \ldots \vee IgnoreOthers$$

This *ConsoleOp* operation is invoked whenever the user provides any input (keypress). The control structure is implicit in the preconditions of all the constituent operations. We have tried to define each of these so that the precondition of exactly one is satisfied each time *ConsoleOp* is invoked. *IgnoreOthers* is the default do-nothing operation.

6 State transition table

The *ConsoleOp* operation defines a *finite-state machine* where each of the constituent operation schemas describes a single transition. A *Mealy machine* is a state machine where outputs are associated with transitions (not states), so any state machine constructed from Z operation schemas as we have done here is a Mealy machine. The analyses of Jaffe, et al. [10] use the Mealy model. Table 1 shows the state transition table derived from the Z texts in [9][2]. The table represents nested states that are modelled in Z by schema inclusion. We developed code that can interpret any state machine that is expressed in this tabular form, not just the particular state machine defined by the operations in [9].

We build Table 1 by listing all of the operation schemas that are combined in *Console Op*. These determine the contents (but not the order) of the third column. Then we extract the preconditions of each (we determined the preconditions by inspection, not by performing the Z precondition calculation described in [14]). Preconditions on *input*? are listed in the middle column, while those involving other state variables are listed in the first column. In our table, sequence order and indentation represent the nesting of states that is expressed in Z by schema inclusion.

The precondition states in the first column are indicated by the names of the Z state schemas from [9] where they are defined. Solid double lines separate the mutually exclusive states *key* = *locked*, *Available* and *Engaged*. Solid single lines, together with indentation, indicate that the table entries enclosed between them are *substates* of preceding entries. The full precondition of a substate is formed by conjoining the preconditions of the preceding entries at lesser indentation. For example, the full precondition of the substate on the row with *ListingPatients* in the first column is *Available* ∧ *Setup* ∧ *ListingPatients*. This conjunction is the precondition given in [9] for the *SelectPatient* operation, that appears in the third column of the same row. When the first column in a row is blank, the substate is the same as the last preceding nonblank column: the precondition for the *Logout* operation is *Available* ∧ *Setup*.

Dotted lines separate mutually exclusive substates. Thus *run* = *running* and *Setup* indicate mutually exclusive substates of *Available*, and *ListingPatients* and *ListingFields* indicate mutually exclusive substates of *Setup*.

The inputs in the middle column are expressed as predicates on the variable *input*?. The right column lists the operation schemas that define the next state, as well as any outputs. Each line also applies to any included substates, so *name input*? ∈ *DISPLAY* elicits the *SelectDisplay* operation in the *Available* state, and also in its substates *run* = *running*, *Setup*, and in its sub-substates *ListingPatients* and *ListingFields*.

Our table suggests an efficient implementation. It should only be necessary to test each distinct state precondition once. It should not be necessary to test preconditions of substates that are indented under states whose preconditions have been found to be false. Conditions that are expected to occur frequently can be placed near the beginning of the table to minimize the number of tests.

7 Formalizing the state transition table

To develop a program that interprets the state transition table, we must express its meaning formally. Each entry in the table is described by the *Transition* schema. The inden-

---

[2] The actual Z texts are more complicated than the simplified excerpts presented in Section 4. Table 1 has been simplified to make it fit on one page.

State	Input	Operation
key = locked	(all inputs)	*Locked*
Available	*name input?* ∈ *DISPLAY*	*SelectDisplay*
	input? = *log_message*	*TypeMessage*
run = running	*input?* = *cancel_run*	*SelectCancelRun*
Setup	*input?* = *auto_setup*	*AutoSetup*
	input? = *expt_mode*	*ExptMode*
	input? = *store_field*	*EditField*
	input? = *edit_setting*	*SelectSetting*
	input? = *override_cmd*	*SelectOverride*
	input? = *logout*	*Logout*
ListingPatients	*name input?* ∈ ran *patients*	*SelectPatient*
ListingFields	*name input?* ∈ dom *fields*	*SelectField*
	(other inputs)	*Ignore*
Engaged	*input?* ∈ *CHAR*	*Get*
¬ *LoggedOut*	*input?* = *cancel*	*Cancel*
EditingCancel	*input?* = *confirm*	*CancelRun*
TypingMessage	*input?* ∈ *terminator*	*WriteMessage*
EditingField	*input?* ∈ *terminator*	*StoreField*
EditingSetting	*input?* ∈ *terminator*	*StoreSetting*
Overriding	*input?* = *confirm*	*Override*
LoggedOut	*input?* ∈ *terminator*	*Login*
	input? = *cancel*	*NoCancel*
	(other inputs)	*Ignore*

Table 1. User interface state transition table.

tation of each row is indicated by an integer nesting *depth* (the double, solid and dashed lines in the table were included to improve its appearance; they do not require any formal representation). The *state* and *input* preconditions in the first and second columns are indicated by predicates expressed as unary relations on *Console* and *INPUT*, respectively. The *operations* in the third column of the table are modelled by functions on *Console* states. Operations are declared to be functions, not relations, because they are intended to be deterministic: for each state and input, there should be only one successor state. Table 1 is modelled by *t*, a sequence of these records.

```
┌─ Transition ──────────────────────────────────────────────
│ depth : N
│ state : P Console
│ input : P INPUT
│ op : Console ⇸ Console
└────────────────────────────────────────────────────────────
```

$$| \quad t : \text{seq } \textit{Transition}$$

Here are some excerpts from Table 1 in Z syntax. The entry for *t* 6 shows how we model rows in Table 1 where the first column is blank: we set the *state* in the *Transition* record to the empty set, $\emptyset$; the depth in these rows is the same as in the nearest preceding non-empty entry.

$(t\ 1).depth = 0$
$(t\ 1).state = \{\ Console \mid key = locked \bullet \theta Console\ \}$
$(t\ 1).input = \{\ input? : INPUT \mid true\ \}$
$(t\ 1).op = \{\ Locked \bullet (\theta Console, \theta Console')\ \}$

$(t\ 5).depth = 1$
$(t\ 5).state = \{\ Setup \bullet \theta Console\ \}$
$(t\ 5).input = \{\ input? : INPUT \mid input? = auto\_setup\ \}$
$(t\ 5).op = \{\ AutoSetup \bullet (\theta Console, \theta Console')\ \}$

$(t\ 6).depth = 1$
$(t\ 6).state = \emptyset$
$(t\ 6).input = \{\ input? : INPUT \mid input? = expt\_mode\ \}$
$(t\ 6).op = \{\ ExptMode \bullet (\theta Console, \theta Console')\ \}$

8 Interpreting the state transition table

In this section we formally define the core of interpreter: the *transition* function that takes any *Console* state and *input*? into a new *Console* state. The table *t* is a parameter of this function.

We interpret the table by traversing it from top to bottom, searching for an entry which is *enabled*. We say an entry is enabled if its full state precondition is satisfied: this is true when the precondition given in the first column of the entry is satisfied, and all the nearest preceding entries at lesser indentations are enabled as well. If the first column entry is blank, the entry is enabled if the nearest preceding entry at the same depth is enabled.

We define a unary prefix relation e on table entries; $e(i)$ is *true* if table entry i is enabled. The unary prefix relation $edd(i)$ describes the effect of sequence order and indentation; it is *true* if all the nearest preceding entries at lesser indentation are enabled. For $e(i)$ to be *true*, $edd(i)$ must be *true*. Its definition uses the binary prefix relation $ed(i, d)$, which is *true* when the nearest entry indented at depth d that precedes entry i is enabled, or if there is no such preceding entry.

$$
\begin{array}{|l}
e\_, edd\_ : \mathbb{P}(\operatorname{dom} t) \\
ed\_ : \operatorname{dom} t \leftrightarrow \mathrm{N} \\
\hline
\forall i : \operatorname{dom} t \bullet \\
\quad (\forall d : \mathrm{N} \bullet (\textbf{let } ds == \{ j : 1 .. i-1 \mid (t\ j).depth = d \} \bullet \\
\qquad\qquad ed(i, d) \Leftrightarrow e(max\ ds) \vee ds = \emptyset)) \wedge \\[4pt]
\quad (\textbf{let } d == (t\ i).depth \bullet \\
\qquad\quad edd(i) \Leftrightarrow (\forall dj : 0 .. d-1 \bullet ed(i, dj))) \wedge \\[4pt]
\quad (\forall c : Console \bullet \\
\qquad\quad e(i) \Leftrightarrow edd(i) \wedge (((t\ i).state = \emptyset \wedge ed(i, d)) \\
\qquad\qquad\qquad\qquad \vee ((t\ i).state \neq \emptyset \wedge c \in (t\ i).state)))
\end{array}
$$

When we have found an enabled entry, we test whether the input precondition given in the second column is also satisfied. If it is, we say the transition has *triggered*. We then apply the function given in the third column to obtain the new state. If no entries are triggered, the new state is the same as the old state.

$$
\begin{array}{|l}
transition : (Console \times INPUT) \rightarrow Console \\
\hline
\forall input? : INPUT;\ c : Console \bullet \\
\quad transition(c, input?) = \\
\qquad \textbf{if } \exists i : \operatorname{dom} t \bullet e(i) \wedge input? \in (t\ i).input \textbf{ then } (t\ i).op\ c \textbf{ else } c
\end{array}
$$

A single application of the *transition* function is the processing of a single input event. It is expressed by the *ConsoleOp1* operation:

$$ConsoleOp1 \mathrel{\widehat{=}} [\ Event \mid \theta Console' = transition(\theta Console, input?)\ ]$$

If the table t is constructed properly, this operation is the same as the *ConsoleOp* we defined in section 5. In other words, *ConsoleOp1* is a *refinement* of *ConsoleOp*.

The refinement is correct because each entry in t corresponds to one disjunct in *ConsoleOp*. The only subtle point involves the preconditions of the substates (the indented table entries). The precondition of a substate is the conjunction of the condition listed in the substate's own table entry, and all the conditions listed in the nearest preceding entries at lesser indentation. We have expressed this formally by including edd in the definition of e.

9 An efficiency measure

In section 8 the formal definition of the *enabled* condition *e* includes quantification over sets. Implementing these literally might result in a complicated, inefficient program. Can we eliminate them?

The quantified predicates describe the sequence of table entries preceding the entry of interest. As we traverse the table from top to bottom, it should not be necessary to test these predicates again and again. Rather, it should be possible to keep track of them incrementally, doing just a little computation at each new table entry.

When traversing the table, we may encounter sequences of disabled entries, where indented entries are disabled because preceding entries at lesser depth are also disabled. At a given table entry, the depths of all the preceding disabled entries form the set *dds*, and *dd* is the smallest element in the set. If the depth of the current table entry *d* is not greater than *dd*, or if there are no preceding disabled entries (the set *dds* is empty), then the current entry can be enabled.

We can express this argument formally:

$$\forall i : \text{dom}\, t \bullet (\textbf{let } d == (t\ i).depth; \qquad\qquad\qquad\qquad [\text{Define } d]$$

$$dp == (t\ (i-1)).depth \bullet \qquad\qquad\qquad [\text{Define } dp]$$

$$e(i) \Rightarrow edd(i) \qquad\qquad\qquad\qquad [\text{Definition of } e(i)]$$

$$\Leftrightarrow (\forall dj : 0..\, d-1 \bullet ed(i, dj)) \qquad\qquad [\text{Def'n of } edd(i)]$$

$$\Leftrightarrow \neg\, (\exists dj : 0..\, d-1 \bullet \neg\, ed(i, dj)) \qquad\qquad [\text{Predicate calc.}]$$

$$\Leftrightarrow (\textbf{let } dds == \{\ dj : 0..\, dp \mid \neg\, ed(i, dj)\ \} \bullet \qquad [\text{Define } dds]$$

$$(0..\, d-1) \cap dds = \emptyset \qquad\qquad [\text{Indices in def'ns of } ed,\, dds]$$

$$\Leftrightarrow dds = \emptyset \vee d-1 < min\ dds \qquad\qquad [\text{Sets, arithmetic}]$$

$$\Leftrightarrow dds = \emptyset \vee d \leq min\ dds \qquad\qquad\qquad [< \text{ vs. } \leq]$$

$$\Leftrightarrow (\textbf{let } dd == min\ dds \bullet dds = \emptyset \vee d \leq dd))) \qquad [\text{Define } dd]$$

This result might help us write a more efficient program because it provides an alternate way to test the predicate $edd(i)$. We only need to test the final predicate in the preceding argument:

$$dds = \emptyset \vee d \leq dd$$

The $dds = \emptyset$ condition holds when we are not traversing a sequence of disabled entries, and the $d \leq dd$ condition occurs when we have just emerged from the end of such a sequence.

10 Implementing the operations

Our implementation language is a commercially available Pascal dialect [3] adapted to multitasking embedded control applications.

We write Pascal code to implement each Z operation schema (such as *SelectDisplay* or *EditMessage*) that appears as a disjunct in *ConsoleOp* (section 5). Working directly from our state transition table, we implement each operation schema as three separate Pascal program units: a boolean function that tests the preconditions on the *Console* state variables, a second boolean function that tests the precondition involving *input?*, and a procedure that implements the change of state. These latter procedures consist largely of assignment statements; they do not check any preconditions, but rely on the boolean functions to do that.

11   Implementing the interpreter

The interpreter reads the users' input from the console and invokes the functions and procedures described in the preceding section, as directed by the contents of the state transition table.

We implement the state transition table as an array of records. This Pascal code is our implementation of the *Transition* schema and table *t* defined in section 7:

```
type
     Transition = record
          depth: integer;
          state: ^function;
          input: ^function;
          op: ^procedure;
     end;

var
     t: array[1..n] of Transition;
```

The nonstandard language extensions in our Pascal dialect include pointers to functions and procedures. The state, input and op fields in the Transition record are pointers to the Pascal functions and procedures that implement the Z operation schemas[3].

Our implementation of the *transition* function is presented here as a Pascal procedure without parameters; the table *t*, the *input?* and the *Console* state variables are all global. Its local variables derive from the formal development.

```
procedure transition;

var
     i: integer;       { i           index of current table entry        }
     e: boolean;       { e(i)        current entry is enabled            }
     empty: boolean;   { dds = 0     no preceding disabled entries       }
     d: integer;       { d           (t i).depth, depth current tbl entry }
     dd: integer;      { dd          d least indentd prev. disabld entry }
     tr: boolean;      { e(i) ∧      current entry is triggered          }
                       { input? ∈ (t i).input                            }
     ed: array[0..maxdepth] of boolean;
                       { ed(i, d)    e for nearest entry at depth d       }
```

---

[3] The syntax for pointing to functions and procedures is more complicated than indicated here.

The `transition` code tests each table entry in turn. Our Pascal dialect [3] provides a (nonstandard) `invoke` function that executes a function or procedure indicated by a pointer.[4] This makes it easy to translate the formal definitions of e and *transition* from section 8. We implement $e(i) \Leftrightarrow edd(i) \wedge p$ as:

```
if edd(i) then e := p else e := false;
```

instead of `e := ` $edd(i)$ `and` p`;` in order to avoid the expense of testing p when $edd(i)$ is *false*. Instead of testing $edd(i)$ directly, we use the equivalent result derived in section 9 and test $dds = 0 \vee d \leq dd$ instead.

```
procedure transition;
  ...
  while i <= n ...
    ...
  { Invariant: dds = 0 ∨ dd = min dds }
      if empty or (d <= dd) then
          if (t[i].state = nil)
              then e := ed[d]
              else e := invoke(t[i].state)
          else e := false;

      if e then tr := invoke(t[i].input) else tr:=false;
      if tr then invoke(t[i].op);

  { Maintain invariant }
    ...
```

This code is correct if the program variables satisfy their formal definitions when the assignment to e is made. This condition is the *loop invariant*: it must be true each time execution reaches the first `if` statement inside the `while` loop. We first establish the invariant by assigning values to program variables before entering the loop. Then, each time e is assigned in the body of the loop, new values must also be assigned to other variables in order to maintain the invariant.

The only tricky part of the invariant involves the variables we need for our alternate formulation of $edd(i)$: $dds = 0 \vee dd = min\ dds$. The formal derivation of the code to maintain this invariant appears in Appendix A. We only need to write code for the three cases where the new values of dd or `empty` differ from their previous values: when a disabled entry is encountered after a sequence of enabled entries, when an enabled entry is encountered after a sequence of disabled entries, and when a disabled entry at lesser depth is encountered after a sequence of disabled entries at greater depth. Writing out the code for each case, we obtain

---

[4] The syntax of `invoke` is more complicated than shown here.

```
{ Maintain invariant }
if (not e) and empty then begin empty := false;
                             dd := d end;
if (not e) and (not empty) and (d <= dd) then dd := d;
if e and (not empty) and (d <= dd) then empty := true;
```

We only need to add a little loop machinery to complete the development. The completed `transition` procedure appears in Fig. 1. The complete dispatcher is simply a loop that, on each turn, removes one event from the head of the X event queue and calls `transition`.

12 Discussion

We believe that the formal development benefited the organization and correctness of our program. We also report the size of our program.

12.1 Program size

The informal requirements for the console program, including but not limited to its user interface, comprise 45 pages of prose and diagrams (chapter 8 in [8]).

A 17 page section in [9] provides a formal definition for the user interface to every operation described in the prose requirements [8]. This section includes about 450 lines of Z text (expressed as LaTeX source) divided between about 50 schemas and some axiomatic definitions.

In our implementation the executable part of the dispatcher, including the `transition` function and the rest of the X event handling machinery, comprise about 300 lines of Pascal (including plenty of comments and blank lines). The code for table t contributes about 300 lines. The functions and procedures that implement the operation schemas comprise another 1100 lines. By far the largest portion of the user interface code — about 4400 lines — determines the appearance of the displays. It is based on the illustrations in [8] and is not described in any formal specification.

12.2 Program organization

We feel we were successful in partitioning our design and factoring out recurring features.

A typical programming style implements an X application program as a large `case` statement with a branch to handle each X event [11][5]. In contrast, we separate the code that does the work of the application apart from the basic X event handling and dispatching.

An advantage of our style is that it splits the code into an application specific part and a generic part. The state transition table and the functions and procedures that implement

---

[5] Bowen [1] modelled some display aspects of the X window system in Z, but did not consider event handling.

```
procedure transition;

var
    i: integer;        { i            index of current table entry        }
    e: boolean;        { e(i)         current entry is enabled            }
    empty: boolean;    { dds = 0      no preceding disabled entries       }
    d: integer;        { d            (t i).depth, depth current tbl entry }
    dd: integer;       { dd           d least indentd prev. disabld entry }
    dleq: boolean;     { d ≤ dd       begin new seq. of disabled entries  }
    tr: boolean;       { e(i) ∧       current entry is triggered          }
                       { input? ∈ (t i).input                             }
    ed: array[0..maxdepth] of boolean;
                       { ed(i,d)      e for nearest entry at depth d       }

begin
    i := 1; tr := false; empty := true;
    for d := 0 to maxdepth do ed[d] := true;
    while (i <= n) and (not tr) do
    begin
        d := t[i].depth;
        dleq := d <= dd;

        { Invariant: dds = 0 ∨ dd = min dds }
        if empty or dleq then
            if (t[i].state = nil)
                then e := ed[d]
                else e := invoke(t[i].state)
            else e := false;

        if e then tr:=invoke(t[i].input) else tr:=false;
        if tr then invoke(t[i].op);

        { Maintain invariant }
        if (not e) and empty then begin empty := false;
                                        dd := d end;
        if (not e) and (not empty) and dleq then dd := d;
        if e and (not empty) and dleq then empty := true;

        ed[d] := e;
        i := i + 1;
    end;
end;
```

Fig. 1. Completed transition procedure.

the operation schemas are application-specific part; the `transition` procedure and the X event loop are generic.

We anticipate providing control software for other parts of our system — for example, the console that the accelerator operator uses to tune the beam (chapter 4 in [8]) — by replacing only the application-specific portion of the code.

12.3 Correctness

We examined our Z texts with a type checker [13] and corrected the errors it discovered.

We developed different portions of the program with differing degrees of formal rigor. The functions and procedures that implement the Z operation schemas — *Edit-Message* and the rest — were translated from Z to Pascal by inspection, with no further formal development. This code is simple: the functions just test for equality or membership, and the operations are mostly sequences of assignments. We discovered several omissions and other errors in the Z texts at this stage; coding constitutes an intensive inspection of a specification.

We developed the *transition* function with more formality because we recognized that it was the more difficult part. It is at the core of the application — it is executed each time the operator touches a key — and we hope to reuse it elsewhere.

We believe it would have been more difficult to achieve a working `transition` procedure without the formal development. In fact we began writing this code intuitively, but we could not convince ourselves that our initial attempts were correct. Instead of resorting to trial and error debugging, we resolved to defer testing until we had completed a convincing formal derivation. This effort quickly exposed an outright error in our intuitively written code, and also revealed that it was needlessly complicated.

When we began testing our formally derived `transition` procedure, it failed almost immediately. We discovered that we had overlooked a case because we made an unfounded assumption about the ordering of items in the state transition table. We unwittingly made this assumption when we first sketched a sample state transition table. Then we dutifully wrote the assumption into our informal prose description, and carried it through the whole formal development. So the error did not arise in one of the formal development steps, but resulted from basing our formal analysis on a poorly chosen example. It was easy to repair the error and correct each step in the development.

After this single correction we have not found any more errors in the `transition` code. We have exercised it with test cases that visit every table entry and probe every condition described in Appendix A. In each case we have confirmed that we can trigger transitions that should be enabled, and we cannot trigger transitions that should be disabled.

Moreover, the formal derivation increased our understanding. We realized that the variable dd which had seemed so necessary was really just a way to test the predicate $edd(i)$. In fact our table is so small, with so few depths, that using dd may confer no benefits. It might be better to test $edd(i)$ explicitly on each turn of the loop after all. This would obviate the code that maintains dd and its lengthy formal justification in Appendix A. Nevertheless, we report the full development here as a testament to the tension between intuition and formality.

References

1. Jonathan P. Bowen. X: Why Z? *Computer Graphics Forum*, 11(4):221 – 234, October 1992.
2. Digital Equipment Corporation, Maynard, Massachusetts. *Introduction to VAXELN*, October 1991.
3. Digital Equipment Corporation, Maynard, Massachusetts. *VAXELN: Pascal Programming Guide*, December 1991.
4. Jonathan Jacky. Formal specifications for a clinical cyclotron control system. In Mark Moriconi, editor, *Proceedings of the ACM SIGSOFT International Workshop on Formal Methods in Software Development*, pages 45 – 54, Napa, California, USA, May 9 – 11 1990. (also in *ACM Software Engineering Notes*, 15(4), Sept. 1990).
5. Jonathan Jacky. Formal specification and development of control system input/output. In J. P. Bowen and J. E. Nicholls, editors, *Z User Workshop, London 1992*, pages 95 – 108. Proceedings of the Seventh Annual Z User Meeting, Springer-Verlag, Workshops in Computing Series, 1993.
6. Jonathan Jacky. Specifying a safety-critical control system in Z. *IEEE Transactions on Software Engineering*, 21(2):99–106, 1995. (also in [15], pages 388–402).
7. Jonathan Jacky, Ruedi Risler, Ira Kalet, and Peter Wootton. Clinical neutron therapy system, control system specification, part I: System overview and hardware organization. Technical Report 90-12-01, Radiation Oncology Department, University of Washington, Seattle, WA, December 1990.
8. Jonathan Jacky, Ruedi Risler, Ira Kalet, Peter Wootton, and Stan Brossard. Clinical neutron therapy system, control system specification, part II: User operations. Technical Report 92-05-01, Radiation Oncology Department, University of Washington, Seattle, WA, May 1992.
9. Jonathan Jacky and Jonathan Unger. Formal specification of control software for a radiation therapy machine. Technical Report 94-07-01, Radiation Oncology Department, University of Washington, Seattle, WA, July 1994. (Draft).
10. Matthew S. Jaffe, Nancy G. Leveson, Mats P. E. Heimdahl, and Bonnie E. Melhart. Software requirements analysis for real-time process control systems. *IEEE Transactions on Software Engineering*, 17(3):241 – 258, March 1991.
11. Adrian Nye. *Xlib Programming Manual*. O'Reilly and Associates, Inc., Sebastopol, CA, 1988.
12. R. W. Scheifler and J. Gettys. The X window system. *ACM Transactions on Graphics*, 5(2):79–109, 1986.
13. J. M. Spivey. *The fUZZ Manual*. J. M. Spivey Computing Science Consultancy, Oxford, second edition, July 1992.
14. J. M. Spivey. *The Z Notation: A Reference Manual*. Prentice Hall, Hemel Hempstead, second edition, 1992.
15. J. C. P. Woodcock and P. G. Larsen, editors. *FME '93: Industrial-Strength Formal Methods*, Odense, Denmark, April 1993. First International Symposium of Formal Methods Europe, Springer-Verlag. Lecture Notes in Computer Science 670.

A   Maintaining the invariant

To maintain the invariant $dds = 0 \vee dd = min\ dds$, three conditions are pertinent: $dds = 0$ and $d \leq dd$ of course, and also $e(i)$. Three binary conditions suggest eight possible cases, but impossible combinations and "don't care" conditions limit the actual number to five:

1. An enabled entry is encountered, not following a sequence of disabled entries: $e(i) \wedge dds = 0$

2. An enabled entry is encountered after a sequence of disabled entries:
   $e(i) \wedge dds \neq 0$

3. A disabled entry is encountered after a sequence of enabled entries:
   $\neg\, e(i) \wedge dds = 0$

4. A disabled entry at greater depth is encountered after a sequence of disabled entries:
   $\neg\, e(i) \wedge dds \neq 0 \wedge d > dd$

5. A disabled entry at lesser depth is encountered after a sequence of disabled entries at greater depth: $\neg\, e(i) \wedge dds \neq 0 \wedge d \leq dd$

The correct assignments to dd and $dds = 0$ can be determined by the case analysis shown in Table 2. The unprimed variables dd and dds on the left side of the table represent the values at the top of the loop when e is assigned, and the primed variables dd' and dds' on the right represent the new values assigned at the bottom of the loop, that will apply when e is assigned in the next turn of the loop. The boxed entries on the right side are the only cases where the old and new values differ; only these cases require assignment statements in the implementation. The numerals in parentheses refer to numbered paragraphs in the following discussion.

$e(i)$	$dds = 0$	$d \leq dd$	$dds' = 0$	dd'
false	false	false	false(2)	dd (3)
true	false	false	XX (4)	XX (4)
false	true	X (1)	$\boxed{false}$ (2,5)	$\boxed{d}$ (5)
true	true	X (1)	true(6)	X (1)
false	false	true	false(2)	$\boxed{d}$ (7)
true	false	true	$\boxed{true}$ (8)	X (1)

Table 2. Maintaining the invariant $dds = 0 \vee dd = min\ dds$

Before working through the table, we establish some generally useful results. It is convenient to package up pertinent state variables, definitions and the loop invariant in the *Loop* schema. The *Step* operation schema models the effect of executing the code at the end of the loop in Fig 1 (after the comment, "Maintain invariant"). Here t is the transition table; i is the index of the current table entry, and d, dd and dds are indentation depths of table entries.

```
┌─ Loop ─────────────────────────────────────────────
│ i : dom t
│ d, dd : N
│ dds : P N
├────────────────────────────────────────────────────
│ d = (t i).depth
│ dds = { dj : 0 .. (t (i − 1)).depth | ¬ ed(i, dj) }
│ dds = ∅ ∨ dd = min dds
└────────────────────────────────────────────────────
```

$$Step \cong [\, \Delta Loop \mid i' = i + 1 \,]$$

We consider all the formal text that follows to be within the scope of the definitions in *Step*. The primed "after" variables are defined by the usual Z convention. Now we can derive some results.

A. The range of dds' is $0 .. d$.

$$
\begin{aligned}
max \; dds' &\leq (t \, (i' − 1)).depth && \text{[Def'n of } dds \text{ range, ' naming convention]} \\
&= (t \, ((i + 1) − 1)).depth && [i' = i + 1] \\
&= (t \, i).depth && \text{[Arithmetic]} \\
&= d && [d = (t \, i).depth]
\end{aligned}
$$

B. The enabled state of the current entry becomes the enabled state of the most recent entry at its own depth: $e(i) \Leftrightarrow ed(i', d)$

$$
\begin{aligned}
(\text{let } ds' &== \{ \, j : 1 .. i' − 1 \mid (t \, j).depth = d \, \} \bullet && \text{[Def'n } ed(i, d), \text{ sect 8]} \\
e(i) &\Leftrightarrow e(max \; ds') && [i' − 1 = i, \, i = max \; 1 .. i, \text{ so } i = max \; ds'] \\
&\Leftrightarrow ed(i', d)) && \text{[Def'n of } ed(i, d), \text{ sect 8]}
\end{aligned}
$$

C. When an entry is not enabled, its depth becomes one of the preceding disabled depths: $\neg \, e(i) \Leftrightarrow d \in dds'$.

$$
\begin{aligned}
\neg \, e(i) &\Leftrightarrow \neg \, ed(i', d) && \text{[Result B]} \\
&\Leftrightarrow d \in dds' && \text{[Result A, definitions of } d, \, dds']
\end{aligned}
$$

D. When an entry is enabled, for the next entry the set of preceding disabled entries is empty: $e(i) \Rightarrow dds' = \emptyset$.

$$
\begin{aligned}
e(i) &\Rightarrow edd(i) && \text{[Definition of } e(i)] \\
&\Leftrightarrow (\forall \, dj : 0 .. d − 1 \bullet ed(i, dj)) && \text{[Definition of } edd(i)] \\
&\Leftrightarrow (0 .. d − 1) \cap dds' = \emptyset && \text{[Indices in defn's of } edd(i) \text{ and } dds'] \\
&\Leftrightarrow dds' = \emptyset && [\text{A: } dds' \subseteq 0 .. d, \text{ but C: } d \notin dds']
\end{aligned}
$$

Now we can work through the entries in Table 2.

1. When $dds = 0$ the value of dd is undefined and does not contribute to the result. X indicates this "don't care" condition. We only need six rows in the table, not eight.

2. When an entry is disabled, its own depth must be one of the elements of dds', which cannot be empty.

$$\neg\, e(i) \Rightarrow dds' \neq 0 \qquad\qquad\qquad [\text{C: } d \in dds']$$

3. When an entry is disabled but its own depth is greater than dd, then dd remains the smallest element in dds'.

$$
\begin{aligned}
\neg\, e(i) &\wedge dds \neq 0 \wedge d > dd & [\text{Entry in Table 2}]\\
&\Rightarrow dds' = dds \cup \{d\} & [\text{C: } d \in dds', \text{ definition of } dds']\\
&\Rightarrow dd = min\ dds' & [dd < d]\\
&\Leftrightarrow dd' = dd & [\text{Definition of } dd']
\end{aligned}
$$

4. Here the entry is enabled but its depth is greater than dd. This condition is impossible because it contradicts the result about $e(i)$ we derived in section 9.

$$
\begin{aligned}
e(i) &\wedge dds \neq 0 \wedge d > dd & [\text{Entry in Table 2}]\\
&\Leftrightarrow false & [\text{Section 9: } e(i) \Rightarrow dds = 0 \vee d \leq dd]
\end{aligned}
$$

Inspection of the code confirms that this condition is prevented by the `if ...then ...else ...` statement in the body of the loop. XX indicates this impossible condition.

5. A disabled entry is first encountered when dds is empty. The depth of this entry becomes the first — and necessarily the smallest — element in dds'

$$
\begin{aligned}
\neg\, e(i) &\wedge dds = 0 & [\text{Entry in Table 2}]\\
&\Leftrightarrow dds' = \{d\} & [\text{C: } d \in dds']\\
&\Rightarrow dds' \neq 0 \wedge d = min\ dds' & [\text{Set theory, definition of } min]\\
&\Leftrightarrow dd' = d & [\text{Definition of } dd']
\end{aligned}
$$

6. If the entry is enabled and dds is empty, it remains so; dds' is empty, by result D.

7. This disabled entry lies at lesser indentation depth than any nearest preceding disabled entry, so dd must be reassigned to maintain the invariant.

$$
\begin{aligned}
\neg\, e(i) &\wedge d \leq dd & [\text{Entry in Table 2}]\\
&\Rightarrow d = min\ dds' & [\text{C: } d \in dds', d \leq min\ dds]\\
&\Leftrightarrow dd' = d & [\text{Definition of } dd']
\end{aligned}
$$

8. An enabled entry follows a disabled entry that was indented to an equal or greater depth. This signals the end of a sequence of nested disabled entries. By result D, dds' should be emptied.

This concludes the case analysis.

The French population census for 1990

Pascal Bernard[†] and Guy Laffitte[‡]

[†] Institut de Recherche Informatique de Nantes & IUT de Nantes, 3 rue du Ml Joffre, F-44041 NANTES

[‡] INSEE, 105 rue des Français Libres, F-44200 NANTES

Abstract. Specification design techniques and methods are illustrated through a real example: the development of a system supporting the French population census for 1990. These techniques are sufficiently generic to find a wide use even if some of them do not seem at first sight to be obvious. We demonstrate the benefits of using B for the specification of the core of the system which supported the census. This is illustrated on a small part of the system: the administrative geography.

1 Introduction

The census consists of taking a snapshot of France at a certain date. The information is gathered through smaller censuses covering the French territory. The information concerns the people living in France. It provides the primary data which are eventually treated to obtain statistics. For example, it permits the comparison of proportions of bachelors in urban and rural areas.

A computer system had to be designed in order to support the collection of data and to organize the data for statistical analysis. Every application program which carries out specific statistics has to access data from the census. This organization was a critical step in the whole census. It involved the collection procedure as well as every eventual application program providing specific statistical analysis. It was decided that both the collection and the access to the data were to be based on the structure of the French administration (such as counties, regions,etc): *the administrative geography* (hereafter called geography). This structure is large and complex. J.R. Abrial was a consultant at the INSEE[1]. He suggested adopting the B notation to describe this structure and to extract its properties. Moreover, it was essential that collectors and programmers associate the same meanings with the same words. It is this stage which is the subject of our article. The B method has been followed for the implementation. We will also give an overview of the implementation.

One problem was to specify the organization for programmers in a clear and unambiguous way. Some properties must hold for the structure to make sense while others are simply expected to hold. For instance, it was important to know if there was a hierarchy covering the whole country. Another problem, therefore, was checking such properties. We had to specify verification procedures in order to check that these properties actually hold. Last but not least, the organization had to be completed for data access performance.

---

[1] Institut National de la Statistique et des Etudes Economiques

Our solution was to define two sets of procedures. The first set was given to the programmers in order to retrieve information about the geography. We carried out the specification of these procedures in order to explain to the programmers what they were for and to know what had to be implemented. The second set of procedures enabled the construction of a representation of the geography from the data which were provided and was used to check the properties. We carried out the specification of these procedures in order to make a reliable implementation. Reliability is very important since the results of statistical analyses entail political decisions. If the system was not correct, all the other applications developed by the programmers would give meaningless results. In short, a failure of our design would have cost a lot of money.

This paper is about the method followed and choices made for the design. It gives an example of an actual development using B [1, 2] as the specification language. We have included a short section describing the implementation of the procedures. In conclusion we evaluate the benefits obtained from our design. The computer system which was designed from these specifications has been in use for a few years now. We will also draw conclusions about the contribution of specifications to the solution of the original problems. This work has already been quoted in [4], and Henri Habrias encouraged us to present it in detail.

Section 2 presents the French geography informally.

Section 3 defines the set constructions we used in order to give a set theoretic model of the geography. Notations are taken from B as we will use B for the formal specification.

Section 4 presents the specification of a part of the geography. This is the specification of an abstract machine which provides a set theoretic modelling of the geography, states the properties of the geography and the semantics of data retrieval procedures in terms of this modelling. This specification, together with the comments associated with it, serves as a very precise manual of the procedures provided to the programmers for retrieving information about the geography.

In section 5 we give the specification of an abstract machine which takes as its input the existing description of the geography and builds up another representation. This representation follows the previous modelling. The properties are checked in tandem with the construction process.

Section 6 describes how the specification was used to build actual implementations.

In conclusion, we give our answer to the question : "Does this experiment qualify the B method for industrial use?".

2   The case study

We give a simplified version of the geography. However, it is not clear whether the informal description will help the reader to understand the formalization, or if the converse is true. We do not believe that the only reason for this is that our English is very coarse. The structure is complex enough to make it difficult to describe it in an informal way. But the informal description relates the dummy identifiers of the formal specification to real life entities. Figure 1 is intended to help in reading this section. Section 4 gives the formalization.

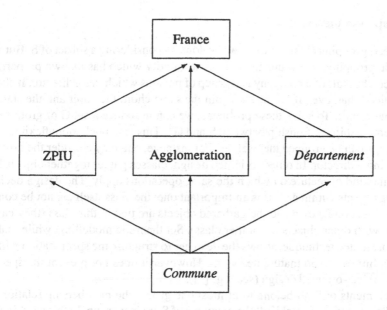

Fig. 1. Structure of French geography

For this presentation, we will consider that the small censuses cover what is called a *commune* in French. A *commune* is a part of the territory administred by the council of a village or a town. Wherever you are in France, you are on some territory administered by one and only one *commune* council. In other words, the set of the *communes* forms a partition of France. This is the finest partition. *Communes* are grouped into *départements* which are like British counties. The set of *départements* also forms a partition of the territory. There is, in addition, the set of French regions which is discarded for the sake of clarity. Finally, there is the country which covers the whole territory.

Communes may also be grouped into agglomerations. Agglomerations may spread over several *départements*, but they generally do not contain all the *communes* of a *département*. Statistics sometimes refer to larger groupings called ZPIU[2]. When belonging to a ZPIU, all the *communes* of an agglomeration must be in the same ZPIU.

There are two kinds of *départements*: metropolitan and overseas *départements*. All the *communes* which are part of an agglomeration are also part of a ZPIU if and only if they belong to a metropolitan *département*. Moreover, no *commune* of an overseas *département* belongs to a ZPIU.

3 Structuring the data: some useful techniques

The set constructions which were used to formalize the geography are explained here. They are introduced in a section of their own because the same techniques may be used in many other cases. The remaining sections illustrate their use.

---

[2] for Zone de Peuplement Industriel et Urbain (economic interest)

3.1 Groups and hierarchies

Let S be a set; grouping elements of S can be done by considering a subset of S. But very often, while grouping, one wants to create a new entity which has its own properties[3]. For instance, the staff of a company is a group of persons which has a director. It should also be noticed that, even if all the staff join the same choir, the choir and the staff are not the same things. To solve these problems, we can introduce a set G of groups over S and a membership relationship between S and G[4]. This is a much more flexible technique. Groups can be grouped themselves. For instance, one may consider the group of groups having a director. In object-oriented design, classes gather together objects having the same kind of feature on which the same operations apply. The design decision of grouping objects within a class is an important one: the class itself cannot be considered as an object on its own and the gathered objects are tied in this class (they cannot be grouped with other objects in another class). Set theoretic modelling, while making use of the presented technique, allows the designer to structure the specification without committing himself to premature decisions. However it does not prevent the specifier from using object-oriented design (see e.g. [5]).

If the elements of S all belong to at most one group, the membership relation is a function. This function is total if all the elements of S are in a group. In this case, it gives rise to a partition of S. Two elements of S are in the same part if they belong to the same group.

We call trees whose leaves are at the same level **hierarchies**. Each level is a set of nodes and each non-terminal node (a node which is not a leaf) groups nodes (the sons) of the level following its level. The membership relation of these groupings is a function (is_child_of) which is total from the set of nodes of one level to the set of nodes of the preceding level. Except for the root of the tree, a childhood relationship gives rise to a partition of the set of nodes of one level. The previous technique applies for modelling hierarchies. This will be illustrated with the country–*département*–*commune* hierarchy.

3.2 Direct product of functions

Given two functions f and g, it is possible to construct a new set by taking the range of their direct product ([2, I.2.4.2 and I.2.6]). Given three sets $S, \mathcal{U}$ and $\mathcal{V}$, two binary relations f and g such that "$f \in S \leftrightarrow \mathcal{U} \land g \in S \leftrightarrow \mathcal{V}$", the direct product of f and g, denoted $f \otimes g$, is defined by:

$$\{(x,(y,z)) \mid (x,(y,z)) \in S \times (\mathcal{U} \times \mathcal{V}) \land (x \mapsto y) \in f \land (x \mapsto z) \in g\}$$

If f and g are functions, then $f \otimes g$ is also a function. Here we are only concerned with the direct product of functions. From this definition, we can deduce that:

$$(u \mapsto v) \in \mathbf{ran}(f \otimes g) \Leftrightarrow \exists s.(s \in S \land u = f(s) \land v = g(s))$$

---

[3] Please note that what is called a "group" here has nothing to do with the well-known algebraic structure

[4] in fact, this is how "users" are organized under UNIX systems for access rights over the files

This equivalence is quite interesting for two reasons. The first one is a matter of style. It provides a means of avoiding quantified formulae which are difficult to understand. The second reason is related to the technique given for grouping. Let $\mathcal{U}$ and $\mathcal{V}$ be sets of groups of elements of S. If f is the membership relation between elements of S and groups of $\mathcal{U}$ and g is the membership relation between elements of S and groups of $\mathcal{V}$, then $u \mapsto v$ is in $\mathbf{ran}(f \otimes g)$ if and only if u and v have a common element of S. Moreover, we have:

$$\mathbf{dom}(f \otimes g \rhd \{u \mapsto v\}) = \{s \mid s \in S \wedge u = f(s) \wedge v = g(s)\}$$

This gives a construction of the intersection of groups for our modelling technique. It can be noticed that $\mathbf{ran}(f \otimes g)$ is a relation between $\mathcal{U}$ and $\mathcal{V}$. This relation is satisfied for pairs of groups which have a non-empty intersection. The set of groups of $\mathcal{U}$ that have a non-empty intersection with a group $g2$ of $\mathcal{V}$ is constructed by:

$$\mathbf{dom}((\mathbf{ran}(f \otimes g)) \rhd \{g2\}).$$

Similarily, the set of groups of $\mathcal{V}$ that have a non-empty intersection with a group $g1$ of $\mathcal{U}$ is constructed by:

$$\mathbf{ran}(\mathbf{ran}(\{g1\} \lhd (f \otimes g))).$$

The direct product will be used for stating the invariant of our system.

4   Formalization of the geography

In the previous description, we did not state the actual geography, but only what it looks like. Categories were introduced for the different entities (*communes*, *départements*,...), and properties governing these entities were stated (e.g., every *commune* belongs to a *département*). The formalization will give a set theoretic modelling of administrative geography together with procedures for retrieving information about an actual geography. As we are using B for this specification, it means that we define an abstract machine whose invariant states what the geography looks like and whose operations consist of those necessary to retrieve information about the actual geography.

The informal specification will be extended during the formalization process. In this way Section 2 serves as a good introduction while details are introduced progressively.

4.1   The context

The informal presentation of the geography is a rather refined presentation compared to what was originally obtained. Nevertheless, modelling the geography requires further analysis. Let us take the example of the *départements*. They are introduced as groupings of *communes*. But they are also entities in their own right which may be named and have properties such as being metropolitan or overseas. They cannot just be sets of *communes*. Figure 2 gives the context part of the abstract machine: it introduces the deferred sets (called basic types in [6]).

```
SETS
    COM⁰₉
    DEP⁰₉
    AGG⁰₉
    ZPIU
```

Fig. 2. The context (first machine)

We use our grouping technique of section 3 by introducing sets of groups for *départements*, agglomerations and ZPIU. They are called DEP, AGG and ZPIU respectively. The set of *communes* is called COM.

As the root of the country–*département*–*commune* hierarchy is a singleton, it is not worth representing the grouping of all the *départements* into the one and only country. One layer of the hierarchy therefore disappears.

4.2 The variables

We have the sets of elements of the actual geography and the functions for grouping. The B introduction of variables is presented in figure 3. *Com* is the set of actual *communes*. *Dep* is a set of groups, the set of actual *départements*. Similarly, *Agg* is the set of actual agglomerations and *Zpiu* is the set of actual ZPIUs. *Com2Dep*, *Com2Agg* and *Com2Zpiu* are the actual membership relations between the *communes* and the different sets of groups. *OverSeas* is the subset of *Dep* which contains overseas *départements*.

The variables *KeyCom*, *KeyDep*, *KeyAgg* and *KeyZpiu* are functions relating the entities of the geography to keys which will be used to identify them. To simplify we will replace keys by natural numbers.

```
VARIABLES
    Com, Dep, Agg, Zpiu,
    Com2Dep, Com2Agg, Com2Zpiu,
    OverSeas,
    KeyCom, KeyDep, KeyAgg, KeyZpiu
```

Fig. 3. The variables

4.3   The invariant

We give some definitions which will simplify the expression of the invariant. Then we state the invariant for an actual geography.

Definitions   Definitions are like macros. They are composed of two parts separated by the symbol $\widehat{=}$. The part on the left-hand side is an identifier and the part on the right-hand side is an expression. In the invariant, the identifiers which are bound to a definition are to be taken as abbreviations for the corresponding expressions.

DEFINITIONS
$$DefInZD \widehat{=} \mathbf{ran}(Com2Zpiu \otimes Com2Dep)\,\S$$
$$DefInAD \widehat{=} \mathbf{ran}(Com2Agg \otimes Com2Dep)\,\S$$
$$DefAgg2Zpiu \widehat{=} \mathbf{ran}(Com2Agg \otimes Com2Zpiu)$$

Fig. 4. The definition of group intersections

The definitions for our case study are given in figure 4. $DefInZD$ is a relation between $ZPIU$ and DEP whose instances are pairs $z \mapsto d$ where z is a ZPIU and d is a *département* having some *commune*(s) in common (Cf section 3). The other definitions are similar. $DefInAD$ is for the intersection of agglomerations with *départements*. $DefAgg2Zpiu$ is for the intersection between agglomerations and ZPIUs. The invariant will force this set to be the graph of a function which associates a *commune* to its ZPIU.

The Invariant   The invariant states the constraints that should satisfy the values of the variables in order to represent a geography. These constraints are the properties of the geography. They are presented in figure 5. They can be read as follows:

- *Com* is a subset of the set of all possible *communes*, etc,

- *départements* keys are natural numbers, each *département* is identified by one key (the function *KeyDep* is total and injective),

- all the overseas *départements* are *départements*,

- agglomerations are identified by their keys,

- ZPIUs are identified by their keys,

- all the *communes* have a key,

- the *communes* of a *département* have different keys. The key of a *commune* is local to its *département*,

---

INVARIANT
$Com \subset COM \land Dep \subset DEP \land Agg \subset AGG \land Zpiu \subset ZPIU \land$
$KeyDep \in Dep \rightarrowtail \textbf{NAT} \quad \land$
$OverSeas \subset Dep \quad \land$
$KeyAgg \in Agg \rightarrowtail \textbf{NAT} \quad \land$
$KeyZpiu \in Zpiu \rightarrowtail \textbf{NAT} \quad \land$
$KeyCom \in Com \rightarrow \textbf{NAT} \quad \land$
$Com2Dep \otimes KeyCom \in Com \rightarrowtail (Dep \times \textbf{NAT}) \quad \land$
$Com2Agg \in Com \twoheadrightarrow Agg \quad \land$
$DefInAD \rhd OverSeas \in Agg \nrightarrow Dep \quad \land$
$Com2Dep \in Com \twoheadrightarrow Dep \quad \land$
$Com2Zpiu \in Com \twoheadrightarrow Zpiu \quad \land$
$DefInZD \rhd OverSeas = \varnothing \quad \land$
$DefAgg2Zpiu \in Agg \nrightarrow Zpiu \quad \land$
$\textbf{dom}(DefAgg2Zpiu) = \textbf{dom}(DefInAD \rhd OverSeas)$

Fig. 5. The invariant of an actual geography

- some *commune*s are in agglomerations, but a *commune* can be in at most one agglomeration (*Com2Agg* is a partial function). Each agglomeration contains at least one *commune* (the function is on to),

- all the *commune*s belonging to overseas *département*s and belonging to the same agglomeration are in the same *département*. Overseas *département*s are islands or group of islands. Agglomerations are sets of contiguous towns. This is why it is impossible to have agglomerations spreading over two overseas *département*s,

- every *commune* is in a *département* and every *département* has a *commune* belonging to it (constraint between two levels of a hierarchy),

- *commune*s belong to at most one ZPIU, and each ZPIU contains at least one *commune*,

- there is no ZPIU spreading over overseas *département*s,

- agglomerations cannot spread over two ZPIUs,

- all of the agglomerations spreading over metropolitan *département*s spread over a ZPIU.

4.4 Initialization

The initialization is a procedure which must establish the invariant. The machine we are defining does not modify the geography. The geography will be constructed through the refinement of the initialization, importing the second machine. The constructed geography will thus satisfy the invariant. The initialization of this machine consists in setting the variables with the actual geography (figure 6). The proofs that the provided geography exists and satisfies the invariant are carried out by the refinement of the machine importing the second machine.

INITIALIZATION
    ANY
        $ActCom, ActDep, ActAgg, ActZpiu,$
        $ActCom2Dep, ActCom2Agg, ActCom2Zpiu,$
        $ActOverSeas,$
        $ActKeyCom, ActKeyDep, ActKeyAgg, ActKeyZpiu$
    IN
        $ActCom \subset COM \wedge ActDep \subset DEP \wedge$
        $ActAgg \subset AGG \wedge ActZpiu \subset ZPIU \wedge$
        $ActKeyDep \in ActDep \rightarrowtail \mathbf{NAT} \quad \wedge$
        $ActOverSeas \subset ActDep \quad \wedge$
        $ActKeyAgg \in ActAgg \rightarrowtail \mathbf{NAT} \quad \wedge$
        $ActKeyZpiu \in ActZpiu \rightarrowtail \mathbf{NAT} \quad \wedge$
        $ActKeyCom \in ActCom \rightarrow \mathbf{NAT} \quad \wedge$
        $ActCom2Dep \otimes ActKeyCom \in ActCom \rightarrowtail (ActDep \times \mathbf{NAT}) \quad \wedge$
        $ActCom2Agg \in ActCom \twoheadrightarrow ActAgg \quad \wedge$
        $ActDefInAD \rhd ActOverSeas \in ActAgg \nrightarrow ActDep \quad \wedge$
        $ActCom2Dep \in ActCom \twoheadrightarrow ActDep \quad \wedge$
        $ActCom2Zpiu \in ActCom \twoheadrightarrow ActZpiu \quad \wedge$
        $ActDefInZD \rhd ActOverSeas = \varnothing \quad \wedge$
        $ActDefAgg2Zpiu \in ActAgg \nrightarrow ActZpiu \quad \wedge$
        $\mathbf{dom}(ActDefAgg2Zpiu) = \mathbf{dom}(ActDefInAD \rhd ActOverSeas)$
    THEN
        $Com := ActCom \parallel Dep := ActDep \parallel$
        $Agg := ActAgg \parallel Zpiu := ActZpiu \parallel$
        $Com2Dep := ActCom2Dep \parallel$
        $Com2Agg := ActCom2Agg \parallel$
        $Com2Zpiu := ActCom2Zpiu \parallel$
        $OverSeas := ActOverSeas \parallel$
        $KeyCom := ActKeyCom \parallel KeyDep := ActKeyDep \parallel$
        $KeyAgg := ActKeyAgg \parallel KeyZpiu := ActKeyZpiu$
    END

Fig. 6. The initialization

4.5 The operations

The specification of data retrieval procedures can be done in terms of the modelling.
Figure 7 gives the example of the specification of an operation which fetches the set of
*commune*s of overseas *département*s. The specification of the other procedures is of no
interest for the purpose of this paper.

```
OPERATIONS
    scod ⟵ Over_Com
        BEGIN
            scod := Prj1(Com2Dep ▷ Overseas)
        END
    ...
```

Fig. 7. Example of data retrieval operation

5 Building a representation

The second machine specifies the procedures for constructing a representation of the geography. The refinement of the first machine will import the specification of this machine. The data retrieval procedures access data in a structure which is built by procedures specified in this section. This cooperation is enforced by taking the same modelling of the geography and by taking the same implementation for the structure with
both machines.

The geography is given by several files containing respectively:

1. the keys of the *département*s,

2. the keys of the agglomerations,

3. the keys of the ZPIU,

4. the local keys of the *commune*s associated with the key of their *département*s,

5. the association of *commune*s to agglomeration by the triplets (key of the *département*, keys of the *commune*, key of the aggomeration),

6. the association of *commune*s to ZPIU by the triplets (key of the *département*, keys
   of the *commune*, key of the ZPIU)

To each file we associate the task (with the same number) of loading the file in order to
construct the geography. From the contents of the files, it can be seen that tasks 1,2 and

3 can be done in any order or even in parallel. They should precede task 4. Task 4 should precede tasks 5 and 6.

In order to improve the performance of the sytem, the intersections defined in 4.3 are computed. The state of the machine contains three variables to hold these values. To help the design of a procedure which computes incrementally the intersection of agglomerations with ZPIU, we have to choose an order for tasks 5 and 6. Our choice is to perform task 5 before task 6.

5.1 The context

The properties of a completed geography may be stronger than those of pieces of geography being constructed. We introduce a variable "Stage" which gives the stage of the construction. The greater the value of this variable, the stronger the constraints of the invariant over the other variables. The significant stages, given the tasks and given the invariant of the first machine are:

- the keys of all groups are loaded. The groups are created and it makes it possible to attach *communes* to them,

- the *communes* are attached to their *départements* and their local keys are defined. *Communes* can be grouped into agglomerations and ZPIU,

- the contents of agglomerations are defined. It is possible to define the contents of ZPIU while constructing the intersection of ZPIUs and agglomerations,

- the contents of ZPIUs are defined. The actual geography is ready for use.

GREATER introduces an order over stages. The other constants are stages. It should be noticed that our specification may be extended easily. It may be enriched by new stage constants which can be "inserted" in between the existing ones. One common mistake is to take numbers instead of such constants. It makes it impossible to enrich the specification with new inserted stages. The problem can be overcome by taking symbolic constants instead of actual values, and by stating the orderings to which they must comply.

OVERKEYS is a set of natural numbers which contains the keys of overseas *départements*.

5.2 The variables

Stage is the current stage of construction of the geography by the machine. The variables *InZD*, *InAD* and *Agg2Zpiu* are introduced to hold the values of the intersections.

5.3 The invariant

During the construction of a geography, the variables of the machine will satisfy a weaker invariant. We state a general invariant which has to hold at all stages of the construction.

On the last block of the invariant (Figure 10), the values of *InZD*, *InAD* and *Agg2-Zpiu* are constrained to correspond to the respective definitions. The rest of the invariant

```
    SETS
        ...
        STAGE = {Scratch, GroupsLoaded, CDepLoaded,
                    CAggLoaded, CZpiuLoaded}

        GREATER, OVERKEYS

        GREATER = {(GroupsLoaded ↦ Scratch),
                    (CDepLoaded ↦ GroupsLoaded),
                    (CAggLoaded ↦ CDepLoaded),
                    (CZpiuLoaded ↦ CAggLoaded)}* ∧
        OVERKEYS ⊂ NAT
```

Fig. 8. The context (second machine)

```
    VARIABLES
        ...,
        Stage, InZD, InAD, Agg2Zpiu
```

Fig. 9. The variables

concerns the same variables modelling the geography as for the first machine. It is almost a carbon copy of the invariant machine. But it has been altered in order to take into account the temporary states of the representation under construction.

5.4 Initialization

It consists in setting an "empty" geography to which will be added the entities given by the files. The first line of the initialization (see figure 11) states that there is no entity of any kind. The variable $Stage$ is set to $Scratch$ in order to allow the addition of new entities. The rest of the variables are set to be empty in order to satisfy the invariant[5].

5.5 The operations

For each task, we specify a procedure for introducing an atomic datum. For instance the procedure Add_Dep takes as input a key and creates a *département* associated with this

---

[5] In a similar specification with Z, these settings would have been enforced by the invariant. We prefer the B policy. In Z, if settings are not enforced by the invariant, they have to be explicitly constrained. With B, no setting changes unless otherwise specified. In order to make sure the invariant is established, the setting of the other variables has to be constrained to the only possible value

INVARIANT

$Com \subset COM \land Dep \subset DEP \land Agg \subset AGG \land Zpiu \subset ZPIU \land$

$KeyDep \in Dep \rightarrowtail \mathbf{NAT} \quad \land$

$OverSeas \subset Dep \quad \land$

$KeyAgg \in Agg \rightarrowtail \mathbf{NAT} \quad \land$

$KeyZpiu \in Zpiu \rightarrowtail \mathbf{NAT} \quad \land$

$KeyCom \in Com \rightarrow \mathbf{NAT} \quad \land$

$Com2Dep \otimes KeyCom \in Com \rightarrowtail (Dep \times \mathbf{NAT}) \quad \land$

$Com2Agg \in Com \twoheadrightarrow Agg \quad \land$

$DefInAD \rhd OverSeas \in Agg \twoheadrightarrow Dep \quad \land$

$Com2Dep \in Com \rightarrow Dep \quad \land$

$(GREATER(Stage, CDepLoaded) \Rightarrow Com2Dep \in Com \twoheadrightarrow Dep \quad) \land$

$(GREATER(Stage, CAggLoaded) \Rightarrow Com2Agg \in Com \twoheadrightarrow Agg \quad) \land$

$Com2Zpiu \in Com \twoheadrightarrow Zpiu \quad \land$

$DefInZD \rhd OverSeas = \varnothing \quad \land$

$DefAgg2Zpiu \in Agg \twoheadrightarrow Zpiu \quad \land$

$\mathbf{dom}(DefAgg2Zpiu) \subset \mathbf{dom}(DefInAD \rhd OverSeas) \quad \land$

$(GREATER(Stage, ZpiuLoaded)$

$\quad \Rightarrow \mathbf{dom}(DefAgg2Zpiu) = \mathbf{dom}(DefInAD \rhd OverSeas)$

$\quad\quad \land Com2Zpiu \in Com \twoheadrightarrow Zpiu) \land$

$InZD = DefInZD \land$

$InAD = DefInAD \land$

$Agg2Zpiu - DefAgg2Zpiu$

Fig. 10. The invariant of a geography for all stages

INITIALIZATION
BEGIN
$\quad Com, Dep, Agg, Zpiu := \varnothing, \varnothing, \varnothing, \varnothing \parallel$
$\quad Com2Dep, Com2Agg, Com2Zpiu := \varnothing, \varnothing, \varnothing \parallel$
$\quad OverSeas := \varnothing \parallel$
$\quad KeyCom, KeyDep, KeyAgg, KeyZpiu := \varnothing, \varnothing, \varnothing, \varnothing \parallel$
$\quad InZD, InAD, Agg2Zpiu := \varnothing, \varnothing, \varnothing \parallel$
$\quad State := Scratch$
END

Fig. 11. The initialization

key. The precondition of these procedures states that they must be called at the right stage of the construction. For each stage, we specify a validation procedure which checks the progression of the invariant, prevents the calls of procedures of the current stage and authorizes the calls of procedures for the following stage.

A selection of specifications of these procedures is given in figures 12 and 13. It is representative of the whole set.

When f is a function $f(x) := u$ means $f := f <+ \{(x \mapsto u)\}$. The operator $<+$ represents the overriding of one function by another.

Add_Dep(dk) creates a new *département* identified by the key dk. This procedure may only be called at the first stage of the construction: the loading of groups. The key must be a natural number which has not already been associated with an existing *département*. This ensures that *KeyDep* remains injective.

Set_GL validates the stage of the loading of groups. The variable *Stage* is set to *Groups-Loaded*. Therefore the preconditions of procedures for adding groups will no longer hold.

Add_Com(clk, d) creates a new *commune* identified by the local key clk within *département d*. This procedure may only be called at the second stage of the construction. d must be an existing *département*, and clk is not associated with another *commune* of the same *département*.

Add_CZ(c, z) incorporates the *commune c* in the ZPIU z. The *commune c* is not already in a ZPIU, it does not belong to an overseas *département*. If c is in an agglomeration and if a *commune* of this agglomeration already is in a ZPIU, then this ZPIU must be z.

Set_ZL is the last validation. It succeeds when the last part of the invariant is satisfied.

6 Implementation

The presentation of the implementation merits a paper on its own. The following is just an overview.

As shown in Figure 14 the system was specified with three abstract machines. The first and second machines are the machines of the previous sections. The third machine, the DB machine, relates the logical view of the relations introduced so far to their physical representation. It sets the tables used for the representation and their maximum cardinalities. The third machine was refined automatically with a home made tool. The second machine was refined by hand using the operations provided by the third machine. The first machine was also refined by hand, using the operations provided by the second machine. Part of the refinement is shown in Figure 15[6]. The translation of the refined specifications into Pascal programs was straightforward. B allows refinements which are very close to programs.

---

[6] This employs a FOR loop instead of a WHILE loop provided by B. This is just to make it more readable.

OPERATIONS

$Add\_Dep(dk) =$
    PRE
        $Stage = Scratch \wedge$
        $dk \in \mathbf{NAT} \wedge dk \notin \mathbf{ran}(KeyDep)$

    THEN
        ANY d **IN** $c \in (DEP - Dep)$
        THEN
            $Dep := Dep \cup \{d\} \parallel$
            $KeyDep(d) := dk \parallel$
            IF $dk \in OVERKEYS$ **THEN**
                $Overseas := Overseas \cup \{d\}$
            END
        END
    END

$\dots$

$Set\_GL =$
    PRE
        $Stage = Scratch$
    THEN
        $Stage := GroupsLoaded$
    END

$Add\_Com(clk, d) =$
    PRE
        $Stage = GroupsLoaded \wedge$
        $d \in Dep \wedge clk \in \mathbf{NAT} \wedge$
        $(d \mapsto clk) \notin \mathbf{ran}(Com2Dep \otimes KeyCom)$
    THEN
        ANY c **IN** $c \in (COM - Com)$
        THEN
            $Com := Com \cup \{c\} \parallel$
            $Com2Dep(c) := d \parallel KeyCom(c) := clk$
        END
    END
$\dots$

Fig. 12. Examples of construction operations

```
      ...
      Add_CZ(c, z) =
          PRE
              Stage = AggLoaded ∧
              c ∈ Com ∧
              z ∈ Zpiu ∧
              c ∉ dom(Com2Zpiu) ∧ Com2Dep(c) ∉ Overseas ∧
              (c ∈ dom(Com2Agg) ∧ Com2Agg(c) ∈ dom(Agg2Zpiu)
                  ⇒ Agg2Zpiu(Com2Agg(c)) = z)
          THEN
              Com2Zpiu(c) := z ‖
              InZD := InZD ∪ {(z ↦ Com2Dep(c))} ‖
              IF c ∈ dom(Com2Agg) THEN
                  Agg2Zpiu(Com2Agg(c)) := z
              END
          END

      ...
      Set_ZL =
          PRE
              Stage = AggLoaded ∧
              ran(Com2Zpiu) = Zpiu ∧
              dom(DefInAD ▷ OverSeas) ⊂ dom(DefAgg2Zpiu)
          THEN
              Stage := ZpiuLoaded
          END
```

Fig. 13. Examples of construction operations (continued)

Because of a lack of time, most of the proof obligations have not been carried out. Nevertheless, the construction resulted in a code in which we could have great confidence. In fact, the system has been used intensively for several years now, and no problem has arisen due to the specification. In any case, all the ingredients are available for performing the proofs (the specifications, their refinements and the logical system of [2]).

The tool for refining data bases was designed with respect to the philosophy of the B method. The main features of this software are:

- each data base system is the implementation of an abstract machine,

- idiosyncrasies of the operating systems have been encapsulated in technical abstract machines,

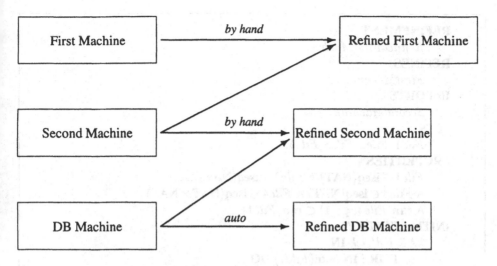

Fig. 14. Implementation

- the lowest level of the implementation consists in the implementations of the abstract machine which is a portable virtual memory. It offers an example of several refinements of the same specification. It provided an opportunity to test the whole system on small databases during the development phase on the workstation,

- the system has been designed for saving storage space and to be portable.

The system was developed and tested on a SUN workstation under UNIX. The target machine was an IBM under MVS. The programming language was Pascal.

The specification of the whole system resulted in 65 pages of B. The invariant occupies 8 pages. It was achieved within two weeks by a B expert and an expert of the geography[7]. The refinement and the implementation took 2 months and resulted in 14K lines of Pascal. The system was then tested and transferred to the mainframe. This took another two weeks. The size of the resulting database is 40Mbytes. These figures do not take into account any training in B nor the time spent in developing the tool generating databases.

The time spent to improve the expression of the invariant was not wasted. This resulted in a relatively easy specification of the required procedures which, in turn, resulted in a quick and reliable implementation. Indeed, we believe that spending extra time for improving the invariant is *always* a good policy. This extra time is easily compensated for by the rapidity with which the implementation is achieved. A carefully designed invariant makes it unlikely that special cases are omitted during the design of operations. Therefore there is a gain of both time and reliability.

---

[7] namely G. Laffitte and J.-P. Faur

```
REFINEMENT
    RefinedFirstMachine
REFINES
    FirstMachine
IMPORTS
    SecondMachine
SETS
    File1, File2, File3, File4, ...
PROPERTIES
    File1 ∈ iseq(NAT) ∧ File2 ∈ iseq(NAT)
    ∧ File3 ∈ iseq(NAT) ∧ File4 ∈ iseq(NAT × NAT)
    ∧ ran(File4; prj1) ⊂ ran(File1)...
INITIALIZATION
    VAR i, dl, lck IN
        FOR i IN dom(File1) DO
            Add_Dep(File1(i))
        END⨟
        FOR i IN dom(File12) DO
            Add_Agg(File2(i))
        END⨟
        FOR i IN dom(File3) DO
            Add_ZPIU(File3(i))
        END⨟
        Set_GL⨟
        ...
        Set_ZL⨟
END
```

Fig. 15. Part of the refinement

7 Conclusion

A significant example of specification for program development has been given. It is based on a *real-life application*. The specification was not carried out a posteriori. Part of it has been extracted for the purpose of this paper. It constitutes a case study for specification development method and for specification language evaluation.

We have presented the method followed in order to achieve the specification:

- a technique for modelling groupings of different kinds and hierarchies. The use of direct product for stating the relationship among the groups,

- the modelling of local keys using direct product,

– the specification of a construction. We started from the modelling of the final object to design the dynamics of the system: stage decomposition is obtained by taking the final invariant as a target and by considering the atomic steps of the construction. We have tried to be as close as possible to the final invariant during the construction. It resulted in well defined and minimal tests.

The system supporting the census has been fully specified and implemented. It has been running for a few years and is still in use. It proves that our solution was adequate. The benefits of our approach are:

for the system development a quick and reliable implementation. The ultimate goal was stated clearly right from the beginning,

for the users of the system a clear and concise definition of the services provided by the system. The specification supplemented by comments, as it is presented in this paper, proved to be a very effective way to explain the geography. This was important since the whole geography is very complex.

The B method proved to be effective for the development of the system supporting the census. The method requires that all the people involved in the project have or get some background in mathematics (set theory and first order predicate calculus). This is the price to be paid in order to make wide and complex projects manageable.

Although it is difficult to draw general conclusion from one example, it seems that the development of different kinds of systems should also benefit from the B method (for instance for concurrent systems see [3, pp 366–380]).

References

1. Jean-Raymond Abrial. *Introduction à la méthode de spécification formelle B (lecture on 6 video tapes with a print of the transparencies).* teknea, Toulouse, 1994.
2. Jean-Raymond Abrial. *The B Book: assigning programs to meanings.* Cambridge University Press, to appear in 1995.
3. Andy S. Evans. Specifying & verifying concurrent systems using Z. In Maurice Naftalin, Tim Denvir, and Miquel Bertran, editors, *FME'94: Industrial Benefit of Formal Methods*, volume 873 of *Lecture Notes in Computer Science*, pages 366–380. Springer-Verlag, 1994.
4. Henri Habrias. *Introduction a la Spécification.* Methodologies du logiciel. Masson, 1993.
5. Anthony Hall. Specifying and interpreting class hierarchies in Z. In J. P. Bowen and J. A. Hall, editors, *Z User Workshop, Cambridge 1994*, Workshops in Computing, pages 120–138. Springer-Verlag, 1994.
6. J. Mike Spivey. *The Z Notation: a Reference Manual.* International Series in Computer Science. Prentice Hall, 1992.

Animation

Implementing Z in Isabelle

Ina Kraan and Peter Baumann

Department of Computer Science, University of Zürich
8057 Zürich, Switzerland
inak,baumann@ifi.unizh.ch

Abstract. We present work in progress on a deep semantic embedding of the specification language Z and its deductive system (as defined by the draft Z standard) in the theorem prover Isabelle. Z is based on Zermelo-Fraenkel set theory and first-order predicate logic, extended by a notion of schemas. Isabelle supports a fragment of higher-order predicate logic, in which object logics such as Z can be encoded as theories. This paper gives an overview of Z-in-Isabelle, including some example proofs, and discusses some of the major issues involved in formalizing Z in Isabelle.

1 Introduction

We are implementing the language Z and a deductive system for Z in the generic theorem proving tool Isabelle. The implementation is based on Z and its deductive system as defined in the draft Z standard [2][1]. It is a deep embedding of Z, i.e., both the syntax and a deductive system are formalized within Isabelle's meta-logic.

Implementations of Z can be roughly divided into two categories. First, there are the direct implementations, including, for instance, the tool Balzac [4, 6]. Second, there are the embeddings of Z in other logics and in logical frameworks. Most such implementations have been shallow rather than deep embeddings, including ProofPower [5], Z in HOL [1], and Z in LEGO [7]. The distinction between shallow and deep is not always clear-cut, but in principle, the deeper the embedding, the more properties of the language can be proven in the embedding. Deep embeddings of Z in logical frameworks include Jigsaw [8] and our system Z-in-Isabelle. Jigsaw is based on the deductive system of Woodcock and Brien [13] and implemented in 2OBJ [3]. Despite being deep embeddings within logical frameworks, the two systems are quite different. A comparison can be found in section 7.

An embedding within a logical framework has the advantage that one can prove faithfulness and adequacy more easily (for an example of such proofs for first-order logic in Isabelle, see [9]). A deep embedding, while involving much more effort than a shallow one, has a number of advantages. First, it can be used to rigorously derive, as theorems in the meta-logic, new rules of inference for the object logic (one of the example proofs presented is the derivation of such a rule). These may be more desirable because they are simpler to use, or because they facilitate the automation of proofs. Second, one can prove properties of the language being embedded, e.g., properties such as commutativity or associativity of operators. Our experience with Isabelle has shown that it is a suitable

---

[1] Where it is incomplete, we have referred to the logic of Woodcock and Brien [13].

tool to implement such an embedding, and that a deep embedding is useful in itself for the reasons mentioned above.

Z has been designed as a powerful specification language that enables writing large specifications concisely. Mechanization was not an uppermost priority, and unsurprisingly, some of the design decisions taken in the development of the language to ease the writing of specifications have proven difficult in implementing the language. The difficulties we encountered and solutions we propose are a central topic of this paper.

We do not assume any knowledge of Isabelle; we do, however, assume some familiarity with the draft Z standard, in particular the deductive system in Appendix F. The paper starts with a brief introduction to Isabelle. Then, some of the pitfalls we encountered in the course of implementing Z in Isabelle are discussed. There follows an overview of the implementation. Subsequently, we present some example proofs to provide an idea of the implementation in action. Finally, we draw conclusions and point out directions for future work.

2 The Generic Theorem Prover Isabelle

Isabelle [10] is a generic theorem prover. Its distinctive feature is that it represents logics in a fragment of higher-order logic. The implemented (object) logic, here Z, is then implemented as a theory in Isabelle's (meta-)logic, where each inference rule of the object logic is an axiom in the meta-logic. Isabelle comes with a number of default logics, including first-order logic, higher-order logic, first-order sequent calculus, Zermelo-Fraenkel set theory, and a constructive type theory. It also has a powerful mechanism to define theories, with elaborate mechanisms to translate between concrete and abstract syntax. Furthermore, Isabelle provides a powerful tactic[2] mechanism, which can be used to automate large parts of proofs. Finally, Isabelle can manage multiple proofs: the current state of the active proof can be saved, and work on it resumed later. This is useful when experimenting and when, in the middle of a proof, one realizes that a lemma is needed.

Isabelle represents syntax using the simply typed lambda calculus. It has polymorphic types and uses type classes to control polymorphism. Variable binding is expressed using λ-abstraction, a technique also known as higher-order abstract syntax [11]. Thus, for example, $\forall x.\ P$ is represented as $\forall(\lambda x.\ P)$. Also, $F(t)$ represents the application of the function F to the argument t. If F has the form $\lambda x.\ P$, Isabelle will β-reduce $(\lambda x.\ P)(t)$ to $P[t/x]$. This enables us to express inference rules involving substitution for bound variables in such a way that the usual side conditions are automatically respected. Thus, for instance, the inference rule

$$\frac{P}{\forall x.\ P}\quad [\forall I]$$

with the side condition that x does not occur free in the assumptions, can be formalized in Isabelle as

$$(\textstyle\bigwedge x.\ \mathit{Trueprop}(P(x))) \implies \mathit{Trueprop}(\forall(\lambda x.P(x)))\ .$$

---

[2] Tactics are programs that apply inference rules.

The function *Trueprop* coerces object-level truth to meta-level truth, and $\bigwedge$ is meta-level universal quantification. Thus, we can read the above as: "if $P(x)$ is true for all x, then $\forall x.\ P(x)$ is true". Fortunately, Isabelle hides much of the meta-logic "overhead" from the user. The above rule, for instance, would actually be entered as something like[3]

```
(!! x. P(x)) ==> ALL x. P(x)
```

Isabelle allows schematic variables or unknowns, which can be instantiated via unification. Schematic variables are distinguished from other variables by a leading question mark and are particularly useful when applying inferences that introduce new parameters or variables. For instance, the rule

$$\frac{P[t/x]}{\exists x.\ P}\quad [\exists I]$$

can be expressed in Isabelle as

```
P(?x) ==> EX x. P(x) .
```

The rule introduces a new schematic variable, ?x, representing the witness for the existential variable x. Thus, the rule can be applied without having to provide the actual witness. The schematic variable ?x will become instantiated to the witness via unification later in the proof.

Isabelle's meta-logic is a fragment of higher-order logic, with universal quantification, implication, and equality. The inference rules of the logic are almost never used directly in proofs, however. Normally, a variety of derived resolution rules are used. These rules are tailored in such a way that a resolution step in Isabelle's meta-logic mimics the application of a rule in the object logic. Thus, the use of a meta-logic is transparent to quite a large extent. Isabelle allows both forward-chaining proof, i.e., two theorems can be resolved against each other to produce a new theorem, and backward-chaining proof, i.e., a conjecture (goal) is resolved against a theorem, which yields one or more new conjectures (subgoals). In backward proof, the original conjecture is a theorem when all subgoals have been reduced to *true*. The proofs in this paper are all backward proofs.

3 The Inference Rules

Some of the inference rules in Appendix F of the draft standard were straightforward to implement. Thus, for instance, the rule $(\vee\vdash)$ is represented in Z-in-Isabelle as:

```
disjL        "[| $H, P, $G |- $E; $H, Q, $G |- $E |]
             ==> $H, P V Q, $G |- $E"
```

---

[3] The symbol ! ! stands for meta-level universal quantification ($\bigwedge$) and ==> for meta-level implication ($\Longrightarrow$).

The list of premises is enclosed in " [| |] ", and premises are separated by " ; ". The symbol "$" signifies that the variable it precedes may unify with a list of formulae, not just a single formula. This Z-in-Isabelle rule corresponds to

$$\frac{\Gamma, P, \Delta \vdash \Lambda \quad \Gamma, Q, \Delta \vdash \Lambda}{\Gamma, P \vee Q, \Delta \vdash \Lambda} \quad [(\vee \vdash)]$$

which is the rule $(\vee \vdash)$ in the draft Z standard (before "rule-lifting").

A number of the inference rules in the draft Z standard, however, posed problems for the implementation, either because they could not be implemented directly, or because they were awkward to use. This is discussed in the following subsections.

3.1 Enumeration

The rules for tuples, Cartesian products, tuple selection, binding extensions, binding selection, simple declarations, and theta expressions are all expressed via enumeration, and thus needed to be reformulated. For all but theta expressions, the rules were rewritten to be recursive. We illustrate this using the draft standard rule for simple declarations on the left:

$$\frac{n_1 \in s \wedge \ldots \wedge n_m \in s \vdash}{[n_1, \ldots, n_m : s]} \quad [([n : s] \vdash)]$$

This rule is replaced by the two rules

$$\frac{m \in s \wedge [n : s] \vdash}{[m, n : s] \vdash} \quad [([m, n : s] \vdash)]$$

and

$$\frac{n \in s \vdash}{[n : s] \vdash} \quad [([n : s] \vdash)]$$

The variables m in the rule $([m, n : s] \vdash)$ and n in the rule $([n : s] \vdash)$ are prevented from being compound themselves by attaching corresponding (Isabelle-)type information to them. The recursive rules are derivable from the original ones.

For theta expressions, a recursive formulation was not possible, since the order of the individual bindings in a binding extension is irrelevant and the representation of a binding extension not unique. We have not yet been able to come up with a formulation of a rule for theta expressions that is sufficiently simple, clean, and efficient. For the time being, therefore, we unfold all theta expressions in a preprocessing step.

3.2 Substitution and Variable Binding

One major design decision was to exploit higher-order abstract syntax as offered by Isabelle. Making higher-order abstract syntax applicable required a number of adaptations.

Many of the inference rules of Z, in particular those that involve binding constructs (universal and existential quantification, and set comprehension) have side conditions to avoid variable capture. Isabelle allows no such side conditions, which means that they must already be enforced directly in the way the inference rule is expressed. As explained above in section 2, the use of higher-order abstract syntax enables us to do this. What this means is that we lift the variable capture problem from the object level to the meta-level.

This is illustrated by the set comprehension rule *compre* in Z-in-Isabelle:

$$\frac{\vdash \exists x : T \mid P(x) \bullet t = F(x)}{\vdash t \in \{x : T \mid P(x) \bullet F(x)\}} \quad [\, compre \,]$$

Variables bound by set comprehension and existential quantification are represented as λ-bound variables at the meta-level. Therefore, applying the *compre* rule (in backwards proof) to a conclusion

$$\vdash y \in \{y : T \mid Q(y) \bullet G(y)\} \,,$$

where y occurs both free and bound, will cause the variable corresponding to the existential variable x in the premise of the rule to be renamed.

While higher-order abstract syntax has the advantage of being simple, elegant and efficient, it does pose some restrictions. First, inference rules can apply to one bound variable at a time only. Second, any possible occurrence of a bound variable must be explicitly represented. The latter means that we must take care when using schema references. Assume, for instance, that the goal to be proven is

$$\vdash x \in \{x : T \mid S \bullet F(x)\} \,,$$

where S is:

$$\begin{array}{|l}
\hline
S \underline{\hspace{8cm}} \\
x : T \\
\hline
P(x) \\
\hline
\end{array}$$

If we apply the *compre* rule, we get

$$\vdash \exists x_0 : T \mid S \bullet x = F(x_0) \,,$$

and if we then expand the schema, we obtain

$$\vdash \exists x_0 : T \mid x \in T \land P(x) \bullet x = F(x_0) \,,$$

which is nonsense. Thus, for higher-order abstract syntax to work correctly, we must expand all schema references from the outset. This is done in a preprocessing step. For large specifications, expanding schema references results in very long formulae and affects efficiency noticeably. The problem can be alleviated at least visually for the user by making the concrete syntax of a schema its reference. Then, the expansion is only visible in the abstract syntax.

Another difficulty are schemas as expressions versus schemas as predicates. We need to decide whether the variables declared in schemas are bound or free. When using higher-order abstract syntax, one represents bound variables at the object level by λ-bound variables at the meta-level and free variables at the object level as constants or possibly variables at the meta-level. At first glance, variables declared in schemas as predicates ought to be free, those declared in schemas as expressions bound. However, this will not work for two reasons. First, since λ-bound variables are anonymous, schema inclusion and the schema calculus could not be used in schemas as expressions, as we need to know the names of the variables to merge them properly. Second, it would mean that we could not use the same representation for a schema whether it is used as a predicate or as an expression. On the other hand, if we let the variables declared in schemas as expressions remain free at the meta-level, a meta-level application may inadvertently substitute some term for such a variable, and thus compromise correctness. Our solution is to eliminate all schemas used as expressions: We replace all occurrences of schemas as expressions S with $\{S \bullet \theta S\}$ and expand any occurring schema references. The elimination of schemas as expressions is also a preprocessing step and must precede the expansion of schema references described in the previous paragraph.

If we represent variables declared in schemas as free variables, we must ensure that they cannot be confused with global variables. To this end, we take the same approach as Woodcock and Brien [13] and uniquely rename all global variables at the outset.

Eliminating schemas as expressions has a few consequences on the deductive system. First, the rule *schemaexp* is superfluous, since we no longer have schemas as expressions. Second, other rules that contain schemas as expressions, i.e., $(\forall \vdash)$ and $(\vdash \exists)$, must be reformulated. For instance, $(\forall \vdash)$ becomes

$$\frac{b \odot [x : S \mid P(x)], \forall x : S \mid P(x) \bullet Q(x), b \odot Q(x) \vdash}{b \odot [x : S \mid P(x)], \forall x : S \mid P(x) \bullet Q(x) \vdash} \quad [(\forall \vdash)]$$

The new rules are derivable from the original rules and *schemaexp*.

Finally, given that we are using higher-order abstract syntax, we would like to exploit the conveniences thereof as far as possible. In particular, we would like to use meta-level rather than object-level substitution in the rules that use explicit bindings in the draft Z standard. These are $(\forall \vdash)$, $(\vdash \exists)$, and *Leibniz*. The versions with meta-level substitution are

$$\frac{\forall x : T \mid P(x) \bullet Q(x) \vdash x \in T \wedge P(x)}{\forall x : T \mid P(x) \bullet Q(x), Q(x) \vdash} \quad [(\forall \vdash)_m]$$

$$\frac{\vdash \exists x : T \mid P(x) \bullet Q(x), x \in T \wedge P(x)}{\vdash \exists x : T \mid P(x) \bullet Q(x), Q(x)} \quad [(\vdash \exists)_m]$$

$$\frac{x = y, P(x) \vdash}{P(y) \vdash} \quad [\, Leibniz_m \,]$$

The above rules have the advantage that the substitution is taken care of automatically at the meta-level, whereas, for the ones with explicit bindings, the substitution would have to be "computed" by repeatedly applying the substitution rules. What this amounts to is lifting substitution from the object to the meta-level. Thus, we need to establish that meta- and object-level substitution are equivalent. What we gain by doing this are much shorter and simpler proofs. The example proofs in section 6 use the rules ($\forall \vdash$)$_m$, ($\vdash \exists$)$_m$, and $Leibniz_m$ (or rather rules derived from them), precisely to avoid overly lengthy and tedious proofs. However, two versions of the deductive system, i.e., one with and one without object-level substitution, are available.

3.3 Automating Proofs

As we discovered in the course of our first proof attempts, a number of the rules are formulated in a way that makes them difficult to use in automated proofs. A good example of this is ($\vdash \exists$):

$$\frac{b \odot [x : S \mid P(x)] \vdash \exists x : S \mid P(x) \bullet Q(x), b \odot Q(x)}{b \odot [x : S \mid P(x)] \vdash \exists x : S \mid P(x) \bullet Q(x)} \quad [\, (\vdash \exists) \,]$$

It presupposes that we have a binding b such that $b \odot [x : S \mid P(x)]$ holds. To obtain such a binding, we often had to use the *cut* rule. In automated proofs, one generally tries to avoid the *cut* rule, since it allows too great a degree of freedom (it does not have the subformula property). Since we are working in a meta-logic, we can simply derive rules of inference that are more suitable for automated proofs. For instance, the rule we would like to have for existential quantification on the right is

$$\frac{\vdash b \odot [x : S \mid P(x)] \quad \vdash \exists x : S \mid P(x) \bullet Q(x), b \odot Q(x)}{\vdash \exists x : S \mid P(x) \bullet Q(x)} \quad [\, exR \,]$$

which does not presuppose a binding b such that $b \odot [x : S \mid P(x)]$ holds. In Z-in-Isabelle, we formulate the new rule as a meta-implication and prove it, in this instance simply by applying the *cut* rule appropriately. From this new rule, we can also easily derive the rule

$$\frac{\vdash b \odot [x : S \mid P(x)] \quad \vdash b \odot P(x)}{\exists x : S \mid P(x) \bullet Q(x) \vdash} \quad [\, exRthin \,]$$

by applying the *thin* rule to the second premise. The latter rule is useful for tactics, since it discards the original conjecture $\exists x : S \mid P(x) \bullet Q(x)$, which helps prevent looping. Note, however, that it is "unsafe" in the sense that applying it may make the conjecture unprovable.

We have derived a large set of such rules, in particular with respect to writing tactics to automate large parts of proofs. Using derived rules often makes for clearer, shorter proofs.

An alternative to derived rules are tactics, which simply perform the same sequence of inferences as were used to prove the derived rule. Using a tactic rather than a derived rule of inference, however, has the disadvantage that the sequence of inferences is executed each time the tactic is executed. For the derived rule, this is done only once when the rule is proven. Hence, derived rules are more efficient than tactics. However, derived rules presuppose a fixed sequence of inferences. Tactics are programs that apply inference rules, and they are more flexible than derived rules in that they can take parameters and make choices, backtrack, etc. Thus, for instance, it is not possible to derive an inference rule that will decompose a schema as predicate as a hypothesis using the rules $([|] \vdash)$ and $([D; D] \vdash)$ from the draft Z standard, and $([m, n : s] \vdash)$ and $([n : s] \vdash)$ from section 3.1, since the number of applications of $([D; D] \vdash)$, $([m, n : s] \vdash)$, and $([n : s] \vdash)$ depends on the number of variables declared in the schema. A tactic can easily achieve this by applying $([|] \vdash)$ once and repeatedly applying first $([D; D] \vdash)$, then $([m, n : s] \vdash)$, and finally $([n : s] \vdash)$ until no longer applicable.

3.4 Partial Functions and Undefinedness

Dealing with partial functions is an ongoing concern. While partial functions are certainly desirable to have in a specification language, there are a number of difficulties in dealing with them. The issues of function application and undefinedness have not yet been decided definitively in the Z standard, and we are currently investigating the various approaches.

One issue we have had to deal with in order to do any non-trivial proofs is that of equality. According to the Z standard, two terms are equal iff they are defined and equal (known as existential equality). However, the reflection rule as stated in the Z standard

$$\frac{}{\vdash x = x} \quad [\, reflection \,]$$

is sound only "with respect to a model in which all well-typed expressions have a value, since it does not require that x be defined, as stipulated in the semantics" [2].

The question is whether to change the semantics of equality or the inference rule. The inference rule can be changed by adding a precondition that requires the object to be an element of some set and thus exist.

$$\frac{\vdash x \in T}{\vdash x = x}$$

Changing the semantics of equality means allowing undefined terms to be equal. There are several approaches, and it remains to examine their effects on proofs and on the soundness of other inference rules. In the example proofs below, we have used the modified reflection rule.

4 Schemas

Z revolves around its concept of schemas. Schemas and schema references can be used very flexibly within specifications. The meaning of a schema varies according to its use as a schema proper, an expression, a predicate or a declaration. What it is being used as is normally clear from context. This flexible use causes difficulties in Isabelle, since a schema, like any other object, cannot change its type from one occurrence to another. Therefore, we insert coercion functions which change the type of a schema according to its context. Isabelle's translation mechanism between concrete and abstract syntax allows us to hide these coercion functions from the user during proofs. Thus, for instance, the concrete syntax of conjecture (1) in section 6 is

$$AddBirthday \vdash known' = known \cup \{name?\}$$

but the abstract syntax is

$$pred([\ldots \mid \ldots]) \vdash known' = known \cup \{name?\}\,,$$

where *pred* is a coercion function that coerces an object of Isabelle-type schema to an object of Isabelle-type predicate. The insertion of these coercion functions is done in a preprocessing step.

5 Z-in-Isabelle

Z has been implemented as a family of theories in Isabelle (see Figure 1). The theory *Pure* contains Isabelle's meta-logic. The basic theory underlying Z-in-Isabelle is *Lkz*. *Lkz* defines sequents using the representation of the standard LK theory distributed with Isabelle[4]. It also defines the type system of Z-in-Isabelle, which mirrors the type system of Z to a large extent, but is less detailed (see Figure 2). There are two type constructors, *Set* (a postfix operator) and ∗ (an infix operator), which construct objects of type *Power* and *Cartesian*, respectively, from objects of type *Type*. Theories for the given (built-in) types integers (*Number*) and strings (*String*) extend *Lkz*. The theory *Z* is based on these three theories, and defines the core language and the deductive system of Z. Further theories define substitution and the mathematical toolkit.

Our ultimate goal is a system whose core is Z-in-Isabelle, with a preprocessor that reads Z specifications in LaTeX format and a postprocessor that presents the results of Z-in-Isabelle in LaTeX format. The preprocessor parses the specification, type-checks it, performs the preprocessing steps described in sections 3.1, 3.2[5], and 4, and translates the specification into Z-in-Isabelle syntax.

A specification is implemented as a theory extending the family of theories in Figure 1. An Isabelle theory file consists of a signature, which contains the information necessary for type checking, parsing and pretty printing terms, and of a set of axioms. We

---

[4] We did not implement Z as an extension to the built-in theory ZF for the simple reason that the latter is written in natural deduction style, whereas the deductive system in the Z standard is written in a sequent calculus style.

[5] While expanding schema references, it also expands Δ and Ξ declarations.

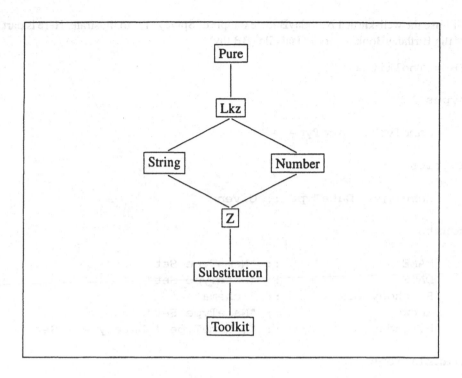

Fig. 1. Overview of Z-in-Isabelle theories.

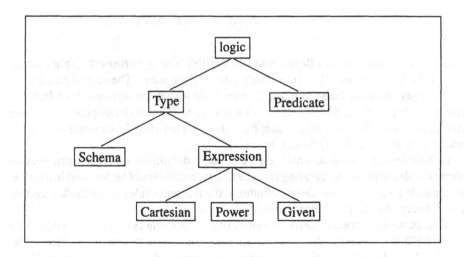

Fig. 2. Overview of Z-in-Isabelle classes.

will use the well-known BirthdayBook example of Spivey [12] to illustrate. Here is part of the BirthdayBook theory in Isabelle (BB.thy):

```
BB = Toolkit +

types

    NameType, DateType 0

arities

    NameType, DateType :: Given

consts

    NAME                  :: "NameType Set"
    DATE                  :: "DateType Set"
    BirthdayBook          :: "Schema"
    known                 :: "NameType Set"
    birthday              :: "( NameType * DateType ) Set"

translations

    "BirthdayBook" == "[
                        known : \power NAME;
                        birthday : NAME \pfun DATE
                      |
                        known = \dom birthday
                      ]"
```

The first part defines the Isabelle types and their arities. The types NameType and DateType belong to the class Given, and they take no parameters. The second part defines the constants and their Isabelle types. The constant NAME, for instance, is of Isabelle-type NameType Set, which means that it is a set of elements belonging to the given type NameType. The last part contains the translation between the concrete and the abstract syntax of the BirthdayBook schema.

Definitions other than schemas, e.g., axiomatic definitions or constraints, become inference rules in the corresponding Isabelle theory, which would be declared in a further section called rules. From these definitions, the inference rules in the toolkit, and the core Z theory, we can prove properties of specifications.

Though we have tried to keep the syntax of Z-in-Isabelle as close as possible to the standard LaTeX style, a number of syntactic requirements on the format of theory files and a few lexical quirks of Isabelle make a certain amount of preprocessing inevitable, as should already be apparent from the BirthdayBook example theory. The preprocessor is currently under development. Until it is usable, Z specifications must be hand-translated into corresponding Z-in-Isabelle theory files.

Z-in-Isabelle currently makes a few restrictions on syntax. As already mentioned, quantifiers and set comprehensions may range over one variable only. Moreover, tuples are currently limited to have two elements only. We plan to lift the latter restriction in the foreseeable future. Barring these exceptions, Z-in-Isabelle supports the full concrete syntax of the draft Z standard.

6 Example Proofs

To give a flavor of Z-in-Isabelle proofs, we present two proofs in some detail. The first derives a new rule of inference, the one-point rule, and the second establishes a property that follows from the AddBirthday schema of the BirthdayBook example.

6.1 Deriving the One-Point Rule

The one-point rule is often used to simplify existentially quantified formulae, particularly when computing the preconditions of schemas. The conjecture for the one-point rule can be formulated as follows:

$$\frac{\vdash y \in T \quad \vdash P(y)}{\vdash \exists x : T \mid x = y \bullet P(x)} \quad [\textit{One-Point Rule}]$$

In an Isabelle session, with the theory Z already loaded, the conjecture for the one-point rule is entered as follows:

```
val [major,minor] = goal Z.thy
      "[| |- y \in T ; |- P(y) |] ==>
       |- \exists x:T | x=y @ P(x)";
```

The identifiers major and minor will be bound to the premises, so that they can be referred to directly in the course of the proof. Variables such as y, T, and P are free, but considered fixed from the point of view of unification[6]. This is in contrast with schematic variables (always preceded by "?"), which are also free, but may become instantiated via unification. The list of premises is enclosed in double square brackets, and premises are separated by semi-colons. Isabelle returns the initial proof state[7]

Level 0
$\vdash \exists x : T \mid x = y \bullet P(x)$
1. $\vdash \exists x : T \mid x = y \bullet P(x)$

---

[6] This prevents unwanted instantiations. For instance, y could inadvertently become instantiated to 0 in the course of the proof. While

$$\frac{\vdash 0 \in T \quad \vdash P(0)}{\vdash \exists x : T \mid x = 0 \bullet P(x)}$$

is certainly a sound rule, it is not particularly useful.

[7] In the following, we present pretty-printed proof states to ease readability.

In each proof state, the original conjecture is followed by the subgoals that remain to be proven. To begin with, the subgoal is the original conjecture. We now need to apply a rule for existential quantification on the right. To this end, we resolve, in Isabelle's metalogic, the subgoal against the rule $exRthin$

$$\frac{\vdash x \in T \quad \vdash P(x) \quad \vdash Q(x)}{\vdash \exists x : T \mid P(x) \bullet Q(x)} \quad [\, exRthin \,]$$

which is derived from the rule $(\vdash \exists)_m$ in section 3.2.

> - by (resolve_tac [exRthin] 1);
> Level 1
> $\vdash \exists x : T \mid x = y \bullet P(x)$
> 1. $\vdash\ ?x \in T$
> 2. $\vdash\ ?x = y$
> 3. $\vdash P(?x)$

This results in three subgoals. It also results in the introduction of a new variable, $?x$, representing the witness for the existentially quantified variable. It is a schematic variable or unknown, meaning that it can be instantiated via higher-order unification in the following proof steps. We can now resolve the first subgoal against the major premise.

> - by (resolve_tac [major] 1);
> Level 2
> $\vdash \exists x : T \mid x = y \bullet P(x)$
> 1. $\vdash y = y$
> 2. $\vdash P(y)$

This solves the subgoal and instantiates the variable $?x$ with y, as is apparent from the new first subgoal. We apply the reflection rule to this subgoal

> - by (resolve_tac [reflection] 1);
> Level 3
> $\vdash \exists x : T \mid x = y \bullet P(x)$
> 1. $\vdash y \in\ ?T1$
> 2. $\vdash P(y)$

which introduces a new subgoal $y \in\ ?T1$. This is solved by resolving against the major premise again.

> - by (resolve_tac [major] 1);
> Level 4
> $\vdash \exists x : T \mid x = y \bullet P(x)$
> 1. $\vdash P(y)$

To solve the remaining subgoal, we appeal to the minor premise.

> - by (resolve_tac [minor] 1);
> Level 5
> $\vdash \exists x : T \mid x = y \bullet P(x)$
> No subgoals!

This leaves us with no subgoals to prove. We bind the theorem to an identifier of our choice, so that we can use it as an inference rule in subsequent proofs.

$$- \text{ val onepoint} = \text{result}();$$

$$\text{val onepoint} =$$
$$[|\ \vdash\ ?y \in\ ?T; \vdash\ ?P(?y)\ |] \Rightarrow\ \vdash \exists x :\ ?T \mid x =\ ?y \bullet\ ?P(x) : \text{thm}$$

Note how the free variables y, T, and P are now turned into schematic variables $?y$, $?T$, and $?P$, so that the rule can be used immediately in resolution. Thus, for instance, if we have a subgoal of the form

$$\ldots \vdash \exists x : \mathbb{Z} \mid x = y \bullet x > 0,$$

we can apply the derived one-point rule directly to obtain the subgoals

$$\ldots \vdash y \in \mathbb{Z} \quad \text{and} \quad \ldots \vdash y > 0.$$

6.2 The BirthdayBook Example

We now show another type of deduction in Z-in-Isabelle, namely proving a theorem that follows from a specification. We prove, from the schemas

```
┌─ BirthdayBook ─────────────────────────────
│ known : ℙ NAME
│ birthday : NAME ⇸ DATE
├────────────────────────────────────────────
│ known = dom birthday
└────────────────────────────────────────────
```

```
┌─ AddBirthday ──────────────────────────────
│ ΔBirthdayBook
│ name? : NAME
│ date? : DATE
├────────────────────────────────────────────
│ name? ∉ known
│ birthday' = birthday ∪ {name? ↦ date?}
└────────────────────────────────────────────
```

the conjecture

$$known' = known \cup \{name?\} \tag{1}$$

It states that after we add a birthday to the birthday book, the known names should include both the previously known names and the new name.

The example is taken from Spivey [12], and the proof here loosely follows the one given there. It makes use of the same lemmas, namely

$$\text{dom}(f \cup g) = (\text{dom} f) \cup (\text{dom} g) \tag{2}$$

$$\text{dom}\{a \mapsto b\} = \{a\} \tag{3}$$

In fact, proving these lemmas is much more difficult than proving (1). The first requires well over one hundred inferences in Isabelle, the second around fifty, which makes the proofs too large to present here in detail. The basic steps are unfolding the definitions of dom, ∪, and in some instances ↦, and decomposing goals and hypotheses as needed. Large stretches of the proofs can be shortened or automated by using derived rules and tactics.

Armed with these lemmas, however, the proof of (1) is relatively straightforward, proceeding by substitutions justified by hypotheses or lemmas. The original (preprocessed) conjecture is

$$goal\ BB.thy\ ``AddBirthday \vdash known' = known \cup \{name?\}";$$

Remember that *AddBirthday* is not a schema reference (since they have all been expanded), but simply the concrete syntax for the *AddBirthday* schema. We apply a tactic that "unpacks" the *AddBirthday* schema as described at the end of section 3.3. This yields the subgoal (for reasons of space, we now show only the subgoals of the main proof steps).

1. $birthday \in NAME \nrightarrow DATE$,
   $birthday' \in NAME \nrightarrow DATE$,
   $known \in \mathbb{P}\,NAME$,
   $known' \in \mathbb{P}\,NAME, name? \in NAME$,
   $date? \in DATE, known = \text{dom}\ birthday$,
   $known' = \text{dom}\ birthday'$,
   $name? \notin known$,
   $birthday' =$
   $birthday \cup \{name? \mapsto date?\}$
   $\vdash known' = known \cup \{name?\}$

We then substitute for *known* using the equation

$$known = \text{dom}\ birthday$$

for *known'* using the equation

$$known' = \text{dom}\ birthday'$$

and for *birthday'* using the equation

$$birthday' = birthday \cup \{name? \mapsto date?\}$$

Substitution involves applying the rule

$$\frac{\vdash x = y \quad \vdash P(y)}{\vdash P(x)}\ [\,subR\,]$$

which is derived from *Leibniz$_m$* in section 3.2. However, the rule unifies in many different ways with the subgoal, most of which are unwanted. We therefore apply a variation of the resolution tactic that pre-instantiates variables in inference rules. In the first substitution, for instance, we instantiate x in the rule with *known*. The substitutions yield the main subgoal

1. $birthday \in NAME \nrightarrow DATE$,
   $birthday' \in NAME \nrightarrow DATE$,
   $known \in \mathbb{P}\,NAME$,
   $known' \in \mathbb{P}\,NAME, name? \in NAME$,
   $date? \in DATE, known = \text{dom}\,birthday$,
   $known' = \text{dom}\,birthday'$,
   $name? \notin known$,
   $birthday' =$
   $birthday \cup \{name? \mapsto date?\}$
   $\vdash \text{dom}(birthday \cup \{name? \mapsto date?\}) = \text{dom}\,birthday \cup \{name?\}$

Using (2), we can substitute

$$(\text{dom}\,birthday) \cup (\text{dom}\{name? \mapsto date?\})$$

for

$$\text{dom}(birthday \cup \{name? \mapsto date?\})$$

which yields

1. $birthday \in NAME \nrightarrow DATE$,
   $birthday' \in NAME \nrightarrow DATE$,
   $known \in \mathbb{P}\,NAME$,
   $known' \in \mathbb{P}\,NAME, name? \in NAME$,
   $date? \in DATE, known = \text{dom}\,birthday$,
   $known' = \text{dom}\,birthday'$,
   $name? \notin known$,
   $birthday' =$
   $birthday \cup \{name? \mapsto date?\}$
   $\vdash \text{dom}\,birthday \cup \text{dom}\{name? \mapsto date?\} =$
   $\qquad \text{dom}\,birthday \cup \{name?\}$

and using (3), we can substitute

$$\{name?\}$$

for

$$\text{dom}\{name? \mapsto date?\}$$

which then yields

1. $birthday \in NAME \nrightarrow DATE$,
   $birthday' \in NAME \nrightarrow DATE$,
   $known \in \mathbb{P}\,NAME$,
   $known' \in \mathbb{P}\,NAME, name? \in NAME$,
   $date? \in DATE, known = \operatorname{dom} birthday$,
   $known' = \operatorname{dom} birthday'$,
   $name? \notin known$,
   $birthday' =$
   $birthday \cup \{name? \mapsto date?\}$
   $\vdash \operatorname{dom} birthday \cup \{name?\} = \operatorname{dom} birthday \cup \{name?\}$

This can be solved by applying reflection, and the proof is complete.

7 Conclusions

In this paper, we have discussed some of the central issues that arose in the course of implementing a deep embedding of the specification language Z in the meta-logic of the generic theorem prover Isabelle. We have given a brief overview of the implementation and provided two examples of proofs carried out in the system. The first proof demonstrates one of the advantages of a deep embedding: The rigorous derivation of a new rule of inference, in this case the well-known one-point rule. The second proof gives an idea of what is involved in proving a property of a given specification.

The two major deep embeddings of Z in logical frameworks are Jigsaw, based on the deductive system of Woodcock and Brien [13] and implemented in 2OBJ [3], and Z-in-Isabelle. They are, in fact, quite dissimilar, wherefore a comparison is worthwhile. Some of the differences between the two embeddings arise from the logical frameworks themselves. For instance, 2OBJ has no notion of schematic variables, which is essential when proving general properties of a language, e.g., the associativity of an operator, and when deriving new rules of inference within the system. Related to the lack of schematic variables in 2OBJ is the fact that 2OBJ uses pattern matching rather than unification. Of the two logical frameworks, Isabelle seems the more flexible.

Other differences between the two systems lie in the approaches taken to the embedding itself. Jigsaw is very faithful to the presentation of the deductive system in [13], directly encoding not only the premises and conclusions of each rule, but also the side conditions. Z-in-Isabelle takes care of side conditions at the meta- rather than the object level by exploiting higher-order abstract syntax. Furthermore, Jigsaw does rule-lifting explicitly via tactics, while Z-in-Isabelle exploits unification to simulate rule-lifting. Jigsaw is thus the deeper embedding of the two.

Jigsaw, however, is slow. One reason for Jigsaw's lack of efficiency is that where Z-in-Isabelle uses derived rules, Jigsaw must rely on tactics. A further loss of efficiency is caused by the large amount computation involved in explicitly computing side conditions and substitutions. In Z-in-Isabelle, this corresponds to much simpler computations at the meta-level. One main emphasis of our implementation was precisely the use of higher-order abstract syntax, an efficient and elegant approach to dealing with variable binding and substitution. While employing higher-order abstract syntax has required us to sacrifice some deepness of embedding for the sake of correctness, we believe that the benefits, i.e., efficiency and simplicity, are worth the cost.

As mentioned above, one major advantage of using a logical framework for an embedding is that faithfulness and the adequacy of the embedding is easier to prove. Such a proof provides users with a high degree of confidence in the system. We are in the course of establishing these properties. From a theoretical point of view, implementing Z in a meta-logic has been useful in pointing out various issues that require further investigation. From a practical point of view, the deep embedding of Z-in-Isabelle has been useful in that we were able to derive inference rules that were simpler to use.

In future work, beyond investigating theoretical issues, we need to build up a library of lemmas such as (2) and (3) based on the mathematical toolkit. Without these, the second example proof would have become orders of magnitude longer and more complex. We plan to build a library of lemmas such that a direct appeal to any toolkit definition will rarely be necessary. Furthermore, we intend to expand the set of derived inference rules and to organize them in tactics in such a way that large parts of proofs can be automated.

The results of the implementation of Z-in-Isabelle we have achieved so far are promising, and we believe that with the resolution of outstanding theoretical issues and the development of a large set of lemmas, derived rules of inference, and tactics, we will have a reliable, useful, and flexible system in which to reason about Z specifications.

References

1. J.P. Bowen and M.J.C. Gordon. Z and HOL. In J.P. Bowen and J.A. Hall, editors, *Z User Workshop, Cambridge 1994*, Workshops in Computing, pages 141–167. Springer-Verlag, 1994.
2. S. M. Brien and J. E. Nicholls. Z base standard. Technical Monograph PRG-107, Oxford University Computing Laboratory, Wolfson Building, Parks Road, Oxford, UK, November 1992. Accepted for ISO standardization, ISO/IEC JTC1/SC22.
3. J. Goguen, A. Stevens, H. Hilberdink, and K. Hobley. 2OBJ: A metalogical theorem prover based on equational logic. *Philosophical Transactions of the Royal Society, Series A*, 339:69–86, 1992.
4. W. T. Harwood. Proof rules for Balzac. Technical Report WTH/P7/001, Imperial Software Technology, Cambridge, UK, 1991.
5. R. B. Jones. ICL ProofPower. *BCS FACS FACTS*, Series III, 1(1):10–13, Winter 1992.
6. L. E. Jordan. The Z syntax supported by Balzac-II/1. Technical Report LEJ/S1/001, Imperial Software Technology, Cambridge, UK, 1991.
7. S. Maharaj. Implementing Z in Lego. Master's thesis, Department of Computer Science, University of Edinburgh, September 1990.
8. A. Martin. *Machine-Assisted Theorem-Proving for Software Engineering*. PhD thesis, University of Oxford, 1994.
9. L.C. Paulson. The foundation of a generic theorem prover. *The Journal of Automated Reasoning*, 5(3):363–397, 1989.
10. L.C. Paulson. *Isabelle. A Generic Theorem Prover*. Lecture Notes in Computer Science. Springer-Verlag, 1994.

11. F. Pfenning and C. Elliot. Higher-order abstract syntax. In *Proceedings of the ACM-SIGPLAN Conference on Programming Language Design and Implementation*, 1988.
12. J. M. Spivey. *The Z Notation: A Reference Manual*. Series in Computer Science. Prentice Hall International, 2nd edition, 1992.
13. J. C. P. Woodcock and S. M. Brien. W: A logic for Z. In J. E. Nicholls, editor, *Z User Workshop, York 1991*, Workshops in Computing, pages 77–96. Springer-Verlag, 1992.

The Z-into-Haskell Tool-kit:
An Illustrative Case Study

Howard S. Goodman

School of Computer Science, University of Birmingham,
Birmingham, England, B15 2TT
Email: h.s.goodman@cs.bham.ac.uk

Abstract. This paper presents an illustrative case study in the use of a new method of writing Haskell programs from Z specifications. Whereas previous methods of writing functional programs from formal specifications have relied either too little or too much on the intellectual skills of the programmer, this new approach aims to strikes a balance, by providing what is essentially a tool-kit, consisting of a framework of code together with a set of rules and guidelines for the programmer.

1   Introduction

The use of formal methods in software engineering has been observed to increase both reliability of code and programmer productivity [1]. The use of a popular dedicated specification language such as Z [2], which is basically just typed set theory, is well documented, to the extent that it could reasonably be argued that writing and reading Z specifications is relatively easy. Indeed, the specification of a system is invariably easier to comprehend than the implementation, because one need not be concerned with the details of execution.

However, if formal methods are to gain more widespread use, it is important that the relative ease of expressing computational problems in Z be extended to the rest of the software development process. At present, it is the case that the techniques by which programs are written from specifications – often termed *refinement* or *animation* of specifications into code – are difficult to use in practice. Such traditional formal methods [3, 4] typically use a small imperative programming language of just five essential constructs. It is intended that programs which have been developed into this language then be re-expressed in a real imperative language.

The end result of such a process is programs which are substantially removed in terms of style from the specifications which they are intended to satisfy, a factor which can, no doubt, be attributed to the inherent differences of purpose between specification languages and procedural programming languages. Whereas the former are declarative, in that they are designed to state *what* is to be computed, with as little execution detail as possible, the latter are imperative, in that they are designed to state *how* to compute the desired result. Thus, when using imperative languages, the intended meaning of programs is obscured, with negative implications for the maintainability of software.

It has long been argued that functional programming languages, with their foundations in the mathematics of the lambda calculus, offer a solution to this problem [5]. Programs written in such languages are referentially transparent, facilitating formal reasoning; higher-order functions enable programs to be constructed elegantly from re-usable smaller units; lazy evaluation enables infinite structures to be represented; and the programmer is freed from having to take care of memory management. Because they have to contain sufficient detail to be executable, such languages are not as abstract as dedicated specification languages, but, nevertheless, the level of abstraction possible in a functional language is considerably higher than that which can ever be attained in an imperative language. Moreover, while, in the past, functional languages have been slow and cumbersome, the tremendous research effort that is currently being put into the development of Haskell [6] continues to yield an increasingly more advanced language and better compilers, so that, at last, a functional language can be used to write real, deliverable programs.

Previous work in animating specifications in functional languages has, on the whole, been geared to automatic translation of specifications into functional prototypes, and crude tools have been produced to this end [7, 8]. In striving for automatic translation, however, it is easy to go too far in simulating Z's mathematical structures in a functional language: for example, the infinite list [1 ..] has been proposed as a model of N_1 [9], when a built-in integer type would clearly be better. If, on the other hand, we accept that, within the limits of present-day technology, programs are better written by people than by computers, we find ourselves at the opposite extreme: the similarities between the specification and programming languages are not exploited to the full [10], and the programmer may even be expected to write programs almost by magic from specifications, as in the example functional animation in chapter 19 (pages 271–8) of [11]. Furthermore, it is a criticism of all such methods to date, automatic or otherwise, that state is handled clumsily, being passed explicitly from function to function, and that input/output, if it is tackled at all, is done so using a technique of *continuations*, which is widely regarded as less intuitive than the sequential, imperative style.

Therefore a new approach to the problem of developing functional programs from Z specifications is proposed in this paper, whose goals are as follows:

- The method should be easy to use, provided that the programmer has been suitably versed in it and is familiar with Z and Haskell.

- The resultant code should be of good quality, with a balance being struck between clarity on the one hand and executability on the other.

- Because it is fairly universally agreed that specifications are easier to understand and manipulate than programs, it is intended that the specification document, rather than the code itself, should be the primary resource as far as program maintenance is concerned.

The key to all three of these goals is to bridge the gap between the specification and code domains. The challenge lies in finding exactly where and how to do this.

Throughout this paper, the use of the new method of writing functional programs from Z specifications will be illustrated by means of a single example specification. A good choice of specification for this role is that of the Internal Telephone Directory, as it is presented in chapter 4 (pages 41–70) of [11]. It is a short, simple specification of a system which is typical of a whole class of systems whose specifications abound in the Z literature.

The structure of this paper, then, is as follows. A brief overview of the 'Z-into-Haskell tool-kit' is presented in section 2. Roughly adhering to the structure of the specification document, the implementation of the state space is discussed in section 3, that of the operations in section 4, and the provision of a user interface in section 5. Finally, the method is reviewed and discussed in section 6.

2 The Z-into-Haskell Tool-kit

The new method of writing Haskell programs from Z specifications whose use is illustrated in this paper can be regarded as a kind of 'tool-kit'. This tool-kit, which is presented in its own right in [12], consists essentially of three discrete parts.

The first part is a high-level strategy to guide the programmer in writing Haskell programs from Z specifications. Essentially, the strategy tells the programmer how to proceed when faced with a specification document: for example, what to do with the system state, how to expand and analyse operation schemas so that interactive functions can be produced from them, and how to structure the program.

The second part of the Z-into-Haskell tool-kit is the framework for handling input/output and state. These features, though absolutely essential parts of Z, are frequently glossed over by functional programmers, as functional languages do not have a state per se, as this would violate referential transparency, and hitherto have not provided sufficiently intuitive mechanisms for expressing interaction. Recently, however, the application of elements of category theory to practical functional programming has yielded the monad [13], an elaborate form of syntactic sugar with which the workings of 'difficult', imperative-style features can be concealed in a purely functional language. For animating Z specifications, a re-usable program module entitled *IOS* has been developed [14]. In this module a monad is defined which incorporates input, output and state, allowing the programmer to deal with these features in a new way which is considerably simpler – and closer in style to the specification language – than before. Further, a framework is provided, consisting of a *main* function and a skeleton command loop which the programmer can adapt to provide a simple, command-line-based interface to any program.

The third part of the tool-kit, known as the 'rules', describes how to implement individual Z types and expressions in Haskell. Some of the rules, where the implementation of the Z construct in question is straightforward, such as the translation of ⌢ as ++, are direct and one-to-one, and are presented in simple tables. Others, which describe how to implement certain Z characteristics where there is a choice of possible implementations, such as the choice between two representations of relations, may be presented in tables with conditions attached to each of the choices. The more complex rules may also be accompanied by textual explanation, and it is this which is presented in boxes throughout this paper.

It is into the 'rules' component of the tool-kit that the majority of the work in bringing the code domain closer to the specification domain has expended. The translations relate to a Haskell implementation of Z's 'mathematical tool-kit' [2], which is provided in the form of a Haskell module, entitled *ZPrelude*, which is designed to be incorporated by default into all programs which are written from Z specifications. A rich collection of data types and operations upon them, which are defined as standard in Z but not in Haskell, has been provided, including sets, and, based on these, bags, and a variety of relation and function types. Sets, of course, cannot contain duplicated elements, whereas Haskell's built-in lists can. The decision to implement sets explicitly, rather than to model them trivially as lists, was primarily the result of the observation that it is not always easy to tell when a list might contain repeated elements. A consistent approach to the use of sets, in which repetitions are filtered out by default, spares the programmer some unnecessary, error-prone work. Furthermore, by using an abstract data type of sets, we are not committed to one particular implementation.

Given this brief overview of the method, its use will now be illustrated by means of a simple case study. Though not essential for the comprehension of this paper, the reader is referred to chapter 4 (pages 41–70) of [11] for the full exposition of the specification in hand.

3 The State Space

3.1 Given Sets

As usual, the first stage in the specification process is the description of the basic types, or *given sets*, which are used in the specification. The 'strategy' part of the Z-into-Haskell tool-kit states that the first stage in the implementation process is to implement these types, typically in a separate *State* module; the 'rules' part dictates how actually to implement each type.

Just two given sets are introduced in this specification, representing the types of people's names and internal telephone numbers respectively:

[*Person, Phone*]

The following rule points the way to an implementation of these types:

> *String* is the most general type to use in such circumstances, the appropriate values being instantiated as and when required; but, where some structure is apparent, a tuple may be more suitable, such as the implementation of a *Name* type by a combination of surname and initials. In either case, it is recommended that a synonym type be defined for clarity's sake.

That is to say, though we may well choose to represent *Person* as, say, a tuple of surname and initials, for the purposes of writing this specimen implementation, in which we are not biased towards any particular implementation, the most appropriate type to choose is simply *String*. The same can be applied to *Phone* as well: although *Int* might be a more obvious choice, telephone numbers may have non-numeric characters in them, and, besides, strings are more efficient to process.

Therefore, in the *State* module of the Haskell implementation, we write:

$$\text{type } \textit{Person} = \textit{String}$$
$$\text{type } \textit{Phone} = \textit{String}$$

As would be expected, a *Report* type is also used in the specification, for the messages which are output to the user of the system. Although in Z such types often have all their possible values enumerated in a free type definition, this is quite unnecessary in the program. Rather, we apply the following rule:

> Where a type is designed primarily for the purposes of textual output, such as the common *Report* or *Message* types, *String* is the only sensible choice.

3.2   The System State

The state schema of the internal telephone directory is defined thus:

$$
\begin{array}{l}
\underline{\quad PhoneDB\quad\qquad\qquad\qquad\qquad\qquad\qquad} \\
members : \mathbf{P}\ Person \\
telephones : Person \longleftrightarrow Phone \\
\hline
\mathrm{dom}\ telephones \subseteq members
\end{array}
$$

According to the strategy, the state schema of a system should be implemented as a tuple in Haskell, comprising all the essential data objects, and omitting both auxiliary data objects and invariants. The former are unnecessary baggage, and can be calculated on demand from the essential data objects. The latter are lost in translation because conditions cannot be attached to a tuple, but this is not a problem, as any well-written specification will ensure that the system invariants are established at the outset and are never violated. This is just one aspect of how the Z document is to be the primary reference for maintenance, and the program itself is reduced to second-class status.

Inspection of the predicate part of the above schema makes it clear that

$$\mathrm{dom}\ telephones \subseteq members$$

is merely an invariant, and that neither *members* nor *telephones* is auxiliary, as neither is wholly defined in terms of any other variables. Therefore the state tuple will simply be a (*members*, *telephones*) pair.

To decide upon appropriate types for *members* and *telephones* – essentially a question of translating Z into Haskell – we return to the rule-base. Sometimes in Z a set is used as a unary relation, merely for testing membership: such a set can be termed a *passive* set [15]. It is easy to see that this is not the case for *members*, which is both *active* – that is to say, used as a complete data structure in its own right – and, more crucially, *mutable*, in that its value is not constant throughout the specification. Hence its implementation is dictated by the rule:

> Where a variable in a Z specification has the type $\mathbf{P}\ X$, for some underlying type X, the default type in an implementation is the active $P\ x$. It is only where a set is obviously used just as a test, i.e. passively, and, furthermore, that it is immutable, that the passive set, or unary relation, type, $Rel1\ x$, should be used.

The type $P\ x$ – a set of objects of type x – is defined in the *ZPrelude*. The unary relation type *Rel1 x* is defined as $x \rightarrow Bool$, using Haskell's built-in function mechanism: thus it can only be used for *immutable* objects, as the behaviour of a Haskell function cannot be modified during a program's execution.

A similar choice of two paths exists for the relation *telephones*, and a similar observation leads us to the correct implementation. It is not a relation in the sense that it is used for testing, like $\leq$, for example, and so the following rule applies:

> If a relation or function is either active or mutable, it *must* be implemented using the Z-style, primed types, $Rel'\ x\ y$ and $Fun'\ x\ y$.

Again, $Rel'\ x\ y$, a set of pairs of type (x, y), is defined in *ZPrelude*. The other type of relation, a 'true Haskell' relation, is called $Rel\ x\ y$, and is equivalent to $x \rightarrow y \rightarrow Bool$.

The underlying simple types *Person* and *Phone* have already been given their implementations, so we can move directly on to define the state tuple itself:

$$\text{type } PhoneDB\ =\ (P\ Person,\ Rel'\ Person\ Phone)$$

4   The Operations

This being a trivial specification, there is no 'general theory' [16] or any other intermediate, static definitions that would warrant inclusion in an intermediate module. Instead, we move straight on to the interactive operations themselves, whose implementations will form the bulk of the *Main* module.

It is common Z style first to define partial operations, which work if the necessary preconditions are satisfied, then to define any number of error conditions, and finally to unify the two into a total operation. That of adding an entry to the database is expressed thus:

$$DoAddEntry\ \hat{=}\ AddEntry \land Success$$
$$\lor$$
$$NotMember$$
$$\lor$$
$$EntryAlreadyExists$$

Because the formation of conjunction and disjunction of operations is not possible in Haskell, and is cited as such as a non-implementable specification construct [17], the strategy says that we must keep all the calculation in the Z domain before moving into the Haskell domain. The normal procedure is, therefore, to expand the total operation into schema form:

```
┌─ DoAddEntry ──────────────────────────────────────────
│  members, members' : P Person
│  telephones, telephones' : Person ⟷ Phone
│  name? : Person
│  newnumber? : Phone
│  rep! : Report
├───────────────────────────────────────────────────────
│  dom telephones ⊆ members
│  dom telephones' ⊆ members'
│
│  ((name? ∈ members ∧
│  name? ↦ newnumber? ∉ telephones ∧
│  members' = members ∧
│  telephones' = telephones ∪ {name? ↦ newnumber?} ∧
│  rep! = "Okay")
│         ∨
│  (name? ∉ members ∧
│  members' = members ∧
│  telephones' = telephones ∧
│  rep! = "Not a member")
│         ∨
│  (name? ↦ newnumber? ∈ telephones ∧
│  members' = members ∧
│  telephones' = telephones ∧
│  rep! = "Entry already exists"))
└───────────────────────────────────────────────────────
```

The next stage, according to the strategy, is 'variable classification', in which the total operation schema is examined so that the variables named in its declaration part may be classified according to their use. Sometimes it will be the case that there are auxiliary variables, which we shall define locally in the Haskell code, or even that apparent input and output variables (designated by ? and ! respectively in Z, but by _in and _out in Haskell) are actually auxiliary, having their values entirely constrained by other variables. In the case of *DoAddEntry*, though, it is clear that *members* and *telephones* are true pre-state variables, *members'* and *telephones'* are their equally necessary post-state counterparts, *name?* and *newnumber?* are true inputs which will be required from the user of the system, and *rep!* will have to be output to the user.

Following on from the variable classification is the 'predicate analysis', in which the actual structure of the schema is examined. In this case, the first two lines of the predicate part are simply the invariant from the state schema, which is lost in translation to Haskell. The remainder is a standard case analysis, expressed as the disjunction of conjunctions. In each of the three cases, there is a test, involving input and pre-state variables only; a change of state, which is easily recognisable by equality expressions involving post-state variables; and some output, for which *rep!* is used.

Now, from this expanded total operation schema and the accompanying analysis, we move straight to an interactive Haskell function, which we shall call simply *addEntry*. This function will take the necessary input from the user and then carry out a case analysis completely analogous to that in the Z. As explained in section 2, interaction and state – aspects which have previously been messy in this kind of development – are handled very neatly by means of a monad [13], in this case the dedicated *IOS* monad [14] which makes provision for input from the user, output to the user, and a system state.

The principal type defined in the *IOS* module is *IOS x*, where *x* is a type variable. An object of this type represents a computation, involving input/output and state, in which the result is of type *x*. The dummy type $()$ is used where, notionally, there is no result: thus the type of a typical interactive function, such as that which will implement *DoAddEntry*, is *IOS* $()$. Sequential composition is achieved by means of either *thenIOS*, which binds the value of the expression on its left to the variable named in the lambda expression on its right, or *seqIOS*, which composes two monad-based operations when the output of the first is uninteresting, i.e. $()$. In addition to these, the following monad-based functions, defined in the *IOS* module, are used in this example: *fetchState*, which binds the current value of the state to the variable named in the lambda expression on the right-hand side of the *thenIOS* operator; *setState*, which sets the state to the value of its argument, returning $()$; *write*, which writes a given string to the standard output, returning $()$; and *read1*, which writes a given prompt to the standard output, then reads a string from the standard input and binds it to the variable named in the lambda expression on the right.

To express the conditional in a function such as *addEntry*, we sugar Haskell's built-in if–then–else construct in a way which allows a notation of guards, and – since it is assumed that the cases are exhaustive, as is normal in well-written Z – does not require a default case. (The conditional *cond*, along with the choice and guard operators ||| and ⊩, are defined in the *ZPrelude*.)

In translating the Z expressions themselves to Haskell, the *ZPrelude* and its accompanying translation rules provide all that is necessary for this simple example. The following table summarises the rules which are required in this instance, and provides a flavour of the use of the rules in general:

Z	Haskell
$\in$	*inset*
$\notin$	*notin*
$x \mapsto y$	(x, y)
$\cup$	*union*
$\{x_1, \ldots, x_n\}$	*set* $[x_1, \ldots, x_n]$

Note that the infix functions, underlined when typeset, are enclosed in opening single quotation marks in ASCII, viz. `'inset'`. (There are, of course, other differences between typeset and ASCII programs, such as the replacement of λ by \.) Also notice the translation of the set display $\{x\}$ into the list $[x]$, with the constructor function *set* applied to yield a set.

The function, then, is as follows:

$addEntry =$
$\quad fetchState \qquad\qquad \underline{thenIOS}\ \lambda(members,\ telephones) \rightarrow$
$\quad read1\ \text{``Name? ''} \quad \underline{thenIOS}\ \lambda name\_in \qquad\qquad \rightarrow$
$\quad read1\ \text{``Number? ''} \quad \underline{thenIOS}\ \lambda newnumber\_in \qquad \rightarrow$
$\quad cond$
$\qquad |||\ (name\_in\ \underline{inset}\ members\ \wedge$
$\qquad\qquad (name\_in,\ newnumber\_in)\ \underline{notin}\ telephones)$
$\qquad\qquad \mapsto (setState\ (members,\ telephones\ \underline{union}$
$\qquad\qquad\qquad\qquad\qquad set\ [(name\_in,\ newnumber\_in)])$
$\qquad\qquad\qquad\qquad \underline{seqIOS}$
$\qquad\qquad\quad write\ \text{``Okay\textbackslash n''})$
$\qquad |||\ name\_in\ \underline{notin}\ members$
$\qquad\qquad \mapsto write\ \text{``Not a member\textbackslash n''}$
$\qquad |||\ (name\_in,\ newnumber\_in)\ \underline{inset}\ telephones$
$\qquad\qquad \mapsto write\ \text{``Entry already exists\textbackslash n''}$

This was arrived at by following the guidelines laid down in the Z-into-Haskell strategy, which can be summarised as follows. A total operation in Z is implemented as a function of type $IOS\ ()$, in which no input is taken and no output is returned. This function should, firstly, fetch the state and bind it to an appropriately named tuple; secondly, the necessary input variables should be received from the user; thirdly, the actions should be defined in a guarded $cond$ expression, any changes to the state being made with a $setState$ expression.

Even to a reader unfamiliar with the use of monads in Haskell, the close correspondence between the $DoAddEntry$ schema and the $addEntry$ function should be apparent. The inputs are mentioned quite explicitly, as are outputs and changes of state, and the body of the operation is clearly a three-case conditional. The primary differences, apart from the stark change in notation and the imposition of sequential ordering, are the loss of the invariants and the variable signatures, and the separation between the guards and their guarded actions, a distinction which is not made explicit in this particular style of Z specification. Whilst the obligatory conversion from Z to Haskell notation is unfortunate in the sense that, for example,

$$telephones \cup \{name? \mapsto newnumber?\}$$

reads so much more clearly than

$$telephones\ \underline{union}\ set\ [(name\_in,\ newnumber\_in)],$$

the great similarity between the Z schema and the Haskell function, and the fact that the latter can be written simply and directly from the former, further illustrates how the specification document can remain the main resource for program maintenance. If the specification of an operation is changed, a new function can easily be produced from it almost as a by-product of the specification.

The implementation of the other operations specified in [11] is omitted here for the sake of brevity.

5   A User Interface

Z specifications are traditionally vague as far as 'non-functional' requirements such as user interfaces are concerned. While attempts have been made to remedy this [18], it is typically the case that the form of the intended program will be quite obvious from the Z document as it stands – for example, there will be a suite of operations defined quite explicitly as schemas – so further elaboration in Z is not necessary.

As to the actual form that the user interface might take, for the sake of simplicity, and following the lead set by the functional animation illustrated in chapter 19 (pages 271–8) of [11], the current incarnation of the Z–into–Haskell tool-kit makes provision for simple text-based input/output. The top level of a program consists of a loop in which the user types short commands to access the various operations offered by the system; the appropriate command is then run, input being taken from the keyboard and output being sent, as ASCII text, to the screen; and the loop is terminated and the program exited by the user typing the command 'end'.

In the case of the Internal Telephone Directory, six operations are specified, from which six Haskell functions can be written as illustrated in the preceding section. The command names, from pages 62–3 of [11], are shown in the following table, alongside the names of the corresponding total Z schemas and the Haskell functions:

Command	Z schema	Haskell function
am	*DoAddMember*	*addMember*
rm	*DoRemoveMember*	*removeMember*
ae	*DoAddEntry*	*addEntry*
re	*DoRemoveEntry*	*removeEntry*
fp	*DoFindPhones*	*findPhones*
fn	*DoFindNames*	*findNames*

As with the individual interactive functions themselves, here, once again, the Haskell code is written directly from the specification. The Z–into–Haskell tool-kit includes a skeleton for the construction of user interfaces of the form described above, of which the following command loop is the main part:

$$
\begin{aligned}
&commandLoop :: IOS\ () \\
&commandLoop = \\
&\quad read1\ \text{``Command?\ ''} \qquad \underline{thenIOS}\ \lambda cmd \rightarrow \\
&\quad if\ cmd == \text{``end''}\ then \\
&\qquad write\ \text{``Bye-bye.\textbackslash n\textbackslash n''} \\
&\quad else \\
&\quad (case\ cmd\ of \\
&\qquad cmd_1 \rightarrow action_1 \\
&\qquad \quad\ \vdots \\
&\qquad cmd_n \rightarrow action_n \\
&\qquad \_\ \ \rightarrow write\ \text{``Unknown command\textbackslash n''}) \\
&\qquad\qquad\qquad \underline{seqIOS} \\
&\quad commandLoop
\end{aligned}
$$

All that is necessary is to fill in the gaps thus –

$$\text{case } \textit{cmd} \text{ of}$$
$$\text{"am"} \rightarrow \textit{addMember}$$
$$\text{"rm"} \rightarrow \textit{removeMember}$$
$$\text{"ae"} \rightarrow \textit{addEntry}$$
$$\text{"re"} \rightarrow \textit{removeEntry}$$
$$\text{"fp"} \rightarrow \textit{findPhones}$$
$$\text{"fn"} \rightarrow \textit{findNames}$$
$$\text{"?"} \rightarrow \textit{showInfo}$$
$$\_ \quad \rightarrow \textit{write } \text{"Unknown command\textbackslash n"}$$

– and to add both a *showInfo* function, which outputs instructions to the screen, and a top-level *main* function, which displays an introduction before launching into the command loop:

$$\textit{mainIOS} :: \textit{IOS } ()$$
$$\textit{mainIOS} =$$
$$\quad \textit{showInfo} \quad \underline{\textit{seqIOS}}$$
$$\quad \textit{commandLoop}$$

Finally, because the current edition of Haskell [6] dictates that a program must have a function *main* in the module *Main*, and that its type must be the continuation-based interactive type *Dialogue*, it is necessary to apply a conversion function (from the *IOS* module) to this:

$$\textit{main} :: \textit{Dialogue}$$
$$\textit{main} = \textit{runIOS mainIOS}$$

This concludes the basic user interface to the program.

6   Discussion

6.1   Summary

In this paper, the use of a new method of writing functional programs (in Haskell) from Z specifications has been demonstrated. The method, which is known as the 'Z-into-Haskell tool-kit', comprises three fairly distinct parts. The part known as the 'strategy' has been seen to guide the programmer, through the specification document in question, towards an implementation. Along this path, the part of the method known as the 'rules' has been called upon to assist in the implementation of individual Z constructs: this consists of translation rules, expressed wherever possible in the form of tables, which show how to translate expressions in Z into expressions in Haskell, augmented by an additional *ZPrelude* module, in which most of Z's mathematical tool-kit is implemented. Finally, further re-usable code, mostly resident in the *IOS* module, has been utilised in the construction of interactive functions and in bringing the disparate parts of the system together.

6.2 Limitations of the Method

Although this paper only refers to one small example, the Z-into-Haskell tool-kit has been developed by animating the specifications of several small and medium-size systems which have been found in the Z literature [11, 16, 19]. By necessity, the number of systems examined so far has been rather small, but it is intended that the method be made as broad as possible, by studying further examples, and thus it will be applicable to an even wider variety of specifications.

This is particularly true of those of the rules which show how to implement 'difficult' Z expressions whose Haskell equivalents are not obvious. These rules represent the generalisation of individual design decisions that have been made during the animation of individual systems, and, as such, are currently few in number. Although the groundwork has been laid, in that the method for discovering and generalising rules has been established, and several useful rules have been created in this way, the rule-base is still in the early stages of development, and further work is necessary for its expansion.

While the rule-base includes the relatively general guidelines as to which of two implementations of, say, relations to choose, it is also home to more specific rules, such as those which deal with Z's binary composition operator for relations, ;. The implementation of ; in Haskell is particularly problematic, because its operands are relations, but, while relations in Z are all of a single kind, and this kind includes sequences and all the varieties of functions as well, in Haskell there are two different implementations of relations alone, another two for functions, and another still for sequences. This leads to a complex set of rules for the one operation alone, of which the following table, which shows how to implement $s \, ; f$ where s is a sequence and f a function, gives a flavour:

Z expr.	Z signature	Haskell signature	Haskell expr.
$s \, ; f$	$s : \text{seq} \, X; f : X \rightarrowtail Y$	$s :: [X]; f :: Fun'\ X\ Y$	$map \, (f \, \#) \, s$
		$s :: [X]; f :: X \rightarrow Y$	$map \, f \, s$

(# is the 'application' operator for functions which are implemented as sets of pairs, while *map* is the standard Haskell function that maps a function onto each member of a list. Also note the condition that f does not necessarily have to be a total function, but it does have to be defined for every point in the range of s in order for the above result to hold.)

Rules like this relate to expressions which arise only occasionally, perhaps just once during the course of this research project. There may be any number of such expressions where a direct translation either does not exist or is not obvious, but, without a rule in the tool-kit, the programmer will have to rely on his/her expertise to produce an implementation, thereby stumbling against the limitations of this method. Even when a solution is found to a difficult problem, without other instances of that problem having been seen, one would be justifiably cautious in generalising this solution into a rule.

6.3 Questions of Formality

Unlike in the earlier formal methods [3, 4], whose reliance on imperative programming languages necessitated rigorous proof that programs fulfilled their specifications, here no mention has been made of proof. Instead the emphasis has been on translation from Z to Haskell in such a way that appears to obviate the need for proof. But to what extent is this the case? How can we be sure that correct use of the Z-into-Haskell tool-kit yields programs which satisfy their specifications?

This is difficult to answer, as we are calling into question the 'bridge' that has been constructed between two language domains. Because these domains are quite separate, with different syntax and different semantics, the construction of proofs that programs meet their specifications is dependent on some postulated correspondences. For example, let us consider the proof that

$$telephones \ \underline{union} \ set \ [(name\_in, newnumber\_in)]$$

is a correct implementation of

$$telephones \cup \{name? \mapsto newnumber?\},$$

which was mentioned in section 4. This translation was carried out using the rules which were tabulated in that section, but it is these translation rules themselves whose 'proofs' are a little elusive.

It is proved in the *ZPrelude* module that *union* achieves the same effect as $\cup$ according to the latter's informal specification. The same holds for the translation of a set display $\{x_1, \ldots, x_n\}$ into $set \ [x_1, \ldots, x_n]$: any duplicated elements in the list are eliminated by the application of *set*. As for the representation of a mapping $x \mapsto y$ as a pair (x, y), this is defined for Z on page 95 of [2], and it is assumed that a pair in Haskell has, to all intents and purposes, the same behaviour as a pair in Z. Thus it appears reasonable to accept that these rules preserve correctness across the linguistic divide, but to what extent this can be proved, and to what extent the rules are axioms, is, as yet, unclear.

In the case of more complex rules, such as that relating to the composition of a sequence and a function, above, these can typically be proved wholly within Z before making the change into Haskell. In the case of certain others, such as those describing how to choose between two varieties of relations and functions, again there appears to be an axiomatic element, although, as long as the chosen variety can be seen to be a correct model, this is adequate for the proof. The same can be said of the input/output and state mechanism, which, moreover, implements aspects of systems which are not even specified in formal Z text. Finally, there is nothing to prove regarding the 'strategy', as this is simply a set of instructions for the user, and does not itself yield anything concrete.

In conclusion, although the hypotheses discussed above have yet to be made more formal, the foundations for carrying out program proofs have been laid, and, with some further work, the Z-into-Haskell tool-kit can be made respectable and robust as a means of writing programs from specifications. Thus, notwithstanding the limitations postulated above, it is already apparent, from work with examples such as the one presented in this paper, that this new method, with its roots in two very usable, real languages, its sensible exploitation of the similarities and differences between them, its pragmatic provision of guidelines for the programmer and its clean handling of input/output and state, constitutes a bold step in the direction of being able to write correct programs from specifications, and, consequently, should help contribute to the more widespread use of formal methods.

Acknowledgements

The author is grateful to Dr Antoni Diller of the University of Birmingham School of Computer Science for his invaluable comments and advice during the course of this research project.

The author is supported by the Engineering and Physical Sciences Research Council (U.K.), grant number 92313752.

References

1. Jonathan P. Bowen and Michael G. Hinchey. Ten commandments of formal methods. *IEEE Computer*, 28(4):56–63, April 1995.

2. J. M. Spivey. *The Z Notation – A Reference Manual*. International Series in Computer Science. Prentice Hall, Hemel Hempstead, 2nd edition, 1992.

3. David Gries. *The Science of Programming*. Texts and Monographs in Computer Science. Springer-Verlag, New York, 1981.

4. Carroll Morgan. *Programming from Specifications*. International Series in Computer Science. Prentice Hall, Hemel Hempstead, 1990.

5. John Backus. Can programming be liberated from the Von Neumann style? A functional style and its algebra of programs. *Communications of the ACM*, 21(8):613–41, 1978.

6. Paul Hudak, Simon Peyton Jones, Philip Wadler, et al. Report on the programming language Haskell, a non-strict, purely functional language. *SIGPLAN Notices*, 27, May 1992. Version 1.2.

7. G. O'Neill. Automatic translation of VDM specifications into Standard ML programs. Technical Report DITC 196/92, National Physical Laboratory, Teddington, 1992.

8. Paulo Borba and Silvio Meira. From VDM specifications to functional prototypes. *Journal of Systems and Software*, 21(3):267–78, June 1993.

9. Michael Johnson and Paul Sanders. From Z specifications to functional implementations. In John E. Nicholls, editor, *Z User Workshop, Oxford 1989*, Workshops in Computing, pages 86–112. Springer-Verlag, 1990.

10. Linda B. Sherrell and Doris L. Carver. Experiences in translating Z designs to Haskell implementations. *Software – Practice and Experience*, 24(12):1159–78, December 1994.

11. Antoni Diller. *Z: An Introduction to Formal Methods*. John Wiley and Sons, Chichester, 2nd edition, 1994.

12. Howard S. Goodman. The Z-into-Haskell tool-kit. Technical Report CSR-95-1, School of Computer Science, University of Birmingham, April 1995. Available from the author's WWW home-page, URL http://www.cs.bham.ac.uk/~hsg/.

13. Philip Wadler. The essence of functional programming. In *19th ACM Symposium on Principles of Programming Languages*, January 1992.

14. Howard S. Goodman. Animating Z specifications in Haskell using a monad. Technical Report CSR-93-10, School of Computer Science, University of Birmingham, August 1993. Revised with corrections, April 1995. Available from the author's WWW home-page, URL http://www.cs.bham.ac.uk/~hsg/.

15. Samuel H. Valentine. Z−−, an executable subset of Z. In John E. Nicholls, editor, *Z User Workshop, York 1991*, Workshops in Computing, pages 157–187. Springer-Verlag, 1992.

16. Ben F. Potter, Jane E. Sinclair, and David Till. *An Introduction to Formal Specification and Z*. International Series in Computer Science. Prentice Hall, Hemel Hempstead, 1990.

17. Ian J. Hayes and Cliff B. Jones. Specifications are not (necessarily) executable. *Software Engineering Journal*, 4(6):330–8, 1989.
18. Antoni Diller. Specifying interactive programs in Z. Technical Report CSR-90-13, School of Computer Science, University of Birmingham, August 1990.
19. Ian J. Hayes, editor. *Specification Case Studies*. International Series in Computer Science. Prentice Hall, Hemel Hempstead, 2nd edition, 1993.

Types and Sets in Gödel and Z

Margaret M. West

School of Computer Studies, University of Leeds
e-mail: mmwest@scs.leeds.ac.uk
URL: http://csirisb.leeds.ac.uk/
ftp: agora.leeds.ac.uk

Abstract. An *animation* of a formal specification language is seen as a useful aid for validating specifications. This paper describes differences between the type systems of the programming language Gödel and the Z notation and the implications this has for devising a set of rules for animation of Z via Gödel. A set of rules are outlined and future work is discussed whereby the correctness of the Gödel implementation could be established.

1 Introduction

The use of a formal specification produces significant advantages for system development and an animated system model (or interpretation) provides an aid for developing and understanding such a specification. The animation may then be used for experiments and the results of the experiments can be used to assess and refine the design of the system. The draft IEC standard on industrial safety-related systems [1] states that the aim of prototyping/animation is:

> To check the feasibility of implementing the system against the given constraints.
> To communicate the specifier's interpretation of the system to the customer, in order to locate misunderstandings.

Other authors have argued against the desirability of executing a specification [2]. However, their objections are to prototyping from specifications rather than animating them. A prototype is for demonstrating the functionality of the required system, whereas an animation should make apparent the logical relationships within the specification [3, 4]. The objections of [2] to executable specifications include the fact that a prototype does not respect properties of a specification such as non-determinism. However this is not necessarily true of an animation; it should be possible to deduce the possibility of multiple outputs from a given input by animating the specification. A further objection (to executable specifications) is that they do not explicitly include specification variables (such as time). A time variable provides information about the performance of the system and since an animation reflects the logical structure of the specification, *all* specification variables should have their representation. An animation of a Z specification possesses many of the advantages of the (non-executable) original and in addition the specification is explained. A small model environment can be provided in order to overcome large domain problems, as suggested by [5].

There are advantages in using a logic programming language for animation purposes [6]. It enables a "what" approach of specification to be modelled via constraint predicates. Queries of the "what if" variety can then be posed: given predicate $Pred(x, y, z, w)$ the value(s) of w can be established where x, y, z are ground or (in principle) the value(s) of x, where y, z, w are ground.

A set of rules for translating a subset of Z to Prolog code was established in [3]. Two strategies were explored: *Formal Program Synthesis* and *Structure Simulation*. The former would have provided a correct execution of a subset of Z, but was abandoned as infeasible. (The reasons are given in Section 4.) The latter, Structure Simulation was more successful. In this, characteristics of Z schemas are identified and adapted so that the logical structures of the specification are preserved as far as possible in the resulting Prolog interpretation. In the simulation, Z schemas are represented by predicates and every variable within a schema is represented by a Prolog variable. Thus specification variables such as time can be explicitly represented. (See [7] for an case study involving time.) However [3] also identifies some *disadvantages* to using Prolog for such a purpose; for example the clauses are selected in the order they appear. This limits the "what if" facility of animation; the deduction of any variable from instantiated variables is not straightforward unless the code is ordered in an appropriate manner. Secondly, Z is based on set theory and Prolog lacks support for sets. Any implementation of Prolog is unsound with respect to the semantics of first order logic and this invalidates any attempt to prove the correctness of a Prolog implementation of Z.

It is argued here that a better technique is to use the Gödel language [8]; the language has been carefully designed and implemented, and supports a set data type. Section 2 of this paper gives a brief introduction to the Gödel language, and Section 3 of the paper will develop rules for translating Z specifications to Gödel. Section 3 also includes a demonstration of the software: a schema from the Unix File System [9] is translated to Gödel. The possibility of proving the correctness of Gödel implementation of Z using the technique of *Abstract Interpretation* [10, 11] is then discussed in Section 4.

2 The Gödel Programming Language

2.1 Overview

A brief introduction to the Gödel language is now presented; a fuller (and formal) exposition is in [8]. The logic programming language Gödel has a greatly improved declarative semantics compared with Prolog, and supports a set data type in a manner described below. (The Prolog predicate "setof" has no declarative meaning.) Prolog has no occur check in the unification algorithm, and unsafe negation. Apart from some well-defined exceptions, a program in Gödel is defined as a theory in first order logic and an implementation must be sound with respect to this semantics. Gödel has a flexible computation rule that can be constrained by user-defined control declarations and ensures that all calls to negative literals are ground.

A predicate definition consists of a declaration, specifying the type(s) of its arguments, and a set of statements of the form

```
Head <- Body.
```

The head is an atom with the defining predicate. The body is a formula in first order logic and may be absent. (Note that <- in Gödel means "if" and in contrast to Prolog, upper case is used for constants and lower case for variables.) Language declarations include the categories: base, constructor, constant, function, proposition and predicate. To allow for program structuring, the language is *modular* and modules can be exported (to modules) and can themselves import (other modules). A Gödel program is a collection of modules. There are a number of system modules and these include modules for integer, rational and floating point arithmetic. Further system modules provide structured data types such as Set, List, String, Table. Gödel provides special facilities for meta programming and system modules are provided for reasoning about other Gödel programs.

2.2 Types and Sets

Many sorted first order logic provides the basis for Gödel, whose sorts are then *types*. Gödel is strongly typed; each constant, function and predicate must have its type(s) specified. The BASE declaration declares the *base types* of the module. For example a base FileId can be declared to represent file identifiers.

In addition there are type constructors such as List/1, so that List(a) is a generic list with parametric type a. Each constructor must be declared but parameters and variables are not declared. A Gödel program is checked for type correctness during compilation. The goal is subsequently checked at run-time before execution.

In Gödel, sets are supported by the system module Sets. Thus Set(a) constructs a term whose *intended meaning* is a set of elements of type a. In order for sets to be manipulated mechanically, it is necessary to present them in a form closer to the level of a computer (such as a list). The axioms for such sets and set operations are presented in detail in a text by Manna and Waldinger [12]. This presents the theory of some fundamental structures of computer science, and includes the "theory of finite sets". Axioms include multiplicity:

$$\forall u, x : u \circ (u \circ x) = (u \circ x)$$

and exchange:

$$\forall u, x, v : u \circ (v \circ x) = v \circ (u \circ x)$$

where $(u \circ v)$ denotes a binary insertion function, the result of inserting element u in set x.

In Gödel Inc is a mapping *Include* with the property $Include(d, S) = \{d\} \cup S$, where d are an element and S a set. *Include* can thus be compared with $\circ$. Braces for sets provide some notational sugar. For example {1, 2, 3} stands for

```
Inc(1, Inc(2, Inc(3, Null)))
```

where Null is the empty set ({ }). The order of elements in a set is irrelevant and sets do not have duplicate elements. Thus the equality of extensional sets takes this into account:

{5, 6, 7} = {7, 7, 6, 5}.

There is no automatic facility for obtaining a set of all elements of a particular type and such a set must be coded explicitly.

Sets terms can also be intensional:

s = {x : 1 < x < 5}.

The precise meaning of this example of an intentional set (term) is

ALL [x] (x In s <-> SOME [y](x = y & 1 < y < 5)).

This semantics for intentional sets was first given in [13].

The Sets module also provides functions for set union (+) and intersection (*). Set difference is represented by (\). Predicate In represents set membership and Subset represents subset. Both are provided by Sets. Further features of the language will become apparent. The definition of types in Gödel and system support for sets has implications for the animation of Z, and this is discussed in the next section.

3 Animation

This section describes a preliminary set of rules for animation in a small model environment. An animated system model provides an aid for developing and understanding a Z specification; a review of work in this field is contained in [14]. In contrast to other researchers [15] who use predicates in the schema signature to generate data, the approach here is that animation fulfills the role of testing. It is akin to conventional testing of software through the use of test cases. However test cases can be *derived* from a formal specification [16, 17] and in addition the test cases can be preserved and used for validating the final system [18]. The results of the animation are compared with the results of the final tests.

Breuer and Bowen [14], have identified requirements for an animation. They include *coverage* of Z, *sophistication* (less likely to go into an infinite loop), *efficiency* (performance) and *correctness* . Inevitably there is a trade-off between some of these requirements.

An aim of the approach presented here is that as much coverage of Z as possible is attempted, and that the code should be *sophisticated*. However correctness should take precedence over coverage, and provided the performance is reasonable, both should take precedence over efficiency.

In this paper schemas are represented by predicates and a query to a schema should result in a single binding satisfying that schema. If backtracking is initiated, further bindings are generated and thus the set of bindings satisfying the schema in the model environment is presented incrementally. Schema taken from the Z specification of the Unix File Store are translated to Gödel to illustrate the method [9] and can be seen in Section 3 and in the Appendix.

3.1 Types in Z

Rules for animation need to take account of the type system for Z. In Z the type of an expression is either *given* or derived using type constructors *Power set, Cartesian product*. A further derived type is a *schema*, a typed set each of whose members is a *binding*, a collection of named (bound) type variables. All these features are implemented in a manner which will be described. Z schemas are interpreted in a single Gödel module UnixFiles and this module imports predicates from a Gödel module Lib which models Z relations and functions [19]. These *characteristic* predicates define partial function, relation etc and are generic to a set of any type. "Given sets" are displayed but sets such as power sets and sets (of ordered pairs) which represent functions are never generated. In general the philosophy is to *test* a set to see if it satisfies a particular characteristic predicate. Schemas are also represented by characteristic predicates, defining the relationship between schema variables. The next subsection describes how various features of the Z notation are implemented in Gödel. The representation in Gödel of given sets, schema variables and schemas is now presented in a tutorial manner. An example of a schema translation is then presented.

3.2 Givensets and Bindings

1. *Given Sets*: In order to provide a small model environment, the current policy is to provide elements of the given sets via axiomatic description in Z. (The prototype Z^{--} interpreter [20] allows only structured free types for this.) The given sets of the specification are declared as one of the base types (in Gödel) and some constants of the base types introduced to model the environment. These constants are coded into extensional sets.

2. *Bindings*: A further type is BindVar, used to facilitate the binding formation of schemas. *Lists* of this type form a binding of a schema. Variable names are bound to their values via functions over the appropriate types. Binding functions always have SetNames as their first argument. (Variable names have the base type SetNames.) The names are "decorated" using Gödel functions: DSet primes a variable name, IN decorates an input variable, OUT an output. Thus function DSet has as its argument a type of SetNames and returns the same type. The other decoration functions are defined in a similar manner.

An example is given to illustrate. The integers are supplied by the system and supposing FileId to be a declared base type, then Bind1, Bind2 can be defined:

```
CONSTANT   F1, F2  : FileId;
           Fid, Posn : SetNames.
FUNCTION   Bind1 : SetNames * FileId ->  BindVar.
FUNCTION   Bind2 : SetNames * Integer ->  BindVar.
```

The following are examples of elements of type BindVar:

```
Bind1(Fid, F1)        Bind2(Posn, 5)
Bind1(DSet(Fid), F2)  Bind2(OUT(Posn), 6)
```

In the next subsection it will be shown how the binding function device ensures that types and identifiers associated with a given schema are completely fixed by function declarations and schema statement. They also explain how schema declarations and predicates are modelled. (This does not include axiomatic descriptions and generic definitions which are explained later.)

3.3 Schema

1. *Schema Bindings*: A schema is modelled by:

$$Schema(binding, name) \Leftarrow Signature(binding\_variables) \wedge$$
$$Predicate(binding\_variables)$$

where *binding* is of type List(BindVar) and *name* is of type SchNames. *Schema(binding, name)* is thus defined:

PREDICATE SchemaType: List(BindVar) * SchNames.

Schema names are also decorated via Gödel functions, thus DSch primes a schema name. (Note: Although, strictly, schema and variable names are the same Z type, *NAME*, it is convenient here to ascribe them to different types.)

2. *Variable Declarations*: Variables are also checked for additional constraints, for example if they represent a partial function.

3. *Schema Predicate*: *Predicate(binding_variables)* constrains the declared variables using predicates and functions provided by the system. Existential and universal quantification are both provided by Gödel, as are set operations. The Lib module provides other parts of the Z notation, such as function override, domain and range restriction. However λ expressions require separate treatment; they will need expansion as an intensional set term, according to the function which is being defined. Such an expansion requires the facility of meta programming and is discussed in Section 4.

For example suppose Bind1, Bind2 are declared as above. A schema named *Chan* with variables named Fid, Posn of types FileId and Integer respectively would have schema clause head as follows:

SchemaType([Bind1(Fid, f), Bind2(Posn,posn)], Chan) <-
··· *signature and predicate.*

Given the binding function declarations Bind1, Bind2, the list inside the predicate is thus equivalent to θ*Chan*.

Declared variables are additionally constrained, for example for functionality. The predicate PF (presented next) is contained in module Lib and is generic to sets of any type. As can be seen the code is strikingly similar to the definition of partial function in ZF.

```
%% Type constructor OP allows the definition of Goedel
%% function OrdPair. This is a device to provide for
%% ordered pairs and functions (a set of ordered
%% pairs) in Z.
CONSTRUCTOR OP/2.
FUNCTION    OrdPair : a * b -> OP(a,b).


%%%%% declaration of partial function %%%%%
PREDICATE   PF  : Set(OP(a,b)) * Set(a)  * Set(b).


%%%%% definition of partial function %%%%%
%%%%% This characteristic predicate determines whether
%%%%% a set is a function from s1 to s2         %%%%%
 PF(pf, s1, s2) <- ALL [z,x,y] (z In pf &
                             (z = OrdPair(x,y))
                 ->  (x In s1) & (y In s2)   &
             ALL [u] (OrdPair(x, u) In pf -> u = y)).
```

The next subsection describes how schema calculus and schema referencing are modelled.

3.4   Schema Calculus and Schema Referencing

1. *Schema Declaration in a Signature*: If a schema B is declared in the signature of schema A, then the clause head for B is conjoined to the signature of A. Variables of B are merged with the other signature variables.

2. *Schema Calculus*: Conjunction and disjunction of schemas A, B is modelled by conjunction and disjunction of the Gödel predicates of schemas A, B. The variables are merged; the lists are appended and duplicates removed. Schema composition and piping is accomplished by conjunction, with appropriate adjustment to the variable list in the new schema.

3. *Schema Referencing*: As explained the constructed type List (BindVar) of *Schema* is equivalent to the binding formation θ*Schema*. Code is provided to form the set $\{S \bullet \theta S\}$. An example of its use is given in Section 3.6 and the code can be seen in the Appendix.

3.5   Axiomatic Descriptions and Generic Definitions

These require special treatment; in the case of the former they are modelled in the same way as a schema, and suitable *names* must be generated for them Axiom1, Axiom2 ... They must then be conjoined to the schemas which refer to them, otherwise there is no check on the constraints introduced. Generic definitions are treated in the same way as the parametrised definitions of functions etc, i.e. by using parameters a, b, ... instead of generic sets $X_1, X_2, \ldots$

3.6 Animation Example

The example given is a schema modelling an operation on a *channel storage* system [9].
A fuller version is given in the Appendix and the library of functions and relations is
in [21]. *CS* is a schema consisting of a partial function between channel identifiers (the
set *CID*) and a schema *CHAN* representing a channel. *CHAN* consists of a file identifier, *fid* and a position within a file, *posn*. The schema *openCS* denotes the opening of
a new channel. The old channel store is updated by the addition of a new channel whose
file position is zero, but *fid* is unconstrained.

$$
\begin{array}{|l}
\hline \_openCS _____ \\
\Delta CS \\
CHAN \\
cid! : CID \\
\hline
cid! \notin \mathrm{dom}\, cstore \\[4pt]
posn = 0 \\[4pt]
cstore' = cstore \oplus \{ cid! \mapsto \theta CHAN \} \\
\hline
\end{array}
$$

This is represented by the following predicate:

```
%%The bindings are expanded out
SchemaType([Bind1(Fid, f), Bind2(Posn,posn),
           Bind3(OUT( ChId), outc),
           Bind4(Cstore, cs),
           Bind4(DSet(Cstore), cs1) ], OpenCs )   <-
%%declared schema
  SchemaType([Bind1(Fid, f), Bind2(Posn,posn)], Chan) &
  SchemaType([Bind4(Cstore, cs),
      Bind4(DSet(Cstore), cs1)], Del(Cs)) &
        GivenSets( _, _,cid) &
          outc In cid &
%%schema predicate
        %% cid not in dom cstore
        DomContents(cs, domcs) &
        ~(outc In domcs) &
          posn = 0 &
 FunOveride(cs1, cs, {OrdPair(outc, [Bind1(Fid, f),
        Bind2(Posn,posn)])} ) .
```

(NB. The predicate FunOveride represents function overriding, $\oplus$ in Z.) A query shows
the effect on a channel store consisting of a single channel, of the opening of a further
channel. The existing channel has file identifier F1 and position 2. Thus before a new
channel is opened, schema *openCS* has component

$$
Cstore = \{ Cid1 \mapsto \langle (Fid, F1), (Posn, 2) \rangle \}
$$

where the angled brackets have been appropriated to denote a binding. The query provides this information and finds the value of CS', the value after the new channel is opened, and information about the new channel. A new channel is added. This can be any of the remaining (unopened) ones and the file identifier can be any. The position in the file is 0.

```
[UnixFiles] <- cs = {OrdPair(Cid1,[Bind1(Fid,F1),
                                    Bind2(Posn,2)])} &
               SchemaType([b1,b2,b3,Bind4(Cstore, cs),
                  Bind4(DSet(Cstore), cs1) ], OpenCs ).
```

```
b1 = Bind1(Fid,F1),
b2 = Bind2(Posn,0),
b3 = Bind3(OUT(ChId),Cid2),
cs = {OrdPair(Cid1,[Bind1(Fid,F1),Bind2(Posn,2)])},
cs1 = {OrdPair(Cid1,[Bind1(Fid,F1),Bind2(Posn,2)]),
       OrdPair(Cid2,[Bind1(Fid,F1),Bind2(Posn,0)])} ? ;
```

Initiating a back track gives a further answer:

```
b1 = Bind1(Fid,F5),
b2 = Bind2(Posn,0),
b3 = Bind3(OUT(ChId),Cid3),
cs = {OrdPair(Cid1,[Bind1(Fid,F1),Bind2(Posn,2)])},
cs1 = {OrdPair(Cid1,[Bind1(Fid,F1),Bind2(Posn,2)]),
       OrdPair(Cid3,[Bind1(Fid,F1),Bind2(Posn,0)])} ?
Yes
```

Thus amongst many possible values

$$Cstore' = \{\, Cid1 \mapsto \langle (Fid, F1), (Posn, 2) \rangle, Cid2 \mapsto \langle (Fid, F1), (Posn, 0) \rangle \}$$

The new channel can be any except $Cid1$ and the file remembered by the new channel, any.

4 Proposed Demonstration of Correctness

4.1 Formal Program Synthesis

A previous paper [3] examined the possibility of deriving a logic program from a specification using formal program synthesis [22]. The basis of the derivation is the axioms for the *Theory of Finite Sets* [12]. The proposed method was to transform specifications to clausal form [23] and to synthesise logic programs using inference procedures such as resolution. (This use of resolution to synthesise programs contrasts with its more usual use of *running* programs.)

However there are difficulties in determining a suitable inference step as there is (in general) no algorithm and the method was abandoned. It is worth noting that a recent overview and survey of logic program synthesis [24] interprets synthesis in a much broader way than in the work just described. This survey paper is intended to provide a framework for future research as well as presenting an assessment of existing work.

4.2 Abstract Interpretation

A different method of proof for animations of Z is proposed by Breuer and Bowen [14], who suggest a set of criteria for correctness of an animation based on the notion of *abstract interpretation*. The seminal work on abstract interpretation was done by Cousot and Cousot [25], who used it for static analysis of imperative programs. Cousot and Cousot discuss an abstraction function, alpha, relating the concrete semantics to an approximate one that explicitly exhibits an underlying structure implicitly present in the richer concrete structure associated with program executions. The work has since been extended to declarative languages. (See [10, 11].) In [26] a procedure based on abstract interpretation is developed that will detect at compile time (of a logic program) most places where the occur check is *not* needed. Abstract interpretation is utilised in *strictness analysis* [10] in functional programming. The objective is to improve program efficiency by detecting when it is possible to pass arguments "by value" rather than "by need". The abstraction must be *safe*, so that an argument must never be detected as strict when it is not. On the other hand there may be strict arguments which are undetected by the abstraction. In this sense the abstracted interpretation holds less information than the concrete one: the concrete interpretation *refines* the abstract. The idea can be generalised to one where an exact or "concrete" semantics is abstracted by an approximate one, whose domain of interpretation is simpler than the standard concrete domain. A brief tutorial in some of the concepts involved in approximation is given in the next subsection.

4.3 Approximation

A useful introduction to the concepts of semantic domains and approximation is contained in [27], while good summaries of abstract interpretation are in [10, 11]. Sets of value spaces for a programming application are known as semantic domains; compound domains are built from primitive domains (Integers, Booleans, Characters etc.). In order to accommodate non termination or incomplete information, the sets that are used as value spaces are 'lifted" by the introduction of a partial element $\perp$ that denotes non-termination. For example if the set of Booleans $Bool = \{tt, ff\}$ then:

$$Bool_\perp = \{tt, ff\} \cup \{\perp\}$$

A *partial ordering* relation $\sqsubseteq$ is introduced, and $\perp$ is the least element. We say $\perp$ *approximates* other elements of the domain. In the case of lifted sets, elements not equal to $\perp$ approximate themselves, i.e. given a set S, if $S_\perp = S \cup \{\perp\}$ then an ordering $\sqsubseteq$ on S is defined by:

$$a \sqsubseteq b \Leftrightarrow a = \perp \; or \; a = b$$

In order to capture the underlying structure of a (richer) concrete domain D_{conc}, an abstraction function α is constructed which maps between D_{conc} and an abstract domain D_{abs}:

$$\alpha : D_{conc} \to D_{abs}$$

D_{abs} is simpler than D_{conc} and is said to *approximate* it in a manner described below. The abstraction function α and concretisation function γ:

$$\gamma : D_{abs} \to D_{conc}$$

are adjoined, so that

$$\forall d : D_{abs} \bullet d = \alpha(\gamma(d))$$
$$\forall d : D_{conc} \bullet d \sqsubseteq (\gamma(\alpha(d))$$

Thus

$$\alpha \circ \gamma = IdentityFunction_{abs},$$
$$\gamma \circ \alpha \sqsupseteq IdentityFunction_{conc}$$

An approximation diagram (Figure 1) is shown below. (An approximation diagram is more general in concept than a commutative diagram, for $\sqsubseteq$ subsumes equality. In order to distinguish them, "dashed" lines are used as in [10].) Env_{conc} and Env_{abs} are concrete and abstract environments respectively. In [11], Cousot and Cousot describe them as *contexts* and they denote the bindings of program identifiers to values at a particular stage in a program. This diagram must then be shown to commute, where f_{conc} and f_{abs} are functions on the concrete and abstract domains respectively. These concepts are discussed further in Section 4.4.

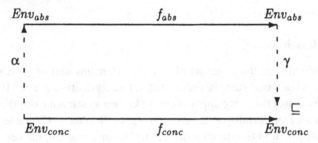

Fig. 1. Approximation Diagram

4.4 Strategy for Proof

Criteria for establishing the correctness of a Z animation have been produced by [14]: the animation is an abstract interpretation of the Z it models. An implementation has been produced, a Miranda script. Proofs are treated as follows: Z is expanded to its constituent schemas and associated variables, and these are treated directly by the authors. A domain *ideal* ZF is defined in order to form comparisons between the abstraction (executable) and the concrete (Z). Thus in order to be correct the operators (e.g. union, intersection, power set) of an interpretation must approximate the corresponding ZF operators from below. The reason for this can be expressed informally: an animation of a

Z specification must not contain any more information than the original Z specification, as it may mislead. The formalisation differs in detail from that illustrated in Fig 1. In order to obtain a relation between abstract and concrete domains, a concretisation function (denoted *rep*) is introduced and there is no explicit α.

The endpoint of the work presented here is to prove the correctness of the Gödel animation of Z via abstract interpretation. As can be seen, the approach to interpretation presented here is somewhat different to [14], where *schemas* and associated variables are treated directly. Other parts of Z (e.g. schema calculus) are expanded *before* interpretation. Whereas in the translation method presented here, parts of the schema calculus are directly interpreted in Gödel. This may inhibit proof of correctness and have to be revised.

The (proposed) concrete domain, D_Z is akin to the domain *ideal*. It consists of expressions and predicates. Expressions are (typed) values, sets and bindings of named variables. Schemas are sets of bindings of the named variables. The typed values include integers. The concrete domain thus consists of the *world of sets*, W_A, of Z (that are to be covered by the animation technique). However in order to accommodate incomplete computations, the concrete domain should also contain $\perp_Z$. Also sets can be incomplete: if set s belongs to W_A, the set denoted $s_{\perp_Z}$ (in Section 4.3) also belongs to the concrete domain.

The proposed abstract domain, D_G, includes (typed) constants integer values and predicates. Sets are also included and values are either displayed or can be decided using characteristic predicates. Distinguished types are names of schema, names of variables and bindings, i.e. SetNames, SchNames, BindVar respectively. The partial element $\perp_G$ represents an incomplete computation, e.g. if the program flounders. A characteristic predicate (for a set) can also flounder given certain inputs so (as with the concrete domain) sets can be incomplete. Schemas are represented by characteristic predicates (as described in Section 3) and can be composed of other schemas.

An abstraction function, α, would map elements of D_Z to elements of D_G, thus expressions of D_Z would be mapped to expressions of D_G, and predicates of D_Z to predicates of D_G. The element $\perp_Z$ of D_Z would be mapped to the element $\perp_G$ of D_G. The concretisation function maps correspondingly from D_G to D_Z. In particular $\perp_G$ of D_G maps to $\perp_Z$ of D_Z. The evaluation of a numerical expression in both D_Z and D_G will be used to illustrate. Suppose an abstraction function, $\alpha : D_Z \rightarrow D_G$, maps natural numbers as follows:

$$\alpha(n) \quad = n, n \leq MaxInt,$$
$$= \perp_G, n > MaxInt$$

where n is a natural number, *MaxInt* is the largest number available on the machine. Suppose x and y are identifiers with natural number values n_1, n_2 and z denotes the product of x and y in D_Z. In D_Z z has value $n_3 (= n_1 n_2)$ However in D_G, $\alpha(n_3) = n_3$ for $n_3 \leq MaxInt$, but $\alpha(n_3) = \perp_G$ for $n_3 > MaxInt$. Applying the concretisation function, $\gamma(\alpha(n_3)) = n_3$ for $n_3 \leq MaxInt$ and $\gamma(\alpha(n_3)) = \perp_Z$ for $n_3 > MaxInt$.

4.5 Finite Sets

The previous example provided an α for mapping between integers and integer expressions in D_Z, specified by Z, to Gödel output domain D_G. However as was demonstrated in the example, Gödel sets are finite, so it remains to determine the best mapping between the (infinite) sets of Z and the finite sets of Gödel. Three possible solutions are outlined:

1. Specifications could be restricted to finite sets. This would reduce coverage and might seem overly restrictive, however in the specification examined, the Unix File system, the authors note that file size, directory size "Inodes", device capacity and position within a file are *bounded*. Many specifications require only finite sets.

2. The mapping could be such that Z variables outside a certain bound map to $\perp$. This was demonstrated in the example and is not very informative.

3. A further possibility is that variables outside a certain bound map to an error message or alternatively to the character ∞. The latter is suggested by the IEEE floating point standard, which allows output of ∞ for division of a natural number by zero. (∞ divided by ∞ is then a *nan*, "not a number".)

Whatever the solution, the next task would then be to prove the approximation result for all the Z features (these are represented by f_{conc}) which are directly translated to Gödel (these are represented by f_{abs}). At present this includes schema bindings, schema calculus, as well as set operations and relations.

4.6 Meta Programming

Gödel provides the facility of meta programming and system modules are provided for this. It is proposed to use this facility for automation of the Z to Gödel translation. As can be seen, some parts of the Z notation are capable of direct translation to Gödel, while others require expanding. For example, the reference to a schema with primed name means the introduction of a Gödel program with appropriate variables renamed as described above. Lambda expressions and axiomatic definitions would also require some special treatment of this type. Meta programming would be used to expand specific constructs to their constituent expressions.

5 Discussion and Conclusion

It was stated previously that an aim for the animation technique presented here is that as much as possible of the Z notation should be covered. The results so far seem promising, i.e. it seems that a substantial part of the Z notation is capable of translation to Gödel. The simpler constructs are capable of direct translation; the more complex constructs require the aid of the meta programming facility. The main restriction is that sets are *finite*. A further aim (and main philosophy) is that animation is equivalent to a *test*. Thus it is not intended that variables should be generated by their declaration (apart from the simple device of generating from a given set); they must be instantiated, or a set of possible values given. A possible strategy would be to derive test values from the specification, as described by other authors [16, 17]. Although not the prime aim, a "reasonable" performance of the code has been the case so far. The longest response to a query is of the order of a second.

The typing system of Gödel has forced a particular approach (in contrast to previous work in Prolog). For example to model the "ordered pair" type, it was necessary to define a type constructor. (In Prolog a list of two elements sufficed!) In a similar manner, it was necessary to define binding functions (Bind1, Bind2, ...) and a new type BindVar, to attach variables of a specific type to their names. The Prolog equivalent was a predicate which took variables of any type. Gödel has proved a much superior language for animation (to Prolog); the support for sets and the provision of types has facilitated the work. The flexible computation rule has meant that there are many more possible queries to the code. It allows for control declarations to be inserted by the programmer to delay certain calls until variables are sufficiently instantiated. However, so far, it has never been necessary to use this facility. The Unix file example has used a single module (for the sake of simplicity in the demonstration). However in order to exploit Gödel facilities more fully, it would be necessary to allow each schema (or related group) to be coded into a module.

A set of proposed rules for translation from Z specifications to the Gödel language has been presented here, with the intention of *animating* the Z. The aim has been to provide as much coverage of the Z as possible, and so far it seems that a large subset of Z is capable of animation. However an important proviso is that the Gödel should correctly represent the Z, and this should be achievable given the notion of "abstract interpretation". For a given set of rules, it will be necessary to find a suitable abstraction function (or its equivalent) between the Z and Gödel domains and show that the approximation diagrams (of the animated features of Z) commute.

6 Acknowledgement

I would like to acknowledge the help and support given by Pat Hill, of the School of Computer Studies, Leeds University. (The Gödel code encorporates many of her suggestions.) I would also like to thank the referees for a number of helpful comments.

References

1. IEC Draft Standard. Software for Computers in the Application of Industrial Safety-related Systems (IEC reference 65a (Secretariat) 122). To be superceded by IEC 1508 in 1995.
2. I J Hayes and C B Jones. Specifications are not (Necessarily) Executable. *Software Engineering Journal*, 4(6):330–338, 1989.
3. M M West and B M Eaglestone. Software Development: Two Approaches to Animation of Z Specifications Using Prolog. *Software Engineering Journal*, 7(4):264–276, July 1992.
4. D W Bustard and T Harmer. PEEP: An Experiment in Software Animation. In P Barnes and D Brown, editors, *Software Engineering 86*, pages 2–18. IEE, Savoy Place, London, UK, 1986.
5. N E Fuchs. Specifications are (Preferably) Executable. *Software Engineering Journal*, pages 323–334, September 1992.

6. B A Wichmann, editor. *Software in Safety-Related Systems (IEE/BCS Joint Study Report)*. John Wiley and Sons, UK, 1992.

7. M M West, T F Buckley, and P H Jesty. A Formal Expression of the Safety and Functional Requirements of a Safety-Critical System. Technical Report 93.38, School of Computer Studies, University of Leeds, 1993.

8. P M Hill and J W Lloyd. *The Gödel Programming Language*. MIT Press, 1994.

9. C Morgan and B Sufrin. Unix Filing system. In I Hayes, editor, *Specification Case Studies (Second Edition)*, pages 45–78. Prentice Hall International (UK) Ltd, 1993.

10. S Abramsky and C Hankin. *Abstract Interpretation of Declaritive Languages*. Ellis Horwood Limited, Chichester, West Sussex, PO19 1EB, England, 1987.

11. P Cousot and R Cousot. Abstract Interpretation and Application to Logic Programs. *The Journal of Logic Programming*, 13:103–179, 1992.

12. Z Manna and R Waldinger. *The Logical Basis for Computer Programming, Vol 1: Deductive Reasoning*. Addison-Wesley, USA, 1985.

13. K L Clark, F G McCabe, and S Gregory. IC-Prolog Language Features. In K L Clark and S A Tarnlund, editors, *Logic Programming*, pages 253–266. Academic Press, London, 1982.

14. P T Breuer and J Bowen. Towards Correct Executable Semantics for Z. In *Z User Workshop, Cambridge, June 1994*, pages 185–209. Springer-Verlag, 1994.

15. A J Dick, P J Krause, and J Cozens. Computer Aided Transformation of Z into Prolog. In *Proceedings of the 4th Annual Z Users Meeting, Oxford University Computing Laboratory PRG*, pages 71–85. Springer-Verlag, December 1989.

16. J Dick and A Faivre. Automating the Generation and Sequencing of Test Cases from Model-Based Specifications. In J C P Woodcock and PG Larsen, editors, *FME '93: Industrial-Strength Formal Methods, First International Symposium of Formal Methods Europe (Proceedings), April 19-23 1993, Odense, Denmark*, pages 268–284. Lecture Notes in Computer Science 670, Springer-Verlag, Germany, 1993.

17. D. Carrington and P. Stocks. A Tale of Two Paradigms: Formal Methods and Software Testing. In *Z User Workshop, Cambridge, June 1994*, pages 51–68, 1994.

18. B A Wichmann. A Development Model for Safety-Critical Software. In B A Wichmann, editor, *Software in Safety-Related Systems (IEE/BCS Joint Study Report)*, pages 211–223. John Wiley and Sons, UK, 1992.

19. J M Spivey. *The Z Notation: A Reference Manual*. Prentice Hall International (UK) Ltd, UK, 1989.

20. S H Valentine. The Programming Language Z - -. *Information and Software Technology*, 37(5), May 1995. to appear.

21. M M West. Types and Sets in Gödel and Z: Unix File System Case Study. Technical Report 94.33, School of Computer Studies, University of Leeds, December 1994.

22. C Hogger. *Introduction to Logic Programming*. Academic Press, London, 1984.

23. R A Kowalski. *Logic For Problem Solving*. North Holland Artificial Intelligence Series, New York, 1979.

24. Y Deville and K K Lau. Logic Program Synthesis. *J. Logic Programming*, 20:321–350, July 1994.

25. P Cousot and R Cousot. Abstract Interpretation: A Unified Lattice Model for Static Analysis of Programs by Construction or Approximation of Fix-points. In *Proc. 4th ACM Symposium on the Principles of Programming Languages*, pages 238–252, 1977.

26. P. M. Hill. Logic Programming with the Occur-Check. Technical Report 94.32, School of Computer Studies, University of Leeds, 1994.

27. D A Schmidt. *Denotional Semantics, A Methodology for Language Development*. Allyn and Bacon, Inc, 7 Wells Avenue, Newton, Massachusetts, US, 1986.

A Example: File System in Unix

A.1 Z schemas

The example is explained briefly, and only a fragment is presented here. A fuller account of the Unix File System Case Study is in [9] and the library necessary for functions and relations is in [21]. The file storage system allows files to be stored and retrieved using file identifiers; the set of all file identifiers is called FID. Further given sets are channel identifiers CID. A channel identifier is a Unix file descriptor.

$$[FID, CID]$$

In order to support random access to files for reading and writing, a channel is defined which remembers a file and the current position in the file:

```
__ CHAN _____
  fid : FID
  posn : N
_____
```

A channel storage system allows channels to be stored and retrieved using channel identifiers taken from CID. The channel storage system is denoted $cstore$ and is defined by the schema CS, where $cstore$ denotes a partial function between CID and the schema $CHAN$.

```
__ CS _____
  cstore : CID ↦ CHAN
_____
```

The following schema shows the result of updating a channel store.

```
__ openCS _____
  ΔCS
  CHAN
  cid! : CID
_____
  cid! ∉ dom cstore

  posn = 0

  cstore' = cstore ⊕ {cid! ↦ θCHAN}
_____
```

A.2 Gödel Code for the schema

The following shows the base types and functions necessary to produce bindings within the types.

```
%%code for schema
EXPORT UnixFiles.
IMPORT  Lib.
```

```
BASE SetNames, SchNames, BindVar, FileId, Cid .


%%Bind1..3 are functions associated with FileId's,
%%integers and Channel ID's
FUNCTION    Bind1 : SetNames * FileId ->  BindVar.
FUNCTION    Bind2 : SetNames * Integer ->  BindVar.
FUNCTION    Bind3 : SetNames * Cid ->  BindVar.


%% In this case, the second component is a binding
FUNCTION    Bind4 : SetNames * Set(OP(Cid, List(BindVar)))
                            -> BindVar.
%%variable and schema decoration
%% on set names, i.e. priming, imput, output
FUNCTION    DSet : SetNames  ->   SetNames.
FUNCTION    OUT  : SetNames  ->   SetNames.
FUNCTION    IN   : SetNames  ->   SetNames.
%% on Schema names, i.e. priming, delta
FUNCTION    DSch : SchNames  ->   SchNames.
FUNCTION    Del : SchNames   ->   SchNames.


%%Generic to Schema types.
PREDICATE SchemaType: List (BindVar ) * SchNames.


%%For instantiating the given sets
PREDICATE  GivenSets : Set(FileId) * Set(Integer)
                                    * Set(Cid).


%%Schema referencing
PREDICATE SThetaS: Set(List(BindVar)) * SchNames.
```

The code for schemas *CHAN*, *CS*, and *openCS* is presented.

```
LOCAL UnixFiles.
CONSTANT    F1, F2, F3, F5, F7 : FileId;
            Cid1, Cid2 , Cid3 : Cid;
            Fid, PosFile, Posn,
            Cstore , ChId : SetNames;
            OpenCs, Chan, Cs : SchNames.
%%%Data%%%

    GivenSets({F1,F2,F3,F5,F7}, {0,1,2,3,4,5},
                            {Cid1, Cid2, Cid3} ).


%%%%%%Schema %%%%%%%

SchemaType(binding, Chan) <-
    binding = [Bind1(Fid, f), Bind2(Posn,posn) ] &
```

```
        GivenSets(fid, posfile,_) &
%%Types of variables, position is a natural number
            posn >= 0 &

            posn In posfile &
            f In fid .
```

```
%%Forms the set of bindings for a schema in simple cases
%%  where functions are not involved
SThetaS(val, schname) <-
    val = {binding : SchemaType(binding, schname)}.
```

```
SchemaType(binding, Cs) <-
        binding = [Bind4(Cstore, c)] &

        GivenSets( _, _,cid) &
        PF(c, cid, schtyp) &
        SThetaS(schtyp, Chan).
```

```
%%Bind1, Bind2 are from delaration of CHAN
%%Bind4 vars are from declaration of Cs and Cs'
SchemaType([Bind1(Fid, f), Bind2(Posn,posn),
  Bind3(OUT( ChId), outc),
  Bind4(Cstore, cs), Bind4(DSet(Cstore), cs1)], OpenCs )
        <-
    SchemaType([Bind1(Fid, f), Bind2(Posn,posn)], Chan) &
    SchemaType([Bind4(Cstore, cs),
                Bind4(DSet(Cstore), cs1)], Del(Cs)) &
        GivenSets( _, _,cid) &
        outc In cid &
        DomContents(cs, domcs) &
        ~(outc In domcs) &
        posn = 0 &
%%Old channel store is updated by addition of a new
%%channel whose file position is zero, but FileId is
%%unconstrained
  FunOveride(cs1, cs, {OrdPair(outc, [Bind1(Fid, f),
                                Bind2(Posn,posn)])} ) .
```

A.3 Possible Query

An example of a query is given in the main sections. Members of the sets are tried one be one. If a backtrack had been forced further bindings would have been produced.

Schema CS is provided with a possible binding equivalent to

$$Cstore = \{\,Cid1 \mapsto \langle (Fid, F1), (Posn, 2)\rangle\}$$

```
[UnixFiles] <- SchemaType ( binding, Cs ) &
    binding = [ Bind4 ( Cstore, { OrdPair ( Cid1,
          [ Bind1 ( Fid, F1 ), Bind2 ( Posn, 2 ) ])}}].

binding = [Bind4(Cstore,{OrdPair(Cid1,[Bind1(Fid,F1),
                              Bind2(Posn,2)])}}] ? ;
```

There are no other values which satisfy the predicate.

Exploring Specifications with Mathematica

Colman Reilly

Department of Computer Science
Trinity College, Dublin

Abstract. The use of *Mathematica* to "explore" specifications is examined. Here exploration is the process by which a user can encode very abstract specifications in the style of Irish VDM and examine their behaviour on sample data.
It is hoped that such exploration can lead to an improved understanding of the semantics and properties of the model and can suggest theorems. We examine the nature of an implementation in *Mathematica* which provides suitable structures and operations to allow the direct translation of specifications from our normal notation to a form which can be executed. The example of a temperature chart, given in [1], is used to illustrate the system suggested.

1 Introduction

Mathematics is not a straightforward pursuit. New results are not discovered in the linear way in which they are presented, but instead as a result of intuition built up by experience with the objects with which one is working. The first step in understanding a problem is to build a mental picture of it, so that we can see it in our mind's eye and predict its behaviour. When we are working with a model in VDM or Z, we would like to be able to build up such a picture of the behaviour and properties of the model we are using. There are two ways of doing this: by examining the formal properties of the model and by calculating its behaviour when given examples drawn from the problem domain.

Calculating the effect on examples can do several things: it can increase our confidence in the correctness of the model, it can demonstrate that our model is not correct, it can suggest properties of the model and it can provide counter-examples for suggested properties.

When first faced with a new class of mathematical object, our instinct is to work through some concrete examples in order to form an understanding of what is happening, and of what sort of behaviour we can expect from the objects in question. We may draw some diagrams, do a few calculations or try to write down an example of the structure we are talking about. When we are building a mathematical model of something it seems natural to "test" the model on some sample input. If our model acts as we expect it to, then this experimental mathematical work gives us some confidence that it is not completely wrong. If we give it input and the result differs from what we expect then there are two possibilities: either our model is incorrect or our expectations are wrong.

We want to provide the worker in the Irish School of VDM ($VDM^{\clubsuit}$) [1] with a sand-pit in which he may build and play with models.

Often, the general properties of objects are suggested by our experience with a number of concrete examples. For example, much of the theory of modern algebra arises from a study of the general properties of the number systems we use. Conversely, if we suspect the existence of a property then experimenting with sample inputs can reinforce that suspicion or discount it when a counter-example is found. In a similar way an exploration of the same model on different test data should suggest theorems about that model.

Given the complexity of all but the simplest specifications it is unlikely that we can carry out very many experiments in reasonable time by hand, so that software tools able to support our work are indispensable. The existing tools are based on formal logic methods or formally defined specification languages, which in our experience do not provide a natural notation for someone working in the Irish style of VDM, where mathematical notation and proofs are the norm. We must therefore create our own tools.

The style of the school is constructive, in that we will normally write down specifications in the form of recursive functions rather than using post-conditions to specify the behaviour of the function. This means that we seldom have problems with the executability of the specification.

We require that our tool supports directly the normal notation and methods of the school. We also require that it should allow us encode the pre-conditions which reflect the assumptions our models make about their arguments.[1]

Since $VDM^{\clubsuit}$ is founded on the classical algebra of monoids and their homomorphisms it is to be expected that *Mathematica* will provide us with an environment suitable for the development of theorems and the construction of proofs. The functional programming style of *Mathematica* supports directly the recursive functional style we are so fond of and its rewriting facilities allow for a variety of notations to be easily supported, including currying, which we often use. In addition its symbolic manipulation facilities seem to be superior to those of the other widely available packages. Another advantage is that we can bring to bear on our modelling the wealth of standard mathematics for which packages exist in *Mathematica*.

We examine first some of the operations defined in the IVDM package for *Mathematica*, and then examine its use when applied to the temperature chart problem.

2 The IVDM Package

The IVDM package provides an implementation of the operators and domains used in $VDM^{\clubsuit}$. We will focus here on the operations provided for maps, but

---

[1] Some work along these lines, involving translating from a textual form similar to our normal usage to Miranda has been done by Dara Gallagher and Andrew Butterfield, also in Trinity.

definitions are also provided for sets, sequences and bags.

When implementing the package, one of our overriding concerns was that we should be able to map our normal notation onto the structures used with only cosmetic changes — we do not wish to alter the structure of our work to accommodate a tool. Instead the tool should accommodate our style of working. One of the advantages of *Mathematica* is that a user can easily specify rewrite rules that will allow her to describe models using her own notation, while still using the same underlying definitions.

In most cases the operations map quite naturally onto the facilities available in *Mathematica*. In particular, the list operations available allow for natural and reasonably efficient implementations. We will describe the format as a user would see it, without delving into the details of the implementation.

A partial map $\mu : X \to Y$ is represented by a term of the form map[...]. For example, the partial map $[a \to 1, b \to 2, c \to 3]$ is represented in *Mathematica* by map[{a,1},{b,2},{c,3}], where {x,y} is the *Mathematica* notation for the list of two elements, or the ordered pair. Expressions break down into two components: a *head*, which we can use to tag the type of the expression, or as a function name, and a *body*, which is the "argument" of the expression. We can assign such an expression a name: mu=map[{a,1},{b,2},{c,3}]. The = operator is similar to the standard assignment operation: the right hand side is evaluated when the assignment is made, and the symbol on the left hand side takes that value. This should be contrasted with the delayed assignment operator :=, in which the right hand side defines an re-write rule to be used each time the value of the right hand side is called for.

A set is represented in a similar way: to represent the set $\{a, b, c\}$ we use set[a,b,c]. Again the set keyword acts to tag the object we are using. We wish to tag types so that we can use *Mathematica*'s pattern matching facilities to constrain functions to act on arguments of given types. The pattern F[x_set] will only be matched if the argument to F is in fact a set.

The next thing to do is to define operations on the structures we are using. We will examine several of them in some detail: map domain, map extend, map application and set membership.

The domain function is Dom[mu_map], which takes a map and returns a set. Dom[x] will only be matched if x is a map structure, so that, for example, Dom[1] is undefined. As an example of the use of Dom, we have

```
In[4]:= mu
Out[4]= {a->1, b->2, c->3}
In[5]:= Dom[mu]
Out[5]= {a, b, c}
In[6]:= Dom @ mu
Out[6]= {a, b, c}
```

We can see two forms of applying the Dom function. The first is the normal functional form, while the second is the prefix notation that *Mathematica* allows us to use. We have defined an *output form* for set which means that when an object with head set is printed it will conform as closely to our normal notation as possible[2] We have also defined appropriate output forms for our other objects, as shown below.

In a similar way we denote the map extend operation, $\mu \sqcup [x \mapsto y]$ by ExtendMap[mu_map,nu_map], which is a function of two maps. We define extend in the strict sense, and require that for $\mu \sqcup \nu$, dom $\mu \cap$ dom $\nu = \emptyset$. This is part of the definition of ExtendMap.

For example:

```
In[9]:= mu ~ExtendMap~ map[{d,4}]
Out[9]= {a->1, b->2, c->3, d->4}
In[10]:= nu=map[{d,4},{e,5}]
Out[10]= {d->4, e->5}
In[11]:= mu ~ExtendMap~ nu
Out[11]= {a->1, b->2, c->3, d->4, e->5}
```

We see here the infix form that *Mathematica* provides for operations: mu ~ExtendMap~ map[{d,4}] is equivalent to ExtendMap[mu, map[{d,4}]]. This is one of the advantages of *Mathematica* over other tools we have used: it allows the easy use of infix and prefix notation, without requiring any special definitions.

Map application is done using the ApplyMap[mu_map,a_] function (a_ matches a single term of any type). Syntactic sugar has been provided to allow this to be done in a more natural fashion, so that mu[a] is equivalent to ApplyMap[mu,a]:

```
In[14]:= ApplyMap[mu,a]
Out[14]= 1
In[15]:= mu[a]
Out[15]= 1
```

This sort of syntactic sugaring allows the writing of expressions in as close a form as possible to the form we use in our work, and is an important part of the package, in that it allows us to indulge to a certain extent in our usual abuses of notation.

The final operation we will examine here is set membership. We normally write this using the curried form of the characteristic function: so $\chi[\![x]\!]S$ is true if and only if $x \in S$. Again we can write this directly — the above becomes Chi[x][S], with almost no change in notation necessary.

---

[2] We can also define TEXoutput forms. For instance, the function μ defined above renders as $\mu = \{[a \mapsto 1], [b \mapsto 2], [c \mapsto 3]\}$ in TEX. Such output can be *spliced* into a TEXdocument. This has been done with the current paper.

```
In[18]:= S=set[a,b,c]
Out[18]= {a, b, c}
In[19]:= Chi[a][S]
Out[19]= True
In[20]:= Chi[d][S]
Out[20]= False
```

3 Implementing the Temperature Chart

The temperature chart is presented in some detail in [2], and we borrow liberally from that source.

A temperature chart is a tabulation of cities and temperatures in the form

Dublin	12
Cork	12
Limerick	9
Belfast	10

We can model this as mapping from cities to temperatures:

$$C \to T$$

where the domain of cities is finite and consists of a set of names:

$$C = \{Dublin|Chicago|Limerick|London|\cdots|NewYork|Paris|Tokyo\}$$

and the domain of temperatures is taken to be the set of integers:

$$T = \mathbf{Z}$$

We ignore the issues of which temperature scale we are using and of any constraints on the temperature.

Our table given above becomes $\mu \in (C \to T)$ defined by $\{[Belfast \mapsto 10], [Cork \mapsto 12], [Dublin \mapsto 12], [Limerick \mapsto 9]\}$

We can directly model this chart using the package defined earlier:

```
chart=map[{Belfast,10},{Cork,12},{Dublin,12},{Limerick,9}]
```

What is the temperature recorded for Dublin? `chart[Dublin]`, or 12.

There are some reasonably obvious operations we need to define on the chart. First, of course, we need to be able to create a chart:

$$Create : nil \to (C \to T)$$

$$Crea[nil] \stackrel{\triangle}{=} \theta$$

This is implemented directly in *Mathematica* as

```
Crea[]:=map[]
```

The := operation defines a re-write rule for **Crea** that will be substituted when **Crea** is evaluated, and not before.

The next thing we need to be able to do is enter a city into the chart:

$$Ent : (C \times T) \rightarrow ((C \rightarrow T) \rightarrow (C \rightarrow T))$$

$$Ent[\![c, t]\!]\mu \stackrel{\triangle}{=} \mu \sqcup [c \mapsto t]$$

$$pre - Ent[\![c, t]\!]\mu \stackrel{\triangle}{=} \neg\chi[\![c]\!]\mathrm{dom}\,\mu$$

This requires us to be able to encode pre-conditions. We can do this using *Mathematica*'s constraint mechanism. First we implement the pre-condition:

```
preEnt[c_,t_][x_]:=Not[Chi[c][Dom[x]]]
```

Then the function can be written as:

```
Ent[c_,t_][x_]:=x "ExtendMap" map[{c,t}] /; preEnt[c,t][x]
```

The use of the conditional operator /; restricts the rule to matching if and only if the condition on the right hand side — in this case our pre-condition — evaluates to true. This means that our definition for **Ent** will not be expanded unless the pre-condition is true.

Lookup, which finds the temperature associated with a city, is specified by:

$$Lkp : C \rightarrow T$$

$$Lkp[\![c]\!]\mu \stackrel{\triangle}{=} \mu(c)$$

and implemented by

```
Lkp[c_][x_]:=x[c]
```

The delete operation, $Del[\![c]\!]\mu \stackrel{\triangle}{=} \twoheadleftarrow [\![c]\!]\mu$, which removes a city from a chart, can be implemented by:

```
Del[c_][x_]:=RemoveMap[set[c]][x] /; preDel[c][x]
preDel[c_][x_]:=Chi[c][Dom[x]]
```

At this stage it is perhaps worthwhile to give an example of this in action: in Figure 1 we show a sample session using the definitions for *Crea*, *Lkp*, and *Ent*.

Such a session can be logged to file and reloaded later.

This corresponds to the way we would have written these operations if we were working through this example by hand.

The preceding is just the basic implementation of a temperature chart. In order to let us examine the way in which one might attempt to explore a specification, let us examine the *Min* operator, which we wish to yield the city or cities which have minimum temperature. First, we know that we can find the temperatures that exist in the chart, using **Rng[mu]**, and that *Mathematica* has a built-in function **Min**.

```
In[24]:= mu=Crea[]
Out[24]= {}
In[25]:= mu=Ent[Dublin,10][mu]
Out[25]= {Dublin->10}
In[26]:= mu=Ent[Cork,12][mu]
Out[26]= {Cork->12, Dublin->10}
In[27]:= mu=Ent[Limerick,11][mu]
Out[27]= {Cork->12, Dublin->10, Limerick->11}
In[28]:= mu=Ent[Belfast,9][mu]
Out[28]= {Belfast->9, Cork->12, Dublin->10, Limeri
ck->11}
In[29]:= Lkp[Dublin][mu]
Out[29]= 10
In[30]:= Lkp[Cork][mu]
Out[30]= 12
In[31]:= mu=Del[Cork][mu]
Out[31]= {Belfast->9, Dublin->10, Limerick->11}
In[32]:= Lkp[Cork][mu]
Out[32]= bottom
```

Fig. 1. The Temperature Chart Definitions at work.

```
In[35]:= mu=Crea[]
Out[35]= {}
In[36]:= mu=Ent[Dublin,10][mu]
Out[36]= {Dublin->10}
In[37]:= mu=Ent[Cork,12][mu]
Out[37]= {Cork->12, Dublin->10}
In[38]:= mu=Ent[Limerick,9][mu]
Out[38]= {Cork->12, Dublin->10, Limerick->9}
In[39]:= mu=Ent[Belfast,9][mu]
Out[39]= {Belfast->9, Cork->12, Dublin->10, Limeri
ck->9}
In[40]:= Rng[mu]
Out[40]= {9, 10, 12}
In[41]:= Min[Rng[mu]]
Out[41]= 9
```

So we now have the minimum temperature in the chart, using the minimum function which is acceptable *a priori* as a basic function.

Now we need to find the set of cities which have this temperature. For this we can use the InverseMap function from the IVDM package, which yields the pre-image of an element of the range of a map.

```
In[44]:= InverseMap[mu][Min[Rng[mu]]]
Out[44]= {Belfast, Limerick}
```

And so we have a definition for our *Min* function, which we can write as

```
min[x_]:= InverseMap[x][Min[Rng[x]]]
```

This then translates into mathematics as:

$$Min : (C \to T) \to T$$

$$Min(\mu) \triangleq \mu^{-1}(\min(\text{rng } \mu))$$

which will clearly do the job we want, assuming that the min function applied to a set returns the least element of that set.

4   Extending the Model

Up until now we have been considering a single temperature chart. Let us consider a system of charts, one for each country. We model this by the domain $N \to (C \to T)$ where N is the set of countries, or nations.

```
In[47]:= mu:=map[{Ireland,map[{Dublin,10},{Cork,11
}]},{Britain,map[{London,11}]}]
In[48]:= isTempSys[mu]
Out[48]= True
In[49]:= mu:=map[{Ireland,map[{Dublin,10},{Cork,11
}]},{Britain,map[{London,11},{Dublin,2}]}]
In[50]:= isTempSys[mu]
Out[50]= False
In[51]:= mu:=map[{Ireland,map[{Dublin,10},{Cork,11
}]},{Britain,map[{London,11}]},{France,map[{Paris,
14},{Dublin,2}]}]
In[52]:= isTempSys[mu]
Out[52]= True
```

Fig. 2. Playing with the invariant on temperature systems.

The invariant is that each city in the system appears in at most one country. For a map μ in $N \to (C \to T)$ we can write this as

$$(\bigcap_{a \in \text{rng } \mu} (\text{dom } a) = \emptyset) \wedge (|\text{dom } \mu| = |\text{rng } \mu|)$$

```
In[63]:= mu=CreaP[]
Out[63]= {}
In[64]:= mu=Add[Ireland][Add[Britain][mu]]
Out[64]= {Britain->{}, Ireland->{}}
In[65]:= isTempSys2[mu]
Out[65]= False
In[66]:= mu=EntP[Ireland,Dublin,10][EntP[Ireland,C
ork,9][mu]]
Out[66]= {Britain->{}, Ireland->{Cork->9, Dublin->
10}}
In[67]:= LkpP[Ireland,Cork][mu]
Out[67]= 9
In[68]:= mu=EntP[Britain,London,10][mu]
Out[68]= {Britain->{London->10}, Ireland->{Cork->9
, Dublin->10}}
In[69]:= isTempSys2[mu]
Out[69]= True
```

Fig. 3. Testing the *Ent'* operation.

i.e. that the domains of the maps in the range are disjoint.

In *Mathematica* an example of such a chart[3] is

```
map[{Ireland,map[{Dublin,10},{Cork,11}]},{Britain,map[{London,11}]}]
```

which we normally write as

$$
\left[
\begin{array}{l}
\text{Ireland} \mapsto \begin{bmatrix} \text{Dublin} \mapsto 10 \\ \text{Cork} \quad \mapsto 11 \end{bmatrix} \\[2ex]
\text{Britain} \mapsto \begin{bmatrix} \text{London} \mapsto 10 \end{bmatrix}
\end{array}
\right]
$$

In order to allow us test that the invariant is satisfied for a given chart we code a predicate, `isTempSys[x_]` which will return true if the structure given as an argument is a temperature system.

```
isTempSys[x_]:=(Intersection[Sequence @@ Dom /@ Rng[mu]]===set[]) &&
(Card[Dom[x]]==Card[Rng[x]])
```

---

[3] This structure is an example of a bundle, similiar to the Dail model discussed in "An Algebraic Proof in $VDM^{\clubsuit}$",[?] in this volume.

We now check that this invariant meets our expectations - see Figure 2. Of course this is nonsense, since the final map

$$\begin{bmatrix} \text{Ireland} \mapsto \begin{bmatrix} \text{Dublin} \mapsto 10 \\ \text{Cork} \quad \mapsto 11 \end{bmatrix} \\ \\ \text{Britain} \mapsto \begin{bmatrix} \text{London} \mapsto 10 \end{bmatrix} \\ \\ \text{France} \mapsto \begin{bmatrix} \text{Paris} \quad \mapsto 14 \\ \text{Dublin} \mapsto 2 \end{bmatrix} \end{bmatrix}$$

has the same city in two countries. The problem is of course the use of the intersection operator in the invariant, which will yield $\emptyset$ so long as any consecutive pair of its arguments have null intersection.

We need to define a new invariant, and a corresponding new isTempSys predicate.

Let us use the following invariant: we require that for all $c \in C$

$$(|\{x \in \text{dom } \mu : \chi[\![c]\!]\text{dom } (\mu(x))\}| \leq 1) \wedge (|\text{dom } \mu| = |\text{rng } \mu|)$$

How does this translate to *Mathematica*?

```
isTempSys2[x_map]:=Module[{cities,sets},
        cities=Union[Sequence @@ Dom /@ Rng[x]];
        sets=Function[city,Select[Rng[x],!#[city]===bottom&]]/@cities;
        ok=Select[sets,Length[#]>1&];
        Return[(List@@ok==List[]) && (Card[Dom[x]])==Card[Rng[x]])]
]
```

The @@ operator acts to change the head of its argument, in this case from a set to a Sequence, so that the Union operator will produce the union of all the sets in its argument.

Now we test our new invariant:

```
In[55]:= mu:=map[{Ireland,map[{Dublin,10},{Cork,11
}]},{Britain,map[{London,11}]}]
In[56]:= isTempSys2[mu]
Out[56]= True
In[57]:= mu:=map[{Ireland,map[{Dublin,10},{Cork,11
}]},{Britain,map[{London,11},{Dublin,2}]}]
In[58]:= isTempSys2[mu]
Out[58]= False
In[59]:= mu:=map[{Ireland,map[{Dublin,10},{Cork,11
}]},{Britain,map[{London,11}]},{France,map[{Paris,
14},{Dublin,2}]}]
In[60]:= isTempSys2[mu]
Out[60]= False
```

Moving on, we need to define similar operations to the previous case, which we will call $Crea', Ent', Lkp'$ and Del'. We also need an operation to add a country, which we call Add.

The $Crea$ operation is the same as before, with signature $Crea' : nil \rightarrow (N \rightarrow (C \rightarrow T))$, $Crea'(nil) \triangleq \theta$.

```
CreaP[]:=map[]
```

The $Add : N \rightarrow ((N \rightarrow (C \rightarrow T)) \rightarrow (N \rightarrow (C \rightarrow T)))$ operation is used to add a new country into a map.

$$Add[n]\mu := \mu \sqcup [n \mapsto \theta]$$

$$pre - Add[n]\mu := \neg\chi[n]\mathrm{dom}\,\mu$$

This can be implemented by

```
Add[n_][x_]:=x ~ExtendMap~ map[{n,map[]}] /; preAdd[n][x]
```

with the precondition

```
preAdd[n_][x_]:=Not[Chi[n][Dom[x]]]
```

We can obviously write Lkp' in the form $LkpP[n_,c_][x_]$, and implement it as:

```
LkpP[n_,c_][x_]:=(x[n])[c]
```

The enter operation, which enters a new city into the map, can be written in terms of the previous one:

$$Ent' : N \rightarrow (C \rightarrow T)$$

$$Ent'[n, c, t] \triangleq \mu \dagger [n \mapsto Ent[c, t](\mu(n))]$$

with the pre-conditions that $\chi[n]\mathrm{dom}\,\mu$ and $\neg\chi[c]\bigcup_{a\in\mathrm{rng}\,\mu}(\mathrm{dom}\,a)$. In *Mathematica* the precondition encodes as

```
preEntP[n_,c_,t_][x_]:= Chi[n][
                Dom[x]] && Not[Chi[c][Union[Sequence @@ Dom /@ Rng[x]]]
        ]
```

where /@ is the **Map** operator which applies **Dom** to each of the elements of the object on the right hand side and returns a set of results. The expression Union[Dom /@ Rng[x]] is clearly equivalent to $\bigcup_{a\in\mathrm{rng}\,\mu}(\mathrm{dom}\,a)$. In fact, if $\mathrm{rng}\,\mu = \{\nu_1, \ldots, \nu_n\}$ then Dom /@ Rng[x] is equal to $\{\mathrm{dom}\,\nu_1, \ldots, \mathrm{dom}\,\nu_n\}$, and the union of those sets is $\bigcup_{a\in\mathrm{rng}\,\mu}(\mathrm{dom}\,a)$.

The Ent' operation now becomes:

```
EntP[n_,c_,t_][x_]:=x ~OverrideMap~ map[{n,Ent[c,t][x[n]]}] /;
    preEntP[n,c,t][x]
```

An obvious question to ask is does this operation satisfy the invariant? Rather than proving here that it does, we will do the thing that is most natural: we will check that it does not immediately breach it. We will still need to present a proof that the invariant it satisfied, but for the moment we will be satisfied if it works for the obvious cases.

In Figure 3 we can see that our EntP function, at least, does not obviously breach the invariant. This sort of immediate feedback gives us confidence in our model as we are building it, but does not substitute for formal proofs, although we hope that these proofs will be easier to arrive at when we are able to test conjectures interactively.

5 Future Directions

There are several directions to be followed in future work. Obviously we need to build up experience with the application of the tool to larger and more complex specifications, in order to examine how well it scales to larger projects. *Mathematica* is also capable of simplifying and expanding expressions: making use of this facility would allow the tool to act as a "mathematician's assistant", allowing us to do symbolic calculations interactively.

As an example from another domain, let us examine how *Mathematica* allows us to manipulate polynomials.

If we take a polynomial $f(x) = (x - 3)(x - 2)(x - 1)$, we can enter it as f[x_]:=(x-3)(x-2)(x-1) and we can use operations such as Simplify to expand it to the form $-6 + 11\,x - 6\,x^2 + x^3$, or if it is in simplified form we can apply the Factor operator to factor it.

```
In[72]:= f[x_]:=(x-3)(x-2)(x-1)
In[73]:= Simplify[f[x]]
                     2     3
Out[73]= -6 + 11 x - 6 x  + x
In[74]:= f[x]/(x-1)
Out[74]= (-3 + x) (-2 + x)
In[75]:= f[x](x+2)
Out[75]= (-3 + x) (-2 + x) (-1 + x) (2 + x)
In[76]:= Simplify[f[x]]
                     2     3
Out[76]= -6 + 11 x - 6 x  + x
In[77]:= Factor[%]
Out[77]= (-3 + x) (-2 + x) (-1 + x)
```

It is possible to write one's own transformation rules, which will allow these sort of operations to be carried out on other structures. It is our hope that we can use these capabilities of *Mathematica* to allow us carry out symbolic manipulations on the expressions we use in specifications. We could then leave the details of calculations in the operator calculus to the machine and the user can concentrate on finding a valid path from assumptions to proof.

As shown above, we can use this package to test exactly what objects satisfy an invariant and what do not. It would be useful to provide a facility to generate a sample set of test cases so that we could check our invariant.

One of the main tools of the mathematician is the diagram - we draw graphs, Venn diagrams, sketch out partial maps from domain to range and so on. We hope to be able to use *Mathematica*'s graphics facilities to automate some of this process, so that a user can "see" the structures he is playing with and the effects of operations on them, in the same way he would draw the effect of a matrix on a vector in 3-space. For example, when checking an invariant we could have the system draw the structures which passed and those which failed, which might facilitate a human checking the consistency of the formal invariant and his informal intuition.

6 Conclusions

Mathematica provides us with an easy to use and natural environment for the execution and testing of specifications written in the style of the Irish School of VDM. It seems that further development of this sort of package could provide those of us working in this style with the support tools we need to do real work.

Acknowledgements

Thanks to Mícheál Mac an Airchinnigh for his editing and invaluable suggestions — not to mention the original idea for this work, to the Formal Methods Group for their comments , and to the Dublin University Mathematical Society for a supply of coffee and distraction.

References

1. Conceptual Models and Computing, Mícheál Mac an Airchinnigh, Phd Thesis, Department of Computer Science, Trinity College Dublin, 1990.
2. Mícheál Mac an Airchinnigh. Tutorial Lecture Notes on the Irish School of the VDM. *VDM'91: Formal Software Development Methods, Volume 2: Tutorials, Lecture Notes in Computer Science, Volume 552*, pages 141–237. Springer-Verlag, 1991.
3. Mathematica: A System for doing Mathematics by Computer, Stephen Wolfram, 2nd edition, Addison-Wesley, 1991.
4. Programming in Mathematica, Roman Maedar, 2ed, Addison-Wesley, 1991.
5. An Algebraic Proof in $VDM^{\clubsuit}$, A.P. Hughes and A.A. Donnelly, *1995 Z User Workshop*, this volume.

Mathematica is a registered trademark of Wolfram Research, Inc.

Method Integration

Using Z to Rigorously Review a Specification of a Network Management System

Tony Bryant[1], Andy Evans[1], Lesley Semmens[1]
Rajko Milovanovic[2,3], Sinclair Stockman[2], Mark Norris[2] and Clive Selley[2]

[1] Faculty of Information & Engineering Systems
Leeds Metropolitan University

[2] British Telecom

[3] University of Sarajevo

Abstract. The specification of a Network Management System (NMS) comprising a data dictionary and capability descriptions has been rigorously reviewed. The review revealed a number of errors and inconsistencies in the original specification. The review was performed using a technique adapted from one originally developed to review the outputs of structured analysis. The context in which the technique was developed is outlined. The NMS and its specification style are described. The adapted review technique is outlined and examples of its application to the NMS specification and the results of the review including the types of errors and inconsistencies revealed are given.

1 Introduction

RRT, a technique for rigorously reviewing the products of structured analysis, was developed during a lengthy collaboration between BT and LMU [1]. This collaboration looked at ways in which formal notations could be incorporated into existing soft ware development methods. In particular we looked at SSADM, Yourdon and Information Engineering (IE). Some of the work involved defining formal semantics for graphical and text-based notations used within these methods while another more pragmatic strand involved looking at ways in which formal notations could best be utilised within the software development process.

We found it useful to distinguish between four main aspects of the developing project which we termed rigorous review, enhancement, convergence and automated support.

Rigorous Review involves provision of a platform for the use of formal notations as a quality review mechanism in systems projects; an additional component to existing methods

Enhancement concerns the incorporation of formal notations within development methods, aiming to enhance and augment the quality and decrease the production costs of systems products; developing from rigorous review, providing a more cohesive use of formal models

Convergence involves using both formal review and enhancement as a basis to assist the transfer of the technology of formal notations and methods into wider use among systems practitioners; an overall methods framework fully incorporating formal models and practices

Automated support involves the development of a model for assessment and specification of tools designed to support integrated methods

Since the project started in 1987, we have become aware of a number of related developments, [2] provides a review of many of these. In order to derive tangible benefit for system practitioners (and budget holders) the immediate areas upon which we concentrated were those of quality review and enhancement. The objective of augmenting existing approaches with formal methods develops from this, in the sense that once the benefit of formal reviews becomes established, the next step is to consider which aspects of the system should be formally modelled in the first instance, rather than as a later review activity. Once these two facets have been established, proven and accepted they will provide a platform for the remaining two. Other work at LMU[3] involving the integration of Yourdon and Z has begun to address these areas.

It is also important to note that the benefit and impact of formal approaches is seen to be portable and extendable. Formal methods are not there to replace existing ones, but rather can be introduced in a variety of ways at specified stages of system development. In some cases formal methods may be seen as complementing existing, structured ones; in other cases they may be offered as alternatives. Eventually a specific set of guidelines will be produced for deciding which methods should be selected for use at a particular stage within identified systems development contexts.

The RRT then represents a first step in harnessing the power of formal notations. It is also, critically, a way of applying context and method to formal approaches themselves.

The operation of the technique can be used to clarify aspects of the specification, expose inconsistencies and contradictions, highlight omissions and ambiguities. The translation process can be carried out by a small number of specialists, who need not be fully involved in the specific project itself. On the other hand, project personnel (with relatively common skills) can quickly be trained to 'read', comprehend and remedy, criticise or elaborate the output from RRT. As well as serving to trap errors early in the life cycle, such output can be used to confirm the specification produced to date, and it can also form the basis for later test specifications of the implemented design.

One of the main weaknesses of formal notations is their lack of method [4]. The RRT partially overcomes this by locating the application of such notations within a specific process model. At BT this has been done within the context of the TELSTAR[4] method which incorporates RRT as a distinctive technique available to practitioners. In addition to its inclusion in the reference manuals, the team from LMU and BT have undertaken field trials on a number of live projects, and the LMU team have developed a range of training materials and seminar presentations to encourage and support use of the technique.

---

[4] TELSTAR[5] is the BT in-house development framework which incorporates methodologies such as SSADM, Information Engineering, and Yourdon.

2 The Rigorous Review Technique

2.1 The technique

RRT was originally developed as a method of reviewing specifications produced using Structured Analysis Techniques of the sort used in SSADM.

The purpose of the technique is to reveal inconsistencies, omissions and redundancy in the structured specification. The effectiveness of the technique has been demonstrated by its application to a number of BT system specifications[1]. The use of integrated techniques for quality assurance has since been recognised by other authors, for example the potential of the SAZ method for quality assurance of SSADM specifications was reported at ZUM'94[6].

The technique comprises three main stages:

1. The derivation of the system state (in Z) from the Entity Relationship Model and underlying data dictionary. Any informally specified constraints on the state can be specified formally at this stage.

2. The derivation of the signatures of the Z operations from the Data Flow Diagrams, Data Flow definitions and entity/datastore cross reference.

3. An attempt to specify the predicates of the Z operations using any available process or function descriptions.

The above sets out the way the technique would be applied given an ideal set of documentation. However, it is recognised that not all projects produce exactly this set of documents and it is possible to use the technique without these specific deliverables provided the same information is contained in any alternative documentation provided. Each time a new style of documentation is encountered decisions have to be made on how best to adapt the technique to fit. This adaptation has, up to now, been done by members of the team who developed the original technique.

3 The Network Management System and its Specification

The specification we reviewed was one that had been reverse engineered from BT's existing Network Management System (NMS).

The reviewed requirements specification belongs to a set of specifications produced specifically for use in BT's Software Design College - a "Top Gun" school for the company's best designers of software intensive systems. The College has had seven cohorts so far, with over one hundred participants in total, most having used the NMS in their "hands on" design work.

The rigorous review of the College's specifications served a multiple purpose:

- to improve further the concrete specifications used in the College

- to expose leading industrial designers of commercial-strength, and frequently mission-critical systems, to beyond toy-size applications of formal methods

– to provide a hint on how and when formal methods might be introduced into large scale industrial practice.

The evidence gathered through the results of the rigorous review described herein confirms that, regarding the last point above, a good deal remains to be desired by the industrial community.

Despite this, BT is looking at some point in future when formal methods might be more helpful on the wider scale in all stages of development. This is likely to be an integral part of the continuous improvement process which is carried out in BT's ISO 9000 and TickIT certified software development units.

3.1 The NMS

The external context in which the NMS operates is BT's public switched telephone network (PSTN). The PSTN provides communication between customers. Internally, the PSTN consists of nodes, and routes which link any two nodes. There are a number of different types of node. Customer's telephones are connected to a subset of these.

Among other things, the System should allow

1. The display of data held on individual nodes and routes

2. Computation and display of traversable nodes and routes by a single call initiated from a telephone linked to a given node, all the way to the called party's node. There is more than one path linking these two end nodes

3. For any such call, and any node traversable by it, identification and display of the node resources traversable by the call

3.2 The specification

The core of the NMS specification comprised

– a set of capability requirements which were essentially natural language descriptions of the central aspects of each capability.

The structure of the documentation for each capability requirement was as follows:

ID < A unique identifier >
Name < A one sentence long description of the capability >
Functioning < A natural language description of the capability >
Anomalous situation behaviour < a natural language description >
Measurable performance < performance requirements, volumes etc - all given as numbers >

For the purposes of the review we concentrated on the **Functioning** associated with each **ID**.

– a data model comprising:

1. An information model

The **information model** was specified using an EBNF notation:

=	*is composed of*
+	*and*
[|]	*one of the following*
{ }	*some number of*
0{ }	*zero or more of*
1{ }	*one or more of*
{ }1	*optionally one of*

2. A data dictionary of types and quantities

The **data dictionary** contained natural language descriptions, including types and quantities for the information items. All information model items whose name ended in ID were strings uniquely identifying the relevant object. The descriptions included parts of the processing associated with the information item giving the algorithm and how to compute it.

3.3 Adapting the technique

As can be seen from the description of the central elements of the specification (see section 3.2) we did not start with a conventional structured specification. There were none of the usual diagrams, however there was a Data Dictionary specifying the Information Model and there were capability descriptions. The formality of the Data Dictionary meant that it was relatively straightforward to specify the "Entity Types" and their instances (an example of this can be seen in section 4.3).

It was less obvious how to derive the operation schemas. One problem was that the details of the interfaces to the capabilities were embedded in their descriptions and another was that some of the details of the capabilities were specified as part of the Data Dictionary descriptions of Information Items. Thus the specification of the capabilities was nearer to writing a formal specification from scratch, with only an informal requirements document and the constraint that the very complex data structure had already been specified. An example of the specification of part of a capability can be seen in section 4.4

4   The Review

4.1   Logistics of the Review

The rigorous review we carried out on the NMS system took approximately three hundred person hours overall. A significant proportion of this time however was spent in gaining an understanding of the system, notation and style used to document it. The translation of the data model into a Z specification was quite straightforward due to the formal style of the model, whereas the translation of the operations took significantly longer due to their greater complexity and terseness of description. Overall, forty schemas were generated for the data model and over fifty (quite complex) schemas were generated for the process model. This compared with three pages for the data model and eight pages for the process model in the original NMS specification document.

4.2 Results of the Review

The review successfully identified a number of ambiguities and omissions in the NMS specification document, which might have otherwise been missed by an informal review. Most of the errors/problems were found during the translation process or by *type checking* the resulting specification (*f*UZZ was particularly useful here). A discussion of the general types of problem identified is given below. Although typical errors are listed in the section to which they relate, many of them were noticed at a later stage, for example during the specification of operations required to access the data model.

4.3 The Data Model

The NMS data model was already formal, being expressed in terms of an extended BNF notation. This was supplemented by a descriptive Data Dictionary. An example of a typical data model fragment and its translation into Z is given below:

$$
\begin{aligned}
\text{codetranslation} &= \text{codetranslationID} + \\
&\quad \text{replacementdigitstring}
\end{aligned}
$$

$[CodeTranslationID]$

$$
\begin{array}{l}
\hline
\text{CodeTranslation} \\
\hline
Code\_Translation\_ID : CodeTranslationID \\
ReplacementDigitString : (\text{seq}\,\mathbb{N}) \\
\hline
\end{array}
$$

$$
\begin{array}{l}
\hline
\text{CodeTranslation\_ds} \\
\hline
CodeTranslations : \mathbb{P}\,CodeTranslation \\
CodeTranslation\_id : CodeTranslationID \rightarrowtail CodeTranslation \\
\hline
CodeTranslation\_id = \{\, c : CodeTranslations \bullet \\
\qquad\qquad (c.Code\_Translation\_ID, c)\} \\
\hline
\end{array}
$$

Here, a Z basic type, *CodeTranslationID* is used to represent the type of the code translation id attribute. Next, the schema *CodeTranslation* is declared with the attributes *Code_Translation_ID* and *ReplacementDigitString* to represent an NMS code translation. Although a data store of code translations is not explicitly declared in the NMS data model, access to the instances of *Code Translation* is required by the NMS process model. Thus, the schema *Code Translation_ds* is declared containing a set of Code-Translations, and a function *Code_Translation_id*, which maps a unique code translation identifier to a Code Translation belonging to the data store.

Due to the formal structure of the data model the task of transforming the formal model into Z was relatively straightforward. However, a number of problems were still identified during the review:

Expressiveness of notation: The extended BNF notation used in the Information Model (akin to a semantic network) of the NMS specification captured a part of the data requirements of the system. In particular, this notation had no means to express constraints on the number of instances of a repeating attribute or data group. Instead, it only contained information on (loosely construed) existential quantification: zero or more, one or more, optionally one.

A rigorously formalised or theoretical option might be that the notation be modified to enable known limits on the number of instances of repeating attributes to be stated. The suggested form for this notation would then be:

$$MIN\{ \}MAX$$

where MIN and MAX are integers ($MIN \leq MAX$). In the case where $MIN = MAX$ this would mean a fixed number of occurrences.

The industrial practice environment for the original specification has instead driven this kind of (not always readily available) quantitative information into the Data Dictionary. Pragmatically based industrial needs influenced this choice:

- it is easier for contemporary software engineers and customers for development - two major camps of readers of the requirements specification - to read EBNF without MIN and MAX visual noise

- frequently, quantities are expressed in the relative form. Distribution information, when (not always) available is a significant help for designers too. E.g. for the part of the Information Model looking like:

$$A = B + \{C + \{D\}\}$$

which reads "A is composed of B and few combinations of 'C and a few of D' ", information on the number of instances is frequently expressed in real industrial setting as something like:

"number of instances of D is between 190 and 870 per one C, with more than 830 in 95 % of cases", and

"number of instances of C is 30K per one B but not more than 40K for any two As combined".

Data Omissions: Another, more obvious problem identified were omissions within the data dictionary of data referred to by other parts of the data dictionary or by NMS operations. These omissions prevented the complete specification of data accessed by NMS.

Data Ambiguities and Inconsistencies: The review also identified a number of ambiguities and inconsistencies in the data model. For example, there were a number of attributes that were either misnamed or whose structure was not fully defined.

4.4 Processing

The process model for NMS consisted of a number of informal natural language descriptions of the overall functionality of each capability of the operations in the system. Precise details of the operations that made up a particular capability were given in an object-oriented style throughout the data description document. The most complex capability we formalised was the *Micro Trace Forward* capability which described the traversal of the internal structure of a PSTN node by a "model excitation" (e.g telephone call). This required over thirty pages of Z to describe fully its functionality. An example of the type of description used to describe a capability along with the Z resulting from its formalisation is given below:

CodeTranslation - *A resource of a PSTNnode. Replaces incoming value of LeadingDigits with ReplacementDigitString and forwards the transformed ModelExcitation to the originating DecodeSysXID.*

CodeTranslationOp
$\Delta Modex$
$\Xi CodeTranslation\_ds$
$decodesysx?, decodesysx! : DecodeSysX$
$node?, node! : PSTNnode$
$modex : ModelExcitation$
$c : CodeTranslationID$
$new\_leading\_digits : LeadingDigits$

$modex \in model\_excitation$
$(modex.LeadingDigits, CodeTrans(c)) =$
$\quad (\mu\, d : decodesysx?.DigitsAndTransOrCat)$
$new\_leading\_digits = (\mu\, ct : CodeTranslations \mid$
$\quad ct.Code\_Translation\_ID = c \bullet ct.ReplacementDigitString)$

$model\_excitation' = model\_excitation \setminus \{modex\} \cup$
$\quad modify\_modex(modex, new\_leading\_digits, modex.TerminationStatus,$
$\qquad\qquad modex.AutomaticAlternativeRoutingFlag,$
$\qquad\qquad modex.AutomaticReRoutingFlag,$
$\qquad\qquad modex.DynamicAlternativeRoutingFlag,$
$\qquad\qquad modex.AutomaticAlternativeRouteingAtNextExchangeFlag)$

$decodesysx! = decodesysx?$
$node! = node?$

The schema *CodeTranslationOp* describes fully the successful code translation operation of a PSTN node (error conditions are not considered). The schema above is a striking example of the simultaneous advantage and drawback of formal methods as they currently stand.

The advantage is that all the "illities" (completeness, coherence, etc) of the requirements specification are brought to bear in full by this segment of Z - which need not be the case in the original requirements specification text. For today's industrial use, drawbacks seem to stand out in a more pronounced way.

The first drawback is that the text is significantly longer. As the requirements specification also serves the purpose of the (virtual or actual) contract between the customer for development and the developer, both - and especially the former like this "contract" to be as short as possible, albeit not at the expense of completeness. The second drawback is a much larger set of concepts (capitalised names) used in the Z text above, than in the original. As each of the concepts from the original text is further associated - in the Information Model part of the requirements specification - with others, the Z equivalent expands it of necessity into a much more frequent referencing pattern for those concepts. That obviously also leads the specification expressed in Z partly into the design domain - what might be relevant for developers, but is most frequently the realm where customers for development do not like to venture on their own. One should keep in mind, however, that the aim of the RRT was not to provide a specification document for customers but to clarify and improve on the informal NMS requirements specification.

The mechanics of the Z text above are as follows: The leading digits of an incoming model excitation (a number, etc) are modified according to the value stored by an associated code translator in the data base. The operation inputs the decoder associated with the PSTN node currently being traversed and the current PSTN node. Provided there exists an associated leading digits and code translation in the decoder the new leading digits is given by the replacement digit string in the code translation. The new model excitation is then updated with the new leading digits.

Overall, a greater number of problems was found in the process model than the data model, probably due to its greater complexity: Some of the general ones were:

Terse-specification: Use of terse informal natural language in the operation descriptions caused some initial problems. As a consequence, some assumptions had to be made about their meaning. The example of the code translation operation described above clearly illustrates this! Furthermore, little detail was available in the original NMS requirements specification on how the system would behave if an operation pre-condition was not satisfied and further assumptions had to be made.

Concrete vs Abstract: As a consequence of the necessary reverse engineering of the NMS specification from its implementation, the data and operations of the NMS system were very close to physical implementation. This placed a greater burden on the reviewers in terms of representing this greater detail in Z, and also made understanding of the functionality of the system more difficult.

Structure: The descriptions of the processing were split between the "Capabilities" and the "Data Dictionary", and this sometimes made the operations quite difficult to follow. The parts of these descriptions that were included in the data dictionary had a very object-oriented feel to them and the fact that the target language is C++ explains this. Here, Z showed its lacking of a object-oriented structuring capabilities.

Although each of the individual operation requirements could be readily specified as a schema it was difficult to specify the communication between these operations. Some progress was made using Z's piping operator[7], however in some cases this prohibited us from specifying recursively communicating operations due to the fact that operations are required to be "declared before use" in Z.

4.5 Recommendations from the Review

Although the review identified a number of ambiguities and omissions in the NMS specification document, their number was no more than we would have expected for a specification of this size.

Many ambiguities revealed in the specification of the data model could have been resolved with additions to the notation. In terms of the process specification, the most common problem with specification was the terseness of the informal natural language descriptions, which often resulted in the under-specification of operations. Thus, once the meaning of the NMS specification had been clarified, the Z specification was generated fairly quickly. However, quite a number of 'holes' in the specification were made apparent during the translation.

Thus it was recommended that greater emphasis be placed on both clarifying the operation descriptions used in the NMS documentation and correcting the ambiguities we had identified. An added benefit of the review process was that the Z specification generated would greatly aid in this process.

5 Conclusions

The application of the Rigorous Review Technique to a system as large and as complex as NMS has clearly confirmed the practical benefits of the technique, and indeed of incorporating formal methods in systems development in general. By applying RRT to the NMS system, we were forced by the precise nature of the Z notation to examine the specification of the system closely and rigorously. Problems, which might easily have escaped the attention of a non-formal review were quickly brought to our attention. In the case of NMS, RRT highlighted problems of underspecification, data inconsistencies and ambiguous statements.

Another benefit of the technique was that it highlighted deficiencies in the structuring of the NMS specification. If required, the Z specification we produced could be used to help in its rationalisation and better organisation.

It is important to recognise that RRT can require a great deal of time and effort on the part of the reviewers. It is not a task that should be lightly undertaken, as the formality of the technique requires that some, critical aspects of the system be fully understood. However, we believe that the benefits of the review (especially in terms of a possible reduction in long-term maintenance costs) clearly outweigh the initial effort. In the case of NMS we were asked to review the entire specification. It is often more cost effective to identify the more critical and/or more complex aspects of a system and subject just those to review.

Large specifications can be generated from seemingly small system descriptions. NMS was a particularly good example of this. However this is probably only a reflection of the system's true complexity.

RRT shows its greatest strength in highlighting inconsistencies and ambiguities in an existing system specification, i.e. verification. It is not aimed to validate the system against its requirements (although the revised specification generated after RRT will go a long way in helping to clarify the functionality of the system).

Finally, RRT works best when there is good contact and feedback between the reviewers and the original developers of the system. Although the reviewers of NMS worked very successfully with the writers of the NMS specification at BT, it could be envisaged that some developers might see RRT as a review of the quality of their work. Instead, it is important that RRT be promoted as *an additional tool for software engineers in the production of quality systems*.

References

1. Sukhvinder Aujla, Tony Bryant & Lesley Semmens, *A Rigorous Review Technique: Using Formal Notations within Conventional Development Methods*, SESS'93, IEEE Journal on Selected Areas in Communication, February 1994.
2. L.T. Semmens, R.B. France & T.W.G. Docker, *Integrated Structured Analysis and Formal Specification Techniques*, The Computer Journal, Vol 35, No 6 pp600-610, December 1992.
3. Lesley T. Semmens, *Methods Integration: Rigorous Systems Specification Using Structured Analysis and Formal Notations*, PhD thesis in preparation, Leeds Metropolitan University, 1995.
4. A. Bryant, *Structured and Formal Methods*, Z User Workshop Proceedings, 1989, J. Nicholls, editor, Springer-Verlag 1990.
5. *TELSTAR, Verification, Validation and Testing: Techniques, Data and Forms*, BT, 1992.
6. Fiona Polack & Keith C. Mander, *Software Quality Assurance using the SAZ Method*, In J.P. Bowen & J.A. Hall (Eds.) *Z User Workshop, Cambridge 1994*, Springer-Verlag, 1994.
7. John B. Wordsworth, *Software Development with Z: a practical approach to formal methods in software engineering*. Addison-Wesley 1992.

A Two-Dimensional View of Integrated Formal and Informal Specification Techniques

R. B. France and M. M. Larrondo-Petrie

Department of Computer Science & Engineering
Florida Atlantic University
Boca Raton
FL 33431-0991
Tel: Robert (407)367-3857 Maria (407)367-3899
robert@cse.fau.edu maria@cse.fau.edu

Abstract. It is often felt that the use of *formal specification techniques* (FSTs) precludes the use of *informal, structured specification techniques* (ISTs). Research on *integrated FSTs and ISTs* (FISTs) has shown that this is not necessarily the case, and that the use of formal techniques can enhance the use of informal specification techniques and vice versa.

In this paper we describe the applicability of FISTs to requirements engineering along two dimensions: process support, and formal-informal transformations. We illustrate aspects of these dimensions with results from student requirements engineering projects that involved the use of a Structured Analysis (SA) and Z FIST.

1 Introduction

During the late seventies/early eighties research on systematic approaches to software development resulted in a number of graphically-based specification methods collectively referred to as *structured specification methods* (SSMs) (e.g., see [3, 8, 9]). These methods were considered revolutionary at the time mainly because of their use of simple, visually-appealing concepts and notation. The specifications produced by SSMs were often more concise and sometimes more precise than the purely natural language specifications that were prevalent then. The evolution of SSM-related research and practice has resulted more recently in the creation of *object-oriented methods* (OOMs) (e.g., see [11, 2, 14]). Most SSMs and OOMs are *informal* in the sense that their applications are likely to produce ambiguous specifications that are not amenable to rigorous semantic analyses. In this paper, we will refer to the techniques used in SSMs and OOMs as *informal specification techniques* (ISTs).

The need to specify problems and their computer-based solutions precisely has led to interest in *formal specification techniques* (FSTs). FSTs use mathematical concepts and notation to precisely define theories and models of application behavior. Precise specifications facilitate effective communication among persons with a stake in the development of the software, and the ability to reason about properties captured in formal specifications allows developers to predict the behavior of implementations during the specification phases of software development.

Despite their potential, FSTs are rarely used in industrial software development environments. A reason often stated is the perceived complexity of the mathematical concepts and notations they utilize, which, to some, makes the creation, reading and understanding of formal specifications difficult. Furthermore, most FSTs are based on textual languages with relatively primitive complexity management mechanisms. There is also the perception that FSTs are revolutionary alternatives to industrially favored ISTs. This has led some practitioners to the conclusion that applying FSTs requires throwing out IST experiences and tools. Given such perceptions it is not entirely surprising that industries with significant investment in ISTs resist adopting FSTs. Recent research on integrating FSTs and ISTs is concerned with bridging the perceived gap between them. A survey of some integrated FSTs and SSMs can be found in [13].

Integrating FSTs and ISTs (FISTs) can be beneficial in the following respects:

- *FISTs enable an evolutionary approach to the use of FSTs in industry.* An FIST enables the use of FSTs in the context of the ISTs. This allows specifiers to become familiar with FSTs while still taking advantage of their IST skills. FISTs allows an organization to preserve, and even enhance, its investment in ISTs while taking advantage of FST-related benefits.
- *FSTs and ISTs can complement each other.* The relatively simple and graphical nature of IST specifications makes them more presentable than the more detailed, often textual, formal specifi-

cations. Furthermore, the flexibility resulting from the use of simple, intuitively-defined concepts and visual constructs makes ISTs suitable for the early, probing phases of software development, since they are more likely to result in abstract models that can be used to gain a high-level understanding of the problem or solution. The abstract models produced through the use of FSTs are often more difficult to understand. On the other hand, the lack of a firm semantic basis for ISTs inhibits their use in rigorous specification and analysis of behavior. FSTs are needed for such activities. FSTs can also enhance the applications of ISTs. Formalizing an informal specification often reveals ambiguities, inconsistencies, and gaps in the informal specification.

- *An FIST can be used as a basis for systematic transformation of vague specifications to precise statements of need.* Requirements engineering is concerned with transforming vague statements of requirements into more precise statements of the problem. During this transformation a number of intermediate specifications, each varying in degree of rigor and detail, may be produced. FISTs allows one to impose structure on the transformation of vague requirements to formal specifications. FISTs can play a similar role in design, where design is viewed as the creation of a series of decreasingly abstract solution specifications.

In this paper we discuss the applicability of FISTs along two dimensions: process support and formal-informal transformations. The process dimension addresses issues related to processes that support the use of FISTs, and the transformation dimension addresses technical aspects of moving from informal to formal notations and concepts, and vice-versa. In section 2 we discuss FISTs in terms of the dimensions, and in section 3 we illustrate aspects of the dimensions using a requirements engineering problem concerned with modeling a student advising system. We give our conclusions in section 4.

2 Dimensions of FISTs

In assessing the practicality of FISTs issues raised by the following categories of questions need to be examined:

1. *Justification for use*
   What benefits can FISTs bring to software development? What are the limitations? In what situations should FISTs be applied/not applied? What costs are involved?
2. *Process concerns*
   What processes can be used to guide the application of FISTs? What resources and development roles are required? What capabilities and responsibilities are associated with the roles?
3. *Technical: formal-informal transformation concerns*
   How does one go from informal to formal specifications, and vice-versa? What aspects of the transitions can be automated, and are there tools that offer non-trivial assistance? Are transitions reversible? How is integration accomplished? The latter question can be broken down into further questions, including: At any stage of the development, is the system represented entirely using formal or informal notation, or can a system representation consist of both formal and informal notation? Is a significant shift in perspective required as one goes from formal to informal and vice-versa? How does one establish the semantic consistency of formal and informal models with respect to each other?
4. *Technical: analysis concerns*
   What types of semantic analyses are possible with FISTs? What advantages do FISTs offer over purely formal or purely informal techniques with respect to analysis?

Attempts at answers to questions in category 1 are necessary if one is going to make a case for the use of FISTs within an organization. In the conclusion we present our views on some of the issues in this category. A deeper treatment of this dimension requires empirical data that we do not have at this time. This paper focuses on some of the issues in categories 2 and 3. The discussion is based on observations of the use of FISTs, on our experiences with the use of both formal and informal specification techniques, and on our current perception of software specification problems.

Our approach to learning more about the applicability of FISTs involves developing models of FIST usage based on current perceptions of the specification process as developed through practice and research, and improving them using feedback obtained and experienced gained through use in industry and academia. Here, a model of usage consists of elements from the process, transformation, and analysis dimensions, that is, a usage model defines a particular process to follow, and provides a model relating formal and informal concepts and notations that guides transformation and analysis activities. The models of FIST usage that we suggest here require validation through experimentation.

2.1 The Process Dimension

In this section we describe three classes of requirements engineering processes that guide the application of FISTs: the *Parallel Interaction Model* (PIM), the *Loosely Integrated Model* (LIM), and the *Tightly Integrated Model* (TIM).

The process models we describe here represent our initial attempts at providing process support for the use of FISTs. The underlying rationale for each model is based on our evolving views of how FISTs can contribute to good software engineering practice. We continue to test our perceptions through experimentation, the results of which will be used to refine the models.

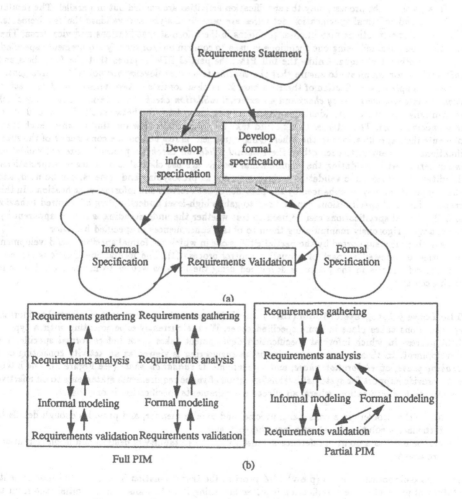

Fig. 1. The Parallel Interaction Model

The Parallel Integrated Model (PIM) The *Parallel Integrated Model* (PIM), shown in Figure 1(a), is a characterization of processes in which the development of formal and informal specifications proceeds in parallel from a common requirements statement. A PIM process consists of two major subactivities: *requirements specification development* (RSD) and *requirements validation* (RV). The RSD phase consists of requirements elicitation, analysis, and specification activities, and the RV phase consists of requirements validation activities. Realizations of the PIM may vary from processes in which the formal and informal specifications are developed and validated in parallel (full PIM) to processes in which only the specification activities are carried out in parallel (partial PIM). Figure 1(b) illustrates the two PIM process class extremes.

Organizationally, a PIM process would require two teams of developers: a team of *formalists*, responsible for developing formal specifications, and a team of *informalists*, responsible for developing informal specifications.

In a full PIM process the formalists and informalists work independently to produce validated requirements specifications. An organization that wants to assess the relative impact of formal and informal specification methods on their development environment can do so, if resources are available, using this model. This approach to comparative analysis can be effective, but it may also require resources that cannot be cost-effectively allocated to a single project. The benefit of doing the comparative analysis needs to be weighed against the cost of the extra resources needed.

In a partial PIM process, only the specification activities are carried out in parallel. The results of formal and informal specification activities are used to analyze and validate the requirements. The formal specifications may uncover problems in the informal specifications and vice versa. The specification obtained using one technique is used to confirm and/or modify requirements specified using the other technique. Unlike the full PIM, the partial PIM requires that the formalists and informalists communicate to ensure that the specifications they develop are not widely divergent.

An example of a realization of the RV phase is one that consists of two types of validation activities: an *internal consistency check* and an *external validation check*. The internal consistency check is primarily concerned with identifying and resolving inconsistencies between the formal and informal specifications. The internal check could take the form of a review meeting in which each team presents their specifications to the other. Once agreement is reached on the consistency of the specifications with respect to each other, the *external validation check* can proceed. External validation is concerned with validating the requirements models against clients' and/or users' expectations. Traditional requirements validation techniques, such as prototyping and reviews, can be used, and their application may be enhanced by the presence of both formal and informal specifications. In this respect, informal specifications can be used to gain a high-level understanding of required behavior and the formal specifications can be used to test whether the understanding is real or apparent by, for example, rigorously manipulating them to obtain consequences of specified behavior.

Another view supported by the partial PIM is one in which the formal specification development activities are viewed as part of a quality control process [1]. The formal specifications serve to identify deficiencies in the products developed using the ISTs as well as to increase confidence in their accuracy.

The Loosely Integrated Model (LIM) In a LIM process the development of informal and formal specifications takes place in some specified order. We are currently experimenting with a type of LIM process in which informal specification development takes place before formal specification development. In these processes, requirements engineering is viewed as an activity consisting of a *probing phase*, an *elaboration phase*, and a *requirements validation phase* (see Figure 2). The intent is to provide support for systematic transformation of vague requirements statements to an effective requirements specification, where, an effective requirements specification is one that:

- states the problem in a consistent, precise, and concise manner, and provides enough details to formulate a solution (or show a lack thereof), and
- can be used as a point of reference against which the suitability of designs and implementations are assessed.

In a probe-elaborate-validate (p-e-v) LIM process, the transformation from an initial requirements statement to an effective specification involves first using ISTs to transform the initial statement to more precise (but informal) specifications, and then using the informal specifications to help build more precise specifications using FSTs.

In the probing phase, the specifier attempts to impose a structure on the problem. Given the exploratory nature of this phase, the tool support should be flexible, that is, allow for the easy incorporation of changes, and should allow the specifier to view the problem in an abstract, but insightful, manner. ISTs from methods such as Yourdon's *Structured Analysis* (SA) [17] and Shlaer-Mellor's *Object-Oriented Requirements Analysis* (OORA) [14], are well-suited to this phase since they mostly employ simple, intuitive modeling concepts and notation, and provide visual abstract views of the structure of a problem from a variety of perspectives. Formal specification techniques, on the other hand, are not well-suited to this phase because they may produce abstract specifications of structure that are not as easy to understand and modify.

When the specifier is satisfied that an adequate structure has been imposed, then he proceeds to the elaboration phase where details pertaining to elements in the structure and their relationships are more rigorously specified. In this phase, formality should be favored over informality because the objective is to create a specification that is analyzable and precise. During the elaboration phase, ambiguities, inconsistencies and incomplete requirements may be identified, and attempts at resolving them may lead to modifications of the structure initially imposed on the problem. Problem

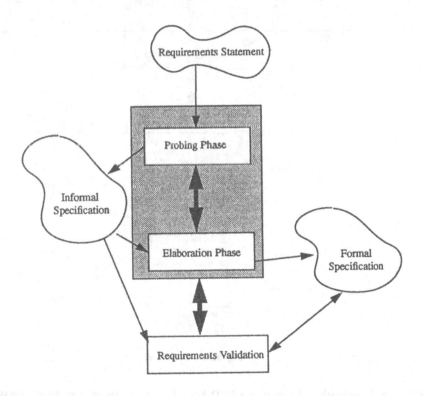

Fig. 2. A LIM Process

analysis in this model is thus an iteration of probing and elaboration phases activities, as indicated by the double headed arrow between the two phases in Figure 2.

The validation activity in the LIM is concerned with establishing the validity of the requirements specification. This phase has the same objectives as the external validation activity discussed earlier.

Tightly Integrated Models (TIMs) Processes in which the development strategies of the informal and formal techniques are intertwined are said to be *tightly integrated*. An example of a TIM process, in which both the formal and informal specifications are developed incrementally, is shown in Figure 3. In this TIM process, specification development starts off with some informal activity which produces a partial structure that the specifier feels is ready for formalization. Formalization of the partial specification then proceeds as in the LIM process described earlier. Once the specification is formalized then additional structure is informally modeled and formally elaborated. This continues until the specification is deemed adequate. The TIM process just described supports a compositional approach to requirements modeling.

An example of a TIM process that supports a decompositional approach to requirements modeling is one in in which a formal specification produced at a particular level of description constrains the subsequent informal refinement activities that produces the next level of description. For example,

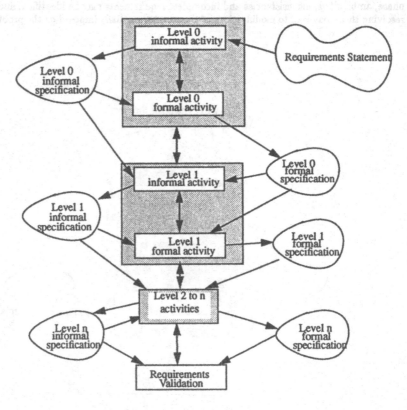

Fig. 3. A TIM Process

in SA, the formalization of a level in a DFD hierarchy can be used to semantically constrain the decomposition of data transforms at that level. The structure obtained as a result of decomposition must be shown to be consistent with the formally defined functional requirements of the parent level. The work described in [7] supports the use of a TIM process in this manner.

2.2 The Transformation Dimension

In this section we classify the types of relationships and transformations that are possible between formal and informal specifications created using an FIST. The manner in which FSTs and ISTs are integrated in an FIST is usually based on one or more of the following considerations:

- *Preservation of the 'intuitive' interpretation of the informal specification*: The integration is such that the formal specifications provide interpretations of the informal specifications that are consistent with intuitively-held interpretations of the informal specifications. An FIST that does not support intuitively-held interpretations of its informal specifications may require that the informal specifications be developed in ways that are not consistent with practices associated with the use of the ISTs. In such cases, a specifier learning and using the FIST is not likely to take full advantage of his prior experience with the ISTs.

- *Level of automated support for moving between formal and informal specifications*: Developing formal specifications can be a very arduous task. Well-defined relationships between FST and IST constructs and concepts can provide the basis for mechanical generation of some parts of the formal specification. In the other direction, it is often possible, given a well-defined relationship between formal and informal specification elements, to mechanically generate informal specifications from formal specifications. This transformation is often easier to automate because it usually entails simply choosing what information to hide.

- *Degree of integration*: In some cases it may not be worthwhile to formally interpret all aspects of an informal specification. It may be sufficient to formalize only those parts that can benefit from more rigorous specification and analysis.

In the following subsections we describe types of transformations and FST/IST relationships.

Cognitive vs. generative transformations The manner in which a specifier goes from formal to informal specifications and vice-versa during the development of an integrated specification can be classified as being *cognitive* or *generative*. A cognitive transformation is characterized by its lack of well-defined, repeatable transformation rules and procedures, and the resulting heavy reliance on human ingenuity in making the transition. A generative transformation is characterized by the existence of well-defined, repeatable transformation rules and/or procedures that provide specifiers with non-trivial assistance during formal to informal, and informal to formal transitions.

In a *cognitive informal to formal specification transformation*, the informal specification is used by the developer to guide the creation of more formal specifications. The specifier does not utilize any formal mappings, making this type of transformation the most flexible in the sense that the specifier is free to develop the formal specification any way he sees fit. The structure of the formal specification may or may not reflect the structure imposed on the problem (or solution) by the informal specification, but it may be desirable to do so to ease the task of demonstrating consistency between the formal and informal specifications.

A *cognitive formal to informal specification transformation* utilizes no formal transformation rules in going from the formal specifications to informal specifications, that is, the specifier is free to determine the type of abstractions he feels are applicable.

Creating a formal specification can be a difficult and arduous task. To reduce the effort required to create formal specifications, ways of automatically generating parts of formal specifications from informal specifications are being investigated (e.g., see [12]). Complete automation of the informal to formal specification transformation is unlikely, given that informal specifications are inherently incomplete. Human interaction is needed during generative transformations to at least simplify and optimize generated formal specifications. Most *generative informal to formal specifications transformations* generate a formal specification skeleton using information available from the informal specifications, then require that the specifier provide additional details in the formal notation. It follows that the more expressive the informal specifications, the more one can generate from them. One can enhance expressiveness by using a variety of informal specification languages to express different aspects, or views, of a problem. For example, complementing DFDs with Entity Relation Diagrams (ERDs), and control models provide information that can not be readily expressed by DFDs alone.

Complete automation of the generation of informal specifications from formal specification is more likely, since the information needed to create an informal specification is usually present in a formal specification; one need only decide what information to abstract out. *Generative formal to informal transformations* can be viewed as abstraction mechanisms.

Note that given an informal specification and a formal specification that was supposedly derived from it, it is not always possible to determine whether the formal specification was derived with the assistance of generative or cognitive transformations. The primary difference between cognitive and generative transformations is that the generative transformations provide mechanical aids to transforming informal specifications to formal specifications and/or vice versa. Both types of transformations are usually guided by a model of a relationship that ties the semantics of the formal notation to the concepts symbolized by the informal notation; a relationship that we will refer to as an *interpretation*[1]. The generative transformations, unlike the cognitive transformations, utilize codified knowledge of well-defined aspects of the interpretations they are based on.

Supplemental vs. interpretive relationships The relationship between the formal specification and informal specification elements of an FIST can be classified as either *supplemental* (or partially interpretive) or (fully) *interpretive*.

In a supplemental coupling, the informal and formal specifications are complementary, that is, when viewed separately they both provide incomplete views of specified properties, but when viewed together they form a more complete specification. An example of a supplemental DFD/Z FIST is one in which the informal data transform (process) specifications (or PSpecs) are replaced by Z specifications. The Z specifications state the goals of the processes, but say little about the sources and destinations of input and output data; such information is provided by the DFD.

When a formal specification provides a formal interpretation of an informal specification then we say that the relationship between them is interpretive. For example, in [7] an interpretive relationship between DFDs and algebraically defined state transition systems (ASTSs) is described. An ASTS associated with a DFD in this approach provides a characterization of a class of formal interpretations for the DFD. Like a supplemental relationship, the interpretive relationship provides meanings to the parts of a DFD (data transforms, datastores, and data). Unlike supplemental relationships, the interpretive relationship also provides meanings for the relationships among DFD parts (e.g., interpretations of the dynamic aspects of data and control flows, and data store accesses).

3   Example Applications of SA-to-Z FISTs

In this section we give samples of the results of SA-to-Z transformations that illustrate some of the major aspects of the dimensions discussed in the previous section. The examples are based on the results of projects assigned to software engineering undergraduate students at Florida Atlantic University. The projects were concerned with the development of an automated student advising system for the Department of Computer Science and Engineering. A group of students in the software engineering project class (referred to here as the *RE group*), developed a formal requirements specification for the system, using a LIM process, in which an entity-relation diagram (ERD), a data dictionary, and control-extended data flow diagrams (DFDs) were produced in the probing phase, and Z specifications [15] produced in the elaboration phase. The transformation of the informal SA models to Z specifications was guided by some rigorously expressed transformation rules, thus it can be classified as generative.

Later, another group of students (referred to here as the *quality group*), were given the SA models developed by the RE group and told to identify any inconsistencies, gaps, and/or ambiguities in the informal models and to improve them accordingly. In this project, the FSTs were used to check the quality of the informal models. The approach taken can be likened to a partial PIM process in which informal and formal specification were developed by separate teams. The quality group used a cognitive transformation approach.

The underlying interpretation used to drive the generative and cognitive transformation approaches outlined here is one in which the ERD captures the state of a system (i.e., the ERD depicts the collections, and relationships among collections, that are needed to describe the required functional behavior), and the DFDs identify processing entities and show their data dependencies, as well as depict their relationships with elements of the state. Elements of the state are depicted by data stores in the DFDs.

A portion of the requirements statement for the advising system is given below:

> The project is concerned with providing automated support for undergraduate advising in the Department of Computer Science and Engineering. An important component of the department's undergraduate advising system is the student evaluation worksheet, which is used to keep track of courses students have taken, are currently enrolled in, and plan to

---

[1] Not to be confused with the mathematical notion of interpretation.

take. Your specific task concerns developing a system that supports information retrieval and updating activities associated with student worksheets.

The system should allow users to do the following:

1. Consistently create and destroy student worksheets.
2. Consistently update worksheet data. The system should be capable of informing users of updates that violate departmental rules, e.g., the system should disallow the recording of an enrollment when the student does not have the necessary pre-requisites.
3. Retrieve worksheets.
4. Generate course plan options from a given semester to graduation.
5. Generate the grade point average.
6. Generate a list of courses that can serve as electives.
7. Check that students satisfy requirements for graduation.

The system will be used by two types of users: *students* and *faculty/staff*. Students will only be allowed to carry out activities 3,4,5,6, and 7, where access is restricted to their worksheets, that is, a student should not be able to access another student's worksheet. Students must not be able to modify their worksheets on the system.

Faculty and staff will be allowed to carry out all the above activities. Facilities should be provided for determining who modified a worksheet and when. This can take the form of a history file that records the originators and dates of modifications.

There are two types of worksheets: the *BS four year student worksheet* and the *Transfer student worksheet*. The Four year worksheet is used for students who are enrolling as freshmen, while the Transfer worksheet is used for students who are transferring from Florida

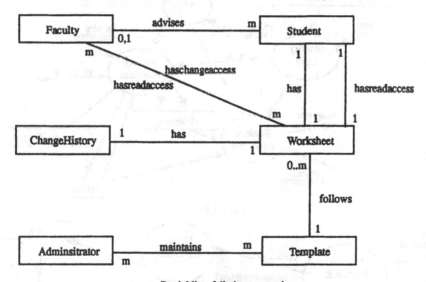

Partial list of dictionary entries

Faculty = @Id
Student = @Id
Template = @Temptype + {Section}1..m
Section = section-type + [Minimum-GPA] + Totalcredits + {Course}1..m
Course = @Course-id + [Pre-Requisites] + Credits + [Minimum-Grade] + [Co-Requisites]
Pre-Requisites, Co-requisites = {{Course}1..m}1..m
Worksheet = @Id + {Section-Status} + Total-Credits + GPA
Section-Status = Total-Credits + [GPA] + {Course-Taken}1..m + {Course-Enrolled}1..m
Course-Taken = Course-id + {(Grade, Term)}1..m
Course-Enrolled = Course-id + Term

.....

Fig. 4. Entity-Relationship Diagram for the Advising System

Community Colleges, (with or without AA degrees) or those who are enrolling for a second Bachelors. Your system should *at least* support four year students and students transferring from community colleges with AA degrees. ...

We show here only some of the results of the informal specification activity (results of the RE group's probing activity). The models presented here are scaled down versions of what the students actually produced. The scaling down of the problem was necessary to keep the paper within reasonable size. The ERD is shown in Figure 4, and a sample of the DFDs produced is shown in Figure 5. In the following sections we outline the transformations to Z specifications, as carried out by the RE and quality groups.

3.1 A generative transformation of the advising problem

In the generative transformation approach used by the RE group, rigorously expressed transformation rules were used to derive a large part of the Z state schema, and information contained in DFDs and the state schema are used to obtain the declaration parts of the operation schemas. The transformations they used were based on ideas and rules expressed by Semmens and Allen in [12].

In the transformation, each entity in the ERD is described by a schema. The components of the schema are the components described in the dictionary associated with the ERD. For example, an entity described in the dictionary as:

$entity = comp1 + comp2 + \{comp3\} + \ldots$

is transformed to a schema:

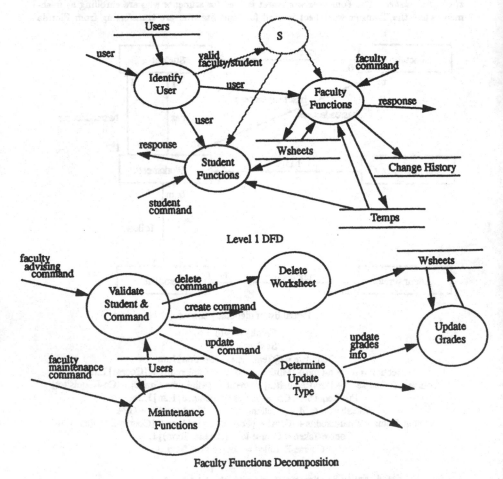

Level 1 DFD

Faculty Functions Decomposition

Fig. 5. Partial DFDs for the Advising System

```
┌─ Entity ─────────────────────────────────────────────────────────────────────
  comp1 : t1
  comp2 : t2
  comp3s : P t3
  ...
└──────────────────────────────────────────────────────────────────────────────
```

In the above schema, ti $(1 \leq i \leq 3)$ is the type of $compi$ (type information is contained in the dictionary). If a component is further defined in the data dictionary, then a separate schema is defined for it and the schema name used as its type name.

Below are some of the schemas derived from the advising system ERD.

```
┌─ Faculty ────────────────────────────────────────────────────────────────────
  id : IDENTIFIER
└──────────────────────────────────────────────────────────────────────────────
```

```
┌─ Template ───────────────────────────────────────────────────────────────────
  Section
  ttype : TEMPTYPE
  sections : F₁ Section
└──────────────────────────────────────────────────────────────────────────────
```

```
┌─ Section ────────────────────────────────────────────────────────────────────
  Course
  stype : SECTTYPE
  totalcredits : N
  courses : F₁ Course
└──────────────────────────────────────────────────────────────────────────────
```

```
┌─ Course ─────────────────────────────────────────────────────────────────────
  courseid : CIDENTIFIER
  credits : N
└──────────────────────────────────────────────────────────────────────────────
```

During the derivation of the Course schema, the students discovered that pre-requisites and co-requisites could be better modeled as relationships among courses rather than as attributes (components) of courses, thus avoiding the more complex recursive definitions. The students also chose to model optional attributes (e.g., minimum GPA and minimum course grade) as partial mappings from instances of entities to instances of the attributes in the final state schema.

The application of the transformation rules sometimes led to cumbersome structures. For example, in transforming the definitions for *Section_Status* the following schemas were obtained:

```
┌─ Section_Status ─────────────────────────────────────────────────────────────
  Course_Taken
  Course_Enrolled
  totalcredits : N
  coursestaken : F Course_Taken
  coursesenrolled : F Course_Enrolled
└──────────────────────────────────────────────────────────────────────────────
```

```
┌─ Course_Taken ───────────────────────────────────────────────────────────────
  courseid : CIDENTIFIER
  termgrade : TERM ↦ GRADE
└──────────────────────────────────────────────────────────────────────────────
```

```
┌─ Course_Enrolled ────────────────────────────────────────────────────────────
  courseid : CIDENTIFIER
  term : TERM
└──────────────────────────────────────────────────────────────────────────────
```

The schemas *Course_Taken* and *Course_Enrolled* can be eliminated by rewriting *Section_Status* as:

```
┌─ Section_Status ─────────────────────────────────
│ totalcredits : N
│ coursestaken : CIDENTIFIER ↠ (TERM ↠ GRADE)
│ coursesenrolled : CIDENTIFIER ↠ TERM
└──────────────────────────────────────────────────
```

In the schema defining the state, entities are represented by variables that have types that are power sets of the schemas describing their structure (i.e., an entity is specified as a set of elements, where the elements represent entity instances) and relationships are modeled as functions or relations, depending on their cardinality (e.g., see [10, 12]). The following state schema is obtained from the advising system ERD:

```
┌─ Advising_State ─────────────────────────────────────────────────
│ Faculty
│ Student
│ Course
│ Worksheet
│ Template
│ ...
│ faculty : F Faculty
│ students : F Student
│ wsheet : F Worksheet
│ temps : F Template
│ hasreadaccess_f, haschangeaccess : Faculty ↠ F Worksheet
│ hasreadaccess_s : Student ↠ F Worksheet
│ courses : P Course
│ prereqs, coreqs : Course ↠ F Course
│ ...
├──────────────────────────────────────────────────────────────────
│ ∀ f : faculty • hasreadaccess_f(f) = wsheets ∧ haschangeaccess(f) = wsheets
│ ∀ s : students • hasreadaccess_s(s) = {w : Worksheet | w.id = s.id ∧ w ∈ wsheets}
│ ...
└──────────────────────────────────────────────────────────────────
```

In formalizing the DFD, only the data transforms that have not been further decomposed are associated with Z schemas. The Z schemas associated with a data transform describes how it affects the state (represented by data stores), and how input data is transformed to output data. The approach used is not compositional because the Z schemas for a system of data transforms are not built-up from schemas associated with the parts. For this reason, the approach used here is supplemental.

The declaration part of an operation schema is derived by examining its input and output flows. If there are flows directed from a data transform to a data store this indicates (in the underlying interpretation) that the data transform will change the state, in which case the associated operation schema contains the declaration ΔS, where S is the schema defining the state. If there are no such flows from a data transform, then the execution of the operation represented by the data transform will not change the state. In such cases, the schemas defining the behavior of the data transforms will contain the declaration ΞS. Data flows directed to a data transform from another data transform and/or external entity are declared as input variables, and data flows emanating from data flows to other data transforms and/or external entities are declared as output variables. For example, the following schema shell is generated for the *UpdateGrades* operation:

```
┌─ UpdateGrade ─────────────────────────────────────────────
│ ΔAdvising_State
│ newgrades? : Course ↦ GRADE
│ student? : Student
│ ...
├───────────────────────────────────────────────────────────
│ ...
│ hasreadaccess_f' = hasreadaccess_f
│ haschangeaccess' = haschangeaccess
│ hasreadaccess_s' hasreadaccess_s
│ temps' = temps
│ courses' = courses
│ prereqs' = prereqs
│ coreqs' = coreqs
└───────────────────────────────────────────────────────────
```

The specifier is required to complete the specification so that it reflects required behavior. In doing so the specifier may choose to associate more than one schema with the data transform, each schema representing a different behavioral situation. For example, the students used two schemas to describe the *UpdateGrade* schema, one defining a successful update, the other defining an unsuccessful update.

As reported elsewhere [6], the students were able to identify problems with their initial SA models using this approach. All the students in the RE group were novice Z specifiers so they found the generative approach useful in the sense that it provided them with starting points and some guidance in the transformation to Z specifications.

3.2 A cognitive transformation of the advising problem

The quality group differed from the RE group in mathematical maturity. The students in the quality group had good backgrounds in discrete mathematics and its application to software engineering (though not necessarily in Z). Their task was to develop a Z specification of the advising system from the SA models created by the RE group, using cognitive transformations.

```
┌─ Worksheet ───────────────────────────────────────────────
│ id : ID
│ temptype : TEMPTYPE
│ hasgrade : COURSE ↦ (TERM ↦ GRADE)
│ enrolledin : COURSE ↦ F TERM
│ gpa : ℝ
│ ...
└───────────────────────────────────────────────────────────
```

```
┌─ wktemplate ──────────────────────────────────────────────
│ core, req, electives : F COURSE
│ prereq : COURSE ↦ F COURSE
│ mingrade : COURSE ↦ GRADE
│ tempty : TEMPTYPE
│ ...
├───────────────────────────────────────────────────────────
│ dom mingrade ⊆ core ∪ req
│ dom prereq ⊆ ((core ∪ req) ∪ electives)
│ ...
└───────────────────────────────────────────────────────────
```

```
AdvisingState
faculty, student, admin, staff : F PERSON
stwksheets : F Worksheet
advisedby : PERSON ↛ PERSON
haswriteaccess : PERSON ↛ F Worksheet
hasreadaccess : PERSON ↛ Worksheet
haschangeaccess : PERSON ↛ F wktemplate
temps : F wktemplate
UserId : PERSON ↣ ID

dom advisedby ⊆ student
ran advisedby ⊆ faculty
faculty ∩ student = ∅
admin ⊆ faculty
faculty ∩ staff = ∅
∀f : faculty • haswriteaccess(f) = stwksheets ∧
     hasreadaccess(f) ∈ stwksheets
∀s : student • haswriteaccess(s) = ∅ ∧
     hasreadaccess(s) = {w : Worksheet | w.id = UserId(s)}
∀a : admin • haschangeaccess(a) = temps
...
```

The advising state schema takes advantage of information not directly reflected in the ERD, for example, the fact that faculty, staff, and students are actually subclasses of another entity, PERSON, is not shown in the ERD. Using the formal specifications developed for the state, the students came up with suggestions for improving the ERD, such as, extending it to reflect generalization-specialization relationships, elaborating the worksheet and template entities so that relationships among courses, grades and students could be made more explicit.

An interesting observation we made is that the Z specifications obtained by this second group is more abstract than the informal models. The students were able to identify situations where details in the SA model (in particular, information in the data dictionary) were not necessary in defining required behavior. For example, in specifying templates, the students abstracted out details pertaining to sections, and instead focused on details related to courses (in other words, the sections were defined in terms of the courses). This resulted in formal specifications that were clearer, and more abstract than the informal models. Rather than adding details, 'elaboration' may actually result in the elimination of some detail, specifically, detail that is not essential to the modeling of required behavior. Given the procedural and descriptive nature of the notations often used in SA it is relatively easy to specify more than is required. Using formal specifications such as Z, which focus on defining relationships, rather than procedural detail, one is less likely to specify such details. Given this view of FSTs, the elaboration phase of the p-e-v LIM should also be concerned with distilling the informal models, so that a specification of required behavior can be obtained at an adequate level of abstraction.

4 Conclusion: Towards Practical Use of FISTs

Much attention has been paid to developing FISTs based on generative informal to formal transformations. The objectives of such approaches may vary from providing support that shield specifiers from the intricacies of the formal specifications, to merely providing some automated support for the more mundane tasks of formal specification creation. While shielding specifiers from the specifics of FSTs may enhance the attractiveness of FISTs, it is very difficult to do currently, and we feel that this is not necessarily a good thing to do. We do not view FISTs as permanent solutions to the problem of applying FSTs in industry, rather, we view them as interim solutions that provide specifiers with a means to progress towards more mature use of FSTs. In this sense, the intent should not be to completely shield the specifier from FST concepts but rather, to introduce FST concepts in a non-intrusive manner. The use of sophisticated FST tools will be enhanced if the specifiers have a good understanding of the tools' underlying concepts. Such skills are less likely to be acquired if the specifiers are shielded from the specifics of the FSTs.

While the emphasis in most FISTs is on the informal to formal specification transformation, the transformation from formal to informal specifications is also useful from a presentation viewpoint. Not only can one present more abstract models from the formal specification using such transformations, but, given well-defined semantic relationships between the informal and formal specifications, one can develop mechanisms that reflect formal analysis activities at the informal level. For example, an operational interpretation of the formal specification may be the basis for simulations at the informal level. To date we know of one major project that exploits this aspect of FISTs (the IPTES FIST [4]), and have seen some independent work on using Petri Nets as a basis for simulating behavior with DFDs (e.g., see [16]).

Experimentation is needed to assess the adequacy of FISTs and to provide feedback that can lead to a deeper understanding of the interplay between FSTs and ISTs. Interpretive relations between informal and formal techniques depend on the consistency, in some sense, between the interpretations provided by formal specifications and intuitively-held interpretations of the informal specifications for practical success. From an education standpoint it may be useful to have early experiments take the form of classroom development projects in software engineering courses. The integrated approaches can be used to present software engineering students with a more unified view of the specification process, in particular, an appreciation of the role of formal and informal modeling and reasoning in the process [6]. While the results and feedback from classroom experimentation may be of some use, the size of the projects may make it difficult to determine whether the results apply to larger projects. From an industry standpoint it is necessary to identify problems with scalability.

To support our experimental work we are also extending CASE tools based on SSMs and OOMs with formal notations and appropriate checking facilities. To date, we have developed an extension to a CASE tool that allows students to generate Z specifications from Entity-Relation Diagrams [5], and have designed an extension that will allow them to generate Z templates from DFDs.

We are currently conducting classroom experiments that employ implementations of the PIM and LIM. We realize that such experiments are limited in their scope, but they should at least provide us with some insight into some of the problems associated with the use of FISTs.

References

1. S. Auijla, A. Bryant, and L. Semmens. A rigorous review technique: Using formal notations within conventional development methods. In *Software Engineering Standards Symposium, U.K.* IEEE Computer Society Press, 1993.
2. D. Coleman, P. Arnold, S. Bodoff, C. Dollin, H. Gilchrist, F. Hayes, and P. Jeremaes. *Object-Oriented Development: The Fusion Method.* Prentice Hall, 1994.
3. T. DeMarco. *Structured Analysis and System Specification.* Prentice-Hall, 1978.
4. R. Elmstrom, R. Lintulampi, and M. Pezze. Giving semantics to SA/RT by means of high-level timed Petri nets. *Real-Time Systems*, 5, 1993.
5. R. France, T. Horton, M. Larrondo-Petrie, and S. Reeves. Process support for rigorous structured analysis. In *Software Engineering Research Forum '93.* SERF'93, 1993.
6. R. France and M. M. Larrondo-Petrie. Understanding the role of formal specification techniques in requirements engineering. In *to appear in Conference on Software Engineering Education.* Springer-Verlag, 1995.
7. R. B. France. Semantically Extended Data Flow Diagrams: A formal specification tool. *IEEE TSE*, 18(4), 1992.
8. C. Gane and T. Sarson. *Structured Systems Analysis: Tools and techniques.* Prentice-Hall, 1978.
9. D. Hatley and I. Pirbhai. *Strategies for Real-Time System Specification.* Dover Press, 1987.
10. F. Polack. Integrating formal notations and system analysis: using entity relationship diagrams. *Software Engineering Journal.*
11. J. Rumbaugh, M. Blaha, W. Premerlani, F. Eddy, and W. Lorensen. *Object-Oriented Modeling and Design.* Prentice Hall, 1991.
12. L. Semmens and P. Allen. Using Yourdon and Z: An approach to formal specification. In *Z User Workshop.* Springer-Verlag, 1991.
13. L. Semmens, R. B. France, and T. W. G. Docker. Integrated structured analysis and formal specification techniques. *The Computer Journal*, 35(6), 1992.
14. S. Shlaer and S. J. Mellor. *Object lifecycles: Modeling the world in states.* Prentice Hall, 1992.
15. J. M. Spivey. *The Z notation: A reference manual.* Prentice-Hall, 1989.
16. T. H. Tse and L. Pong. Towards a formal foundation for DeMarco data flow diagrams. Technical report, Center of Computer Studies, University of Hong Kong, 1986.
17. E. Yourdon. *Modern Systems Analysis.* Prentice-Hall, 1989.

Viewpoints and Objects

John Derrick, Howard Bowman and Maarten Steen

Computing Laboratory, University of Kent, Canterbury, CT2 7NF, UK.
(Phone: + 44 1227 764000, Email: {jd1,hb5,mwas}@ukc.ac.uk.)

Abstract. There have been a number of proposals to split the specification of large and complex systems into a number of inter-related specifications, called viewpoints. Such a model of multiple viewpoints forms the cornerstone of the Open Distributed Processing (ODP) standardisation initiative. We address two of the technical problems concerning the use of formal techniques within multiple viewpoint models: these are unification and consistency checking. We discuss the software engineering implications of using viewpoints, and show that object encapsulation provides the necessary support for such a model. We then consider how this might be supported by using object-oriented variants of Z.

Keywords: ODP, Viewpoints, Consistency, Object-Orientation, ZEST.

1 Introduction

With the increasing complexity of system specifications, the process of separation of concerns is being applied to the design *process* in addition to the design itself. Consideration of this and related issues has led to a number of proposals for partial specification or *viewpoints* to be used to split the specification of large and complex systems into a number of inter-related specifications, [1, 16, 12]. The purpose of decomposing a system specification into viewpoints is to enable different participants in the requirements and specification process to observe a system from a suitable perspective and at a suitable level of abstraction, [12]. One area where viewpoints are playing a prominent role is in the Open Distributed Processing (ODP) standardisation initiative, which is a natural progression from OSI, broadening the target of standardisation from the point of interconnection to the end-to-end system behaviour. The objective of ODP [12] is to enable the construction of distributed systems in a multi-vendor environment through the provision of a general architectural framework that such systems must conform to. There are five separate viewpoints presented by the ODP model: Enterprise, Information, Computational, Engineering and Technology. Requirements and specifications of an ODP system can be made from any of these viewpoints.

The ODP reference model (RM-ODP) recognises the need for formalism, with Part 4 of the RM-ODP defining an architectural semantics which describes the application of formal description techniques (FDTs) to the specification of ODP systems. Z is likely to be used for at least the information, and possibly the enterprise and computational viewpoints. The first ODP compliant specification, the Trader [13], is being written using Z for the information viewpoint.

Zave considers using partial specifications in [16] for specification of complex tele-phone systems. Viewpoints have also arisen in connection with requirements analysis [1]. Although the terminology used in [16, 1] is different, their work reflects a similar set of concerns to those within ODP.

Whilst it has been accepted that the viewpoint model greatly simplifies the develop-ment of system specifications and offers a powerful mechanism for handling diversity within a particular area of concern (e.g. ODP), the practicalities of how to make the ap-proach work are only beginning to be explored. In particular, one of the consequences of adopting a multiple viewpoint approach to development is that descriptions of the same or related entities can appear in different viewpoints and must co-exist. *Consistency* of specifications across viewpoints thus becomes a central issue.

In [8] we provided a mechanism, called *unification*, to describe the combination of specifications from different viewpoints into a single specification, and to check the con-sistency of them. Although our motivation arises from ODP, the mechanism we describe is a general strategy for unifying two partial specifications of a system written in Z.

Although, there has been some initial work on developing techniques to combine dif-ferent viewpoint descriptions or specifications, there has been little work on the software engineering consequences of adopting a multiple viewpoint approach. In particular, al-though the RM-ODP has adopted an object-oriented approach within viewpoints, there has been no firm evidence that object encapsulation and viewpoint specification lead to a satisfactory development scenario. Our aim here is to discuss some of the issues arising from this, and to provide initial evidence that formal approaches to object encapsulation provide the necessary support for viewpoint specification, their combination and conse-quent consistency checking.

Section 2 of this paper briefly sketches the unification mechanism, and in section 3 we discuss the use of consistency checking within the mechanism. These discussions are needed in order to motivate the material in sections 4 and 5. The former discusses software engineering issues which arise from the use of viewpoint specification and the latter presents a case study showing the application of these techniques to an object based approach.

2 Unification

In a model of multiple viewpoints, descriptions of the same or related entities can appear in different viewpoints and must co-exist. Clearly, the different viewpoints of the same specification must be consistent, i.e. the properties of one viewpoint specification do not contradict those of another. In addition, during the development process there must be some way to combine specifications from different viewpoints into a single implemen-tation specification. This process of combining two specifications is known as *unifica-tion* in ODP (other terms used include amalgamation [1] and composition [16]). Because detail and design decisions in a viewpoint specification should be preserved within the complete system specification, a natural way to view unification is as the least refine-ment of the component viewpoints, indeed this is the approach taken in ODP, [5, 12]. Unification can also be used, because of this least refinement, as a method by which to check consistency. To check the consistency of two viewpoint specifications, we check for contradictions within the unified specification.

Unification of Z specifications depends upon the Z refinement relation, which is given in terms of two separate components - data refinement and operation refinement, [18].

Although ODP has adopted unification as being the least refinement, there are alternatives, for example [1, 2]. However, the motivation behind work described in [1, 2] comes from requirements capture, and consequently there are fundamental differences to the ODP approach. In particular, there is no requirement on unification to be a refinement (let alone the least refinement) of the individual viewpoints. Hence, the techniques developed are different from our work and cannot be adapted to provide consistency checks. Recently, the approach has been formalized in terms of augmented co-refinement [2], which is a weaker relation than the usual Z refinement relation. For the purposes of this paper we shall retain the terminology and usage derived from ODP.

The unification algorithm we describe is divided into three stages: normalisation, common refinement (which we usually term unification itself), and re-structuring. This algorithm can be shown to produce the least refinement of both viewpoints, [8].

Normalisation identifies commonality between two different viewpoint specifications, and re-writes each specification into a normal form suitable for unification. Clearly, the two specifications that are to be unified have to represent the world in the same way (e.g. if an operation is represented by a schema in one viewpoint, then the other viewpoint has to use the same name for its (possibly more complex) schema as well), and that the correspondences between the specifications have to have been identified by the specifiers involved. These will be given by mappings that describe the naming, and other, conventions in force. Once the commonality has been identified, normalisation re-names the appropriate elements of the specifications. Normalisation will also expand data-type and schema definitions into a normal form. Examples of normalisation are given in [18, 20].

Unification itself takes two normal forms and produces the least refinement of both. Because normalisation will hide some of the specification structure introduced via the schema calculus, it is necessary to perform some re-structuring after unification to re-introduce the structure chosen by the specifier.

2.1 State Unification

The purpose of state unification is to find a common state to represent both viewpoints. The state of the unification must be the least data refinement of the states of both viewpoints, since viewpoints represent partial views of an overall system description.

The essence of all constructions will be as follows. We unify declarations rather than types, so non-identical types with name clashes are resolved by re-naming, then we unify declarations as follows. If an element x is declared in both viewpoints as $x : S$ and $x : T$ respectively, then the unification will include a declaration $x : U$ where U is the least refinement of S and T. The type U will be the smallest type which contains a copy of both S and T. For example, if S and T can be embedded in some maximal type then U is just the union $S \cup T$. Given two viewpoint specifications both containing the following fragment of state description given by a schema D:

$$
\boxed{\begin{array}{l} D \\ \hline x : S \\ \hline preds \end{array}} \qquad \boxed{\begin{array}{l} D \\ \hline x : T \\ \hline pred_T \end{array}}
$$

we unify as follows:

$$
\boxed{\begin{array}{l} D \\ \hline x : S \cup T \\ \hline x \in S \implies preds \\ x \in T \implies pred_T \end{array}}
$$

whenever $S \cup T$ is well founded. If S and T cannot be embedded in a single type then the unification will declare x to be a member of the disjoint union of S and T, and the mechanism to describe disjoint unions has to be included in the unification. In these circumstances we again achieve the least refinement of both viewpoints. The correspondence rules will relate the types across the viewpoint specifications, and in particular will describe correspondences within the type hierarchy. Therefore the unification can construct well formed unions for overlapping types as in the example in section 5.1.

Axiomatic descriptions are unified in exactly the same manner as state schemas.

2.2 Operation Unification

Once the data descriptions have been unified, the operations from each viewpoint need to be defined in the unified specification. Unification of schemas then depends upon whether there are duplicate definitions. If an operation is defined in just one viewpoint, then it is included in the unification (with appropriate adjustments to take account of the unified state).

For operations which are defined in both viewpoint specifications, the unified specification should contain an operation which is the least refinement of both, w.r.t. the unified representation of state. The unification algorithm first adjusts each operation to take account of the unified state in the obvious manner, then combines the two operations to produce an operation which is a refinement of both viewpoint operations.

The unification of two operations is defined via their pre- and post-conditions (which can always be derived). Given two schemas A and B representing operations, both applicable on some unified state, then the unification of A and B is:

$$
\boxed{\begin{array}{l} \mathcal{U}(A,B) \\ \vdots \\ \hline pre\ A \vee pre\ B \\ pre\ A \implies post\ A \\ pre\ B \implies post\ B \end{array}}
$$

where the declarations are unified in the manner of the preceding subsection.

We show, in [8], that this construction is the least refinement of the two viewpoint specifications. It is also associative, allowing the natural extension of unification to an arbitrary finite number of viewpoints.

3 A word about Consistency

What is consistency? A specification is consistent if it does not contain specifications of entities which cannot possibly exist. That is, given a proof system for Z, with a validity relation ⊢, a specification is said to be consistent if it is not possible to prove $S \vdash false$. For example, a Z specification of a function will be inconsistent if the predicate part of its axiomatic definition contradicts the fact that it was declared as a function. Another way in which inconsistencies can arise in Z specifications is in the definition of free types. Examples of how such inconsistencies can occur are given in [4, 25, 22]. In general, it is undecidable whether or not a set of axioms given in a Z specification is consistent. Sufficient conditions for the consistency of certain combinations of Z paragraphs, in particular axiomatic definitions, given sets and free types, are discussed in [4].

In addition, consistency usually refers to consistency of the state model, i.e. for a given state there exists at least one possible set of bindings that satisfies the state invariant, [18, 20]. With this consistency condition comes a requirement to prove the Initialisation theorem (see below), which asserts there exists a state that satisfies the initial conditions of the model. Due to an ODP requirement associated with multiple viewpoints, we also require operation consistency, because an ODP conformance statement in Z corresponds to an operation schema(s), [24]. A conformance statement is behaviour one requires at the location that conformance is tested. Thus a given behaviour (i.e. occurrence of an operation schema) conforms if the post-conditions and invariant predicates are satisfied in the associated Z schema. That is, operations defined in two viewpoints are consistent if whenever both operations are applicable, their post-conditions are consistent (in the logical sense). Hence, operations in a unification will be implementable whenever each operation has consistent post-conditions on the conjunction of its pre-conditions.

Thus a viewpoint consistency check in Z involves checking the unified specification for contradictions, and has 5 components: axiom, axiomatic, state and operation consistency in addition to the Initialisation theorem. Assuming the individual viewpoints themselves are consistent, the components then take the following form.

Axiom Consistency Axioms constrain existing global constants. Hence, to check for consistency of the two viewpoints, axioms from one viewpoint have to be checked against the second viewpoint w.r.t. any terms appearing in the axioms which are defined in the second viewpoint. If an axiom contains no terms appearing in other viewpoints, its consistency checking requirements are discharged.

State Consistency Consider the general form of state unification given in Section 2.1:

$$
\begin{array}{l}
\rule{0.4pt}{2.5em}\!\!\!D \underline{\hspace{24em}} \\
\quad x : S \cup T \\
\quad \underline{\hspace{10em}} \\
\quad x \in S \implies pred_S \\
\quad x \in T \implies pred_T \\
\rule{0.4pt}{2.5em}\underline{\hspace{24em}}
\end{array}
$$

This state model is consistent as long as both $pred_S$ and $pred_T$ can be satisfied for $x \in S \cap T$.

Axiomatic Consistency Similar to state consistency.

Operation Consistency Consistency checking also needs to be carried out on each operation in the unified specification. The definition of operation unification means that we have to check for consistency when both pre-conditions apply. That is, if the unification of A and B is denoted $\mathcal{U}(A, B)$, we have:

$$ pre\ \mathcal{U}(A, B) = pre\ A \vee pre\ B, \quad post\ \mathcal{U}(A, B) = (pre\ A \Rightarrow post\ A) \wedge (pre\ B \Rightarrow post\ B) $$

So the unification is consistent as long as the post-conditions agree whenever $(pre\ A \wedge pre\ B)$ is satisfied.

Initialisation Theorem The Initialisation Theorem is a consistency requirement of all Z specifications. It asserts that there exists a state of the general model that satisfies the initial state description, formally it takes the form:

$$ \vdash \exists\ State \bullet InitState $$

For the unification of two viewpoints to be consistent, clearly the Initialisation Theorem must also be established for the unification. [8] discusses how this can be achieved.

4 Software Engineering Issues

The previous section elucidated the consistency checking requirements for unified viewpoint specifications. Given these requirements, it is necessary to seek software engineering strategies that make viewpoint decomposition feasible. By feasible we mean it is possible to describe viewpoints which are consistent and that the effort involved in consistency checking is minimised. In this section we explore some of the software engineering consequences of the consistency checking requirements.

Either viewpoint description and analysis will work with arbitrary specifications and specification styles; or some style guidelines or further methodology is needed for the process to become feasible. In consideration of the former position there are some issues which need to be addressed:

No encapsulation of state and operations When the state is unified, all operations acting on that state are adjusted to take account of the unified state. Hence, during unification of an operation, two adjustments are made: the first due to the unified state (declarations are updated to take account of the unified state) and the second due to the change in pre- and post-conditions. Therefore, to keep track of the consistency checking requirements the operations need to be encapsulated with the state they affect. Without this consistency checking is possible, but unrealistic for larger examples.

No operation set representation As Zave noted in [30], Z provides no means of representing the operation set of a specification (i.e. the set of operations visible by the environment). The consequences of this for unification are that if an operation schema appears in both viewpoints, then it *has* to be unified, since there is no means to tell whether it was defined (in either viewpoint) for internal structuring purposes only. If there was such information available, then internal structuring schemas could be re-named and just operations in the operational set unified.

For example, in the specification of the CME agent in section 5.1 below, the operations *Select* and *CreateSelect* are defined purely for internal structuring purposes. Given another viewpoint which contains a specification of the CME agent, we require that only the external interface represented by the *Create*, *Delete* and *Enrol* operations be unified, and that any internal operations with name clashes are resolved instead of unified.

Correspondence rules In order to describe the relation between viewpoints, the RM-ODP includes the concept of a correspondence rule. Part of their purpose is to identify the commonality between the specifications, and describe any possible renamings between them. Any viewpoint methodology will need to include mappings such as these. The limited structure in an ordinary Z specification makes a succinct naming impossible for correspondences, since, for any non-trivial systems, it is likely that a correspondence will wish to name more than one state/operation.

Viewpoint encapsulation The work of [1] indicates that in a non-object approach a large number of re-namings and re-workings of the viewpoints have to be undertaken *during* the unification process. This appears to be because the boundaries of the decomposition are not well defined, leading to viewpoint specifiers referencing and defining similar aspects of the same entity. Again this is a manifestation of the lack of encapsulation when defining the area of concern for each viewpoint specifier.

From case studies undertaken and consideration of these issues it seems that viewpoint description without any style guidelines is unlikely to be practical for anything other than small examples. Encapsulation and identity are central to the practical realization of the viewpoints model. Both of these facets could be provided by a number of software engineering methodologies, however, object-orientation is an obvious choice. Many of the problems identified above can be resolved if one adopts an object-oriented approach.

Encapsulation of state and operation The over-riding advantage of object-oriented methods is their encapsulation of state and operation. This will clearly delimit the consistency checking requirements within a unification, with each unified object generating local consistency checking requirements which do not escalate to global consistency checking problems.

Operation set representation Some, although not all, object-oriented versions of Z provide the ability to specify an operation set, or visibility list, [17, 9], which partitions all the defined operations into disjoint sets of visible and internal operations. If this is provided, then the issue of operation set representation is completely resolved. Even if such a visibility list is not provided by the language used, the encapsulation that comes with object-orientation still provides the opportunity for partial resolution of the problem.

In an object-based world it is likely that a viewpoint partitioning will include the internal specification of the behaviour of an object in only one of the viewpoints. The other viewpoints will then (possibly) reference objects from viewpoints as parameters, or place constraints on the use of those objects. Hence, in these circumstances the unification of two internal representations is unlikely to occur, and so the issue of operation set representation would not occur. Of course, if the internal specification of an object's behaviour did occur in more than one viewpoint (as in the example below), the need for a visibility list would then arise again.

Correspondence rules Identity is a key property of an object, and will allow correspondence rules to relate suitably complex parts and combination of parts of the viewpoint descriptions in a manner which is not currently supported in Z.

These considerations naturally lead to a choice of an object-based or object oriented language for viewpoint decomposition, where each viewpoint specifies a number of interacting objects. Full object-orientation is not necessarily needed, however, if it is available then object-oriented facilities such as inheritance can be exploited. It is preferable that only one viewpoint specifies the internal representation of a given object, and references to objects from one viewpoint will appear as parameters, either as inheritance within another object, or as an abstraction purely in terms of object or method names. The next section investigates the support available in Z for this approach.

5 Using Object-Oriented Techniques

The previous section indicated that an object-oriented style of specification is particularly suitable for viewpoint descriptions, and indeed the RM-ODP has adopted such an approach. We now consider the use and consequences of using viewpoints with object-oriented styles of Z specification. There have been a number of different approaches proposed for providing Z with object-oriented facilities. These include the provision of object-oriented style guidelines, and extensions to Z to allow fully object-oriented specifications. Examples of using Z in an object-oriented style include: Hall's style [10, 11]; ZERO [29]; and the ODP architectural semantics [12]. Examples of object-oriented extensions to Z include: Object-Z [9]; ZEST [7]; MooZ [17]; OOZE [3]; Schuman & Pitt [23]; and Smith [26]. See [27] for a summary and comparison of several approaches.

Z itself is not object-oriented because it does not provided sufficient support for either encapsulation or inheritance. However, it is also possible for Z to be used in an object based fashion, see [20] for a discussion, although there is nothing to keep the specifier within an object based style in contrast to the style guidelines above.

The ODP standardisation initiative requires the use of (near) standardised formal methods, hence the architectural semantics uses Z as opposed to any object-oriented variant of that language. However, given that ODP has adopted the object-oriented paradigm, there is obvious interest in object-oriented variants that can support the required ODP modelling concepts. In particular, Object-Z and ZEST are receiving attention within the ODP community as a specification medium. However, all the object-oriented extensions to Z have an unstable definition, or lack a full semantics, or both. Therefore, techniques with a flattening (or approximate flattening) into Z are of considerable interest to our work. By using such a technique we can define unification and consistency checking of viewpoints without compromising the necessity of a standardised formal description technique. Object-Z and ZEST are suitable from this perspective, however, it is unfortunate that Object-Z is moving away from a flattening semantics.

Of the Z guidelines the work of Hall is the most general. The style adds no new features to Z, however, there are conventions for writing an object-oriented specification. He also provides conventions for modelling classes and their relationships, and, in addition, there is formal support for inheritance through subtyping, [11]. In order to support encapsulation, the RM-ODP has adopted conventions for the use of Z within ODP. Here, encapsulation is achieved by letting each Z specification denote just one object. This achieves the required encapsulation, but clearly any specification of an aggregate of objects or interaction between objects cannot then be modelled within Z. There is a clear need to extend the framework offered by ODP by considering further style guidelines for the specification of collections of objects.

Here we shall show how the unification techniques can be used with ZEST for the specification of viewpoints. To do so we use ZEST to describe two viewpoints consisting of objects or aggregates of objects. We can then flatten ZEST to Z, in order to generate the unification of the two viewpoints and to check for consistency. The unification can then easily be re-assembled into a ZEST specification if further object oriented development is required. Other object-oriented variants of Z could equally have been used for the basis of this example, in particular, Object-Z would have provided a similar set of facilities as those we have called upon. Although we have applied unification by first flattening the ZEST, it is important to note that we do not lose the benefits of using object-orientation by doing this. The encapsulation can be recovered, and the consistency checking requirements still lie within the boundaries defined by the object encapsulation.

5.1 Example

The application of Z in the ODP information viewpoint to the modelling of OSI Management has been investigated by a number of researchers, see [21, 28] for introductions. We show here how the object-based approach in [28] can be used within viewpoint specifications by considering viewpoint specifications of sieve managed objects and their controlling Common Management Environment (CME) agent. We shall consider one viewpoint containing a specification of an event reporting sieve object together with a second viewpoint which describes both a sieve object and a CME agent and its manipulation of the sieve objects.

In ZEST a managed object class is described by a ZEST class, and a managed object is an instance of a managed object class. An instance of a class is created through initialisation of the class specification, which assigns values to all fixed attributes of the class. The initialisation schema provides predicates that must be satisfied on initialisation.

A managed object definition cannot include a *Create* operation, since before it is created a managed object cannot perform any operation, including *Create* itself. However, by including a *Create* operation in the CME agent viewpoint as we do below, we can describe formally the interaction between *Create* and the sieve managed object definition.

Viewpoint 1 To describe the sieve object, we first declare the types. *ObjectId* represents a set of object identifiers, and *SieveConstruct* is used in the event reporting process, its internal structure is left unspecified at this stage. The remaining types are declared as enumerated types.

$$[SieveConstruct, ObjectId]$$
$$Operational ::= disabled \mid active \mid enabled \mid busy$$
$$Admin ::= locked \mid unlocked \mid shuttingdown$$
$$Event ::= nothing \mid enrol \mid deenrol$$
$$Status ::= created \mid deleted$$

Status models the life-cycle of the sieve object, and is used as an internal mechanism to control which operations are applicable at a given point within an object's existence.

The ZEST sieve class is then defined by:

```
┌─ Sieve ──────────────────────────────────────────────
│  sieveid : ObjectId
│  destadd : ObjectId
│ ──────────────────────────────────────────────────────
│  opstate : Operational
│  sico : SieveConstruct
│  adminstate : Admin
│  status : Status
│
│  ┌─ INIT ───────────────────────────────────────────
│  │ opstate = active
│  │ adminstate = unlocked
│  └──────────────────────────────────────────────────
│
│  filter : Event ↔ SieveConstruct
│
│  ┌─ Filter ─────────────────────────────────────────
│  │ event? : Event
│  │ notification! : Event
│  │ ──────────────────────────────────────────────────
│  │ status ≠ deleted
│  │ opstate = active ∧ adminstate = unlocked
│  │ (event?, sico) ∈ filter ⇒ notification! = event?
│  │ (event?, sico) ∉ filter ⇒ notification! = nothing
│  └──────────────────────────────────────────────────
└──────────────────────────────────────────────────────
```

Within the class definition we have described just one of the operations available within a sieve object (for a full description of operations see [21]). The relation *filter* represents criteria to decide which events to filter out and which to pass on and *Filter* represents the operation to perform the filtering.

Viewpoint 2 The second viewpoint contains a description of a controlling CME agent together with a sieve object. For our purposes here we present a very simplified version of an agent which consists of a number of sieve managed objects. The CME agent promotes the *Delete* operation defined on individual sieve objects, and defines a *Create* operation to instantiate sieve objects as required. First of all we declare the types (notice that in this viewpoint *Status* includes an additional value, *being_created*)

$$[SieveConstruct, ObjectId]$$

$$Operational ::= disabled \mid active \mid enabled \mid busy$$

$$Admin ::= locked \mid unlocked \mid shuttingdown$$

$$Event ::= nothing \mid enrol \mid deenrol$$

$$Status ::= being\_created \mid created \mid deleted$$

The ZEST sieve object is then specified in this viewpoint as shown in Figure 1.

Within this class definition an operation to delete a sieve is declared. Upon deletion a sieve sends a *deenrol* notification to its environment, and moves into a state where no further operations can be applied. A *CMEagent* is modelled as a collection of sieve objects, where and initially no sieve objects have been created.

```
┌─ CMEagent ──────────────────────────────────
│ id : ObjectId
├─────────────────────────────────────────────
│ sieves : ℙ Sieve
│
│ ┌─ INIT ────────────────────────────────────
│ │ sieves = ∅
│
│ ┌─ Select ──────────────────────────────────
│ │ s? : Sieve
│ ├───────────────────────────────────────────
│ │ s? ∈ sieves
│
│ ┌─ CreateSelect ────────────────────────────
│ │ Δ(sieves)
│ │ s? : Sieve
│ ├───────────────────────────────────────────
│ │ s? ∉ sieves
│ │ sieves' = sieves ∪ {s?}
│
│ Delete ≙ Select • s?.Delete
│ Enrol ≙ Select • s?.Enrol
│ Create ≙ CreateSelect • s?.Init
└─────────────────────────────────────────────
```

```
┌─Sieve──────────────────────────────────────────────
│ ┌──────────────────────────────────────────────────
│ │ sieveid : ObjectId
│ │ destadd : ObjectId
│ ├──────────────────────────────────────────────────
│ │ opstate : Operational
│ │ sico : SieveConstruct
│ │ adminstate : Admin
│ │ status : Status
│ ├──────────────────────────────────────────────────
│ │ opstate ∈ {active, disabled}
│ │ adminstate ∈ {locked, unlocked}
│ │
│ │ ┌─INIT─────────────────────────────────────────────
│ │ │ status = being_created
│ │ │ opstate = active
│ │ │ adminstate = unlocked
│ │ └──────────────────────────────────────────────────
│ │
│ │ ┌─Delete──────────────────┐ ┌─Enrol───────────────────
│ │ │ Δ(status)               │ │ Δ(status)
│ │ │ notification! : Event   │ │ notification! : Event
│ │ ├─────────────────────────┤ ├─────────────────────────
│ │ │ status = created        │ │ status = being_created
│ │ │ notification! = deenrol │ │ status' = created
│ │ │ status' = deleted       │ │ notification! = enrol
│ │ └─────────────────────────┘ └─────────────────────────
│ │
│ │ newfilter : Event ↔ SieveConstruct
│ │
│ │ ┌─Filter───────────────────────────────────────────
│ │ │ event? : Event
│ │ │ notification! : Event
│ │ ├──────────────────────────────────────────────────
│ │ │ status = created
│ │ │ opstate = active ∧ adminstate = unlocked
│ │ │ (event?, sico) ∉ newfilter ⇒ notification! = nothing
│ │ └──────────────────────────────────────────────────
└─┴──────────────────────────────────────────────────
```

Fig. 1. ZEST sieve object specification

Notice that the extra structuring available in ZEST does away with the need to define framing schemas to achieve the promotion of operations from Sieve to the aggregate of sieves used in the agent. The agent operation to delete a specific sieve object can now be defined quite simply by referencing the method of the appropriate sieve object. The other managed object operations are promoted in a similar fashion. It is situations such as these where a visibility list is necessary, because we do not want operations *Select* and *CreateSelect* to be visible to the environment. Ideally they should be declared as internal operations. Recently, a small extension to ZEST included such a mechanism.

Notice that the *Create* operation is not part of the sieve specification, so we have preserved the concept that *Create* must occur before any operation in the sieve specification can be applied.

Unification of Viewpoints To describe the unification of viewpoints, we decorate with subscripts, so for example $Filter_1$ is the schema *Filter* from the first viewpoint. To apply the unification of the two viewpoints we first flatten both ZEST specifications into Z specifications. The flattening used is that defined in [19], and we do not repeat the details here. However, the flattening of the CMEagent need not be calculated because no other viewpoint contains a specification of the CMEagent's behaviour. Therefore it suffices to consider just the flattening of the ZEST sieve specifications. We give this here, and then unify the two Z specifications, however, experience shows that it is straightforward to factor the unification through the ZEST extended structuring in an obvious manner, and that the unification can be derived just as easily from the ZEST specifications.

The flattening of the sieve object in the first viewpoint is:

$$
\begin{array}{|l}
\hline
\_Sieve_____ \\
\hline
sieveid : ObjectId \\
destadd : ObjectId \\
opstate : Operational \\
sico : SieveConstruct \\
adminstate : Admin \\
status : Status \\
\hline
\end{array}
$$

$$
\begin{array}{|l}
\hline
\_InitSieve_____ \\
\hline
Sieve \\
\hline
opstate = active \\
adminstate = unlocked \\
\hline
\end{array}
$$

$$filter : Event \leftrightarrow SieveConstruct$$

$$
\begin{array}{|l}
\hline
\_Filter_____ \\
\hline
\Xi Sieve \\
event? : Event \\
notification! : Event \\
\hline
status \neq deleted \\
opstate = active \wedge adminstate = unlocked \\
(event?, sico) \in filter \Rightarrow notification! = event? \\
(event?, sico) \notin filter \Rightarrow notification! = nothing \\
\hline
\end{array}
$$

The equivalent Z sieve object in the second viewpoint can be derived in a similar fashion. To unify the two specifications we first unify the state. The only conflict in the declarations are due to differing types $Status_1$ and $Status_2$. To resolve this conflict, the type $Status$ in the unification is taken as the least refinement of $Status_1$ and $Status_2$ (i.e. $Status_1 \cup Status_2$, and this is well formed because the correspondence rules collapse the type hierarchy), and state unification is applied to the schema *Sieve*. Hence, in addition to the declarations which are not in conflict, the unification will contain the following:

$Status ::= being\_created \mid created \mid deleted$

```
┌─ Sieve ──────────────────────────────────────────────────
│ sieveid : ObjectId
│ destadd : ObjectId
│ opstate : Operational
│ sico : SieveConstruct
│ adminstate : Admin
│ status : Status
├──────────────────────────────────────────────────────────
│ status ∈ { created, deleted } ⇒ true
│ status ∈ { being_created, created, deleted } ⇒
│      ( opstate ∈ { active, disabled } ∧ adminstate ∈ { locked, unlocked })
└──────────────────────────────────────────────────────────
```

Upon simplification the schema *Sieve* becomes:

```
┌─ Sieve ──────────────────────────────────────────────────
│ sieveid : ObjectId
│ destadd : ObjectId
│ opstate : Operational
│ sico : SieveConstruct
│ adminstate : Admin
│ status : Status
├──────────────────────────────────────────────────────────
│ opstate ∈ { active, disabled }
│ adminstate ∈ { locked, unlocked }
└──────────────────────────────────────────────────────────
```

In a similar fashion we unify *InitSieve*$_1$ and *InitSieve*$_2$, which simplifies to:

```
┌─ InitSieve ──────────────────────────────────────────────
│ Sieve
├──────────────────────────────────────────────────────────
│ status = being_created
│ opstate = active
│ adminstate = unlocked
└──────────────────────────────────────────────────────────
```

The *Delete* and *Enrol* schemas are defined in just one viewpoint. Hence, both these schemas are included in the unification (with adjustments due to the unified state schema *Sieve*). Similarly the unification contains both relations *filter* and *newfilter*.

To unify *Filter*$_1$ and *Filter*$_2$ we first adjust *Filter*$_1$ due to the unified state schema. The predicate part of *Filter*$_1$ is then

$status \in \{ created, deleted \} \Rightarrow$
$(status \neq deleted \wedge opstate = active \wedge adminstate = unlocked)$

Calculation of the pre-conditions *preFilter*$_1$ ∨ *preFilter*$_2$ then simplifies to:

$(status = created \wedge opstate = active \wedge adminstate = unlocked)$

Thus the unification of $Filter_1$ and $Filter_2$ is then given by:

$Filter$ _____

$\Xi Sieve$
$event?: Event$
$notification!: Event$

$status = created \wedge opstate = active \wedge adminstate = unlocked$
$(event?, sico) \in filter \Rightarrow notification! = event?$
$(event?, sico) \notin filter \Rightarrow notification! = nothing$
$(event?, sico) \notin newfilter \Rightarrow notification! = nothing$

Since we were unifying two Sieve objects, the unification can easily be re-written into the equivalent ZEST class shown in Figure 2.

5.2 Conclusions from Case Studies

The above example illustrates the use of object specifications within viewpoints and their unification. Although the example is simple, it clearly shows how viewpoints can reference other objects and their methods, or contain a partial description of an object's behaviour. We have undertaken a number of case studies in order to test the hypothesis that object-oriented description is the preferable viewpoint specification medium, and our conclusions so far support this claim. The studies involving non-object based descriptions were significantly harder to check for consistency and much harder to specify in an independent fashion in the viewpoints, the specification in [1] is another indication of the difficulty of non-object based viewpoint specifications.

Conversely, the object based viewpoint descriptions were much more successful. When the viewpoints contain only references to objects defined in other viewpoints (as opposed to specifying any of their behaviour) consistency checking is relatively straight-forward (although the viewpoints can still be inconsistent). If two viewpoints both contain (partial) descriptions of the same object, then there can be a non-trivial consistency checking process, however, due to encapsulation the boundaries that inconsistency can arise within are well defined.

Examples were undertaken using the following styles: Hall's conventions; encapsulation as defined by the ODP architectural semantics; ZEST and finally using Z specifications produced from the object-oriented methodology described by Smith [26]. The style of the object-oriented variant chosen did not significantly affect the success or otherwise of the viewpoint specification or unification. There are clear merits in using Z without extended syntax, particularly in the use within ISO initiatives. To that extent, there are clear advantages in using the work of Hall and Smith. Hall in particular offers formal and well-defined support for inheritance, which is lacking for some other Z object-oriented variants. The extended syntax approaches have advantages for the developer, who is then not constrained by conventions for embedding object-orientation in Z, but only if a clear semantics, and preferably a flattening into Z, can be given.

```
┌─Sieve──────────────────────────────────────────────────┐
│  ┌──────────────────────────────────────────────────┐ │
│  │ sieveid : ObjectId                                │ │
│  │ destadd : ObjectId                                │ │
│  ├──────────────────────────────────────────────────┤ │
│  │ opstate : Operational                             │ │
│  │ sico : SieveConstruct                             │ │
│  │ adminstate : Admin                                │ │
│  │ status : Status                                   │ │
│  ├──────────────────────────────────────────────────┤ │
│  │ opstate ∈ {active, disabled}                      │ │
│  │ adminstate ∈ {locked, unlocked}                   │ │
│  └──────────────────────────────────────────────────┘ │
│                                                         │
│  ┌─INIT─────────────────────────────────────────────┐ │
│  │ status = being_created                            │ │
│  │ opstate = active                                  │ │
│  │ adminstate = unlocked                             │ │
│  └──────────────────────────────────────────────────┘ │
│                                                         │
│  ┌─Enrol──────────────────┐  ┌─Delete───────────────┐ │
│  │ Δ(status)              │  │ Δ(status)            │ │
│  │ notification! : Event  │  │ notification! : Event│ │
│  ├────────────────────────┤  ├──────────────────────┤ │
│  │ status = being_created │  │ status = created     │ │
│  │ status' = created      │  │ notification! = deenrol│ │
│  │ notification! = enrol  │  │ status' = deleted    │ │
│  └────────────────────────┘  └──────────────────────┘ │
│                                                         │
│  │ filter, newfilter : Event ↔ SieveConstruct          │
│                                                         │
│  ┌─Filter───────────────────────────────────────────┐ │
│  │ event? : Event                                    │ │
│  │ notification! : Event                             │ │
│  ├──────────────────────────────────────────────────┤ │
│  │ status = created ∧ opstate = active ∧ adminstate = unlocked │
│  │ (event?, sico) ∈ filter ⇒ notification! = event?  │ │
│  │ (event?, sico) ∉ filter ⇒ notification! = nothing │ │
│  │ (event?, sico) ∉ newfilter ⇒ notification! = nothing │ │
│  └──────────────────────────────────────────────────┘ │
└─────────────────────────────────────────────────────────┘
```

Fig. 2. Equivalent ZEST class

5.3 Relation between Unification and Inheritance

It is important to recognise that unification is a 'horizontal' rather than 'vertical' development activity. By that we mean it is used to check development at a particular stage (consistency checking) or possibly to combine development specifications (unification), rather than a development activity that serves to define the implementation more closely (as in refinement or inheritance). Given that unification is a horizontal activity, one needs to describe the relationship between it and vertical development activities. The relationship between unification and refinement is well known since unification is based upon (least) refinement. We describe here the relation between inheritance and unification.

To do so we need a formal approach to inheritance and subtyping. In [11], Hall discusses known definitions in terms of both extensional and intensional semantics. Given that we are interested in behaviour of specifications, we shall consider definitions due to intensional semantics here. His intensional meaning of subclass is in terms of subclass instances being valid implementations of the superclass, however, the definition is different from a refinement relation (as one would expect). To exhibit subtyping there must exist a retrieve relation Abs between the superclass and subclass such that the following are true.

S1 $\forall$ $Superstate$; $Substate$ • pre $Superop \land Abs \Rightarrow$ pre $Subop$

S2 $\forall$ $Superstate$; $Substate$; $Substate'$ • pre $Superop \land Abs \land Subop \Rightarrow$
   $(\exists Superstate' • Abs' \land Superop)$

S3 $\forall$ $Substate$ • $(\exists Superstate • Abs)$

Only the third rule differs from the rules for refinement, see [11] for justification of this. Hall also compares his definition with those of Cusack [6], Lano & Haughton [14] and Liskov & Wing [15]. We are interested here in the relation between unification and the individual viewpoints. In these circumstances the retrieve relations will be partial functions (and only total if one viewpoint is degenerate). In this case Hall's subtyping implies subtyping in the sense of Cusack, Lano & Haughton and Liskov & Wing (ignoring the history predicates of the latter two). In particular, the rules S1-3 suffice for subtyping in both Hall and ZEST, and we will thus work with this definition.

It is easy to construct examples to show that the unification of two viewpoints is not in general a subtype of each viewpoint. However, this is unsurprising because one viewpoint is only a partial description of an object's behaviour. Instead the natural result to seek is the following:

Theorem 1. *Let P_i, O_i be objects in viewpoint i. Let P_i be a subtype of O_i. Then $\mathcal{U}(P_1, P_2)$ is a subtype of $\mathcal{U}(O_1, O_2)$, where $\mathcal{U}$ is the unification operator between viewpoints.*

Proof
The full proof involves construction of appropriate retrieve relations between $\mathcal{U}(P_1, P_2)$ and $\mathcal{U}(O_1, O_2)$ in a manner similar to the proof that unification is the least refinement, see [8]. The outline of the proof is as follows:

The subtyping rules S1 and S2 between $\mathcal{U}(P_1, P_2)$ and $\mathcal{U}(O_1, O_2)$ are satisfied because unification is the least refinement.

For S3, note that every state in the $\mathcal{U}(O_1, O_2)$ unification appears in either O_1 or O_2 or both. Thus every state in $\mathcal{U}(P_1, P_2)$ is related to some state in $\mathcal{U}(O_1, O_2)$ via the retrieve relation defined for the least refinement.   □

It is straightforward to construct examples to show that the converse is not true, that is $\mathcal{U}(P_1, P_2)$ being a subtype of $\mathcal{U}(O_1, O_2)$ does not imply that P_i is a subtype of O_i.

The theorem then provides a sound footing for the use of object-oriented techniques in viewpoint descriptions. The relationship between unification and multiple inheritance is clearly of importance (especially w.r.t method consistency), and is currently under investigation.

6 Conclusion

Formal description techniques are being employed extensively in ODP and have proved valuable in supporting precise definition of reference model concepts. The use of viewpoints to enable separation of concerns to be undertaken at the specification stage is a cornerstone of the ODP model. Therefore two issues of importance to ODP and other models of multiple viewpoints are unification and consistency checking.

We discussed the need for encapsulation of the consistency checking boundaries within a unification, and showed how object-oriented methodologies provide the necessary support for such a process. This was illustrated by providing a unification mechanism on an object-oriented variant of Z, where we specified a number of viewpoints of an OSI Management application.

Acknowledgements

This work was partially funded by British Telecom Labs., Martlesham, Ipswich, UK and by the Engineering and Physical Sciences Research Council under grant number GR/K13035. Thanks to David Wolfram for discussing the semantics of ZEST with the authors.

References

1. M Ainsworth, AH Cruickshank, LJ Groves, and PJL Wallis. Viewpoint specification and Z. *Information and Software Technology*, 36(1):43–51, February 1994.
2. M Ainsworth and PJL Wallis. Co-refinement. In D Till, editor, *Proc. 6th Refinement Workshop*, City University, London, 5th–7th January 1994. Springer-Verlag.
3. A. J. Alencar and J. A. Goguen. OOZE: An object oriented Z environment. In P. America, editor, *ECOOP '91 - Object-Oriented Programming*, LNCS 512, pages 180–199. Springer-Verlag, 1991.
4. R. D. Arthan. On free type definitions in Z. In J. E. Nicholls, editor, *Sixth Annual Z User Workshop*, pages 40–58, York, December 1991. Springer-Verlag.
5. G. Cowen, J. Derrick, M. Gill, G. Girling (editor), A. Herbert, P. F. Linington, D. Rayner, F. Schulz, and R. Soley. *Prost Report of the Study on Testing for Open Distributed Processing*. APM Ltd, 1993.
6. E. Cusack. Inheritance in object oriented Z. In P. America, editor, *ECOOP '91 - Object-Oriented Programming*, LNCS 512, pages 167–179. Springer-Verlag, 1991.
7. E. Cusack and G. H. B. Rafsanjani. ZEST. In S. Stepney, R. Barden, and D. Cooper, editors, *Object Orientation in Z*, Workshops in Computing, pages 113–126. Springer-Verlag, 1992.
8. J. Derrick, H. Bowman, and M. Steen. Maintaining cross viewpoint consistency using Z. In *IFIP International Conference on Open Distributed Processing*. Chapman Hall, 1995.

9. R. Duke, P. King, G. A. Rose, and G. Smith. The Object-Z specification language version 1. Technical Report 91-1, Software Verification Research Centre, Department of Computer Science, University of Queensland, May 1991.

10. A. J. Hall. Using Z as a specification calculus for object-oriented systems. In D. Bjorner, C.A.R. Hoare, and H. Langmaack, editors, *VDM '90 VDM and Z - Formal Methods in Software Development*, LNCS 428, pages 290–318, Kiel, FRG, April 1990. Springer-Verlag.

11. J. Hall. Specifying and interpreting class hierarchies in Z. In J. Bowen and J. Hall, editors, *Eighth Annual Z User Workshop*, pages 120–138, Cambridge, July 1994. Springer-Verlag.

12. ITU Recommendation X.901-904 — ISO/IEC 10746 1-4. *Open Distributed Processing - Reference Model - Parts 1-4*, July 1995.

13. ITU/ISO CD ISO 13235/ITU.TS Rec.9tr. *ODP Trading Function*, 1994.

14. K. Lano and H. Haughton. Reuse and adaption of Z specifications. In J. P. Bowen and J. E. Nicholls, editors, *Seventh Annual Z User Workshop*, pages 62–90, London, December 1992. Springer-Verlag.

15. B. Liskov and J. M. Wing. A new definition of the subtype relation. In O. M. Nierstrasz, editor, *ECOOP '93 - Object-Oriented Programming*, LNCS 707, pages 118–141. Springer-Verlag, 1993.

16. P. Mataga and P. Zave. Formal specification of telephone features. In J. Bowen and J. Hall, editors, *Eighth Annual Z User Workshop*, pages 29–50, Cambridge, July 1994. Springer-Verlag.

17. S. L. Meira and A. L. C. Cavalcanti. Modular object oriented Z specifications. In J. E. Nicholls, editor, *Fifth Annual Z User Workshop*, pages 173–192, Oxford, December 1990. Springer-Verlag.

18. B. Potter, J. Sinclair, and D. Till. *An introduction to formal specification and Z*. Prentice Hall, 1991.

19. G. H. B. Rafsanjani. ZEST - Z Extended with Structuring: A users's guide. Technical report, BT, June 1994.

20. B. Ratcliff. *Introducing specification using Z*. McGraw-Hill, 1994.

21. S. Rudkin. Modelling information objects in Z. In J. de Meer, V. Heymer, and R. Roth, editors, *IFIP TC6 International Workshop on Open Distributed Processing*, pages 267–280, Berlin, Germany, September 1991. North-Holland.

22. M. Saaltink. Z and Eves. In J. E. Nicholls, editor, *Sixth Annual Z User Workshop*, pages 223–243, York, December 1991. Springer-Verlag.

23. S. A. Schuman, D. H. Pitt, and P. J. Byers. Object-oriented process specification. In C. Rattray, editor, *Specification and Verification of Concurrent Systems*, Workshops in Computing, pages 21–70. Springer-Verlag, 1990.

24. R. Sinnott. *An Initial Architectural Semantics in Z of the Information Viewpoint Language of Part 3 of the ODP-RM*. ISO/IEC SC21/WG7 N915, July 1994. BSI Input document to the ODP Plenary meeting in Southampton.

25. A. Smith. On recursive free types in Z. In J. E. Nicholls, editor, *Sixth Annual Z User Workshop*, pages 3–39, York, December 1991. Springer-Verlag.

26. G. Smith. An object-oriented development framework for Z. In J. Bowen and J. Hall, editors, *Eighth Annual Z User Workshop*, pages 89–107, Cambridge, July 1994. Springer-Verlag.

27. S. Stepney, R. Barden, and D. Cooper, editors. *Object Orientation in Z*. Workshops in Computing. Springer-Verlag, 1992.

28. C. Wezeman and A. J. Judge. Z for managed objects. In J. Bowen and J. Hall, editors, *Eighth Annual Z User Workshop*, pages 108–119, Cambridge, July 1994. Springer-Verlag.

29. P. J. Whysall and J. A. McDermid. An approach to object oriented specification using Z. In J. E. Nicholls, editor, *Fifth Annual Z User Workshop*, pages 193–215, Oxford, December 1990. Springer-Verlag.

30. P. Zave and M. Jackson. Techniques for partial specification and specification of switching systems. In J. E. Nicholls, editor, *Sixth Annual Z User Workshop*, pages 205–219, York, December 1991. Springer-Verlag.

Education Session

Teaching Programming as Engineering

Prof. David Lorge Parnas

Department of Electrical and Computer Engineering
McMaster University, Hamilton, Ontario, Canada L8S 4K1

ABSTRACT

In spite of unheralded advances in computer hardware and software, most of today's introductory programming courses are much like courses taught 30 years ago. Although the programming languages have changed, we continue to equate teaching programming with teaching the syntax and semantics of programming languages. This paper describes a different approach being taken in the Faculty of Engineering at McMaster University. Our course emphasises program design rather than language syntax, insisting that the program design is something distinct from the detailed code. It allows students a choice of programming languages for use in their laboratory work. Students learn a mathematical model of programming and are taught to use that model to understand program design, analysis, and documentation. Considerable effort is spent on teaching the students how to apply what they see as "theory" in practice.

1 Programming Courses and Engineering

Professional Engineers are expected to use discipline, science, and mathematics to assure that their products are reliable and robust. We should expect no less of anyone who produces programs professionally.

In most jurisdictions, Engineers are expected either to be graduates of carefully accredited University programs or to have passed exams that demonstrate that they have the knowledge that they would have obtained in such a program. It is unfortunate that there is no corresponding registration process for those who design programs.

This paper treats programming as an engineering discipline and describes a course that is intended to teach the fundamental knowledge that should be expected of any professional programmer. Because we believe in teaching good programming design habits from the start, this is our student's first course in programming. Unfortunately, many students come to University having learned to use a computer in High School or elsewhere. With rare exceptions those students have learn to program intuitively and resist learning a new, more disciplined, approach.

2 The Important Characteristics of Programming Courses

For many years, the first question asked about any introductory programming course for Engineers was "What programming language do you teach?" If one examines the many textbooks available for such courses, one finds that *at least* half of the book, and usually more, is simply a description of the syntax and interpretation of a particular language. The situation is exactly as if courses on circuit design were dominated by a description of one model of oscilloscope. We seem to have forgotten that our task is to teach students how to design programs, not the characteristics of one or two human artifacts.

Engineering educators have long known that their students must be prepared to work in rapidly changing fields. We have recognised that the educational programme must stress fundamentals: science, mathematics, and design discipline so that graduates will find their education still valid and useful late in their careers. Most of the books that I used in my own engineering education are still correct and relevant, several decades later. In contrast, many introductory programming books are considered out of date before the students who use them have graduated.

The approach taken by those who teach programming differs greatly from the approach taken by those who teach Engineering. All aspects of computing, have developed very rapidly and will continue to do so. Instead of reacting to rapid change by focusing on fundamentals, programming books and courses try to keep up with the latest developments. Each new language, each new operating system, each new database package, each new windowing system, and each new programming fad, gives rise to another wave of books. The few books that claim to focus on fundamentals are highly theoretical. Instead of learning the latest tools, students are given theories that, while they don't get out of date, do not seem relevant to the task of programming.

The subject matter of most introductory programming courses is material that many Engineers of previous generations learned on their own. Just as we did not have courses devoted to the use of slide-rules, we did not need an academic course to learn FORTRAN. In fact, many students of my generation found that a one-week evening course, offered by the computer centre (without credit), was sufficient as an introduction to the tools that were available.

If we are to include a course on programming as part of a university education, we must focus on programming fundamentals and their application, not on tools.

3 The Role of Mathematics in Engineering

One of the clear differences between typical programming courses and most engineering courses is that the programming courses neither teach, nor make much use of, mathematics.

Those who do not have an engineering education often do not realise how much mathematics is taught to engineers. At my university, approximately 30% of an engineer's education is devoted to things that are explicitly titled mathematics. There is also a great deal of mathematics taught in specialised engineering courses. The ability to use mathematics is one of the things that differentiate professional engineers from technicians.

In engineering education we emphasise the concept of professional responsibility. An Engineer is taught from her first day at University, that her products must be "fit for use". She learns that she cannot rely on intuition alone. Much of their education is devoted to learning how to perform *both* mathematical analysis and carefully planned testing of their products. My own engineering education included approximately as much mathematics as would have been taken by a mathematics major and, at my *alma mater*, many of the courses were the same ones taken by the mathematics majors. However, we did not learn math as an "end" but as a "means". We learned how to use mathematics in developing and analysing product designs.

In spite of the fact that, in every other area of engineering, students are expected to learn the relevant mathematics, most programming courses treat the mathematics of programming as if it were (a) too difficult and (b) not relevant to programming. Moreover, many of those who advocate the use of mathematical methods in analysing programs disparage the use of testing. They treat complementary methods of assuring quality as if they were alternatives. Students should be taught how to use mathematics, *together with testing*, to increase the reliability of their programs.

In fact, the mathematics needed to understand modern programming techniques is elementary and entirely within the capability of anyone who can learn differential and integral calculus. That includes all Engineers and Scientists.

4 The Role of Programming in Engineering, Business, and Science

There is no longer any question about whether or not engineering, business, and science students should take courses in programming. Computers and software are now ubiquitous in those fields. Many engineering products include computers and software; most others are designed and analysed using computers. Hardly a week passes in which we do not hear some anecdote about the failure of an engineering product caused by an error either in the software contained in the product or in the software used to design it. Since people rarely talk loudly about their failures, we can assume that these anecdotes are just the "tip of the iceberg". The question is no longer "Do we teach our students about programming?", but "What do we teach our students about programming?" Educators should insist that courses on programming be more substantive than is currently the case.

5 The Content of Most "Standard" Programming Courses

It is time to question the intellectual content of many courses in computing. We need to ask whether these courses are comparable to other engineering, mathematics or science courses. The typical programming course simply teaches about a programming language, an artifact designed by a few fallible human beings. Most of the time is spent on things that are neither mathematical truths nor facts about the world; they are just human design decisions. Such courses are analogous to teaching about a particular calculator, including the location of its buttons, how to turn it on, how to change the display, etc. I often hear complaints that many of these courses teach almost the same artifacts that were taught 30 years ago; computer scientists think that we should teach more modern languages than FORTRAN. I believe that the teaching of "old" languages is not the real problem; the real problem is that the subject of the course is an artifact, any artifact.

At my University, we were teaching two distinct courses under a single number (to preserve the illusion that we had a common first year for all engineering students). One "section" was a FORTRAN course; the other a course in Pascal. Because these two languages are quite different, the material taught to one section differed greatly from material taught to the other. For example, Pascal users learn about records; FORTRAN users do not. In fact, the concept of a record is an important one that can be used in FORTRAN even though the language has no explicit facilities for it; if one group of students should learn about records, so should the other.

6 Programming Courses are not Science Courses

We must recognise another difference between engineering education and the education of scientists and mathematicians. In Engineering schools there is great emphasis on *design*, i.e. on how to invent useful things. Science courses can focus on science, i.e. on facts about the world. Like Engineering courses, programming courses should emphasize how to apply those facts. This has very practical implications. Every course is limited in the amount of contact and student time; when we chose to stress the application of science, we are choosing not to teach certain facts or theories that might be interesting and elegant for all involved.

As part of a course in programming, students must learn problem solving skills. They must learn how to formulate problems, how to decompose a problem into smaller problems, how to integrate solutions, etc. The emphasis in many courses taught in Computer Science departments is quite different. Computer Scientists are interested in programming languages, (and the science of programming), for their own sake. There is more emphasis on the syntax and semantics of a language, than on how to use the language or how to decide when to use them. Computer Scientists are also interested in models of programming and often teach automata theory, language theory, etc. in introductory courses. The material is elegant and has great intellectual value; there is rarely emphasis on how to use it. In Engineering courses we teach theory and models, but the emphasis is always on the application of what we teach to design problems.

Design and analysis may be viewed as complementary skills. Design is inherently creative and most of the things we can teach about design are heuristics, things that don't always work. Because a heuristic, intuitive, design process often yields designs that are "almost right" (i.e. wrong), solid, disciplined analysis of the results of the design process is essential. In engineering, mathematics is most often used as a means of design analysis. In contrast, many Computer Scientists talk about systematically deriving programs from specifications. Program derivation is analogous to deriving a bridge from a description of the river and the expected traffic. Refining a formal specification to a program would be like refining a blueprint until it turned into a house. Neither is realistic; the creative steps in design are absolutely essential. This is as true in programming as it is in any other area of Engineering.

Engineers have learned to make a clear distinction between the product itself and a description of it. This distinction seems to have been lost in the Computer Science literature on programming. Mathematical tradition, in which formulae are the products, has led computer scientists to view programs and their mathematical descriptions as if they were the same things.

Those who chose engineering as a career path are often people with a fairly pragmatic view of life. They appreciate mathematics that is simple and elegant, but they want frequent assurance that the mathematics is useful. It is important to show them how to use a mathematical concept, not simply to teach them the definitions and theorems. In engineering mathematics, the emphasis has always been more on application of theorems than on proofs. Computer Science has followed the approach taken by mathematicians. When most Computer Scientists design a course, they will discuss proof of correctness more they discuss design.

7 A New Approach to Teaching Programming

We are currently teaching a novel programming course for all first year Engineering Students. It differs from conventional programming courses in two ways:

- The early part of the course teaches the basic mathematics of programming with emphasis on the use of mathematics to describe what a program does, or must do. Programming assignments are expressed as mathematical specifications. Students learn to compare programs with mathematical descriptions.

- We stress that the programming language is not the subject of the course. Students are given a choice of programming languages that can be used in the laboratories. Two of the three lectures per week were taught in an algorithmic notation based on Dijkstra's guarded commands [2,3]. The third, "laboratory", lecture uses a "real" language. Currently we offer FORTRAN and C. Lectures are scheduled so that students who wish to do so can attend both C and FORTRAN lectures. A system of bonus-points rewards students who learn both languages. The lectures present the same algorithms. Students see every algorithm at least twice, once in pseudocode, once in FORTRAN or C.

Our course emphasises both the creative steps in programming and the analytical steps needed to confirm the correctness of a design. Students are taught a "divide and conquer" approach to programming in which problems are systematically reduced to simpler problems and programs can be inspected in a disciplined, systematic way.

8 The Mathematics Needed for Professional Programming

This section describes the mathematical content of this first course on programming. Although the material would be familiar to any Computer Scientist, it will be unfamiliar to most Engineers; most engineers have not kept up with developments in Computer Science or software design methods.

8.1 Finite State Machines

The first step in getting students to take a professional approach to programming is to get rid of the "giant brain" and "obedient servant" views of a computer. It is essential that students see computers as purely mechanical devices, capable of mathematical description. Students learn that "remembering" or "storing" data is just a state-change, and are then taught to analyse simple finite state machines to verify that they accomplish specified tasks. The Moore-Mealy model is used. Students are also taught how to design finite state machines to perform simple tasks. We do not present the usual "automata theory". The emphasis is on understanding how the machines function and on designing them. A simulator is provided so that students can test their designs. The use of the state machine concept allows students to take a disciplined approach to designing systems that process sequences of data.

8.2 Sets, Functions, Relations, Composition

We present a naive set theory in which all sets consist of a finite number of elements selected from previously defined finite universes. It is important to present the students with examples of the use of these concepts and exercises in their use. We want the students to know far more than the definitions and the algebraic laws; we ask them to use the concepts to provide precise models of real-world situations. We show how state machines can be described by a pair of mathematical relations and how to use the operations of union, intersection, negation and relational composition. This is the first step towards describing programs using functions and relations.

8.3 Mathematical Logic Based on Finite Sets

In the first two sections, finite state machines, and sets have been kept not just finite, but small, so that they could all be described by enumeration. The next step is to point out that these are unrealistically small sets, that it is not practical to describe most sets by enumeration, and that we must be able to make general statements about classes of states. We then introduce an interpretation of classical predicate logic in which expression denotations are finite sets and show how to use predicate calculus to characterise sets, including functions and relations. By restricting the definitions to finite sets, we eliminate all the problems that Mathematicians and Computer Scientists find interesting, and can focus on the use of these concepts in program design.

We define the logic formally, but unlike most Computer Science and Mathematics courses, we do not give proof rules. Instead we give the evaluation rules, which are much simpler. We use logic to describe program behaviour, not to prove theorems.

The logic allows use of partial functions (defining all primitive predicates on undefined values to be *false*). It is important to provide numerous examples in which the students use predicate logic to characterise the states of something real. Arrays (viewed as partial functions) provide a rich source of examples such as, "Write a predicate that is *true* if array A contains a palindrome of length 3." We repeatedly use the mathematics to say things about programs, and teach students to *use*, as contrasted to *prove theorems about*, logic[1].

8.4 Programs as "Initial States"

We briefly present an unconventional view of programming as picking the initial state of a finite state machine. This helps to explain such concepts as table driven programs, interpreters, etc. It allows us to explain von Neumann's insight about the interchangeability of program and data and the practical implications of his contributions. We show practical examples of "trade-offs" made by moving decision logic between program and data, e.g. introducing table driven programs.

8.5 Programs as Descriptions of State Sequences

Next, we present a more conventional view of programs as descriptions of a sequence of state changes. This concept is presented abstractly; we do not give any programming language notation for describing the sequences.

8.6 Programs Described by Functions from Starting-state to Stopping-state

After pointing out that programs can be characterised as either *terminating* or *non-terminating* we indicate that our course focuses on programs that are intended to terminate after computing some useful values. We explain how the most important characteristics of programs can be described by a mathematical relation between its starting-states and stopping states. The exact model used is LD-relations[2,3]. We provide examples in which the students use relations to describe distinct sets of sequences that are equivalent in the sense of having the same set of (start-state, final-state) pairs. We show how these relations can be used to describe a class of programs that are equivalent (in the sense of getting the same answers) but may differ in the algorithms that they use. We show that this allows a mathematical description of a program that is simpler, and easier to understand than the program itself because it omits information about the intermediate states in the state sequences.

8.7 Tabular Descriptions of Functions and Relations

We extend the notation of predicate calculus by introducing 2-dimensional tableaux, which we call simply tables, whose entries are predicate expressions or terms. We show that these are equivalent to more conventional notation, but much easier to read. Students are given many examples in which we describe mathematical functions using these tables[4]. Figure 1 is an example of a complete specification of a program.

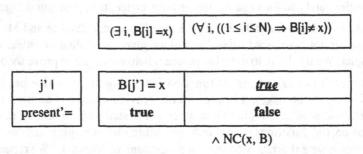

| $(\exists i, B[i] = x)$ | $(\forall i, ((1 \leq i \leq N) \Rightarrow B[i] \neq x))$ |

j'		$B[j'] = x$	*true*
present'=		true	false

$$\wedge \, NC(x, B)$$

Figure 1: Specification of a Programming Assignment.

9 Teaching Programming with this Mathematical Background

Although we teach a lot of mathematics, the purpose of the course is teaching students to program well. We must teach them to use the mathematics. We stress four points: (1) postponing program design until one has a precise statement of the requirements, (2) producing a precise program design before starting detailed coding, (3) constructing programs from components rather than simply, "writing" the lines down in execution order, detailed checking of algorithms using a "divide and conquer" philosophy.

9.1 Programming Professionally

Students are taught that they should never begin coding until they have a precise description of the program that they are trying to produce. All programs are introduced, not just with a natural language description, but with a mathematical description of the required behaviour.

9.2 Program Construction

A main theme of the course is that Engineers should *not* program by "thinking like a computer"; they should not plan the steps to be followed by the computer in the order that the computer will follow them. Instead, they are told that their job is to assemble new programs from previously constructed "building-blocks", simpler programs. They are reminded that, if they are successful, their products will later be used as components of still larger programs. They are also taught that they may have to deal with programs that are thousands, even millions, of lines long. They cannot expect to understand all lines of the program. They must have precise "black box" descriptions of the program building-blocks. They are shown how to use mathematical descriptions instead of examining the code for the programs they will use as building-blocks.

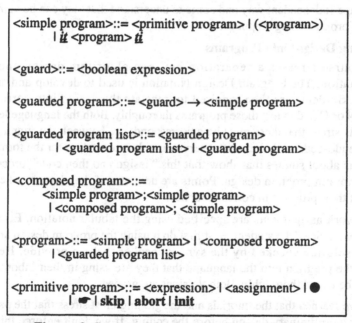

```
<simple program>::= <primitive program> | (<program>)
    | it <program> ti

<guard>::= <boolean expression>

<guarded program>::= <guard> → <simple program>

<guarded program list>::= <guarded program>
    | <guarded program list> | <guarded program>

<composed program>::=
    <simple program>;<simple program>
    | <composed program>; <simple program>

<program>::= <simple program> | <composed program>
    | <guarded program list>

<primitive program>::= <expression> | <assignment> | ●
    | ☞ | skip | abort | init
```

Figure 2: Syntax of the Program Design Notation

9.3 A Simple Language of Program Constructors

To keep the main lectures "language neutral" we introduce a stripped-down notation for describing programs. The syntax is shown in Figure 2. The language is introduced in a way consistent with the idea that we are constructing programs from building-blocks. The language provides four "constructors" or "constructs" that are used to construct programs from components. The simplest, ";", provides sequential execution. The others provide conditional execution, alternatives, and iteration. The programming notation is defined using the mathematical concepts taught earlier[3].We begin with very simple programs and continue, always using the "divide and conquer" discipline, to construct programs that solve more complex engineering problems. The students see every program twice, first developed systematically in the program planning notation, then translated into the programming language of their choice. A discipline of program design is stressed in every example. We systematically decompose a problem into simpler problems, then construct the complete solution from the components. We show how this program design notation, together with the mathematics provide a method of analysis that can be used to validate a design.

Students are shown how to systematically determine whether or not a design covers all cases and does the right thing in each case. Although, we never talk of "correctness proofs", we do use correctness concepts to explain programs. For example, we usually identify an "invariant" when explaining a loop, and demonstrate that the invariant is maintained by an execution of the body of the loop. We demonstrate how thinking in terms of "invariants" makes it unnecessary to try to enumerate all the execution sequences that might arise. Students learn these concepts in a language-independent way and come to understand that they can be applied in any imperative programming language.

9.4 Turning Designs into Programs

The course stresses a separation between algorithm design and detailed implementation. The Program Design Notation is used to develop and analyse the algorithm. Students are then required to translate the design into correct code in FORTRAN or C and to test those programs thoroughly. Both the language lectures and the tutorials stress that there must be a correspondence between the running program and the pseudo-code so that the latter serves as a documentation for the former. We tell the students about studies that show that this "design and then code" process greatly reduces errors in program design. Points are deducted if they have not kept the two versions of their program in correspondence.

Homework assignments are specified using the tabular notation. Each problem must be done twice. The student must first do it using the program design notation and have that solution checked by the syntax checker that we provide. He/She then translates the program into the language that they are using in their laboratory work. This must be tested thoroughly. Both versions are graded.

We have learned that the tutorials and the grading must stress that the pseudo-code is used for a preliminary design before the coding. If we don't enforce this, students fall back to their high-school habits, write the running code using intuitive methods, and then write the "design" as an afterthought.

9.5 Tests

The first test deals entirely with mathematical concepts and state machines. The second tests the student's ability to match pseudocode programs with mathematical specifications. In the third, the student deals with more complex programs and specifications. In the second and third test, and the final exam, students are also asked to match pseudocode programs to programs in their chosen "real" language.

Because it is impossible to grade 400 student programs accurately, we actually ask students to complete programs, using a multiple choice format, so that there is only one correct answer and the tests can be graded mechanically.

10 Experience

The course has been taught three times, each time to about 400 first year engineering students. In the first year there was strong student resistance to the change. Students who turned to upper year students for help found that they were being asked to learn more than the students in previous years. "Why do we have to learn this when they did not?" was a frequent remark. Some of the strongest resistance came from students with previous computer experience; they thought that they already "knew" how to program (often in Basic) and did not need to be taught theory, design methods, or mathematics. However, at the end of the term many confessed that they now understood the material and could not explain their original resistance.

In the second year, there was much less resistance. but students learned the "theory" without applying it to practice. Many students observed that they could learn the program design aspects without actually running C or FORTRAN programs. We discovered a great deal of copying of those by the end of the term.

By the third year, we felt we had the basic material correct and focused on changing the delivery so that students would see the connection between the design lectures and producing working programs. We made several changes:

- We changed our illustrative examples to illustrate the practical relevance of the methods. For example, instead of using recognisers to explain state tables, we had a series of problems on the mode control in a bicycle-computer.

- We provided consultants in the Computer Labs to help people who got stuck because of "mysterious" details about the compiler languages. We did this to remove any excuse for copying another student's work. In previous years students justified copying by claiming that they had no other source of help.

- We introduced a "laboratory exam" in which students were required to debug a simple program under supervision of Teaching Assistants (TAs). Students were given a correct pseudo-code design for the program a week before the exam, but had to find the errors in the C or FORTRAN in "real time".

- We changed our grading procedure to increase the likelihood of detecting copying and treated all copying as academic dishonesty.

- We introduced carefully worked out tutorial sessions in which problem similar to the assignments were solved using exactly the methods called for in the design lectures. The TAs who presented these tutorials were carefully selected for their familiarity with the design method and mathematics.

Some instructors from other courses complained after the second year saying that we were teaching theory not programming. In part, this was justified. The students who had cheated on their homework did not learn how to get a program to run.

However, an instructor whose classes included both students from the old, language-oriented, course, and students from the new course, found that the students from the old course were equally weak on the practical aspects. Computer assignments are far too easy to copy. Busy students will cheat to buy time.

Instructors in other subjects, e.g. Chemical Engineering, Civil Engineering, were also resistant because the material was new to *them*. One Department accused us of teaching "recent research" claiming that "finite state machines" were an example of recent theoretical advances; discussion revealed that they had never seen that half-century old concept. We often forget how little people in other areas understand about the advances made in Computer Science and Programming Methods in the last 30 years. They still program in the same way that people programmed 30 years ago. To be successful in such an environment, those who hope to upgrade programming courses must devote more effort than we did to the education of their colleagues.

Our goal is to improve student's ability to program in any language throughout their careers. It will be years before we can know how well we have succeeded.

11 Conclusions

Programming should be taught as if it were engineering because it is engineering. We should not be satisfied with teaching the detailed characteristics of one of today's tools. To prepare students for the rapidly changing world of computing, we must focus on fundamentals and design discipline. First year engineering students can be taught the mathematics and discipline necessary for professional programming

It is difficult to teach students with programming experience and novices in one class. We must chose between boring the experienced and confusing the novices.

12 Acknowledgements

These thoughts have been strongly influenced by H. D. Mills and N.G. de Bruijn. I am also deeply grateful to Prof. E. M. Williams, former Head of the Department of Electrical Engineering at Carnegie Institute of Technology (now deceased) for having taught me his philosophy of engineering education. Referees have contributed to clarification of an earlier version of this paper. Brian Bauer, Dennis Peters, Ruth Abraham, and Preeti Rastogi, made wonderful contributions in the third year.

13 References

[1]   Parnas, D. L., "Predicate Logic for Software Engineering", *IEEE Transactions on Software Engineering*, Vol. 19, No. 9, Sept. 1993, pp. 856 - 862

[2]   Parnas, D. L., "A Generalized Control Structure and Its Formal Definition", *Communications of the ACM*, Vol. 26, No. 8, August 1983, pp. 572-581.

[3]   Parnas, D. L., Wadge, W.W., "Less Restrictive Constructs for Structured Programs", *Technical Report 86-186*, Queen's, C&IS, Kingston, Ontario, Canada, October 1986, (available from the author)

[4]   Parnas, D. L., "Tabular Representation of Relations", *CRL Report 260*, Communications Research Laboratory, McMaster University, October 1992

A Course on Formal Methods in Software Engineering: Matching Requirements with Design

Paolo Ciancarini and Paolo Ciaccia

Corso di Laurea in Scienze dell'Informazione
University of Bologna - Italy
e-mail: {cianca,ciaccia}@cs.unibo.it

Abstract In this paper we shortly describe the course plan and syllabus used in a course included in the degree in Computer Science of University of Bologna since 1992. We discuss a project developed by students using formal notations for the specification of both requirements and design. The formal methods we use are based on Z notation for requirements specification and Larch for design specifications.
Keywords: Course Plan, Project Assignment.

1 Introduction

Developing a software engineering course in a scientifically oriented Master degree is not an easy task. Our students acquire a strong background on theoretical informatics, on computing theory, on programming language semantics, on logics, and on program verification, and it is not easy to match this kind of background with a practice-oriented software engineering syllabus. Our first idea was in fact to use books like [26] or [18]. These books are very complete, however they are oriented towards an engineering background, because they put emphasis on the design phase and on tools and project management. We felt that this design-oriented approach was not appropriate for our students. Thus, we decided to give a strong formal content to the course, and choose *Formal methods for software specification and design* as title. The course goal is stated as follows:

> *Learning to understand and evaluate the main production methods of real software systems from an architectural perspective, studying and using some importants paradigms of specification and design in-the-large. We will put emphasis on formal models and notations to write and reason with software specifications and architectures, presenting real systems useful to design new systems.*

We decided to follow two guiding principles for structuring the sequence of lectures: a) the notion of software production process should be presented early, to offer a conceptual framework justifying the methods and tools presented, and to suggest the students a method to organize their involvement in the project;

b) the notion of formal specification document should be presented introducing several realistic examples, both at the requirements level and at the design level, aiming at showing the students useful case studies and methods of reasoning on them.

We present to the students several specification and design notations, either formal or informal. Two main examples are discussed in all formalisms presented, namely the Library problem [29], and the specification of a system connected with the project assignment (see below).

The main themes of the course were stated as follows:

– The software production process;
– Informal versus formal specification methods;
– Design and formal properties of sw architectures and sw components;
– Notations, environments and tools for software specification and design.

We inspired ourselves to the course plan described in [16, 14]. This is the course plan we actually followed, together with the papers used for lectures or suggested to students:

1. Introduction to Software Engineering [2, 18]
2. The software production process and software process models [3, 10]
3. Informal and Semi-formal specification [29, 30]
4. Formal specification: theoretical background [6]
5. Formal Notations for declarative specification [21]
6. The Z language and tools (LaTeX, Fuzz) [11, 27]
7. Methods for formal specification based on Z: examples [5, 11]
8. Software design: functional vs. object oriented [4, 9, 23, 32]
9. Software Architectures [1, 15, 24, 25]
10. Algebraic languages as formal notations [6, 12, 13]
11. Larch: Larch Shared Language, Larch Interface Languages, and the Larch tools [20]
12. Example in Larch: the specification of a graphic editor [31, 33]
13. Proving properties of a design document: the Larch prover [17, 19]

The course plan is covered in 70 hours divided in about 35 lectures. The course starts in November and ends in May.

The final examination is intended to test the level of comprehension of software engineering methods and techniques by a single student, and to discuss a project suggested by the teachers and elaborated by each student team. The project quality is evaluated to give the basis for grading (projects were partitioned in three classes: excellent, good, sufficient); each student is required to prove his strong involvement in the project; the individual proficiency on theoretical issues establishes the final grade.

2 The project

The lecture course is integrated by a project that students, organized in teams of four members each, have to elaborate. A clear software development process is enacted since the first lessons to manage the coordination of teams. Teachers, playing the role of customers, propose an initial document describing a desired product. Students are given some weeks to study, discuss and clarify the informal requirements with customers and colleagues. This resulted in a new, final draft of the informal requirements. This is the starting point for the formal requirements specification in Z, to be completed over a period of two months.

The formal documents that were produced are checked with **fuzz** and cross reviewed and evaluated by the teams themselves, and the "best specification" is chosen as input for the next phase, i.e., the formal design specification in Larch/C or Larch/C++. Students are encouraged to reuse the available LSL library of traits (abstract data types) integrating their own modules. Finally, each team is required to provide at least a partial validation of the final design using the Larch prover to prove properties of the system being designed. Such properties should correspond to properties stated in the Z document.

In the following subsections we describe the first project assigned to students.

2.1 Informal requirements

The initial document (informal requirements) provided by the "customers" described *Chessmate*, an application useful for the training of chess players. What follows gives an idea of the contents of such a document:

> Several software packets that support training of chess players are currently available on personal computers:
> - playing programs: these are the most widespread, since they allow to play a game against the computer [8];
> - databases of games: programs used to consult large sets of games, possibly stored on CD-ROM;
> - databases of positions: programs used to build a tree of chess positions, useful to study opening and endgame theory;
> - network or modem interface programs: programs to play remotely against other players;
> - publishing tools for chess documents: chess game editor, chess fonts, chess game formatters.
>
> For an example of graphic interface to playing programs, see [22]. For an introduction to chess-related software, see [8]. What follows is an example of chess text created by the LaTeX formatter with some special macros [28].

```
\newgame
\move e2e4 c7c5
\move g1f3 d7d6
```

```
\move d2d4 c5d4
\move f3d4 g8f6
\move b1c3 a7a6
\move f1c4 e7e6
$$\showboard$$
```

1.	e2–e4	c7–c5
2.	♘g1–f3	d7–d6
3.	d2–d4	c5×d4
4.	♘f3×d4	♘g8–f6
5.	♘b1–c3	a7–a6
6.	♗f1–c4	e7–e6

Most of the programs are stand-alone, i.e., they are not integrated with other packages. We would like to market a new package that offers all the above features in an fully integrated environment: for instance, the playing program should be able to use in the initial phase of a game a position database; the games database should be able to call the player program to analyze a particular variant; when playing by modem a human player should be able to consult the playing program; it should be simple to store a game and print it with the chess fonts, and so on.

Your task is to design such a new package and its user interface.

The informal requirements document was given to students in December; they got three weeks to study the document and "improve" it, namely to clarify its obscurities and ambiguities. Students were also given some software that partially implemented the required functions, to exemplify the intended interface and functionalities.

We got a number of comments and suggestions, that we used to produce the final draft of the informal requirements.

2.2 Formal specification in Z

The final draft containing the informal requirements was given to students at the end of January: they had to produce a formal specification using Z by the end of March. The formal specification in Z had to be written in LaTeX and checked with Fuzz.

We now show a sample of part of a specification produced by a team; its purpose is to define the abstract state of the system.

Given primitive types BOARD, that represents chess boards with pieces (it includes two special values: constant *initboard* represents the initial disposition of pieces, and constant *nilBoard* is a special value useful to deal with error conditions) and PLAYER (enumeration of *white, black*) a game state is defined by schema *POSITION*:

$$
\begin{array}{|l}
\_POSITION \underline{} \\
Board : BOARD \\
ToMove : PLAYER \\
\hline
\end{array}
$$

A game starts from an initial board, and white has the move:

$$
\begin{array}{|l}
\_InitPosition \underline{} \\
POSITION' \\
\hline
Board' = initboard \\
ToMove' = white \\
\hline
\end{array}
$$

There are some functions that are useful to deal with boards. *LegalMoves* returns the moves that are possible in a given position; there is a special move called *showboard*, that will be used to print a diagram.

$$
\begin{array}{|l}
LegalMoves : BOARD \nrightarrow \mathbb{P}\, PLY \\
\hline
\forall\, board : BOARD\, \bullet \\
\quad showboard \in LegalMoves(board) \\
\end{array}
$$

NextBoard returns the new board given a board and a move:

$$
\begin{array}{|l}
NextBoard : (BOARD \times PLY) \nrightarrow BOARD \\
\hline
\forall\, board : BOARD;\ move : PLY \mid \\
\quad move \notin LegalMoves(board)\, \bullet \\
\qquad NextBoard(board, move) = nilBoard \\
\forall\, move : PLY\, \bullet \\
\quad NextBoard(nilBoard, move) = nilBoard \\
\forall\, board : BOARD\, \bullet \\
\quad NextBoard(board, showboard) = board \\
\end{array}
$$

FinalBoard returns an updated board given the initial board and a sequence of moves:

$$\begin{array}{|l}
\hline
FinalBoard : \text{seq } PLY \twoheadrightarrow BOARD \\
\hline
\forall\, moves : \text{seq } PLY \bullet \\
\quad FinalBoard((0\mathrel{..}0) \vartriangleleft moves) = initboard \\
\forall\, i : \mathsf{N}_1;\ moves : \text{seq } PLY \mid i \leq \#moves \bullet \\
\quad FinalBoard((1\mathrel{..}i) \vartriangleleft moves) = \\
\qquad NextBoard(FinalBoard((1\mathrel{..}(i-1)) \vartriangleleft moves), moves(i))
\end{array}$$

The legality check for a sequence of moves is done by function *Legal*:

$$\begin{array}{|l}
\hline
Legal : \text{seq } PLY \rightarrow BOOLEAN \\
\hline
\forall\, moves : \text{seq } PLY \mid \\
\quad (\forall\, i : \mathsf{N}_1 \mid i \leq \#moves \bullet \\
\qquad moves(i) \in LegalMoves(FinalBoard((1\mathrel{..}i) \vartriangleleft moves))) \bullet \\
\qquad Legal(moves) = true \\
\forall\, moves : \text{seq } PLY \mid \\
\quad (\exists\, i : \mathsf{N}_1 \mid i \leq \#moves \bullet \\
\qquad moves(i) \notin LegalMoves(FinalBoard((1\mathrel{..}i) \vartriangleleft moves))) \bullet \\
\qquad Legal(moves) = false
\end{array}$$

Now the schema *GAME* can be defined, that defines the format of games that can be stored.

$$\begin{array}{|l}
\hline
\_GAME\underline{} \\
Wplayer, Bplayer : PLAYERNAME; \\
Tournament : TOURNAMENT;\ Data : DATA; \\
Moves : \text{seq } PLY \\
\hline
(Wplayer \neq Bplayer \lor \\
\quad (Wplayer = noPlayer \land Bplayer = noPlayer)) \\
Legal(Moves) = true
\end{array}$$

Schema *System* defines how games are handled by Chessmate. There are several sets of files, because we have several different representations: ASCII files for text of games, compressed collections of games, position databases

$$
\begin{array}{|l}
\hline
\text{\_\_\_System _____} \\
AsciiFiles : NAMEFILE \twoheadrightarrow GAME \\
GameFiles : NAMEFILE \twoheadrightarrow (IDENTGAME \twoheadrightarrow COMPRESSEDG) \\
BoardFiles : NAMEFILE \twoheadrightarrow BOARDGRAPH \\
Modem : COMMUNICATION \\
\hline
\bigcap\{\text{dom}\, AsciiFiles, \text{dom}\, GameFiles, \text{dom}\, BoardFiles, \} = \emptyset \\
\forall\, Name1, Name2 : NAMEFILE;\ Part1, Part2 : GAME\ | \\
\quad\quad Part1 = Part2; \\
\quad\quad (Name1, Part1) \in AsciiFiles; \\
\quad\quad (Name2, Part2) \in AsciiFiles\, \bullet \\
\quad\quad\quad Name1 = Name2 \\
\forall\, Name : NAMEFILE;\ Identif : IDENTGAME\ | \\
\quad Name \in \text{dom}\, GameFiles; \\
\quad Identif \in \text{dom}(GameFiles(Name))\, \bullet \\
\quad\quad (\exists_1\, Game : COMPRESSEDG\, \bullet \\
\quad\quad\quad Game = (GameFiles(Name))(Identif)) \\
\hline
\end{array}
$$

A typical error is included in the last schema: to say that all file name domains are pairwise disjoint, students say that their intersection is empty!

One of the teams used also a graphical notation to show the main relationships among the schemata they introduced. In Fig. 1 it is shown part of the Z formal requirements document. The picture shows a number of schemata representing operations on the basic abstract states, namely *Editing* and *System*.

At the end of March started the phase in which each team cross refereed the specifications of two other teams. At the end we had a "winner", that was given as input document for the next phase, formal design.

2.3 Formal design in Larch

Given the formal specifications, teams were required to write a formal design with Larch, putting emphasis on the overall structure of the software architecture [1] and on reuse of standard traits given in LSL library. They also were required to use the Larch Prover to interactively prove at least one relevant property of the system being designed. Figure 2 shows the overall design, corresponding to the Z specification given in fig. 1.

We include a LSL *trait* (the basic unit of specification in Larch) where operators and their properties are introduced. These are its main features:

- A ChessBoard is specified (in the included trait *Squares*) as a 2D coordinate system, and a move is a line in this system.
- The included trait *Table* describes tables that store values in indexed places. This is used to model the position of pieces on the chessboard. The operators of the *Table* trait are properly renamed to improve readability. For instance,

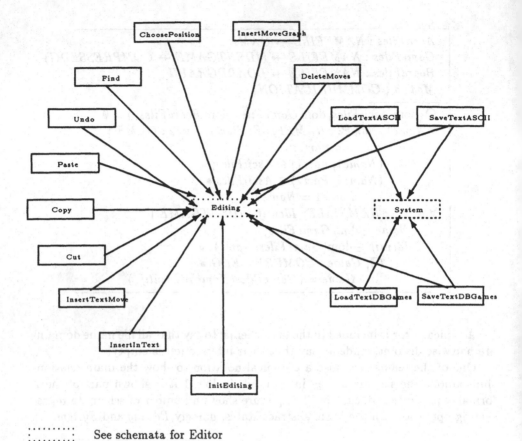

Figure1. Structure of schemata of a Z document; each box is a schema

getPiece is a renaming of *lookup*, and allows indexed access to the chessboard via a *Square* term.

- A *Position* is defined as a tuple with fields *board* and *to_move*, the latter representing the player who is up to move.
- The only new operator introduced in the *Positions* trait is *delPiece*, that removes a piece from the chessboard.
- The *implies* clause claims that removing a piece from the chessboard decrements by 1 the number of pieces on the chessboard. The claimed property was subsequently proved using the Larch Prover.

Positions(Piece, Player) : trait
includes
Squares
Table(Square, Piece, Board, position for add, getPiece for lookup,
noPiece for isEmpty, n_Pieces for size, newBoard for new)
Position tuple of board : Board, to_move : Player

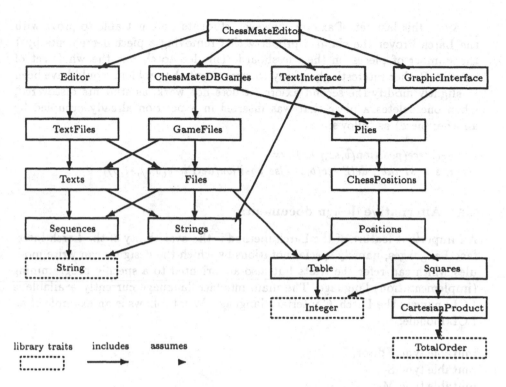

Figure2. A Structure of LSL Traits used in the project

introduces
delPiece : Board, Square → Board
asserts
∀ *b:Board, s, s1:Square, p:Piece*
s ∈ delPiece(b,s) == false;
% A removed piece is not on the chessboard anymore
not(s = s1) ⇒
(s1 ∈ delPiece(b,s) = s1 ∈ b ∧ getPiece(delPiece(b,s),s1) = getPiece(b,s1))
% Removing a piece does not affect the position of the others
implies
∀ *b:Board, s, s1:Square, p:Piece*
s ∈ b ⇒ (n_Pieces(delPiece(b,s)) + 1) = n_Pieces(b)
% Removing a piece decrements by 1 the number of pieces on the chessboard

In a first attempt to specify the semantics of the *delPiece* operator, students used the following alternate set of axioms. They relate *delPiece* to the two *constructors* of the *Board* sort, that is, *newBoard* and *position*:

delPiece(newBoard,s) == newBoard
delPiece(position(b,s1,p),s) == if s = s1 then b else position(delPiece(b,s),s1,p)
% Positioning and then removing a piece does not change the chessboard

With this last set of axioms, however, students were not able to prove with the Larch Prover the claimed property that removing a piece decrements by 1 the number of pieces on the chessboard. This led to change the whole set of axioms. Rather interestingly, a way to circumvent the problem would have been to slightly modify the second axiom (it does not work, as students discovered, when one deletes a piece that was inserted in a position already occupied by another piece) as follows:

delPiece(position(b,s1,p),s) ==
if s = s1 then delPiece(b,s) else position(delPiece(b,s),s1,p)

2.4 Alternative design documents

An important feature of the Larch method is the availability of the Larch Interface Languages, namely special notations by which the designer can write modules which can refer the traits but also are oriented to a specific programming (implementation) language. The main interface language currently available is LCL, namely the Larch/C interface language. What follows is an example of an LCL module:

```
mutable type Editor;
mutable type Sc;
mutable type Ms;
mutable type M;
mutable type I;
mutable type StrMos;
mutable type IntSet;
mutable type FileSys;
mutable type BoardPos;

uses Mossa;
uses StringMoves(char* for AbsStrMos);
uses Board;
uses Heading;
uses ChessEditor;
uses FileSystem(char* for Id);
uses sprint(M, char*);

claims selez_propr(Editor ed) {
    ensures
        (ed•.bSel ⊂ ed•.b) ∨ (ed•.bSel ⊂ ed•.bBkp);
}


void ed_initMod(Editor ed) {
    modifies ed;
```

ensures
 ed'.b = {} ∧
 ed'.bBkp = {} ∧
 ed'.bSel = {} ∧
 ed'.s = initialBoard;
}

Such a module refers a number of LSL traits (clause **uses**). A different design document, say for a C++-based implementation, would have a completely different structure, but could still refer the same traits.

3 Discussion

This discussion of the results of the course concerns the first year only of teaching its contents; in the second year we gave a different project concerning a client-server software architecture. We are now ending the third year and are re-evaluating the approach. Moreover, we have developed a complete description and analysis of the method we have chosen in a companion paper [7].

Although the emphasis was on formal methods, our students used a lot of tools. We feel that this use of tools cannot be over-emphasized in a software engineering course. The students were requested to enact the process using SUN workstations and tools like LATEX, xfig, Fuzz (for checking Z specifications), the LSL checker (for checking Larch traits), the LCL checker and LCLlint (for checking Larch/C modules), the Larch/C++ checker (when it became available), and the Larch Prover (to check properties of the Larch design). This approach was very successful, because the quality of the resulting documents improved and moreover students enjoyed the process.

Probably the most important aspect of our approach was the methodology we suggested, based on a form of contractual development process using strongly formalized notations. The overall process is depicted in Fig.3. Each rounded box is a process phase; each box is a document used as input for lower phases.

The best formal requirements document was used as input for the design phase, where they had to use an algebraic language to provide the software architecture document, paying attention to any occasion of reuse. Such a document had to be at least partially verified, proving with the help of a theorem prover some relevant properties of the system being designed.

We believe we were lucky in the choice of the project assignment: it was neither too small nor too complex. Students initially discussed a lot of chess-related questions, then realized that they were required to develop a system whose main components are an editor and a database, and their final documents were fairly centered upon these features. We also are glad of the use of two different formalisms for requirements and design specification. Students understood quite well the difference in the goals of the different phases and resulting documents.

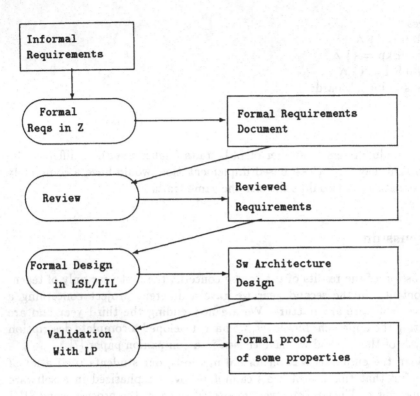

Figure3. The development process used by our students

Most teams tried to follow the format we suggested for specification in Z: top-down development, first specifying data, then states, and finally operations. We were quite satisfied with the level of proficiency in Z demonstrated by the students; they skillfully used advanced features of the language, and also the experience with the type checker Fuzz was positive. All the teams reported that Fuzz was a good help for checking the schemata. A team developed also an original graphical documentation to help examiners in understanding the specification; see fig. 1. However, most teams concentrated much more on algorithmic features than on declarative model-oriented presentation. The schema calculus was scarcely used; comments were almost absent. Almost nobody tried spontaneously to prove properties of the abstract system: we had to force them.

The Larch-based formal design phase was more difficult to assess. Basically, the Larch method requires to write algebraic theories giving a semantics to the basic operators (this is done using the Larch Shared Language). These operators need not to be implemented, but can be used in the definition of module interfaces (written in a Larch Interface Language specific to the implementation language), where the functions that the module makes available to its client are introduced. This so-called *two-tiered* style of specification of modules encourages modularity and separation of concerns, but it challenges specifiers as to how and where split the tiers. The teams were instinctively led to put in the LSL tier most of the required functionalities of the modules. This resulted in scarcely reusable design specifications. Sometimes, also error-handling considerations were present.

We observed that all teams tried to reuse, and possibly extend, some of the traits in the LSL library. According to the Larch style, the design mainly proceeded bottom-up, with traits and operators somehow resembling, respectively, types and operations specified in Z. Finally, the formal proof of properties with the Larch Prover was the most difficult task to perform, however the students appreciated its importance. In at least one case (see above) it led to discover inconsistencies in the specification.

The approach we have chosen generate some important questions, to which we cannot yet provide definitive answers.

- *Do student learn to design better if they start from formal requirements rather than informal ones?*
  Although we observed that in most final documents the separation of concerns between the two phases of requirements and design specification was not so clear-cut (e.g., design considerations were often found in Z documents) the bottom-up design activity was guided by the need of providing adequate support to the high-level functionalities expressed in Z. In some sense, this simplified the activity of designing the overall architecture of the system. It seems that the choice of an appropriate abstraction level is a crucial point for formal requirements to be effectively advantageous.
- *What is the relationship between a model-based (i.e., Z) and an algebraic (i.e., Larch) style of specification?*
  Translating from one language to the other is not trivial. The main difficulty, from the student point of view, appeared to be moving from one language where the concept of *system state* is a basic one, to another where it is completely meaningless.
  A formal link is given by verification: infact, it is possible to specify and prove properties in Z, that then can to be expressed in Larch and finally proved by means of the Larch Prover. However, this was only done with properties expressed in LSL, because the nature of the available tools. We conjecture that the glue between the two styles could be represented by the interface specification, where the state concept is peculiar to the specific implementation language, and algebraic-defined terms are first introduced.

Acknowledgements. First of all, we have to thank our students that had enough patience and enthusiasm to be the first testers of this course, and provided us with interesting project solutions. Wilma Penzo contributed with many ideas and suggestions to the development of the course contents and main design method. We also thank the people that built and made available the tools (Fuzz, Larch Checkers, and Larch Prover) we used in this course. This paper has been partially supported by the Italian Ministry of University and Scientific and Technological Research (MURST).

References

[1] G. Abowd, R. Allen, and D. Garlan. Using Style to Understand Descriptions of Software Architecture. In D.Notkin, editor, *Proc. ACM SIGSOFT 1st Conf. on Fundamentals of Software Engineering*, volume 18:5 of *ACM SIGSOFT Software Engineering Notes*, 1993.

[2] D. Berry. Academic legitimacy of the software engineering discipline. Technical Report CMU-SEI-92-34, Software Engineering Institute, Carnegie Mellon Univ., 1992.

[3] B. Boehm. A spiral model of software development and enhancement. *IEEE Computer*, 21(5):61–72, May 1988.

[4] G. Booch. Object-oriented development. *IEEE Transactions on Software Engineering*, 12(2):211–220, Feb. 1986.

[5] J. Bowen. Formal Specification of Window Systems. Technical Report PRG-74, Oxford University Computing Laboratory, England, June 1989.

[6] B. Chen, W. Harwood, and M. Jackson. *The Specification of Complex Systems*. Addison Wesley, 1986.

[7] P. Ciaccia, P. Ciancarini, and W. Penzo. A Formal Approach to Software Design: The Clepsydra Methodology. In *Proc. 9th Annual Z Users Meeting*, Workshops in Computing, Limerick, Ireland, 1995 (to appear). Springer-Verlag, Berlin.

[8] P. Ciancarini. *Artificial Chess Players (in Italian)*. Mursia, 1992.

[9] P. Coad and E. Yourdon. *Object-Oriented Design*. Yourdon Press, 1991.

[10] B. Curtis, M. Kellner, and J. Over. Process Modeling. *Communications of the ACM*, 35(9):75–90, September 1992.

[11] A. Diller. *Z: An Introduction to Formal Methods*. Wiley, 1990.

[12] H. Ehrig, B. Mahr, I. Classen, and F. Orejas. Introduction to Algebraic Specification. Part 1: Formal Methods for Software Development. *The Computer Journal*, 35(5):460–467, 1992.

[13] H. Ehrig, B. Mahr, and F. Orejas. Introduction to Algebraic Specification. Part 2: From Classical View to Foundations of System Specifications. *The Computer Journal*, 35(5):468–477, 1992.

[14] D. Garlan. Formal Methods for Software Engineers: Tradeoff in Curriculum Design. In C. Sledge, editor, *Software Engineering Education, Proc. SEI Conference*, volume 640 of *Lecture Notes in Computer Science*, pages 131–142, SanDiego, CA, October 1992. Springer-Verlag, Berlin.

[15] D. Garlan and M. Shaw. An Introduction to Software Architecture. In V. Ambriola and G. Tortora, editors, *Advances in Software Engineering and Knowledge Engineering*, pages 1–40. World Scientific Publishing Co., 1992.

[16] D. Garlan, M. Shaw, C. Okasaki, C. Scott, and R. Swonger. Experience with a course on architectures for software systems. In *Proc. Conf. on*

Software Engineering Education, volume 640 of *Lecture Notes in Computer Science*, pages 23–43. Springer-Verlag, Berlin, 1992.

[17] S. Garland and J. Guttag. An overview of LP, the Larch Prover. In B. Springer-Verlag, editor, *Proc. 3rd Int. Conf. on Rewriting Techniques and Applications*, volume 355 of *Lecture Notes in Computer Science*, pages 137–151, 1989.

[18] C. Ghezzi, M. Jazayeri, and D. Mandrioli. *Fundamentals of Software Engineering*. Prentice Hall, 1991.

[19] J. Guttag, S. Garland, and J. Horning. Debugging Larch Shared Language Specifications. *IEEE Transactions on Software Engineering*, 16(9):1044–1075, September 1990.

[20] J. Guttag and J. Horning. *Larch: Languages and Tools for Formal Specification*. Springer-Verlag, Berlin, 1993.

[21] R. Kemmerer. Testing Formal Specifications to Detect Design Errors. *IEEE Transactions on Software Engineering*, 11(1):32–43, January 1985.

[22] A. Kierulf, K. Chen, and J. Nievergelt. Smart game board and go explorer: A study in software and knowledge engineering. *Communications of the ACM*, 33(2):152–166, February 1990.

[23] B. Meyer. *Object-Oriented Software Construction*. Prentice Hall, 1988.

[24] D. Perry and G. Kaiser. Models of Software Development Environments. *IEEE Transactions on Software Engineering*, 17(3):283–295, 1991.

[25] D. Perry and A. Wolf. Foundations for the Study of Software Architecture. *ACM SIGSOFT Software Engineering Notes*, 17(4):40–52, October 1992.

[26] I. Sommerville. *Software Engineering*. Addison Wesley, 4 edition, 1991.

[27] J. Spivey. *The Z Notation. A Reference Manual*. Prentice Hall, 2 edition, 1992.

[28] P. Tutelaers. A Font and a style for Typesetting Chess using LaTeX or TeX. *TUGboat*, 13(1):85–90, 1992.

[29] J. Wing. A Study of 12 Specifications of the Library Problem. *IEEE Software*, 5(4):66–76, July 1988.

[30] J. Wing. A Specifier's Introduction to Formal Methods. *IEEE Computer*, 23(9):8–24, September 1990.

[31] J. Wing and A. Zaremski. A Formal Specification of a Visual Language Editor. In C. Ghezzi and G. Roman, editors, *Proc. 6th IEEE Int. Workshop on Software Specification and Design*, pages 120–129, Como, Italy, October 1991. IEEE Computer Society Press.

[32] E. Yourdon and L. C. Constantine. *Structured Design*. Yourdon Press, 1979.

[33] A. Zaremski. A Larch Specification of the Miró Editor. Technical Report CMU-CS-91-111, Carnegie Mellon Univerity, February 1991.

Hints for Writing Specifications

Jeannette M. Wing*

School of Computer Science, Carnegie Mellon University,
Pittsburgh, PA 15213, USA

Based on my experience in teaching formal methods to practicing and aspiring software engineers, I present some of the common stumbling blocks faced when writing formal specifications. I've broadly categorized the issues along the following dimensions:

- Figuring out *why* you are going through this specification effort. What do you hope to get out of using formalism?

- Figuring out *what* of the system you want *to specify*.

- Figuring out *how* to specify. The most important hurdle to overcome is learning to abstract. I also give specific suggestions on how to make incremental progress when writing a specification.

- Figuring out *what to write down*. Learn a formal method's set of conventions but do not feel constrained by them. Also, we all make logical errors sometimes; I point out some common troublespots in getting the details of a specification right.

I address all these issues by giving hints to specifiers. My talk should be of interest not only to teachers of formal methods but also to their students.

* This research is sponsored by the Wright Laboratory, Aeronautical Systems Center, Air Force Materiel Command, USAF, and the Advanced Research Projects Agency (ARPA) under grant number F33615-93-1-1330. Views and conclusions contained in this document are those of the authors and should not be interpreted as necessarily representing official policies or endorsements, either expressed or implied, of Wright Laboratory or the United States Government.

Mental Models of Z:
I — Sets and Logic

Neville Dean

Anglia Polytechnic University, Cambridge, CB1 1PT, UK
cdean@bridge.anglia.ac.uk

Abstract. Learning any new skills or knowledge involves building mental models expressed in terms of existing internalized knowledge and experience. Developing the formal and mathematical skills required in software engineering should be achieved in stages, with set theory and logic forming the first stage. The mental models for set theory and logic are most effective when they are based upon the practical experience of computer scientists and software engineers, and can be introduced in programming terms.

1 Introduction

It is generally accepted that there is a certain reluctance for industry to adopt formal methods, perhaps more accurately described as 'mathematical methods', in software development. Part of the reason must surely lie with the difficulty that pragmatists find in assimilating these methods, and this in turn arises from the lack of suitable understanding of the underlying concepts.

One view of learning is that we assimilate new ideas and concepts by creating 'mental models' in terms of existing knowledge and experience. Eventually this new learning becomes 'internalized' and may itself be used to build models of new concepts and ideas. Now software development itself entails learning about the system [12], and this learning can be greatly assisted by the use of a mathematically based notation such as Z. This use of Z can only be effective, however, when the notation itself has been internalized by building *appropriate* mental models of its constructs. There has of course been considerable advances in developing a semantics (and a deductive system) for Z. Such semantics, however, are not necessarily the most appropriate way to teach the ideas of Z, nor formal methods in general.

The process of building mental models of Z is considered in two phases:

1. modelling the set theory and logic which underpin Z;

2. modelling the concepts and constructs of Z in terms of set theory and logic.

This paper addresses the first of these two stages. The main issue to be considered is what *is* an appropriate mental model of set theory and logic. Various approaches can be used, and these are considered briefly in section 2. An alternative approach is then described in section 3.

2 Review of some Alternative Models

2.1 Natural Language Approach

In the natural language approach, the mathematical notation is introduced as a shorthand for ordinary language. For example ∀ represents the phrase 'for all'. The explanation of the notation and the concepts is in terms of everyday objects and language. To the student this certainly represents an easy way to learn the subject matter. Unfortunately it invalidates the *raison d'être* of using a formal notation, which is to overcome the shortcomings of natural language and to provide an alternative way of defining a variety of computing concepts and languages. Furthermore, the student sees the use of mathematical hieroglyphics as totally unnecessary — after all they do not seem to be achieving anything that cannot be done in ordinary language.

There is another disadvantage of using everyday objects to explain mathematical concepts. The mental models that are developed often work fine in simple cases but lead to erroneous conclusions in the more complex cases. As an example consider the following elementary approach to defining sets.

In elementary mathematics, a set is sometimes introduced as an 'unordered collection of objects'; essentially it is regarded as a 'bag' of objects in which, for some strange reason, repetitions do not count. Although such a mental image no doubt has its uses, it tends to be a little misleading.

1. The student cannot grasp why including repetitions does not produce a different set. Special rules have to be memorized to cope with this, and frequently forgotten by the beginner. Certainly it is difficult for such apparently arbitrary rules to become internalized.

2. Set membership may be confusing, especially when applied to sets of sets. For example, many students using this model are convinced that 1 is an element of $\{\{1\}\}$ simply because the 1 is 'inside the bag'.

3. Similarly there is much confusion between set inclusion and set membership.

In short, the natural approach to discrete mathematics is fine for giving the impression that something has been learnt, but in my experience this learning alone never became sufficiently internalized to enable the student to *apply* it, and certainly not to *want* to apply it. Of course it is necessary and indeed desirable to have mental models which relate the notation to ordinary concepts and language, but these should be supplementary to more effective models.

2.2 Approach Through Applications

Much effort is frequently spent in getting students to apply set theory and logic to simple systems; this can be a trying and frustrating experience for both tutor and student. A common approach is to demonstrate a few cases, then ask the student to do something very similar or to add to what has already been done. Unfortunately this approach meets with little success. In my own experience, the problem seems to stem from an insufficiently developed mental model; it is only possible for a student to apply discrete mathematics when an *appropriate* model has been *internalized*. Worse! Too much use of 'realistic' applications encourages the student to develop mental models which do not reflect the precision and generality of the mathematics. Trying to push applications too much at an early stage tends to have the opposite effect to that intended; the student is less able to apply the mathematics and is less convinced of its usefulness.

Of course, applications *are* important and should be introduced to motivate the student and to justify the mathematical notation. But equally care should be taken to ensure that the students do not take the easy route and base their mental models of sets and logic on these applications alone; furthermore students cannot be expected to apply discrete mathematics without help — the most that can be expected in a first course is for them to be able to answer highly structured questions about a clearly defined application, for example, sequences in the Z mathematical toolkit.

The ability to apply Z, and mathematical modelling in general, is a high level skill which is frequently underestimated and perhaps undervalued. In my own experience a separate, and specially designed, course needs to be used to develop such skills [3].

2.3 The Rôle and Nature of Logic

There seems to be a certain amount of confusion in the published literature concerning the nature and rôle of logic. Logic seems to exist in a variety of 'flavours'.

1. Classical 'Aristotelian' logic, which uses a restricted (formalised) form of natural language to prove consequences from a given set of premisses.

2. Symbolic logic, which is basically classical logic expressed with symbols (to represent natural language).

3. Circuit logic, which is the application of a Boolean algebra to model digital circuits.

4. Formal logic, which is similar to symbolic logic but the emphasis is on the formal language and the construction of certain types of expression (*wffs* and proofs).

5. Logic for set comprehensions.

To the established practitioner it is clear that these various 'flavours' are all different aspects of logic. But to most students, they appear to be completely different subjects.

Although it is important eventually that students become more comfortable with formalism, it is not appropriate in the early stages to develop formal logic *per se*. Neither is it necessary to introduce logic as a means of arguing or proving conclusions from premisses, even though this too is important. Rather it is best to restrict the use of logic to one particular 'flavour', even though the students should be aware that other uses of logic will follow in due course. For mathematicians, the use of logic for proof would appear to be the most suitable choice; for example see the book by Epp [5]. Yet it is arguable whether this is the best first choice for software engineers and computer scientists.

2.4 Formal Languages

An alternative approach to discrete mathematics is to treat the notation as a formal language. This approach is particularly well exemplified in books such as that by Woodcock and Loomes [11]. Sooner or later the software engineer needs to be able to see the subject in this way, but I question whether most ordinary students have the necessary background to assimilate such an approach to begin with. What is important to the student in the early stages is to lay down the body of experience and understanding upon which a good mental model of formalism can be subsequently built.

2.5 Mathematical Rigour

Another approach, particularly favoured by mathematicians, is to use an abstract axiomatic approach in which sets and other objects are defined solely in terms of the rules that they obey. The difficulty with the traditional mathematical approach is that it is not even formal. It has long been recognized that traditional mathematical proof is even harder than formal proof and should be a skill developed once confidence in formalism has been established; in the past, teachers seemed well aware of this, yet sadly it seems to have become overlooked as the twentieth century has progressed. Thus in the preface to [2] we find:

> In Plane Geometry greater liberty has been assumed as the student will have outgrown formal treatment. But in Solid Geometry formal treatment has been largely retained as it is likely to inspire confidence in a beginner.

Perhaps the failure of Russell and Whitehead [9] to formalize mathematics has led to the belief that learning formalism is an unnecessary and incorrect thing for mathematicians to be doing! Undoubtedly the use of mathematical, semi-formal methods have much to offer the software engineer, but skills such as these can only be developed at a late stage of educational development — once formalism has been established.

3 An Outline of a 'Procedural' Approach

What is required is a mental model which builds on the knowledge and practical experience of computer scientists and software engineers; in particular which builds on the experience of operations, functions, procedures and the like. This section outlines one possible approach. Note that the outline presented in this paper is not in a form that is necessarily suitable for teaching; for the material actually used see [4].

The concepts are described in terms of procedures and relations, with the notation regarded as a programming language. Learning the mathematics is a matter of learning the syntax and 'semantics' of this programming language. The resulting 'semantics' may not be strictly acceptable to pure mathematicians, nor indeed to many theoretical computer scientists, but nevertheless it gives pragmatists an acceptable basis for assimilation.

The approach adopted begins with developing some basic notions of set theory, before considering propositional logic. Subsequently, the more difficult aspects of set theory are developed alongside predicate logic before applying these ideas to relations and functions.

3.1 Set Theory

Set Membership The approach adopted is very close to Cantor's original concept of a set.

Suppose we have a procedure which takes an input value, x, and outputs a Boolean value (.true. or .false.). This procedure defines a set, X. If the output is .true. then we write $x \in X$; if .false. then we write $x \notin X$. Every such procedure corresponds to one set, provided it can produce an output for every input, at least in principle. Some procedures may get locked into infinite loops, something with which most novice programmers soon become acquainted. One example is the set of sets which are not members of themselves; attempting to implement this procedure results in the procedure calling itself and getting into a loop. Russell's antinomy is thus introduced as something very familiar, and perhaps more relevant than barbers who may or may not shave themselves.

Set Equality Usually a formal attempt to define set equality is met with some disbelief: 'surely we don't need to define set equality'. Now however we can talk of two different procedures corresponding to sets A and B; although the procedures are different, the sets may be the same. We say that $A = B$ if, and only if, the corresponding procedures always produce the same outputs for the same input. Again, this relates to a common experience that there is more than one possible solution to a programming problem.

Set Enumeration Set enumeration, or set display, is introduced as corresponding to procedures in which the input object x is compared successively with the items in a list. If a match is found then x is a member; if the end of the list is reached without a match being made then x is not a member. Lists with repetitions and different orderings, although corresponding to different procedures, do correspond to sets which are equal, for example $\{1, 1, 2, 3\} = \{1, 3, 2\}$.

The Empty and Universal Sets A simple procedure can be devised which always gives an output of .false. and which therefore corresponds to the empty set.

Conversely a procedure which always gives .true. corresponds to the universal set, V (the set of *everything*). Although it is often said that permitting a universal set leads to antinomies, such as Russell's antinomy, this is a mistaken belief [6]. Of course ZF set theory does *not* permit a universal set; but it is nevertheless possible to build other theories in which a universal set V does exist (with the defining property that $V \in V$) without giving rise to any antinomies. The argument that a power set has a greater cardinality than the set from which it is derived does not apply to the universal set. The subsequent introduction of types then has the effect of eliminating both the universal set and the possibility of a procedure which can take *any* input; thus the correspondence between sets and procedures is maintained.

In fact the empty set and the universal set are examples of sets which are complements of each other. Two decision procedures which always gives the opposite values are complementary, and the corresponding sets are also complementary.

Operations on Sets Sets can themselves be input to procedures. Thus we can say that $\{3\} \in \{\{3\}\}$ or that $\{1, 2, 3\} \cap \{2, 3, 4\} = \{2, 3\}$. In a sense this is laying the foundations for regarding programs as objects in their own right, and thus to laying the foundations of formal methods. At an early stage of learning however it is perhaps best to play down this concept. Rather it is easier to use a secondary model of sets as lists; this follows from the set enumeration types of decision procedure. If the student has not yet encountered any form of list processing, then they most likely soon will. Furthermore, this can lead on to more practical work involving simple list processing programs and use of a language such as ISETL. The notion of power sets can be introduced most successfully using this approach; given a set enumeration, the student can find the value of the power set by using a simple algorithm. By now the student is beginning to handle the notation as a formal structure which can be manipulated according to well defined rules, and which can be interpreted in terms of procedures.

Set Comprehension Set comprehension can now be introduced in the 'set-logic' language as program code corresponding to the decision procedure for deciding set membership. The special part of the language, indeed the main part, is logic. In this sense, logical notation is introduced as a formal structure, although the student is unaware of this.

3.2 Propositional Logic

A brief overview of the different flavours of logic is given mainly to alert the students to what they need to miss out if read any of the many textbooks on discrete mathematics. Furthermore, many students will have 'done Boolean algebra' or circuit logic, and need to be alerted to the differences in notation and concepts. The review is also a good way of showing the commonalities between all the different approaches to logic, even though it is unlikely that many students will understand the connections fully. Although it is less confusing for beginner to think in terms of several different subjects called 'logic', they should be aware that these subjects are all really different aspects of the same subject.

Logic is thus introduced for two purposes:

1. as a means of defining sets via set comprehension;

2. in order to make statements about what is true for a system, and in particular to the notion of invariants of the system.

Following the usual introduction of propositions, propositional forms are introduced as schemata in which the schematic letters are referred to as propositional letters (p, q, r). A schema such as $p \wedge q$ is referred to as a propositional form rather than as a proposition, although a warning is given that convention tends to use the shorter phrase; again, it is fine for the expert to use language loosely, but it makes life very difficult for the beginner. There is nothing new about this idea, and several texts do make a point of it [10, 7]. This pedantic distinction does seem to give the student a greater sense of security. Exercises are given in which the students instantiate forms with specific values for the propositional letters (using the replace operation of a word processor is suggested here!) Conversely propositions can be grouped into (non-disjoint) classes; in each class all the propositions can be obtained by instantiation of the same propositional form. This helps to introduce the idea of studying the nature of information and its processing; this is an aspect which is picked up later in considering relations and can be easily related to database management, for example.

Having made the careful distinction between proposition and propositional forms, it is not conceptually that difficult to introduce the notion of metastatements about propositional forms; in particular the ideas of equivalence, tautology and contradiction, as well as that of logical implication.

Connectives are initially introduced in the conventional way in the introductory review of logic. After developing truth tables, connectives are reconsidered as a means of constructing invariants to describe systems, and as a means of strengthening of invariants. Modelling the if... then... construct by the conditional connective, $\Rightarrow$, is considered; the definition of the truth table for the conditional connective can then be justified in terms of its usefulness.

3.3 Predicate Logic

Predicate logic is introduced via set comprehension; a predicate with one free variable is used to 'program' the decision procedure which determines set membership. This concept can then be developed into more than one free variable.

Types Students are familiar with specifying the parameter types of a routine; it is not unreasonable then to restrict the inputs to a decision procedure. This is achieved by stochastic assignment of the free variable x to a value taken from the predefined set A, using the $:\in$ operator. (This notation is used in the Abstract Machine Notation of B [13] and also some applications packages such as Derive [8].) At this pre-formal stage of logic, the type of a free variable is regarded as the set of all the possible *values* of the free variable. It needs to be pointed out to the student that this is not the whole story about types in logic, and indeed it is perhaps better to avoid talking of types at this stage and to avoid using the type declaration notation $x : A$. The Z notation can be introduced later as part of a more general development of a mental model for types.

Thus the notation $x :\in A$ is regarded as expressing an operation which selects a value for x from A; this value can then be 'piped' to a predicate which has x as a free variable. The symbol used, $|$, is familiar to many students as a pipe. The result of instantiating a particular value for x is to produce a proposition (assuming that x is the only free variable in the predicate, of course.) The students can experiment with this; thus with the notation $x :\in A \mid p(x)$ where A is a finite set and $p(x)$ they can associate a procedure which first selects all the possible values of x from the set A, then pipes these values to the predicate $p(x)$, which then selects only those values of x which give true propositions. The idea of piping is carried further with the $\bullet$ symbol (a 'fat pipe' seen end on!): this passes the selected values of x to a further process, in which each value of x is substituted in a term, $T(x)$. Thus we may write $x :\in A \mid p(x) \bullet T(x)$ to represent the whole process. Whereas $p(x)$ is always a predicate, $T(x)$ can be any sort of term — perhaps that is why a 'fat pipe' is needed!

Quantifiers A logical expression such as $x :\in A \mid p(x) \bullet T(x)$ can be regarded as an object in its own right, as indeed can a program or just a single of code. It is therefore possible for an appropriate function to map such an expression to a single value, such as a number or a truth value. This is essentially what is achieved by quantifiers, for example: $\sum, \{\ldots\}, \forall$ and $\exists$.

Initially the concept of a quantifier can be developed by considering finite sets. If A is finite, then the value of the quantified expression can be obtained by a loop which allows x to take all the values in A; of course the 'loop' variable x is now bound, a concept which has much in common with loop variables (whereas a free variable can be likened to a parameter of a procedure). In simple cases, for example

$$\forall x :\in \mathbb{N} \mid x \leq 3 \bullet x - 2 \in \mathbb{N}$$

the value of the expression (in this case a proposition) can be found explicitly. Getting the student to work through specific examples helps to establish the concepts much more effectively than the conventional approach of 'rewrite this sentence using quantifiers'.

Subsequently, the notion of a quantifier can be developed for infinite sets. Now instead of a loop variable, the bound variable can be regarded as an input; the universal quantifier, for example, corresponds to the situation in which the result is always `.true.` no matter what input is chosen.

4 Evaluation and Conclusions

An approach has been outlined which constitutes an initial stage of developing mathematical and formal methods in the software engineer. The underlying philosophy is essentially constructivist in which the student builds new mental models in terms of existing ones; these new models in turn become internalized and can then be used as the basis for further learning. What is therefore required in the early stages is to utilize the particular backgrounds and viewpoints that software engineering and computer science students are likely to have; in particular an understanding of procedures and programming.

Such an approach has been gradually developed over the last four years with several hundred students (both full-time and part-time) at the Anglia Polytechnic University in Cambridge studying first degrees in Computing, Mathematics and Real-time Systems; recently the approach has been extended to students studying for an HNC or HND in Computing[1]. Students typically receive just one hour's lecture and one hour's tutorial each week over a nine week period — they are expected to supplement this by six hours a week private study.

The results have been very encouraging with a noticeable improvement in marks attained both in coursework and in examinations. It was observed that as the teaching approach moved further away from the traditional style of presentation to the programme outlined, results continued to improve; this was especially so for students already working in the software industry — this would seem to confirm the usefulness of the approach

---

[1] In the British higher educational system, an HNC or HND is at a level comparable to the first or second year of a degree course, and is often taken as a part-time course usually by students already in employment.

for pragmatists. In the latest delivery of the course, there were no failures out of 43 students. Before beginning the course, many of these students were fearful of mathematics, and generally regarded themselves as being weak mathematically.

The general response of the students is that they enjoy the discrete mathematics, certainly more than they expected; indeed they see the work as something of a challenge to be enjoyed rather than endured. Practically all students acknowledge that the material learnt is relevant and useful, though few can see the connection to actual applications at this stage — that only seems to come when the students meet appropriate applications later.

Although formalism and logical deduction are not treated explicitly, it is noticeable that many (though not all) students do develop these skills to a certain extent. Indeed, the more able students are well able to demonstrate the ability to abstract and generalize particular results. Nearly all students are able to express themselves more clearly, even the few that are still using natural language to present their arguments.

The learning subsequently helps with other courses such as relational databases, object oriented design, artificial intelligence, Prolog and, of course, formal methods and modelling with Z.

5 Summary

- In order to write effective Z, and indeed to read Z effectively, it is necessary to have an understanding which is independent of ordinary language but which is seen to derive from systems concepts.

- Discrete mathematics and subsequently formalism can be developed from the basis of practical experience. Such an approach does works well:

  - the student sees the material as being closely associated with computing concepts;

  - being more closely related to the students' experience, the content is more readily assimilated and more easily applied;

  - the approach provides a good basis for subsequent development of formalism and logical argument.

- Set theory and logic need not be introduced by means of applications. Instead examples may be used to motivate the need for and to justify the mathematics. In the early stages, it is important not to expect too much ability to apply the mathematics since applications only begin to make complete sense once the mathematics itself has been internalized — not the other way round.

- Beginners do not believe they can cope with a highly formal approach, even though such an approach would perhaps easier for them. Nevertheless, formalism and abstraction can developed without the student necessarily being aware. In fact the student is being given the basic building blocks with which to make mental models of abstraction and formalism later.

References

1. Bowen, J.P. & Hall, J.A. (editors); *Proceedings of the Eighth Z User Meeting*, Springer-Verlag, London, 1994.
2. Davidson, J. & Pressland, A.J.; *A Second Geometry*, Clarendon Press & Humphrey Milford, 1926.
3. Dean, C.N. & Hinchey, M.G.; *Introducing Formal Methods Through Rôle-Playing*, SIGCSE Bulletin, **27** (1) : 302-306, 1995.
4. Dean, C.N.; *The Essence of Discrete Mathematics* Prentice Hall Essence of Computing Series, to appear 1996.
5. Epp, S.; *Discrete Mathematics*, International Thomson Publishing Ltd., 2nd edn., 1995.
6. Forster, T.E.; *Set Theory with a Universal Set*, Oxford Logic Guides No. 20, Clarendon Press, 1992.
7. Mattson, H.F.; *Discrete Mathematics with Applications*, John Wiley, 1993.
8. Rich, A., Rich, J. & Stoutemyer, D.; *Derive: A Mathematical Assistant for Your Personal Computer*, Version 3.0, Soft Warehouse, Inc., 1994.
9. Russell, B.A.W. & Whitehead, A.N.; *Principia Mathematica*, Cambridge University Press, 1910-13.
10. Stanat, D.F. & McAllister, D.F.; *Discrete Mathematics in Computer Science*, Prentice Hall, 1977.
11. Woodcock, J.C.P. & Loomes, M.; *Software Engineering Mathematics*, Pitman, 1989.
12. Worden, R; *Fermenting and Distilling* in [1], 1994.
13. Wordsworth, J.B.; *Software Development with B*, To be published 1996.

Equational logic: a great pedagogical tool for teaching a skill in logic

David Gries

Computer Science
Cornell University
Ithaca, NY 14853, USA

Extended Abstract

The text "A Logical Approach to Discrete Math", by Gries and Schneider [1], opens up new possibilities for teaching logic and mathematics. The text advocates spending 6–7 weeks learning equational logic (substitution of equals for equals is the dominant inference rule) and gaining a skill in formal manipulation and the development of proofs. Thereafter, the text *uses* formal equational logic in teaching all the topics of discrete math, from set theory to induction to solving recurrence relations. Everything is more formal than in conventional courses.

The approach is different from others in at least the following ways:

(0) Students gain a skill in formal manipulation, which helps them lose (some of) their fear of mathematics.

(1) Students see that formal logic is *useful*.

(2) Equational logic is not simply a formalization of how people normally prove things (as is natural deduction); instead, it offers a complementary way of proving things.

(3) Equational logic is superior to natural deduction because complexity does not overwhelm when one tries to be completely formal (this is the experience of those who have used equational logic). As David Hilbert said in his famous speech in 1990, rigor is not the enemy of simplicity; instead, the very effort for rigor forces us to develop simpler methods of proof and leads the way to methods that are more capable of development than the old methods of less rigor.

One of the difficulties in getting people to use formal notations like Z is that they have no skill in manipulation and have difficulty translating from English into Z and then back. If they *do* use Z, more than likely they use it only to specify their problems, but rarely to reason about them. Equational logic offers the real chance of overcoming this difficulty. Once people are at ease with and can manipulate predicate logic formulas, there will be much less resistance to learning and using Z.

This work on equational logic is not a threat to Z or in competition with Z. Z is basically predicate calculus together with nice higher-level features for dealing with types and for organizing predicates into very useful units. Z provides higher-level abstractions than does predicate calculus. We are interested only in teaching the basic predicate calculus, and our predicate calculus notation is not too different from that of Z.

We say a few words about equational propositional logic E versus most other propositional logics. Most propositional logics are built around inference rule Modus Ponens:

$$\textbf{Modus ponens:} \quad \vdash P, P \Rightarrow Q \;\longrightarrow\; \vdash Q$$

Implication $\Rightarrow$ assumes real significance, while equivalence $\equiv$ is treated as a second-class citizen —$b \equiv c$ is merely an abbreviation for $b \Rightarrow c \wedge c \Rightarrow b$. And, people begin using $\Rightarrow$ in complicated ways where use of $\equiv$ would be clearer and shorter. Further, the main method of a proof is to break a formula into its little subcomponents and then construct a new formula from them. This is often not the simplest method of proof.

Equational logic E uses substitution of equals of equals (Leibniz) as a main inference rule instead of Modus Ponens. Here are the four inference rules of E.

$$\textbf{Substitution}: \quad \vdash P \;\longrightarrow\; \vdash P[v := Q]$$
$$\textbf{Leibniz}: \quad \vdash P = Q \;\longrightarrow\; \vdash E[v := P] = E[v := Q]$$
$$\textbf{Transitivity}: \quad \vdash P = Q,\; Q = R \;\longrightarrow\; \vdash P = R$$
$$\textbf{Equanimity}: \quad \vdash P, P \equiv Q \;\longrightarrow\; \vdash Q$$

Using Leibniz, manipulations are done by substituting (perhaps large) components of a formula by something equal to them, and this often results in shorter and more understandable proofs. Further, proofs generally follow a form that people have learned in high school: a proof is basically a sequence of substitution of equals for equals.

Note also that $\equiv$ is an associative operator: $(b \equiv c) \equiv d$ and $b \equiv (c \equiv d)$ have the same value in every state. Heavy use of this associativity brings great economies in formulas and their manipulations, just as heavy use of associativity of addition and multiplication in arithmetic does.

Of course, in this short extended abstract, we cannot give further arguments and examples that back up the claims made above. The lecture will do that.

References

1. David Gries & Fred B. Schneider. *A Logical Approach to Discrete Math*. Springer-Verlag, New York, 1993.

Z Browser - Tool for Visualisation of Z Specifications

Luboš Mikušiak, Vladimír Vojtek
Jozef Hasaralejko, Jana Hanzelová
Department of Computer Science and Software Engineering
Faculty of Electrical Engineering and Information Technology
Ilkovicova 3, 812 19 Bratislava, Slovak Republic
E-mail: mikusiak@elf.stuba.sk

Abstract

This article presents Z Browser - Microsoft Windows based tool which helps novices starting to learn Z, even those with little knowledge of Z, to easier comprehend Z specifications. First, Z Browser and its integrated help facility for the Z notation (Z Help) is described. Further, we describe the tests which we carried out in order to find out how much Z Browser reduces the time needed for comprehension of Z specifications and if the quality of comprehension is better. For Z Help, we designed a graphical representation of data types of Z language and the operations on them.

1. Introduction

The usual approach to teaching formal specification languages (not only Z) is to teach the students the notation first and possibly demonstrate individual concepts of the language on small examples. It takes some time until the students are able to read and comprehend more complex specifications where they can see how they can benefit from formal methods. The period of learning might be rather exhausting because they do not have motivation for learning as they do not see any benefits. This might lead to the condition where many students build a pessimistic attitude toward formal methods just from the beginning of their study.

Z Browser is a tool which is able to relieve this problem. It enables even users with small knowledge of Z to read and comprehend Z specifications, and so explore the benefits in a very early stage of studying Z.

Z Browser is a Microsoft Windows based tool for quick and comfortable browsing through formal specifications in Z language. It consists of the Microsoft Windows application, and the integrated help facility for the entire Z language according to The Z Notation, A Reference Manual [1]. Its target audience are mostly novices of Z; like students of computer science taking their formal methods courses, individuals trying to teach themselves Z, and clients being given presentations of Z specifications by software developers.

There are two reasons which motivated us to do this research. First, it is our conviction that formal methods are the means for increasing software quality and reliability. Second, it was the experience that formal methods are often rejected at the beginning by users for their presumed complexity and mathematical background. Having a tool which would enable novice users to understand Z specifications and explore the advantages in the early stages of their study might increase the use of Z.

Z Browser, we concluded, is the tool which is able to reduce essentially the time needed for reading and comprehension of Z specifications. We believe that it is the tool which can help many novices to start to learn Z and focus on the meaning of Z specifications without the previous long study of, for many students quite complicated, Z notation and/or the mathematical background of Z.

As the input format we have chosen LaTeX format and zed and fuzz LaTeX styles because the largest number of Z specifications written so far use these two LaTeX styles.

The rest of this article is organised as follows: Chapter 2 describes in more detail what Z Browser is capable of doing and the reasons why we have designed it in the way it has been designed. In Chapter 3, the Z Browser program is described. Chapter 4 outlines Z Help which is integrated into Z Browser. Chapter 5 is about testing Z Browser. And, Chapter 6 contains conclusions and suggestions for further research.

2. Foundations upon Z Browser is based

At the beginning we tried to find out why Z may seem to be difficult for the novice users and how these difficulties might be overcome. The three main difficulties we identified are:

1. Mathematical background of Z language.
2. Complicated notation with many symbols not frequently used in mathematics nor other formal languages.
3. Complex structure of Z specifications where, for example, schemas contain references to other paragraphs of a Z specification making it difficult to keep track of all these relationships.

For the design of Z Browser, we used human-computer interaction methods as we found the problem of displaying formal notation in a more understandable way similar to the problem of designing comprehensible user interfaces. Modern graphical user interfaces are designed in such a way that they should be comprehensible for the novice and casual users. Also, they should be designed in such a way that most of the users can use them without previous training. Our problem was similar: we wanted to make a Z specification comprehensible to novice users who might have minimal knowledge about Z. We also did not find much literature about visualisation of formal methods [5] so we let ourselves be inspired mostly by the human-computer interaction research [7][11][4][9]. We were also studying visual programming languages [6][12] while we were designing the graphical representation of operations on Z data types. Although visual programming might appear to be a better starting point, the research performed so far is focused on programming languages. We found it difficult to apply this research to visualisation of formal methods.

First of all, the assumption was maintained that the capacity of the human short-term memory is limited to 5±2 chunks of information [14]. When studying a Z specification, this capacity is used for bits and pieces of Z specification as well as for the concepts of Z occurring in the specification. Based on the content of short-term memory, the user makes conclusions and moves toward another part of the specification. He remembers only conclusions, forgets the rest of the information and fills in his short-term memory with new information. In Z, when trying to understand one schema, for example, we need to see also another paragraphs which are referenced in the given schema. These are additional information which need to be held in short-term memory, and mostly for a longer period.

We tried to design Z Browser in such a way that the user needs just one or several mouse clicks in order to display the required information (either another paragraph of Z specification or help for a particular concept) and that it takes another one or two mouse clicks to return to the original place in the specification.

In terms of human-computer interaction: make the screen the visual cache [10] of the users' minds by enabling them to display the needed information in several windows for further reference. In so doing, letting them easily load into their short-term memory information they currently need for reasoning.

To make understanding of the mathematical concepts of Z easier and to simplify the process of searching for the required information about the Z notation, we have designed an integrated help system which enables users to very easily access any information about Z notation. It also explains the mathematical concepts on which Z language is based. We have always used several ways to present the same concept.

3. The Z Browser program

Z Browser program is a Microsoft Windows application. It allows the user to load several Z specifications at the same time. After a Z specification is loaded, the main window of Z Browser looks as follows:

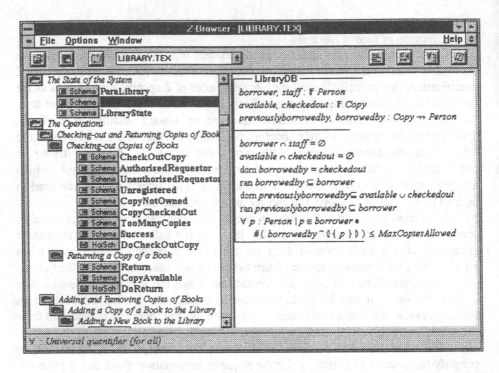

The left side contains a list of all Z paragraphs (paragraph according to Z notation, i.e. schema, generic definition, etc.) occurring in the specification. Each line representing a Z paragraph contains a graphical symbol showing the paragraph type.

A single paragraph can be displayed in the right side of the window. To achieve this, the user has to double-click on the paragraph's name in the paragraph list on the left side. The name of the selected paragraph is highlighted.

The list of paragraphs is interlaced with the headings of chapters, subchapters, sections, etc. in the source LaTeX document. After a double-click on a heading, it is either closed or opened depending on its current state. By closing, we mean hiding all Z paragraphs occurring after the heading to the next heading of the same level. Using this feature, the user can display only those parts of specification which are currently relevant.

This feature is appreciated by users browsing through large Z specifications or large LaTeX documents containing Z specifications. For example, specifications which are part of thesises, where it would accelerate searching for the desired part of the specification.

All the symbols of Z language occurring in the paragraph displayed on the right side are highlighted in a different colour. After a single click by mouse on such a symbol, its name is displayed in the status bar on the bottom of the Z Browser's main window. When the user double-clicks on a highlighted symbol, the appropriate section of Z Help (described in the next section) is displayed in the WinHelp program.

As one paragraph of Z specification usually contains references or symbols introduced in another paragraphs, users need to view these paragraphs and then return to the original paragraph. In order to display, for example, the referenced schema, the user needs just to click on its name. This causes the schema to be displayed. To return to the original place in the specification, the user has to click on the ⬚ button.

Since it is often useful to have open several paragraphs at the same time, the user can open one or more windows where a single paragraph is displayed. In order to display a paragraph in a separate window, the user has to select this paragraph and then push the ⬚ button.

4. Z Help

Z Help is hypertext help for the Z notation which can be displayed using WinHelp (winhelp.exe), the standard program for displaying help texts for Microsoft Windows applications. For each concept of the Z language, it uses a combination of text, graphics and formal Z notation in order to explain this concept. Everything is illustrated on one or several simple examples. We designed Z Help in a way that both novices and advanced users can find here explanation appropriate to their level of knowledge of Z and the mathematical background of Z.

Z Help contains four main sections:

1. *The Z Notation* (types of Z paragraphs, built-in symbols of Z language, schema operations)
2. *The Z Mathematical Tool-kit*
3. *First-order Logic*
4. *Structure of Z Specifications* (small example of the Z specification)

In addition, Z Help contains the Symbol Map, which is a section where all symbols of the Z language are displayed:

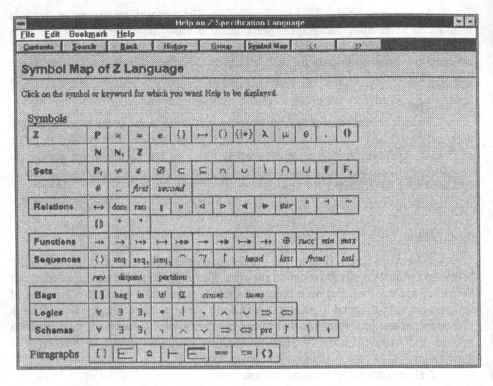

When the user clicks on any of them, the subsection explaining the selected symbol is displayed.

Beside the Symbol Map, there is also a keyword search:

In this way, Z Help can be used separately, when studying printed publications with Z specifications.

We tried to keep the structure all the Z Help sections similar to each other. So when users get used to it, they will be able to search for the required information inside the topic very easily.

The largest section is *The Z Mathematical Tool-kit*. Subsection for each operation contains its description, one ore more examples, graphical representation (about 70% of subsections) and definition using generic schema or axiomatic description.

Further on, we will describe the subsection for relational image. Its first part contains the textual description:

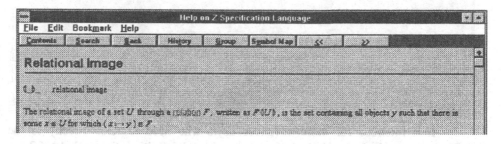

The word relation and symbols $\in$ and $\mapsto$ are dot-underlined which means that a short help for them can be displayed in a pop up window when the user clicks on them.

The next part of the subsection is the graphical representation:

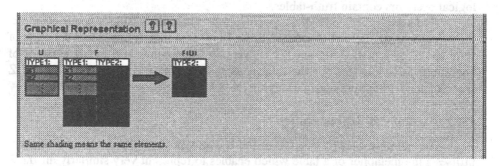

The concepts of the graphical representation are explained in the next subchapter. For the representation of types used in Z we used a metaphor of a table. The same shading of some parts of different tables indicates that they contain the same elements. The two question-mark buttons display pop up windows, one of them showing a graphical representation of relation and the second one explaining the meaning of shading. Not for all operations of the Z *Mathematical Tool-kit* we have found some clarifying graphical representation, but about 70% of subsections describing operations of the Z Mathematical Tool-kit contain also graphical representation.

Next part of this help subsection is the example:

```
Example

F = {(1 ↦ 'A'), (2 ↦ 'C'), (3 ↦ 'B')}
U = {1, 2}

F(U) = {'A', 'C'}
```

The last part of the subsection is the definition of relational image using the generic definition:

```
Definition in Z

[X, Y]
_(_) : (X ↔ Y) × (P X) → (P Y)

∀ F : X ↔ Y; U : P X •
    F(U) = {x : X, y : Y | x ∈ U ∧ (x ↦ y) ∈ F • y}
```

It is intended for more advanced users who need to refer directly to the definition of particular symbols. In both example and definition, the symbols of Z are dot-underlined so the short help text for them may be displayed in a pop up window.

Subsections in the sections *The Z Notation* and *First-order Logic* look similar to those in the *Z Mathematical Tool-kit*. They contain descriptions of the concepts of Z and examples. Subsections on Z paragraphs contain also BNFs and subsections on logical operators contain truth-tables.

The section *Structure of Z Specifications* illustrates, using a small example, the most usual structure of Z specifications. In other words, it introduces concepts of state space, invariant, Delta and Xi conventions, operations and robust versions of operations. It is intended for the total novices in Z who try to understand some Z specification without previously studying the Z language.

4.1 The Graphical Representation

The graphical representation of the Z data types and the operations on them utilises a the metaphor of a table which enables to represent very similarly all the Z data types:

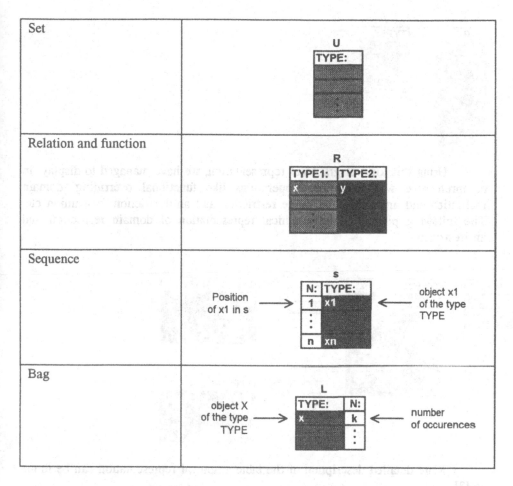

The metaphor of table is also closer to many problems specified in Z than Venn diagrams used frequently for this purpose. For example, in database systems, the user can directly imagine the effects of operations when using this graphical representation

Colour is used to express the fact that some different objects contain equal elements. So the following picture expresses the fact that the set V contains all those elements which are also contained in U, and in addition, U contains some other elements which are not included in V.

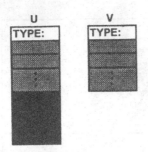

Using this kind of graphical representation, we have managed to display, in comprehensive way, even such operations like functional overriding, domain restriction and antirestriction, range restriction and antirestriction, bag union etc. The following picture is the graphical representation of domain restriction and antirestriction:

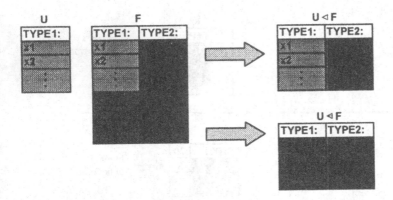

More detailed description of the table graphical representation can be found in [3].

5. Testing of Z Browser

We tested Z Browser to see how comfortable users are with this tool and so we could formulate conclusions for its further development. We also wanted to verify whether or not:

1. Z Browser improves the quality of understanding of Z specifications, and
2. The comprehension of Z specifications is faster when using Z Browser compared to the usual approach of using books for references.

With some members of staff, M.Sc. and Ph.D. students at our Department of Computer Science and Software Engineering we carried out the following tests. Each subject had to read two Z specifications and for each of them fill in the test with questions related to the specification. Subjects studied one of the specifications using

Z Browser and the other specification using books on Z or their notes from classes or seminars. We measured the number of correct answers and the time the subjects needed to fill in the tests.

One half of the subjects wrote their first test using Z Browser and the other half without Z Browser. The period between administering these two tests was at least 3 days in order to "allow the subjects to forget" what they had learned about Z in the first test.

We have chosen two rather simple Z specifications. One of them was the specification of the *phone book*. [2]. It contained one invariant schema with 2 state variables: set of persons and the relation between persons and their phone numbers - the actual phone book. It specified five operations: addition of a new person to the set of persons, addition of a new person - phone number ordered pair to the phone book, finding phone numbers for a given person, deletion of a person - phone number ordered pair from the phone book and removal of a person from the set of persons together with removal of all his/her phone numbers from the phone book.

The other Z specification was a specification of the *security system* of a building where the security system verifies IDs of employees when entering and exiting the building. This was a modified specification used at seminars by our Department. It has one invariant schema which contains 3 state variables: First one is the database of all employees and their IDs - which is a partial function from the set of employees to the set of their IDs, the second state variable is the set of IDs of employees currently outside the building and the third state variable is the set of IDs of employees currently inside the building. The operations specify addition of a new employee and deletion of employee from the database of employees, check-in and check-out from the building, and finding out the names of all employees that are currently inside the building. Each specification embodies 34 schemas and they both contain schemas defining operations and schemas defining robust versions of operations.

One half of the subjects studied at first the specification of the phone book the other half studied at first security system specification.

Both tests contained 37 questions. For every question, there were 4 possible answers from which one was correct. Of the questions, 33 were related directly to the specifications and 4 were related to the Z notation. The first three questions asked about the invariant of the specification, the actual state variables and the initial state. This required the subjects with little to no knowledge of Z to search in Z Help and find out about the basic concepts of Z specification. Next followed questions about each operation. These questions required the user to determine what are the preconditions of operations and what is the effect of the operation on the state space, what the robust versions of operations specify, what error states might occur and what are the responses in those cases. Such structure of the questions required the subjects to understand each schema in the specification. All questions used the least possible amount of Z or mathematical notation. In this way we tried to avoid the situation where by subjects might choose the "most similar looking" answers to what they see in the specifications.

The last 4 questions asked about the operations on Z data types, for example: "Function $F = \{...\}$, function $G = \{...\}$, what is the result of $F \oplus G$?"

In total, 40 subjects were tested. Among them, 3 have never studied Z and 12 passed an exam on Z during the past two years. Three of the tested subjects were members of the staff at the Department, nine were Ph.D. students and the rest were M.Sc. or B.Sc. students. The results showed that the average number of correct answers was 35.82 when using Z Browser and 34.78 without Z Browser. The time which took the subjects to fill in the tests was 26.4 minutes with Z Browser and 33.3 minutes without. This means that subjects not using Z Browser needed approximately 25% more time to fill in the tests.

The separate results for subjects starting without Z Browser (Group A) and starting with Z Browser (Group B) show improvement in both number of correct answers and time needed to fill in the tests even in Group B doing their first test without Z Browser, when the subjects had already opportunity to learn about Z.

Table 1: Average time needed for filling in the tests

	without Z Browser	with Z Browser
Group A	34.1 minutes	25.3 minutes
Group B	32.5 minutes	27.4 minutes
Together	33.3 minutes	26.4 minutes

Table 2: Average number of correct answers

	without Z Browser	with Z Browser
Group A	34.12	35.20
Group B	35.43	36.43
Together	34.78	35.82

The shorter time required to fill in the test when using Z Browser can be explained by the fact, that referring to Z Help was much easier than searching in the books or notes. The subjects used keyword search to find the sections explaining basic concepts of Z specifications like invariant, state space etc. They used mostly Symbol Map in order to quickly access the sections about the operators of the Z Mathematical Tool-kit. The similar structure of all topics enabled the subject to find required parts of sections easily.

Almost equal and quite high number of correct answers resulted from the fact that both specifications were relatively simple. The subjects spent most of the time trying to understand the Z notation in the specification.

After completing the test using Z Browser, subjects were asked which parts of Z Help explaining operations of Z Mathematical tool-kit they were using most. Here, about half used the graphical representation and the other half read examples first. When they did not understand the graphical representation or examples, they turned to the textual description. After that, sometimes they had another look at the graphical representation or examples. This confirms, that each concept should to be explained in several ways, so if users are not sure about one way of explanation, they can verify their assumptions using other ways of explanation.

6. Conclusions

We have designed, implemented and tested Z Browser, the Microsoft Windows based tool, for faster comprehension of Z specifications. Z Browser helps novices to take the first steps in studying the Z language and learn about possible areas of utilisation, by enabling them to read and comprehend Z specifications faster than when using only reference books about Z language.

The Z Browser package consists of the Microsoft Windows program and Z Help - integrated help facility for the Z notation. Help on a particular concept or symbol of Z can be retrieved simply by double-click of a mouse on the symbol in the specification. Z Browser always displays the list of Z paragraphs and, after double-click of the mouse, the whole paragraph is displayed. Properties like this let the user retrieve required information on the specification quickly and comfortably.

After Z Browser has been implemented, we carried out tests in order to find out how useful it is in studying Z specifications and in order to formulate conclusions for further development. Our tests proved that the subjects using Z Browser performed better in both quality of comprehension of Z specifications and time required for this study.

We want to continue in our research on Z Browser. Z Browser can be extended for the possibility to display schemas according to their purpose in the specifications, i.e. invariant, operations, error handling schemas and robust versions operations. This would offer the user another point of view at the specification.

Z Browser, thus far, is not able to display the prose included between Z paragraphs. The time we had for implementation of Z Browser did not allow us to implement this part. As the prose between Z paragraphs is a valuable source of information, we want to implement this in the next version.

We also want to make Z Browser open to cooperate with Z syntax checkers, where Z Browser will be a front-end for syntax checkers working from command line. Z Browser will be able to process the report files generated by syntax checkers and highlight the places in the specifications with errors.

In the tests that we have carried out, we used rather simple Z specifications. It would be interesting to do similar tests with more complex specifications.

References

1.   Spivey J. M.: *The Z Notation A Reference Manual*, Prentice Hall, 1989

2.   Diller A.: *Z an Introduction to Formal Methods*, Wiley & Sons, 1991

3. Mikušiak L., Vojtek V., Morrey I. C.: *Browser for Formal Specifications Written in Z*, 15th International Conference "Information Technology Interfaces", 1993

4. Benbasat I., Todd P.: *An experimental investigation of interface design alternatives: icon vs. text and direct manipulation vs. menus*, International Journal of Man-Machine Studies, Vol. 38, 1993

5. Berztiss A.: *Formal Specification Methods and Visualization*, v Chang: Principles of Visual Programming Systems, Prentice Hall, 1990

6. Bonar J., Liffick B.: *A Visual Programming Language for Novices*, v Chang: Principles of Visual Programming Systems, Prentice Hall, 1990

7. Cansfield D., Irby C., Kimball R., Verplank B.: *Designing Star User Interface*, Human-Computer Interaction, McGraw-Hill, 1989

8. Flynn M., Hoverd T., Braizer D.: *Formaliser - An Interactive Support Tool for Z*, Z User Workshop, Proceedings of the 4th Annual Z User Meeting, 1990

9. Heckel P.: *The Elements of Friendly Software Design*, SYBEX, 1991

10. Johnson K.: *Human-Computer Interaction*, Prentice Hall, 1992

11. Marcus A.: *Principles of Effective Visual Communication for Graphical User Interface of CASE Systems*, Conference Proceedings, NCGA '91, 12th Annual Conference and Exposition Dedicated to Computer Graphics, 1991

12. Matsura K., Tayama S.: *Visual Man-Machine Interface for Program Design and Production*, v Chang: Principles of Visual Programming Systems, Prentice Hall, 1990

13. O'Neil K., Ashoo K.: *The Z Tool*, Z User Workshop, Proceedings of the 4th Annual Z User Meeting, 1990

14. Potter B., Sinclair J., Till D.: *An Introduction to Formal Specification and Z*, Prentice Hall, 1991

15. Skemp R.: *The Psychology of Learning Mathematics*, Lawrence Erlbaum Associates, Inc., 1987

16. Tratz W.: *Computer Programming and the Human Thought Process*, Software - Practice and Experience, Vol. 19, 1979

Appendices

Select Z Bibliography

Jonathan P. Bowen *

Oxford University Computing Laboratory
Wolfson Building, Parks Road, Oxford OX1 3QD, UK
Email: Jonathan.Bowen@comlab.ox.ac.uk
URL: http://www.comlab.ox.ac.uk/oucl/people/jonathan.bowen.html

Abstract. This bibliography contains a list of Z references that are either available as published papers, books or technical reports from institutions, from the author, the Oxford University Computing Laboratory (OUCL), or on-line. The bibliography is in alphabetical order by author name(s).

Introduction

The list of references presented here is maintained in electronic form, in BIBTEX bibliography database format, which is compatible with the widely used LATEX document preparation system [297]. It is intended to keep the bibliography up to date and to issue it to coincide with the regular Z User Meetings. The latest version of BIBTEX source file used for this bibliography [53] is available as a searchable on-line database on the World Wide Web under the following Uniform Resource Location (URL):

http://www.comlab.ox.ac.uk/archive/z/bib.html

The actual BIBTEX source is also available under via anonymous FTP on the Internet ftp.comlab.ox.ac.uk:/pub/Zforum/z95.bib and via electronic mail by sending an electronic mail message containing the command 'send z z95.bib' to the address archive-server@comlab.ox.ac.uk. For more information on accessing the bibliography electronically from the Z archive, see [543].

To add new references concerned with Z to this list, please send details via electronic mail to the address zforum-request@comlab.ox.ac.uk or post them to Jonathan Bowen (address above). It is helpful if you can give as much information as possible so the entry could be included as a reference in future papers concerning Z.

This bibliography has been regularly maintained for Z User Meeting proceedings in the past (e.g., see [58]). For an alternative annotated Z bibliography, see [66], originally started as a result of work on the ZIP project.

Acknowledgements

Ruaridh Macdonald of RSRE, Malvern, initiated the idea of a Z bibliography and helped maintain it for several years. Joan Arnold at the OUCL has previously assisted in maintaining the bibliography as part of her work as secretary to the European ESPRIT ProCoS-WG Working Group (no. 8694) on 'Provably Correct Systems'. Thank you to everybody who has sent in entries over the years.

* Funded by the UK Science and Engineering Research Council (SERC) on grant no. GR/J15186.

References

1. G. D. Abowd. Agents: Communicating interactive processes. In D. Diaper, D. Gilmore, Gilbert Cockton, and Brian Shackel, editors, *Human-Computer Interaction: INTERACT'90*, pages 143–148. Elsevier Science Publishers (North-Holland), 1990.

2. G. D. Abowd. *Formal Aspects of Human-Computer Interaction*. DPhil thesis, Oxford University Computing Laboratory, Wolfson Building, Parks Road, Oxford, UK, 1991.

3. G. D. Abowd, R. Allen, and D. Garlan. Using style to understand descriptions of software architectures. *ACM Software Engineering Notes*, 18(5):9–20, December 1993.

4. G. D. Abowd, J. P. Bowen, A. J. Dix, M. D. Harrison, and R. Took. User interface languages: A survey of existing methods. Technical Report PRG-TR-5-89, Oxford University Computing Laboratory, Wolfson Building, Parks Road, Oxford, UK, October 1989.

5. J.-R. Abrial. The B tool. In Bloomfield et al. [40], pages 86–87.

6. J.-R. Abrial. The B method for large software, specification, design and coding (abstract). In Prehn and Toetenel [399], pages 398–405.

7. J.-R. Abrial. *The B-Book*. Cambridge University Press, To appear.
   Contents: Mathematical reasoning; Set notation; Mathematical objects; Introduction to abstract machines; Formal definition of abstract machines; Theory of abstract machines; Constructing large abstract machines; Example of abstract machines; Sequencing and loop; Programming examples; Refinement; Constructing large software systems; Example of refinement;
   Appendices: Summary of the most current notations; Syntax; Definitions; Visibility rules; Rules and axioms; Proof obligations.

8. J.-R. Abrial, S. A. Schuman, and B. Meyer. Specification language. In R. M. McKeag and A. M. Macnaghten, editors, *On the Construction of Programs: An Advanced Course*, pages 343–410. Cambridge University Press, 1980.

9. J.-R. Abrial and I. H. Sørensen. KWIC-index generation. In J. Staunstrup, editor, *Program Specification: Proceedings of a Workshop*, volume 134 of *Lecture Notes in Computer Science*, pages 88–95. Springer-Verlag, 1981.

10. M. Ainsworth, A. H. Cruikchank, P. J. L. Wallis, and L. J. Groves. Viewpoint specification and Z. *Information and Software Technology*, 36(1):43–51, 1994.

11. A. J. Alencar and J. A. Goguen. OOZE: An object-oriented Z environment. In P. America, editor, *Proc. ECOOP'91 European Conference on Object-Oriented Programming*, volume 512 of *Lecture Notes in Computer Science*, pages 180–199. Springer-Verlag, 1991.

12. A. J. Alencar and J. A. Goguen. Two examples in OOZE. Technical Report PRG-TR-25-91, Oxford University Computing Laboratory, Wolfson Building, Parks Road, Oxford, UK, 1991.

13. P. Ammann and J. Offutt. Functional and test specifications for the Mistix file system. Technical Report ISSE-TR-93-100, Department of Information & Software Systems Engineering, George Mason University, USA, January 1993.

14. P. Ammann and J. Offutt. Using formal methods to mechanize category-partition testing. Technical Report ISSE-TR-93-105, Department of Information & Software Systems Engineering, George Mason University, USA, September 1993.

15. D. B. Arnold, D. A. Duce, and G. J. Reynolds. An approach to the formal specification of configurable models of graphics systems. In G. Maréchal, editor, *Proc. EUROGRAPHICS'87, European Computer Graphics Conference and Exhibition*, pages 439–463. Elsevier Science Publishers (North-Holland), 1987.
   The paper describes a general framework for the formal specification of modular graphics systems, illustrated by an example taken from the Graphical Kernel System (GKS) standard.

16. D. B. Arnold and G. J. Reynolds. Configuring graphics systems components. *IEE/BCS Software Engineering Journal*, 3(6):248–256, November 1988.

17. R. D. Arthan. Formal specification of a proof tool. In Prehn and Toetenel [398], pages 356–370.

18. R. D. Arthan. On free type definitions in Z. In Nicholls [375], pages 40–58.

19. K. Ashoo. The Genesis Z tool – an overview. *BCS FACS FACTS*, Series II, 3(1):11–13, May 1992.

20. S. Aujla, A. Bryant, and L. Semmens. A rigorous review technique: Using formal notations within conventional development methods. In *Proc. 1993 Software Engineering Standards Symposium*, pages 247–255. IEEE Computer Society Press, 1993.

21. P. B. Austin, K. A. Murray, and A. J. Wellings. File system caching in large point-to-point networks. *IEE/BCS Software Engineering Journal*, 7(1):65–80, January 1992.

22. S. Austin and G. I. Parkin. Formal methods: A survey. Technical report, National Physical Laboratory, Queens Road, Teddington, Middlesex, TW11 0LW, UK, March 1993.

23. C. Bailes and R. Duke. The ecology of class refinement. In Morris and Shaw [359], pages 185–196.

24. M. Bailey. Formal specification using Z. In *Proc. Software Engineering anniversary meeting (SEAS)*, page 99, 1987.

25. J. Bainbridge, R. W. Whitty, and J. B. Wordsworth. Obtaining structural metrics of Z specifications for systems development. In Nicholls [373], pages 269–281.

26. J.-P. Banâtre. About programming environments. In J.-P. Banâtre, S. B. Jones, and D. de Métayer, editors, *Prospects for Functional Programming in Software Engineering*, volume 1 of *Research Reports*, chapter 1, pages 1–22. Springer-Verlag, 1991.

27. R. Barden and S. Stepney. Support for using Z. In Bowen and Nicholls [75], pages 255–280.

28. R. Barden, S. Stepney, and D. Cooper. The use of Z. In Nicholls [375], pages 99–124.

29. R. Barden, S. Stepney, and D. Cooper. *Z in Practice*. BCS Practitioner Series. Prentice Hall, 1994.

30. G. Barrett. Formal methods applied to a floating-point number system. *IEEE Transactions on Software Engineering*, 15(5):611–621, May 1989.

    This paper presents a formalization of the IEEE standard for binary floating-point arithmetic in Z. The formal specification is refined into four components. The procedures presented form the basis for the floating-point unit of the Inmos IMS T800 transputer. This work resulted in a joint UK Queen's Award for Technological Achievement for Inmos Ltd and the Oxford University Computing Laboratory in 1990. It was estimated that the approach saved a year in development time compared to traditional methods.

31. L. M. Barroca and J. A. McDermid. Formal methods: Use and relevance for the development of safety-critical systems. *The Computer Journal*, 35(6):579–599, December 1992.

32. P. Baumann. Z and natural semantics. In Bowen and Hall [67], pages 168–184.

33. M. Benjamin. A message passing system: An example of combining CSP and Z. In Nicholls [371], pages 221–228.

34. M. Benveniste. Operational semantics of a distributed object-oriented language and its Z formal specification. Rapport de recherche INRIA 1230, IRISA/INRIA-Rennes, Campus de Beaulieu, 35042 Rennes Cédex, France, May 1990.

35. M. Benveniste. Writing operational semantics in Z: A structural approach. In Prehn and Toetenel [398], pages 164–188.

36. S. Bera. Structuring for the VDM specification language. In Bloomfield et al. [40], pages 2–25.

37. J. Bicarregui and B. Ritchie. Invariants, frames and postconditions: A comparison of the VDM and B notations. *IEEE Transactions on Software Engineering*, 21(2):79–89, 1995.

38. P. G. Bishop, editor. *Fault Avoidance*, chapter 3, pages 56–140. Applied Science. Elsevier Science Publishers, 1990.
Section 3.88 (pages 94–96) provides an overview of Z. Other sections describe related techniques.

39. D. Bjørner, C. A. R. Hoare, and H. Langmaack, editors. *VDM and Z – Formal Methods in Software Development*, volume 428 of *Lecture Notes in Computer Science*. VDM-Europe, Springer-Verlag, 1990.
The 3rd VDM-Europe Symposium was held at Kiel, Germany, 17–21 April 1990. A significant number of papers concerned with Z were presented [105, 161, 191, 145, 203, 209, 290, 424, 458, 500, 530].

40. R. Bloomfield, L. Marshall, and R. Jones, editors. *VDM – The Way Ahead*, volume 328 of *Lecture Notes in Computer Science*. VDM-Europe, Springer-Verlag, 1988.
The 2nd VDM-Europe Symposium was held at Dublin, Ireland, 11–16 September 1988. See [5, 36].

41. D. Blyth. The CICS application programming interface: Temporary storage. IBM Technical Report TR12.301, IBM United Kingdom Laboratories Ltd, Hursley Park, Winchester, Hampshire SO21 2JN, UK, December 1990.
One of a number of reports on the CICS application programming interface. See also [247, 289, 360].

42. A. Boswell. Specification and validation of a security policy model. In Woodcock and Larsen [527], pages 42–51.
Revised version in [43].

43. A. Boswell. Specification and validation of a security policy model. *IEEE Transactions on Software Engineering*, 21(2):99–106, 1995.
This paper describes the development of a formal security model in Z for the NATO Air Command and Control System (ACCS): a large, distributed, multilevel-secure system. The model was subject to manual validation, and some of the issues and lessons in both writing and validating the model are discussed.

44. L. Bottaci and J. Jones. *Formal Specification Using Z: A Modelling Approach*. International Thomson Publishing, London, 1995.

45. J. P. Bowen. Formal specification and documentation of microprocessor instruction sets. *Microprocessing and Microprogramming*, 21(1–5):223–230, August 1987.

46. J. P. Bowen. The formal specification of a microprocessor instruction set. Technical Monograph PRG-60, Oxford University Computing Laboratory, Wolfson Building, Parks Road, Oxford, UK, January 1987.
The Z notation is used to define the Motorola M6800 8-bit microprocessor instruction set.

47. J. P. Bowen, editor. *Proc. Z Users Meeting, 1 Wellington Square, Oxford*, Wolfson Building, Parks Road, Oxford, UK, December 1987. Oxford University Computing Laboratory.
The 1987 Z Users Meeting was held on Friday 8 December at the Department of External Studies, Rewley House, 1 Wellington Square, Oxford, UK.

48. J. P. Bowen. Formal specification in Z as a design and documentation tool. In *Proc. Second IEE/BCS Conference on Software Engineering*, number 290 in Conference Publication, pages 164–168. IEE/BCS, July 1988.

49. J. P. Bowen, editor. *Proc. Third Annual Z Users Meeting*, Wolfson Building, Parks Road, Oxford, UK, December 1988. Oxford University Computing Laboratory.
The 1988 Z Users Meeting was held on Friday 16 December at the Department of External Studies, Rewley House, 1 Wellington Square, Oxford, UK. Issued with *A Miscellany of Handy Techniques* by R. Macdonald, RSRE, *Practical Experience of Formal Specification: A programming interface for communications* by J. B. Wordsworth, IBM, and a number of posters.

50. J. P. Bowen. Formal specification of window systems. Technical Monograph PRG-74, Oxford University Computing Laboratory, Wolfson Building, Parks Road, Oxford, UK, June 1989.
   Three existing window systems, X from MIT, WM from Carnegie-Mellon University and the Blit from AT&T Bell Laboratories are covered.

51. J. P. Bowen. POS: Formal specification of a UNIX tool. *IEE/BCS Software Engineering Journal*, 4(1):67–72, January 1989.

52. J. P. Bowen. Formal specification of the ProCoS/safemos instruction set. *Microprocessors and Microsystems*, 14(10):631–643, December 1990.
   This article is part of a special issue on *Formal aspects of microprocessor design*, edited by H. S. M. Zedan. See also [437].

53. J. P. Bowen. Z bibliography. Oxford University Computing Laboratory, 1990–1993.
   This bibliography is maintained in BIBTEX database format at the Oxford University Computing Laboratory. To add entries, please send as complete information as possible to zforum-request@comlab.ox.ac.uk.

54. J. P. Bowen. X: Why Z? *Computer Graphics Forum*, 11(4):221–234, October 1992.
   This paper asks whether window management systems would not be better specified through a formal methodology and gives examples in Z of X11.

55. J. P. Bowen. Formal methods in safety-critical standards. In *Proc. 1993 Software Engineering Standards Symposium*, pages 168–177. IEEE Computer Society Press, 1993.

56. J. P. Bowen. Report on Z User Meeting, London 1992. *BCS FACS FACTS*, Series III, 1(3):7–8, Summer 1993.
   Other versions of this report have appeared as follows:
   – Z User Meetings, *Safety Systems: The Safety-Critical Systems Club Newsletter*, 3(1):13, September 1993.
   – Z User Group activities, *JFIT News*, 46:5, September 1993.
   – Report on Z User Meeting, *Information and Software Technology*, 35(10):613, October 1993.
   – Z User Meeting Activities, *High Integrity Systems*, 1(1):93–94, 1994.

57. J. P. Bowen. Comp.specification.z and Z FORUM frequently asked questions. In Bowen and Hall [67], pages 397–404.

58. J. P. Bowen. Select Z bibliography. In Bowen and Hall [67], pages 359–396.

59. J. P. Bowen. *Formal Specification and Documentation using Z: A Case Study Approach*. International Thomson Press, London, 1995.

60. J. P. Bowen. Z glossary. *Information and Software Technology*, 37(5):333–334, May 1995.

61. J. P. Bowen, P. T. Breuer, and K. C. Lano. A compendium of formal techniques for software maintenance. *IEE/BCS Software Engineering Journal*, 8(5):253–262, September 1993.

62. J. P. Bowen, P. T. Breuer, and K. C. Lano. Formal specifications in software maintenance: From code to Z^{++} and back again. *Information and Software Technology*, 35(11/12):679–690, November/December 1993.

63. J. P. Bowen, R. B. Gimson, and S. Topp-Jørgensen. The specification of network services. Technical Monograph PRG-61, Oxford University Computing Laboratory, Wolfson Building, Parks Road, Oxford, UK, August 1987.

64. J. P. Bowen, R. B. Gimson, and S. Topp-Jørgensen. Specifying system implementations in Z. Technical Monograph PRG-63, Oxford University Computing Laboratory, Wolfson Building, Parks Road, Oxford, UK, February 1988.

65. J. P. Bowen and M. J. C. Gordon. Z and HOL. In Bowen and Hall [67], pages 141–167.

66. J. P. Bowen and M. J. C. Gordon. A shallow embedding of Z in HOL. *Information and Software Technology*, 37(5):269–276, May 1995.
   Revised version of [65].

67. J. P. Bowen and J. A. Hall, editors. *Z User Workshop, Cambridge 1994*, Workshops in Computing. Springer-Verlag, 1994.
    Proceedings of the Eigth Annual Z User Meeting, St. John's College, Cambridge, UK. Published in collaboration with the British Computer Society. For individual papers, see [32, 65, 58, 57, 81, 103, 106, 151, 172, 174, 189, 210, 211, 215, 222, 298, 329, 393, 446, 504, 526, 531]. The proceedings also includes an *Introduction and Opening Remarks*, a *Select Z Bibliography* [58] and a section answering *Frequently Asked Questions* [57].

68. J. P. Bowen and M. G. Hinchey. Formal methods and safety-critical standards. *IEEE Computer*, 27(8):68–71, August 1994.

69. J. P. Bowen and M. G. Hinchey. Seven more myths of formal methods: Dispelling industrial prejudices. In Naftalin et al. [361], pages 105–117.

70. J. P. Bowen and M. G. Hinchey. Editorial. *Information and Software Technology*, 37(5):258–259, May 1995.
    A special issue on Z. See [60, 71, 66, 78, 190, 310, 326, 331, 497].

71. J. P. Bowen and M. G. Hinchey. Report on Z User Meeting (ZUM'94). *Information and Software Technology*, 37(5):335–336, May 1995.

72. J. P. Bowen and M. G. Hinchey. Seven more myths of formal methods. *IEEE Software*, 12(4), July 1995.
    This article deals with further myths in addition to those presented in [208]. Previous versions issued as:
    – Technical Report PRG-TR-7-94, Oxford University Computing Laboratory, June 1994.
    – Technical Report 357, University of Cambridge, Computer Laboratory, January 1995.

73. J. P. Bowen and M. G. Hinchey. Ten commandments of formal methods. *IEEE Computer*, 28(4):56–63, April 1995.
    Previously issued as: Technical Report 350, University of Cambridge, Computer Laboratory, September 1994.

74. J. P. Bowen and M. G. Hinchey, editors. *ZUM'95 – 9th International Conference of Z Users, Limerick 1995*, Lecture Notes in Computer Science. Springer-Verlag, 1995.

75. J. P. Bowen and J. E. Nicholls, editors. *Z User Workshop, London 1992*, Workshops in Computing. Springer-Verlag, 1993.
    Proceedings of the Seventh Annual Z User Meeting, DTI Offices, London, UK. Published in collaboration with the British Computer Society. For individual papers, see [27, 79, 118, 124, 134, 132, 155, 238, 261, 293, 307, 319, 332, 376, 382, 400, 420, 482, 499]. The proceedings also includes an *Introduction and Opening Remarks*, a *Select Z Bibliography* and a section answering *Frequently Asked Questions*.

76. J. P. Bowen and V. Stavridou. The industrial take-up of formal methods in safety-critical and other areas: A perspective. In Woodcock and Larsen [527], pages 183–195.

77. J. P. Bowen and V. Stavridou. Safety-critical systems, formal methods and standards. *IEE/BCS Software Engineering Journal*, 8(4):189–209, July 1993.
    A survey on the use of formal methods, including B and Z, for safety-critical systems. Winner of the 1994 IEE Charles Babbage Premium award. A previous version is also available as Oxford University Computing Laboratory Technical Report PRG-TR-5-92.

78. J. P. Bowen, S. Stepney, and R. Barden. Annotated Z bibliography. *Information and Software Technology*, 37(5):317–332, May 1995.
    Revised version of [461].

79. A. Bradley. Requirements for Defence Standard 00-55. In Bowen and Nicholls [75], pages 93–94.

80. P. T. Breuer. Z! in progress: Maintaining Z specifications. In Nicholls [373], pages 295–318.

81. P. T. Breuer and J. P. Bowen. Towards correct executable semantics for Z. In Bowen and Hall [67], pages 185–209.

82. S. M. Brien. The development of Z. In D. J. Andrews, J. F. Groote, and C. A. Middelburg, editors, *Semantics of Specification Languages (SoSL)*, Workshops in Computing, pages 1–14. Springer-Verlag, 1994.

83. S. M. Brien and J. E. Nicholls. Z base standard. Technical Monograph PRG-107, Oxford University Computing Laboratory, Wolfson Building, Parks Road, Oxford, UK, November 1992. Accepted for standardization under ISO/IEC JTC1/SC22.
This is the first publicly available version of the proposed ISO Z Standard. See also [457] for the current most widely available Z reference manual.

84. C. Britton, M. Loomes, and R. Mitchell. Formal specification as constructive diagrams. *Microprocessing and Microprogramming*, 37(1–5):175–178, January 1993.

85. D. J. Brown and J. P. Bowen. The Event Queue: An extensible input system for UNIX workstations. In *Proc. European Unix Users Group Conference*, pages 29–52. EUUG, May 1987.
Available from EUUG Secretariat, Owles Hall, Buntingford, Hertfordshire SG9 9PL, UK.

86. D. Brownbridge. Using Z to develop a CASE toolset. In Nicholls [371], pages 142–149.

87. A. Bryant. Structured methodologies and formal notations: Developing a framework for synthesis and investigation. In Nicholls [371], pages 229–241.

88. G. R. Buckberry. ZED: A Z notation editor and syntax analyser. *BCS FACS FACTS*, Series II, 2(3):13–23, November 1991.

89. A. Burns and I. W. Morrison. A formal description of the structure attribute model for tool interfacing. *IEE/BCS Software Engineering Journal*, 4(2):74–78, March 1989.

90. A. Burns and A. J. Wellings. Occam's priority model and deadline scheduling. In *Proc. 7th Occam User Group Meeting, Grenoble*, 1987.

91. A. Burns and A. J. Wellings. A formal description of Ada tasking in Z. Computer Science Report YCS122, University of York, Heslington, York YO1 5DD, UK, 1989.

92. A. Burns and A. J. Wellings. Priority inheritance and message passing communication: A formal treatment. Computer Science Report YCS116, University of York, Heslington, York YO1 5DD, UK, 1989.

93. J. S. Busby and D. Hutchison. The practical integration of manufacturing applications. *Software Practice and Experience*, 22(2):183–207, 1992.

94. P. Butcher. A behavioural semantics for Linda-2. *IEE/BCS Software Engineering Journal*, 6(4):196–204, July 1991.

95. M. J. Butler. Service extension at the specification level. In Nicholls [373], pages 319–336.

96. M. J. Butler. Feature interaction analysis using Z. Åbo Akademi University, Finland, 1994.

97. M. J. Butler. Z specification of the X400 reliable transfer service. Åbo Akademi University, Finland, 1994.

98. D. Carrington. ZOOM workshop report. In Nicholls [375], pages 352–364.
This paper records the activities of a workshop on Z and object-oriented methods held in August 1992 at Oxford. A comprehensive bibliography is included.

99. D. Carrington, D. J. Duke, R. Duke, P. King, G. A. Rose, and G. Smith. Object-Z: An object-oriented extension to Z. In S. Vuong, editor, *Formal Description Techniques, II (FORTE'89)*, pages 281–296. Elsevier Science Publishers (North-Holland), 1990.

100. D. Carrington, D. J. Duke, I. J. Hayes, and J. Welsh. Deriving modular designs from formal specifications. *ACM Software Engineering Notes*, 18(5):89–98, December 1993.

101. D. Carrington, D. J. Duke, I. J. Hayes, and J. Welsh. Deriving modular designs from formal specifications: The analysis phase. Technical Report 93-13, Department of Computer Science, University of Queensland, St. Lucia 4072, Australia, October 1993.

102. D. Carrington and G. Smith. Extending Z for object-oriented specifications. In *Proc. 5th Australian Software Engineering Conference (ASWEC'90)*, pages 9–14, 1990.

103. D. Carrington and P. Stocks. A tale of two paradigms: Formal methods and software testing. In Bowen and Hall [67], pages 51–68.
Also available as Technical Report 94-4, Department of Computer Science, University of Queensland, 1994.

104. D. Carrington and P. Stocks. A tale of two paradigms: Formal methods and software testing. Technical Report 94-4, Department of Computer Science, University of Queensland, St. Lucia 4072, Australia, February 1994.

105. P. Chalin and P. Grogono. Z specification of an object manager. In Bjørner et al. [39], pages 41–71.

106. D. K. C. Chan and P. W. Trinder. An object-oriented data model supporting multi-methods, multiple inheritance, and static type checking: A specification in Z. In Bowen and Hall [67], pages 297–315.

107. W. Chantatub and M. Holcombe. Software testing strategies for software requirements and design. In Proc. EuroSTAR'94, pages 40/1–40/29, 3000-2 Hartley Road, Jacksonville, Florida 32257, USA, 1994. Software Quality Engineering.
The paper describes how to construct a detailed Z specification using traditional software engineering techniques (ERDs, DFDs, etc.) in a top down manner. It introduces a number of notational devices to help with the management of large Z specifications. Some issues about proving consistency between levels are also addressed.

108. A. S. K. Cheng, J. Han, J. Welsh, and A. Wood. Providing user-oriented support for software development by formal methods. Technical Report 92-8, Department of Computer Science, University of Queensland, St. Lucia 4072, Australia, December 1992.

109. S. J. Clarke, A. C. Combes, and J. A. McDermid. The analysis of safety arguments in the specification of a motor speed control loop. Computer Science Report YCS136, University of York, Heslington, York YO1 5DD, UK, 1990.
This report describes some timing extensions to Z.

110. B. Cohen. Justification of formal methods for system specifications & A rejustification of formal notations. IEE/BCS Software Engineering Journal, 4(1):26–38, January 1989.

111. B. Cohen and D. Mannering. The rigorous specification and verification of the safety aspects of a real-time system. In COMPASS '90, 1990.

112. B. P. Collins, J. E. Nicholls, and I. H. Sørensen. Introducing formal methods: The CICS experience with Z. In B. Neumann et al., editors, Mathematical Structures for Software Engineering. Oxford University Press, 1991.

113. J. Cooke. Editorial – formal methods: What? why? and when? The Computer Journal, 35(5):417–418, October 1992.
An editorial introduction to two special issues on Formal Methods. See also [31, 114, 344, 431, 524] for papers relevant to Z.

114. J. Cooke. Formal methods – mathematics, theory, recipes or what? The Computer Journal, 35(5):419–423, October 1992.

115. A. C. Coombes, L. Barroca, J. S. Fitzgerald, J. A. McDermid, L. Spencer, and A. Saeed. Formal specification of an aerospace system: The attitude monitor. In Hinchey and Bowen [245], pages 307–332.

116. A. C. Coombes and J. A. McDermid. A tool for defining the architecture of Z specifications. In Nicholls [373], pages 77–92.

117. A. C. Coombes and J. A. McDermid. Specifying temporal requirements for distributed real-time systems in Z. Computer Science Report YCS176, University of York, Heslington, York YO1 5DD, UK, 1992.

118. A. C. Coombes and J. A. McDermid. Using diagrams to give a formal specification of timing constraints in Z. In Bowen and Nicholls [75], pages 119–130.

119. D. Cooper. Educating management in Z. In Nicholls [371], pages 192–194.

120. V. A. O. Cordeiro, A. C. A. Sampaio, and S. L. Meira. From MooZ to Eiffel – a rigorous approach to system development. In Naftalin et al. [361], pages 306–325.

121. S. Craggs and J. B. Wordsworth. Hursley Lab wins another Queen's Award & Hursley and Oxford – a marriage of minds & Z stands for quality. *Developments, IBM Hursley Park*, 8:1–2, 21 April 1992.

122. I. Craig. *The Formal Specification of Advanced AI Architectures*. AI Series. Ellis Horwood, September 1991.
    This book contains two rather large (and relatively complete) specifications of Artificial Intelligence (AI) systems using Z. The architectures are the blackboard and Cassandra architectures. As well as showing that formal specification *can* be used in AI at the architecture level, the book is intended as a case-studies book, and also contains introductory material on Z (for AI people). The book assumes a knowledge of Z, so for non-AI people its primary use is for the presentation of the large specifications. The blackboard specification, with explanatory text, is around 100 pages.

123. D. Craigen, S. Gerhart, and T. Ralston. Formal methods technology transfer: Impediments and innovation. In Hinchey and Bowen [245], pages 399–419.

124. D. Craigen, S. L. Gerhart, and T. Ralston. An international survey of industrial applications of formal methods. In Bowen and Nicholls [75], pages 1–5.

125. D. Craigen, S. L. Gerhart, and T. J. Ralston. Formal methods reality check: Industrial usage. In Woodcock and Larsen [527], pages 250–267.
    Revised version in [127].

126. D. Craigen, S. L. Gerhart, and T. J. Ralston. An international survey of industrial applications of formal methods. Technical Report NIST GCR 93/626-V1 & 2, Atomic Energy Control Board of Canada, US National Institute of Standards and Technology, and US Naval Research Laboratories, 1993.
    Volume 1: Purpose, Approach, Analysis and Conclusions; Volume 2: Case Studies. Order numbers: PB93-178556/AS & PB93-178564/AS; National Technical Information Service, 5285 Port Royal Road, Springfield, VA 22161, USA.

127. D. Craigen, S. L. Gerhart, and T. J. Ralston. Formal methods reality check: Industrial usage. *IEEE Transactions on Software Engineering*, 21(2):90–98, 1995.
    Revised version of [262].

128. D. Craigen, S. Kromodimoeljo, I. Meisels, W. Pase, and M. Saaltink. EVES: An overview. In Prehn and Toetenel [398], pages 389–405.

129. S. Croxall, P. Lupton, and J. B. Wordsworth. A formal specification of the CPI communications. IBM Technical Report TR12.277, IBM United Kingdom Laboratories Ltd, Hursley Park, Winchester, Hampshire SO21 2JN, UK, 1990.

130. E. Cusack. Inheritance in object oriented Z. In P. America, editor, *Proc. ECOOP'91 European Conference on Object-Oriented Programming*, volume 512 of *Lecture Notes in Computer Science*, pages 167–179. Springer-Verlag, 1991.

131. E. Cusack. Object oriented modelling in Z for open distributed systems. In J. de Meer, editor, *Proc. International Workshop on ODP*. Elsevier Science Publishers (North-Holland), 1992.

132. E. Cusack. Using Z in communications engineering. In Bowen and Nicholls [75], pages 196–202.

133. E. Cusack and M. Lai. Object oriented specification in LOTOS and Z (or my cat really is object oriented!). In J. W. de Bakker, W. P. de Roever, and G. Rozenberg, editors, *REX/FOOL School/Workshop on Foundations of Object-Oriented Languages*, volume 489 of *Lecture Notes in Computer Science*, pages 179–202. Springer-Verlag, 1990.

134. E. Cusack and C. Wezeman. Deriving tests for objects specified in Z. In Bowen and Nicholls [75], pages 180–195.

135. R. S. M. de Barros. Formal specification of relational database applications: A method and an example. Research report GE-93-02, Department of Computing Science, University of Glasgow, UK, September 1993.

136. R. S. M. de Barros. Deriving relational database programs from formal specifications. In Naftalin et al. [361], pages 703–723.

137. R. S. M. de Barros. On the derivation of relational database programs from formal specifications. Research report GE-94-01, Department of Computing Science, University of Glasgow, UK, July 1994.
      An extended version of [136].

138. R. S. M. de Barros and D. J. Harper. Formal development of relational database applications. In D. J. Harper and M. C. Norrie, editors, *Specifications of Database Systems, Glasgow 1991*, Workshops in Computing, pages 21–43. Springer-Verlag, 1992.
      Zc, a Z-like formalism, is used.

139. R. S. M. de Barros and D. J. Harper. A method for the specification of relational database applications. In Nicholls [375], pages 261–286.

140. A. M. L. de Vasconcelos and J. A. McDermid. Incremental type-checking in Z. Computer Science Report YCS185, University of York, Heslington, York YO1 5DD, UK, 1992.

141. C. N. Dean and M. G. Hinchey. Introducing formal methods through rôle-playing. *ACM SIGCSE Bulletin*, 27(1):302–306, March 1995.

142. B. Dehbonei and F. Mejia. Formal methods in the railways signalling industry. In Naftalin et al. [361], pages 26–34.

143. N. Delisle and D. Garlan. Formally specifying electronic instruments. In *Proc. Fifth International Workshop on Software Specification and Design*. IEEE Computer Society, May 1989. Also published in ACM SIGSOFT Software Engineering Notes 14(3).

144. N. Delisle and D. Garlan. A formal specification of an oscilloscope. *IEEE Software*, 7(5):29–36, September 1990.
      Unlike most work on the application of formal methods, this research uses formal methods to gain insight into system architecture. The context for this case study is electronic instrument design.

145. R. Di Giovanni and P. L. Iachini. HOOD and Z for the development of complex systems. In Bjørner et al. [39], pages 262–289.

146. A. J. J. Dick, P. J. Krause, and J. Cozens. Computer aided transformation of Z into Prolog. In Nicholls [371], pages 71–85.

147. A. Diller. Specifying interactive programs in Z. Research Report CSR-90-13, School of Computer Science, University of Birmingham, Birmingham B15 2TT, UK, August 1990.

148. A. Diller. *Z: An Introduction to Formal Methods*. John Wiley & Sons, 1990.
      This book offers a comprehensive tutorial to Z from the practical viewpoint. Many natural deduction style proofs are presented and exercises are included. A second edition is now available [150].

149. A. Diller. Z and Hoare logics. In Nicholls [375], pages 59–76.

150. A. Diller. *Z: An Introduction to Formal Methods*. John Wiley & Sons, Chichester, UK, 2nd edition, 1994.
      This book offers a comprehensive tutorial to Z from the practical viewpoint. Many natural deduction style proofs are presented and exercises are included. Z as defined in the 2nd edition of *The Z Notation* [457] is used throughout.
      Contents: Tutorial introduction; Methods of reasoning; Case studies; Specification animation; Reference manual; Answers to exercises; Glossaries of terms and symbols; Bibliography.

151. A. Diller and R. Docherty. Z and abstract machine notation: A comparison. In Bowen and Hall [67], pages 250–263.

152. A. J. Dix. *Formal Methods for Interactive Systems*. Computers and People Series. Academic Press, 1991.

153. A. J. Dix, J. Finlay, G. D. Abowd, and R. Beale. *Human-Computer Interaction*. Prentice Hall International, 1993.

154. V. Doma and R. Nicholl. EZ: A system for automatic prototyping of Z specifications. In Prehn and Toetenel [398], pages 189–203.

155. C. Draper. Practical experiences of Z and SSADM. In Bowen and Nicholls [75], pages 240–251.

156. D. A. Duce, D. J. Duke, P. J. W. ten Hagen, and G. J. Reynolds. PREMO - an initial approach to a formal definition. *Computer Graphics Forum*, 13(3):C–393–C–406, 1994.
PREMO (Presentation Environments for Multimedia Objects) is a work item proposal by the ISO/IEC JTC11/SC24 committee, which is responsible for international standardization in the area of computer graphics and image processing.

157. D. J. Duke. Structuring Z specifications. In *Proc. 14th Australian Computer Science Conference*, 1991.

158. D. J. Duke. Enhancing the structures of Z specifications. In Nicholls [375], pages 329–351.

159. D. J. Duke. *Object-Oriented Formal Specification*. PhD thesis, Department of Computer Science, University of Queensland, St. Lucia 4072, Australia, 1992.

160. D. J. Duke and R. Duke. A history model for classes in Object-Z. Technical Report 120, Department of Computer Science, University of Queensland, St. Lucia 4072, Australia, 1989.

161. D. J. Duke and R. Duke. Towards a semantics for Object-Z. In Bjørner et al. [39], pages 244–261.

162. D. J. Duke and M. D. Harrison. Event model of human-system interaction. *IEE/BCS Software Engineering Journal*, 10(1):3–12, January 1995.

163. D. J. Duke and M. D. Harrison. Mapping user requirements to implementations. *IEE/BCS Software Engineering Journal*, 10(1):13–20, January 1995.

164. R. Duke and D. J. Duke. Aspects of object-oriented formal specification. In *Proc. 5th Australian Software Engineering Conference (ASWEC'90)*, pages 21–26, 1990.

165. R. Duke, I. J. Hayes, P. King, and G. A. Rose. Protocol specification and verification using Z. In S. Aggarwal and K. Sabnani, editors, *Protocol Specification, Testing, and Verification VIII*, pages 33–46. Elsevier Science Publishers (North-Holland), 1988.

166. R. Duke, P. King, G. A. Rose, and G. Smith. The Object-Z specification language. In T. Korson, V. Vaishnavi, and B. Meyer, editors, *Technology of Object-Oriented Languages and Systems: TOOLS 5*, pages 465–483. Prentice Hall, 1991.

167. R. Duke, P. King, G. A. Rose, and G. Smith. The Object-Z specification language: Version 1. Technical Report 91-1, Department of Computer Science, University of Queensland, St. Lucia 4072, Australia, April 1991.
The most complete (and currently the standard) reference on Object-Z. It has been reprinted by ISO JTC1 WG7 as document number 372. A condensed version of this report was published as [166].

168. R. Duke and G. A. Rose. A complete Z specification of an interactive program editor. Technical Report 71, Department of Computer Science, University of Queensland, St. Lucia 4072, Australia, 1986.

169. R. Duke, G. A. Rose, and A. Lee. Object-oriented protocol specification. In L. Logrippo, R. L. Probert, and H. Ural, editors, *Protocol Specification, Testing, and Verification X*, pages 325–338. Elsevier Science Publishers (North-Holland), 1990.

170. R. Duke and G. Smith. Temporal logic and Z specifications. *Australian Computer Journal*, 21(2):62–69, May 1989.

171. D. Edmond. *Information Modeling: Specification and Implementation*. Prentice Hall, 1992.

172. M. Engel. Specifying real-time systems with Z and the Duration Calculus. In Bowen and Hall [67], pages 282–294.

173. A. S. Evans. Specifying & verifying concurrent systems using Z. In Naftalin et al. [361], pages 366–400.

174. A. S. Evans. Visualising concurrent Z specifications. In Bowen and Hall [67], pages 269–281.

175. P. C. Fencott, A. J. Galloway, M. A. Lockyer, S. J. O'Brien, and S. Pearson. Formalising the semantics of Ward/Mellor SA/RT essential models using a process algebra. In Naftalin et al. [361], pages 681–702.

176. N. E. Fenton and D. Mole. A note on the use of Z for flowgraph transformation. *Information and Software Technology*, 30(7):432–437, 1988.

177. E. Fergus and D. C. Ince. Z specifications and modal logic. In P. A. V. Hall, editor, *Proc. Software Engineering 90*, volume 1 of *British Computer Society Conference Series*. Cambridge University Press, 1990.

178. C. J. Fidge. Specification and verification of real-time behaviour using Z and RTL. In J. Vytopil, editor, *Formal Techniques in Real-Time and Fault-Tolerant Systems*, Lecture Notes in Computer Science, pages 393–410. Springer-Verlag, 1992.

179. C. J. Fidge. Real-time refinement. In Woodcock and Larsen [527], pages 314–331.

180. C. J. Fidge. Adding real time to formal program development. In Naftalin et al. [361], pages 618–638.

181. C. J. Fidge. Proof obligations for real-time refinement. In Till [490], pages 279–305.

182. C. J. Fidge, P. Kearney, and J. Staples. Formally verified real-time software: an integrated development strategy (extended version). Technical Report 93-10, Department of Computer Science, University of Queensland, St. Lucia 4072, Australia, June 1993.

183. M. Flynn, T. Hoverd, and D. Brazier. Formaliser – an interactive support tool for Z. In Nicholls [371], pages 128–141.

184. I. Fogg, B. Hicks, A. Lister, T. Mansfield, and K. Raymond. A comparison of LOTOS and Z for specifying distributed systems. *Australian Computer Science Communications*, 12(1):88–96, February 1990.

185. D. C. Fowler, P. A. Swatman, and P. M. C. Swatman. Implementing EDI in the public sector: Including formality for enhanced control. In *Proc. 7th International Conference on Electronic Data Interchange*, June 1993.

186. N. E. Fuchs. Specifications are (preferably) executable. *IEE/BCS Software Engineering Journal*, 7(5):323–334, September 1992.

187. P. H. B. Gardiner, P. J. Lupton, and J. C. P. Woodcock. A simpler semantics for Z. In Nicholls [373], pages 3–11.

188. D. Garlan. The role of reusable frameworks. *ACM SIGSOFT Software Engineering Notes*, 15(4):42–44, September 1990.

189. D. Garlan. Integrating formal methods into a professional master of software engineering program. In Bowen and Hall [67], pages 71–85.

190. D. Garlan. Making formal methods effective for professional software engineers. *Information and Software Technology*, 37(5):261–268, May 1995. Revised version of [189].

191. D. Garlan and N. Delisle. Formal specifications as reusable frameworks. In Bjørner et al. [39], pages 150–163.

192. D. Garlan and N. Delisle. Formal specification of an architecture for a family of instrumentation systems. In Hinchey and Bowen [245], pages 55–72.

193. D. Garlan and D. Notkin. Formalizing design spaces: Implicit invocation mechanisms. In Prehn and Toetenel [398], pages 31–45.

194. S. L. Gerhart. Applications of formal methods: Developing virtuoso software. *IEEE Software*, 7(5):6–10, September 1990.
This is an introduction to a special issue on Formal Methods with an emphasis on Z in particular. It was published in conjunction with special Formal Methods issues of *IEEE Transactions on Software Engineering* and *IEEE Computer*. See also [144, 208, 362, 455, 508].

195. S. L. Gerhart, D. Craigen, and T. Ralston. Observations on industrial practice using formal methods. In *Proc. 15th International Conference on Software Engineering (ICSE), Baltimore, Maryland, USA*, May 1993.

196. S. L. Gerhart, D. Craigen, and T. Ralston. Experience with formal methods in critical systems. *IEEE Software*, 11(1):21–28, January 1994.
Several commercial and exploratory cases in which Z features heavily are briefly presented on page 24. See also [294].

197. S. Gilmore. Correctness-oriented approaches to software development. Technical Report ECS-LFCS-91-147 (also CST-76-91), Department of Computer Science, University of Edinburgh, Edinburgh EH9 3JZ, UK, 1991.
This PhD thesis provides a critical evaluation of Z, VDM and algebraic specifications.

198. R. B. Gimson. The formal documentation of a Block Storage Service. Technical Monograph PRG-62, Oxford University Computing Laboratory, Wolfson Building, Parks Road, Oxford, UK, August 1987.

199. R. B. Gimson and C. C. Morgan. Ease of use through proper specification. In D. A. Duce, editor, *Distributed Computing Systems Programme*. Peter Peregrinus, London, 1984.

200. R. B. Gimson and C. C. Morgan. The Distributed Computing Software project. Technical Monograph PRG-50, Oxford University Computing Laboratory, Wolfson Building, Parks Road, Oxford, UK, July 1985.

201. J. Ginbayashi. Analysis of business processes specified in Z against an E-R data model. Technical Monograph PRG-103, Oxford University Computing Laboratory, Wolfson Building, Parks Road, Oxford, UK, December 1992.

202. H. S. Goodman. Animating Z specifications in Haskell using a monad. Technical Report CSR-93-10, School of Computer Science, University of Birmingham, Birmingham B15 2TT, UK, November 1993.

203. R. Gotzhein. Specifying open distributed systems with Z. In Bjørner et al. [39], pages 319–339.

204. A. M. Gravell. Minimisation in formal specification and design. In Nicholls [371], pages 32–45.

205. A. M. Gravell. What is a good formal specification? In Nicholls [373], pages 137–150.

206. H. Habrias, S. Dunne, and B. Stoddart. NIAM and Z specifications. Rapport de recherche IRIN 32, Institut de Recherche en Informatique de Nantes, IUT de Nantes, Département Informatique, 3 rue du Maréchal Joffre, 44041 Nantes Cedex 01, France, May 1993.

207. F. Halasz and M. Schwartz. The Dexter hypertext reference model. In *NIST Hypertext Standardization Workshop*, January 1990.

208. J. A. Hall. Seven myths of formal methods. *IEEE Software*, 7(5):11–19, September 1990.
Formal methods are difficult, expensive, and not widely useful, detractors say. Using a case study and other real-world examples, this article challenges such common myths.

209. J. A. Hall. Using Z as a specification calculus for object-oriented systems. In Bjørner et al. [39], pages 290–318.

210. J. A. Hall. Specifying and interpreting class hierarchies in Z. In Bowen and Hall [67], pages 120–138.

211. J. G. Hall and J. A. McDermid. Towards a Z method: Axiomatic specification in Z. In Bowen and Hall [67], pages 213–229.

212. P. A. V. Hall. Towards testing with respect to formal specification. In *Proc. Second IEE/BCS Conference on Software Engineering*, number 290 in Conference Publication, pages 159–163. IEE/BCS, July 1988.

213. U. Hamer and J. Peleska. Z applied to the A330/340 CICS cabin communication system. In Hinchey and Bowen [245], pages 253–284.

214. V. Hamilton. The use of Z within a safety-critical software system. In Hinchey and Bowen [245], pages 357–374.

215. J. A. R. Hammond. Producing Z specifications from object-oriented analysis. In Bowen and Hall [67], pages 316–336.

216. J. A. R. Hammond. Z. In J. J. Marciniak, editor, *Encyclopedia of Software Engineering*, volume 2, pages 1452–1453. John Wiley & Sons, 1994.

217. M. D. Harrison. Engineering human-error tolerant software. In Nicholls [375], pages 191–204.

218. C. L. Harrold. Formal specification of a secure document control system for SMITE. Report no. 88002, RSRE, Ministry of Defence, Malvern, Worcestershire, UK, February 1988.

219. K. Harwood, P. A. Lindsay, and R. Matthews. An approach to constructing verified software. Technical Report 93-8, Department of Computer Science, University of Queensland, St. Lucia 4072, Australia, September 1993.

220. W. T. Harwood. Proof rules for Balzac. Technical Report WTH/P7/001, Imperial Software Technology, Cambridge, UK, 1991.

221. W. Hasselbring. A formal Z specification of ProSet-Linda. Technical report, University of Essen, Fachbereich Mathematik und Informatik – Software Engineering, Schuetzenbahn 70, 4300 Essen 1, Germany, 1992.

222. W. Hasselbring. Animation of Object-Z specifications with a set-oriented prototyping language. In Bowen and Hall [67], pages 337–356.

223. W. Hasselbring. *Prototyping Parallel Algorithms in a Set-Oriented Language*. Dissertation (University of Dortmund, Dept. Computer Science). Verlag Dr. Kovac, Hamburg, Germany, 1994.

    This book presents the design and implementation of an approach to prototyping parallel algorithms with ProSet-Linda. The presented approach to designing and implementing ProSet-Linda relies on the use of the formal specification language Object-Z and the prototyping language ProSet itself.

224. H. P. Haughton. Using Z to model and analyse safety and liveness properties of communication protocols. *Information and Software Technology*, 33(8):575–580, October 1991.

225. H. P. Haughton and K. C. Lano. Three dimensional maintenance. In M. Munro and P. Carroll, editors, *Fourth Software Maintenance Workshop Notes*. Centre for Software Maintenance, Durham, UK, 18–20 September 1990.

    This paper presents an object-oriented extension to Z with the aim to aid software maintenance.

226. I. J. Hayes. Applying formal specification to software development in industry. *IEEE Transactions on Software Engineering*, 11(2):169–178, February 1985.

227. I. J. Hayes. Specification directed module testing. *IEEE Transactions on Software Engineering*, 12(1):124–133, January 1986.

228. I. J. Hayes. Using mathematics to specify software. In *Proc. First Australian Software Engineering Conference*. Institution of Engineers, Australia, May 1986.

229. I. J. Hayes. A generalisation of bags in Z. In Nicholls [371], pages 113–127.

230. I. J. Hayes. Multi-sets and multi-relations in Z with an application to a bill-of-materials system. Technical Report UQCS-176, Department of Computer Science, University of Queensland, St. Lucia 4072, Australia, July 1990.

541

The first chapter is an updated version of [229] and the second and third chapters are updated versions of [233]. The changes are mainly for consistency with [235, 457].

231. I. J. Hayes. Specifying physical limitations: A case study of an oscilloscope. Technical Report 167, Department of Computer Science, University of Queensland, St. Lucia 4072, Australia, July 1990.

232. I. J. Hayes. Interpretations of Z schema operators. In Nicholls [373], pages 12–26.

233. I. J. Hayes. Multi-relations in Z: A cross between multi-sets and binary relations. *Acta Informatica*, 29(1):33–62, February 1992.

234. I. J. Hayes. VDM and Z: A comparative case study. *Formal Aspects of Computing*, 4(1):76–99, 1992.

235. I. J. Hayes, editor. *Specification Case Studies*. Prentice Hall International Series in Computer Science, 2nd edition, 1993.

This is a revised edition of the first ever book on Z, originally published in 1987; it contains substantial changes to every chapter. The notation has been revised to be consistent with *The Z Notation: A Reference Manual* by Mike Spivey [457]. The CAVIAR chapter has been extensively changed to make use of a form of modularization.

Divided into four sections, the first provides tutorial examples of specifications, the second is devoted to the area of software engineering, the third covers distributed computing, analyzing the role of mathematical specification, and the fourth part covers the IBM CICS transaction processing system. Appendices include comprehensive glossaries of the Z mathematical and schema notation. The book will be of interest to the professional software engineer involved in designing and specifying large software projects.

The other contributors are W. Flinn, R. B. Gimson, S. King, C. C. Morgan, I. H. Sørensen and B. A. Sufrin.

236. I. J. Hayes and C. B. Jones. Specifications are not (necessarily) executable. *IEE/BCS Software Engineering Journal*, 4(6):330–338, November 1989.

237. I. J. Hayes, C. B. Jones, and J. E. Nicholls. Understanding the differences between VDM and Z. *FACS Europe*, Series I, 1(1):7–30, Autumn 1993.

Also available as Technical Report UMCS-93-8-1, Department of Computer Science, University of Manchester, UK, 1993.

238. I. J. Hayes and L. Wildman. Towards libraries for Z. In Bowen and Nicholls [75], pages 9–36.

239. He Jifeng, C. A. R. Hoare, M. Fränzle, Markus Müller-Ulm, E.-R. Olderog, M. Schenke, A. P. Ravn, and H. Rischel. Provably correct systems. In H. Langmaack, W.-P. de Roever, and J. Vytopil, editors, *Formal Techniques in Real Time and Fault Tolerant Systems*, volume 863 of *Lecture Notes in Computer Science*, pages 288–335. Springer-Verlag, 1994.

240. He Jifeng, C. A. R. Hoare, and J. W. Sanders. Data refinement refined. In B. Robinet and R. Wilhelm, editors, *Proc. ESOP 86*, volume 213 of *Lecture Notes in Computer Science*, pages 187–196. Springer-Verlag, 1986.

241. D. Heath, D. Allum, and L. Dunckley. *Introductory Logic and Formal Methods*. A. Waller, Henley-on-Thames, UK, 1994.

242. B. Hepworth. ZIP: A unification initiative for Z standards, methods and tools. In Nicholls [371], pages 253–259.

243. B. Hepworth and D. Simpson. The ZIP project. In Nicholls [373], pages 129–133.

244. M. G. Hinchey. Formal methods for system specification: An ounce of prevention is worth a pound of cure. *IEEE Potentials Magazine*, 12(3):50–52, October 1993.

245. M. G. Hinchey and J. P. Bowen, editors. *Applications of Formal Methods*. Prentice Hall International Series in Computer Science, 1995.

A collection on industrial examples of the use of formal methods. Chapters relevant to Z include [115, 123, 192, 213, 214, 246, 330].

246. M. G. Hinchey and J. P. Bowen. Applications of formal methods FAQ. In *Applications of Formal Methods* [245], pages 1–15.

247. I. S. C. Houston. The CICS application programming interface: Automatic transaction initiation. IBM Technical Report TR12.300, IBM United Kingdom Laboratories Ltd, Hursley Park, Winchester, Hampshire SO21 2JN, UK, December 1990.
   One of a number of reports on the CICS application programming interface. See also [41, 289, 360].

248. I. S. C. Houston and M. Josephs. Specifying distributed CICS in Z: Accessing local and remote resources (short communication). *Formal Aspects of Computing*, 6(6):569–579, 1994.

249. I. S. C. Houston and S. King. CICS project report: Experiences and results from the use of Z in IBM. In Prehn and Toetenel [398], pages 588–596.

250. I. S. C. Houston and J. B. Wordsworth. A Z specification of part of the CICS file control API. IBM Technical Report TR12.272, IBM United Kingdom Laboratories Ltd, Hursley Park, Winchester, Hampshire SO21 2JN, UK, 1990.

251. A. D. Hutcheon and A. J. Wellings. Specifying restrictions on imperative programming languages for use in a distributed embedded environment. *IEE/BCS Software Engineering Journal*, 5(2):93–104, March 1990.

252. P. L. Iachini. Operation schema iterations. In Nicholls [373], pages 50–57.

253. M. Imperato. *An Introduction to Z*. Chartwell-Bratt, 1991.
   Contents: Introduction; Set theory; Logic; Building Z specifications; Relations; Functions; Sequences; Bags; Advanced Z; Case study: a simple banking system.

254. D. C. Ince. Z and system specification. In D. C. Ince and D. Andrews, editors, *The Software Life Cycle*, chapter 12, pages 260–277. Butterworths, 1990.

255. D. C. Ince. *An Introduction to Discrete Mathematics, Formal System Specification and Z*. Oxford Applied Mathematics and Computing Science Series. Oxford University Press, 2nd edition, 1993.

256. INMOS Limited. Specification of instruction set & Specification of floating point unit instructions. In *Transputer Instruction Set – A compiler writer's guide*, pages 127–161. Prentice Hall, 1988.
   Appendices F and G use a Z-like notation to give a specification of the instruction set of the IMS T212 and T414 transputers, and the T800 floating-point transputer.

257. A. Jack. It's hard to explain, but Z is much clearer than English. *Financial Times*, page 22, 21 April 1992.

258. D. Jackson. Abstract model checking of infinite specifications. In Naftalin et al. [361], pages 519–531.

259. D. Jackson. Structuring Z specifications with views. Technical Report CMU-CS-94-126, Carnegie-Mellon University, USA, March 1994.

260. J. Jacky. Formal specifications for a clinical cyclotron control system. *ACM SIGSOFT Software Engineering Notes*, 15(4):45–54, September 1990.

261. J. Jacky. Formal specification and development of control system input/output. In Bowen and Nicholls [75], pages 95–108.

262. J. Jacky. Specifying a safety-critical control system in Z. In Woodcock and Larsen [527], pages 388–402.
   Revised version in [263].

263. J. Jacky. Specifying a safety-critical control system in Z. *IEEE Transactions on Software Engineering*, 21(2):99–106, 1995.
   Revised version of [262].

264. J. Jacob. The varieties of refinements. In Morris and Shaw [359], pages 441–455.

265. Jin Song Dong and R. Duke. An object-oriented approach to the formal specification of ODP trader. In *Proc. IFIP TC6/WG6.1 International Conference on Open Distributed Processing*, pages 341–352, September 1993.

266. Jin Song Dong, R. Duke, and G. A. Rose. An object-oriented approach to the semantics of programming languages. In *Proc. 17th Australian Computer Science Conference (ACSC-17)*, pages 767–775, January 1994.

267. C. W. Johnson. Using Z to support the design of interactive safety-critical systems. *IEE/BCS Software Engineering Journal*, 10(2):49–60, March 1995.

268. M. Johnson and P. Sanders. From Z specifications to functional implementations. In Nicholls [371], pages 86–112.

269. P. Johnson. Using Z to specify CICS. In *Proc. Software Engineering anniversary meeting (SEAS)*, page 303, 1987.

270. W. Johnston and G. A. Rose. Guidelines for the manual conversion of Object-Z to C++. Technical Report 93-14, Department of Computer Science, University of Queensland, St. Lucia 4072, Australia, September 1993.

271. C. B. Jones. Interference revisited. In Nicholls [373], pages 58–73.

272. C. B. Jones, R. C. Shaw, and T. Denvir, editors. *5th Refinement Workshop*, Workshop in Computing. Springer-Verlag, 1992.
    The workshop was held at Lloyd's Register, London, UK, 8–10 January 1992. See [435].

273. R. B. Jones. ICL ProofPower. *BCS FACS FACTS*, Series III, 1(1):10–13, Winter 1992.

274. D. Jordan, J. A. McDermid, and I. Toyn. CADiZ – computer aided design in Z. In Nicholls [373], pages 93–104.

275. L. E. Jordan. The kernel Z type checking rules. Technical Report LEJ/TC3/001, Imperial Software Technology, Cambridge, UK, 1991.

276. L. E. Jordan. The Z syntax supported by Balzac-II/1. Technical Report LEJ/S1/001, Imperial Software Technology, Cambridge, UK, 1991.

277. M. B. Josephs. The data refinement calculator for Z specifications. *Information Processing Letters*, 27(1):29–33, 1988.

278. M. B. Josephs. A state-based approach to communicating processes. *Distributed Computing*, 3:9–18, 1988.
    A theoretical paper on combining features of CSP and Z.

279. M. B. Josephs. Specifying reactive systems in Z. Technical Report PRG-TR-19-91, Oxford University Computing Laboratory, Wolfson Building, Parks Road, Oxford, UK, July 1991.

280. M. B. Josephs and D. Redmond-Pyle. Entity-relationship models expressed in Z: A synthesis of structured and formal methods. Technical Report PRG-TR-20-91, Oxford University Computing Laboratory, Wolfson Building, Parks Road, Oxford, UK, July 1991.

281. M. B. Josephs and D. Redmond-Pyle. A library of Z schemas for use in entity-relationship modelling. Technical Report PRG-TR-21-91, Oxford University Computing Laboratory, Wolfson Building, Parks Road, Oxford, UK, August 1991.

282. D. H. Kemp. Specification of Viper1 in Z. Memorandum no. 4195, RSRE, Ministry of Defence, Malvern, Worcestershire, UK, October 1988.

283. D. H. Kemp. Specification of Viper2 in Z. Memorandum no. 4217, RSRE, Ministry of Defence, Malvern, Worcestershire, UK, October 1988.

284. H. Kilov. Information modeling and Object Z: Specifying generic reusable associations. In O. Etzion and A. Segev, editors, *Proc. NGITS-93 (Next Generation Information Technology and Systems)*, pages 182–191, June 1993.

285. H. Kilov and J. Ross. Declarative specifications of collective behavior: Generic reusable frameworks. In H. Kilov and W. Harvey, editors, *Proc. Workshop on Specification of Behavioral Semantics in Object-Oriented Information Modeling*, pages 71–75, Institute for

Information Management and Department of Computer and Information Systems, Robert Morris College, Coraopolos and Pittsburgh, Pennsylvania, USA, 1993. OOPSLA.

286. H. Kilov and J. Ross. Appendix A: A more formal approach. In *Information Modeling: An Object-Oriented Approach*, Object-Oriented Series, pages 199–207. Prentice Hall, 1994.

287. H. Kilov and J. Ross. *Information Modeling: An Object-Oriented Approach*. Object-Oriented Series. Prentice Hall, 1994.

288. P. King. Printing Z and Object-Z LATEX documents. Department of Computer Science, University of Queensland, May 1990.

A description of a Z style option 'oz.sty', an extended version of Mike Spivey's 'zed.sty' [454], for use with the LATEX document preparation system [297]. It is particularly useful for printing Object-Z documents [99, 161].

289. S. King. The CICS application programming interface: Program control. IBM Technical Report TR12.302, IBM United Kingdom Laboratories Ltd, Hursley Park, Winchester, Hampshire SO21 2JN, UK, December 1990.

One of a number of reports on the CICS application programming interface. See also [41, 247, 360].

290. S. King. Z and the refinement calculus. In Bjørner et al. [39], pages 164–188.

Also published as Technical Monograph PRG-79, Oxford University Computing Laboratory, February 1990.

291. S. King and I. H. Sørensen. Specification and design of a library system. In McDermid [337].

292. S. King, I. H. Sørensen, and J. C. P. Woodcock. Z: Grammar and concrete and abstract syntaxes. Technical Monograph PRG-68, Oxford University Computing Laboratory, Wolfson Building, Parks Road, Oxford, UK, 1988.

293. J. C. Knight and D. M. Kienzle. Preliminary experience using Z to specify a safety-critical system. In Bowen and Nicholls [75], pages 109–118.

294. J. C. Knight and B. Littlewood. Critical task of writing dependable software. *IEEE Software*, 11(1):16–20, January 1994.

Guest editors' introduction to a special issue of *IEEE Software* on *Safety-Critical Systems*. A short section on formal methods mentions several Z books on page 18. See also [196].

295. R. D. Knott and P. J. Krause. The implementation of Z specifications using program transformation systems: The SuZan project. In C. Rattray and R. G. Clark, editors, *The Unified Computation Laboratory*, volume 35 of *IMA Conference Series*, pages 207–220, Oxford, UK, 1992. Clarendon Press.

296. M. K. F. Lai. A formal interpretation of the MAA standard in Z. Technical Report DITC 184/91, National Physical Laboratory, Teddington, Middlesex, UK, June 1991.

297. L. Lamport. *LATEX User's Guide & Reference Manual*. Addison-Wesley Publishing Company, Reading, Massachusetts, USA, 1986.

Z specifications may be produced using the document preparation system LATEX together with a special LATEX style option. The most widely used style files are fuzz.sty [456], zed.sty [454] and oz.sty [288].

298. L. Lamport. TLZ. In Bowen and Hall [67], pages 267–268. Abstract.

299. K. C. Lano. Z⁺⁺, an object-orientated extension to Z. In Nicholls [373], pages 151–172.

300. K. C. Lano. Refinement in object-oriented specification languages. In Till [490], pages 236–259.

301. K. C. Lano and P. T. Breuer. From programs to Z specifications. In Nicholls [371], pages 46–70.

302. K. C. Lano, P. T. Breuer, and H. P. Haughton. Reverse engineering COBOL via formal methods. *Software Maintenance: Research and Practice*, 5:13–35, 1993.

Also published in a shortened form as Chapter 16 in [502].

303. K. C. Lano and H. P. Haughton. An algebraic semantics for the specification language Z^{++}. In *Proc. Algebraic Methodology and Software Technology Conference (AMAST '91)*. Springer-Verlag, 1992.

304. K. C. Lano and H. P. Haughton. Reasoning and refinement in object-oriented specification languages. In O. L. Madsen, editor, *ECOOP '92: European Conference on Object-Oriented Programming*, volume 615 of *Lecture Notes in Computer Science*, pages 78–97. Springer-Verlag, 1992.

305. K. C. Lano and H. P. Haughton. *The Z^{++} Manual*. Lloyd's Register of Shipping, 29 Wellesley Road, Croydon CRO 2AJ, UK, 1992.

306. K. C. Lano and H. P. Haughton, editors. *Object Oriented Specification Case Studies*. Object Oriented Series. Prentice Hall International, 1993.

Contents: Chapters introducing object oriented methods, object oriented formal specification and the links between formal and structured object-oriented techniques; seven case studies in particular object oriented formal methods, including:

The Unix Filing System: A MooZ Specification; An Object-Z Specification of a Mobile Phone System; Object-oriented Specification in VDM^{++}; Specifying a Concept-recognition System in Z^{++}; Specification in OOZE; Refinement in Fresco; SmallVDM: An Environment for Formal Specification and Prototyping in Smalltalk.

A glossary, index and bibliography are also included. The contributors are some of the leading figures in the area, including the developers of the above methods and languages: Silvio Meira, Gordon Rose, Roger Duke, Antonio Alencar, Joseph Goguen, Alan Wills, Cassio Souza dos Santos, Ana Cavalcanti.

307. K. C. Lano and H. P. Haughton. Reuse and adaptation of Z specifications. In Bowen and Nicholls [75], pages 62–90.

308. K. C. Lano and H. P. Haughton. *Reverse Engineering and Software Maintenance: A Practical Approach*. International Series in Software Engineering. McGraw Hill, 1993.

309. K. C. Lano and H. P. Haughton. Standards and techniques for object-oriented formal specification. In *Proc. 1993 Software Engineering Standards Symposium*, pages 237–246. IEEE Computer Society Press, 1993.

310. K. C. Lano and H. P. Haughton. Formal development in B Abstract Machine Notation. *Information and Software Technology*, 37(5):303–316, May 1995.

311. K. C. Lano, H. P. Haughton, and P. T. Breuer. Using object-oriented extensions of Z for maintenance and reverse-engineering. Technical Report PRG-TR-22-91, Oxford University Computing Laboratory, Wolfson Building, Parks Road, Oxford, UK, 1991.

312. G. Laycock. Formal specification and testing: A case study. *Software Testing, Verification and Reliability*, 2(1):7–23, May 1992.

313. D. Lightfoot. *Formal Specification using Z*. Macmillan, 1991.

Contents: Introduction; Sets in Z; Using sets to describe a system – a simple example; Logic: propositional calculus; Example of a Z specification document; Logic: predicate calculus; Relations; Functions; A seat allocation system; Sequences; An example of sequences – the aircraft example again; Extending a specification; Collected notation; Books on formal specification; Hints on creating specifications; Solutions to exercises. Also available in French.

314. P. A. Lindsay. Reasoning about Z specifications: A VDM perspective. Technical Report 93-20, Department of Computer Science, University of Queensland, St. Lucia 4072, Australia, October 1993.

315. P. A. Lindsay. On transferring VDM verification techniques to Z. In Naftalin et al. [361], pages 190–213.

Also available as Technical Report 94-10, Department of Computer Science, University of Queensland, 1994.

316. P. A. Lindsay and E. van Keulen. Case studies in the verification of specifications in VDM and Z. Technical Report 94-3, Department of Computer Science, University of Queensland, St. Lucia 4072, Australia, March 1994.

317. R. L. London. Specifying reusable components using Z: Sets implemented by bit vectors. Technical Report CR-88-14, Tektronix Laboratories, P. O. Box 500, MS 50-662, Beaverton, Oregon 97077, USA, November 1988.

318. R. L. London and K. R. Milsted. Specifying reusable components using Z: Realistic sets and dictionaries. *ACM SIGSOFT Software Engineering Notes*, 14(3):120–127, May 1989.

319. M. Love. Animating Z specifications in SQL*Forms3.0. In Bowen and Nicholls [75], pages 294–306.

320. P. J. Lupton. Promoting forward simulation. In Nicholls [373], pages 27–49.

321. A. MacDonald and D. Carrington. Synthesising designs from formal specifications: A case study. Technical Report 93-21, Department of Computer Science, University of Queensland, St. Lucia 4072, Australia, November 1993.

322. R. Macdonald. Z usage and abusage. Report no. 91003, RSRE, Ministry of Defence, Malvern, Worcestershire, UK, February 1991.

This paper presents a miscellany of observations drawn from experience of using Z, shows a variety of techniques for expressing certain class of idea concisely and clearly, and alerts the reader to certain pitfalls which may trap the unwary.

323. B. P. Mahony and I. J. Hayes. A case-study in timed refinement: A central heater. In Morris and Shaw [359], pages 138–149.

324. B. P. Mahony and I. J. Hayes. A case-study in timed refinement: A mine pump. *IEEE Transactions on Software Engineering*, 18(9):817–826, September 1992.

325. B. P. Mahony, C. Millerchip, and I. J. Hayes. A boiler control system: A case-study in timed refinement. Technical report, Department of Computer Science, University of Queensland, St. Lucia 4072, Australia, 23 June 1993.

A specification and top-level design of a steam generating boiler system is presented as an example of the formal development of a real-time system.

326. K. C. Mander and F. Polack. Rigorous specification using structured systems analysis and Z. *Information and Software Technology*, 37(5):285–291, May 1995.

Revised version of [393].

327. A. Martin. Encoding W: A logic for Z in 2OBJ. In Woodcock and Larsen [527], pages 462–481.

328. P. Martin. The formal specification in Z of task migration on the testbed multicomputer. Technical Report ECS-CSG-2-94, Department of Computer Science, University of Edinburgh, Edinburgh EH9 3JZ, UK., February 1994. To appear in *IEEE Software*.

329. P. Mataga and P. Zave. Formal specification of telephone features. In Bowen and Hall [67], pages 29–50.

330. P. Mataga and P. Zave. Multiparadigm specification of an AT&T switching system. In Hinchey and Bowen [245], pages 375–398.

331. P. Mataga and P. Zave. Using Z to specify telephone features. *Information and Software Technology*, 37(5):277–283, May 1995.

Revised version of [329].

332. I. Maung and J. R. Howse. Introducing Hyper-Z – a new approach to object orientation in Z. In Bowen and Nicholls [75], pages 149–165.

333. M. D. May. Use of formal methods by a silicon manufacturer. In C. A. R. Hoare, editor, *Developments in Concurrency and Communication*, University of Texas at Austin Year

of Programming Series, chapter 4, pages 107–129. Addison-Wesley Publishing Company, 1990.

334. M. D. May, G. Barrett, and D. E. Shepherd. Designing chips that work. In C. A. R. Hoare and M. J. C. Gordon, editors, *Mechanized Reasoning and Hardware Design*, pages 3–19. Prentice Hall International Series in Computer Science, 1992.

335. M. D. May and D. E. Shepherd. Verification of the IMS T800 microprocessor. In *Proc. Electronic Design Automation*, pages 605–615, London, UK, September 1987.

336. J. A. McDermid. Special section on Z. *IEE/BCS Software Engineering Journal*, 4(1):25–72, January 1989.
A special issue on Z, introduced and edited by Prof. J. A. McDermid. See also [51, 110, 453, 519].

337. J. A. McDermid, editor. *The Theory and Practice of Refinement: Approaches to the Formal Development of Large-Scale Software Systems*. Butterworth Scientific, 1989.
This book contains papers from the 1st Refinement Workshop held at the University of York, UK, 7–8 January 1988. Z-related papers include [366, 291].

338. J. A. McDermid. Formal methods: Use and relevance for the development of safety critical systems. In P. A. Bennett, editor, *Safety Aspects of Computer Control*. Butterworth-Heinemann, Oxford, UK, 1993.
This paper discusses a number of formal methods and summarizes strengths and weaknesses in safety critical applications; a major safety-related example is presented in Z.

339. M. A. McMorran and J. E. Nicholls. Z user manual. Technical Report TR12.274, IBM United Kingdom Laboratories Ltd, Hursley Park, Winchester, Hampshire SO21 2JN, UK, July 1989.

340. M. A. McMorran and S. Powell. *Z Guide for Beginners*. Blackwell Scientific, 1993.

341. S. L. Meira and A. L. C. Cavalcanti. Modular object-oriented Z specifications. In Nicholls [373], pages 173–192.

342. S. L. Meira and A. L. C. Cavalcanti. The MooZ specification language. Technical report, Universidade Federal de Pernambuco, Departamento de Informática, Recife – PE, Brasil, 1992.

343. B. Meyer. On formalism in specifications. *IEEE Software*, 2(1):6–26, January 1985.

344. V. Mišić, D. Velašević, and B. Lazarević. Formal specification of a data dictionary for an extended ER data model. *The Computer Journal*, 35(6):611–622, December 1992.

345. J. D. Moffett and M. S. Sloman. A case study representing a model: To Z or not to Z? In Nicholls [373], pages 254–268.

346. B. Q. Monahan. Book review. *Formal Aspects of Computing*, 1(1):137–142, January–March 1989.
A review of *Understanding Z: A Specification Language and Its Formal Semantics* by Mike Spivey [452].

347. B. Q. Monahan and R. C. Shaw. Model-based specifications. In J. A. McDermid, editor, *Software Engineer's Reference Book*, chapter 21. Butterworth-Heinemann, Oxford, UK, 1991.
This chapter contains a case study in Z, followed by a discussion of the respective trade-offs in specification between Z and VDM.

348. C. C. Morgan. Data refinement using miracles. *Information Processing Letters*, 26(5):243–246, January 1988.

349. C. C. Morgan. Procedures, parameters, and abstraction: Separate concerns. *Science of Computer Programming*, 11(1), October 1988.

350. C. C. Morgan. The specification statement. *ACM Transactions on Programming Languages and Systems (TOPLAS)*, 10(3), July 1988.

351. C. C. Morgan. Types and invariants in the refinement calculus. In *Proc. Mathematics of Program Construction Conference*, Twente, June 1989.

352. C. C. Morgan. *Programming from Specifications*. Prentice Hall International Series in Computer Science, 2nd edition, 1994.

This book presents a rigorous treatment of most elementary program development techniques, including iteration, recursion, procedures, parameters, modules and data refinement.

353. C. C. Morgan and K. A. Robinson. Specification statements and refinement. *IBM Journal of Research and Development*, 31(5), September 1987.

354. C. C. Morgan and J. W. Sanders. Laws of the logical calculi. Technical Monograph PRG-78, Oxford University Computing Laboratory, Wolfson Building, Parks Road, Oxford, UK, September 1989.

This document records some important laws of classical predicate logic. It is designed as a reservoir to be tapped by *users* of logic, in system development.

355. C. C. Morgan and B. A. Sufrin. Specification of the Unix filing system. *IEEE Transactions on Software Engineering*, 10(2):128–142, March 1984.

356. C. C. Morgan and T. Vickers, editors. *On the Refinement Calculus*. Formal Approaches to Computing and Information Technology series (FACIT). Springer-Verlag, 1994.

This book collects together the work accomplished at Oxford on the refinement calculus: the rigorous development, from state-based assertional specification, of executable imperative code.

357. C. C. Morgan and J. C. P. Woodcock. What is a specification? In D. Craigen and K. Summerskill, editors, *Formal Methods for Trustworthy Computer Systems (FM89)*, Workshops in Computing, pages 38–43. Springer-Verlag, 1990.

358. C. C. Morgan and J. C. P. Woodcock, editors. *3rd Refinement Workshop*, Workshops in Computing. Springer-Verlag, 1991.

The workshop was held at the IBM Laboratories, Hursley Park, UK, 9–11 January 1990. See [434].

359. J. M. Morris and R. C. Shaw, editors. *4th Refinement Workshop*, Workshops in Computing. Springer-Verlag, 1991.

The workshop was held at Cambridge, UK, 9–11 January 1991. For Z related papers, see [23, 264, 323, 512, 522, 507].

360. P. Mundy and J. B. Wordsworth. The CICS application programming interface: Transient data and storage control. IBM Technical Report TR12.299, IBM United Kingdom Laboratories Ltd, Hursley Park, Winchester, Hampshire SO21 2JN, UK, October 1990.

One of a number of reports on the CICS application programming interface. See also [41, 247, 289].

361. M. Naftalin, T. Denvir, and M. Bertran, editors. *FME'94: Industrial Benefit of Formal Methods*, volume 873 of *Lecture Notes in Computer Science*. Formal Methods Europe, Springer-Verlag, 1994.

The 2nd FME Symposium was held at Barcelona, Spain, 24–28 October 1994. Z-related papers include [69, 120, 136, 173, 175, 180, 258, 315]. B-related papers include [142, 415, 472].

362. K. T. Narayana and S. Dharap. Formal specification of a look manager. *IEEE Transactions on Software Engineering*, 16(9):1089–1103, September 1990.

A formal specification of the look manager of a dialog system is presented in Z. This deals with the presentation of visual aspects of objects and the editing of those visual aspects.

363. K. T. Narayana and S. Dharap. Invariant properties in a dialog system. *ACM SIGSOFT Software Engineering Notes*, 15(4):67–79, September 1990.

364. T. C. Nash. Using Z to describe large systems. In Nicholls [371], pages 150–178.

365. Ph. W. Nehlig and D. A. Duce. GKS-9x: The design output primitive, an approach to specification. *Computer Graphics Forum*, 13(3):C–381–C–392, 1994.

366. D. S. Neilson. Hierarchical refinement of a Z specification. In McDermid [337].

367. D. S. Neilson. From Z to C: Illustration of a rigorous development method. Technical Monograph PRG-101, Oxford University Computing Laboratory, Wolfson Building, Parks Road, Oxford, UK, 1990.

368. D. S. Neilson. Machine support for Z: The zedB tool. In Nicholls [373], pages 105–128.

369. D. S. Neilson and D. Prasad. zedB: A proof tool for Z built on B. In Nicholls [375], pages 243–258.

370. J. E. Nicholls. Working with formal methods. *Journal of Information Technology*, 2(2):67–71, June 1987.

371. J. E. Nicholls, editor. *Z User Workshop, Oxford 1989*, Workshops in Computing. Springer-Verlag, 1990.
Proceedings of the Fourth Annual Z User Meeting, Wolfson College & Rewley House, Oxford, UK, 14–15 December 1989. Published in collaboration with the British Computer Society. For the opening address see [384]. For individual papers, see [33, 86, 87, 119, 146, 183, 204, 229, 242, 268, 301, 364, 388, 443, 459, 505].

372. J. E. Nicholls. A survey of Z courses in the UK. In *Z User Workshop, Oxford 1990* [373], pages 343–350.

373. J. E. Nicholls, editor. *Z User Workshop, Oxford 1990*, Workshops in Computing. Springer-Verlag, 1991.
Proceedings of the Fifth Annual Z User Meeting, Lady Margaret Hall, Oxford, UK, 17–18 December 1990. Published in collaboration with the British Computer Society. For individual papers, see [25, 80, 95, 116, 187, 205, 232, 243, 252, 271, 274, 299, 341, 345, 368, 372, 381, 403, 430, 506, 536]. The proceedings also includes an *Introduction and Opening Remarks*, a *Selected Z Bibliography*, a selection of posters and information on Z tools.

374. J. E. Nicholls. Domains of application for formal methods. In *Z User Workshop, York 1991* [375], pages 145–156.

375. J. E. Nicholls, editor. *Z User Workshop, York 1991*, Workshops in Computing. Springer-Verlag, 1992.
Proceedings of the Sixth Annual Z User Meeting, York, UK. Published in collaboration with the British Computer Society. For individual papers, see [18, 28, 139, 98, 149, 158, 217, 369, 374, 394, 422, 444, 483, 498, 525, 541].

376. J. E. Nicholls. Plain guide to the Z base standard. In Bowen and Nicholls [75], pages 52–61.

377. J. E. Nicholls et al. Z in the development process. Technical Report PRG-TR-1-89, Oxford University Computing Laboratory, Wolfson Building, Parks Road, Oxford, UK, June 1989.
Proceedings of a discussion workshop held on 15 December 1988 in Oxford, UK, with contributions by Peter Collins, David Cooper, Anthony Hall, Patrick Hall, Brian Hepworth, Ben Potter and Andrew Ricketts.

378. C. J. Nix and B. P. Collins. The use of software engineering, including the Z notation, in the development of CICS. *Quality Assurance*, 14(3):103–110, September 1988.

379. A. Norcliffe and G. Slater. *Mathematics for Software Construction*. Series in Mathematics and its Applications. Ellis Horwood, 1991.
Contents: Why mathematics; Getting started: sets and logic; Developing ideas: schemas; Functions; Functions in action; A real problem from start to finish: a drinks machine; Sequences; Relations; Generating programs from specifications: refinement; The role of proof; More examples of specifications; Concluding remarks; Answers to exercises.

380. A. Norcliffe and S. Valentine. Z readers video course. PAVIC Publications, 1992. Sheffield Hallam University, 33 Collegiate Crescent, Sheffield S10 2BP, UK.
Video-based Training Course on the Z Specification Language. The course consists of 5 videos, each of approximately one hour duration, together with supporting texts and case studies.

381. A. Norcliffe and S. H. Valentine. A video-based training course in reading Z specifications. In Nicholls [373], pages 337–342.

382. G. Normington. Cleanroom and Z. In Bowen and Nicholls [75], pages 281–293.

383. C. O' Halloran. Evaluation semantics in Z. In Naftalin et al. [361], pages 502–518.

384. B. Oakley. The state of use of formal methods. In Nicholls [371], pages 1–5.
A record of the opening address at ZUM'89.

385. C. O'Halloran. The software repeater (an exercise in Z specification). Report no. 4090, RSRE, Ministry of Defence, Malvern, Worcestershire, UK, 1987.

386. E. A. Oxborrow and H. M. Ismail. KBZ - an object-oriented approach to the specification and management of knowledge bases. Technical Report 51, Computing Laboratory, University of Kent at Canterbury, UK, 1988.
The report describes the important features of KBZ, an extension of Z which should better support the specification of the semantics of conceptual data models. The report then clarifies the use of KBZ by discussing its application in a distributed database environment.

387. C. E. Parker. Z tools catalogue. ZIP project report ZIP/BAe/90/020, British Aerospace, Software Technology Department, Warton PR4 1AX, UK, May 1991.

388. M. Phillips. CICS/ESA 3.1 experiences. In Nicholls [371], pages 179–185.
Z was used to specify 37,000 lines out of 268,000 lines of code in the IBM CICS/ESA 3.1 release. The initial development benefit from using Z was assessed as being a 9% improvement in the *total development cost* of the release, based on the reduction of programmer days fixing problems.

389. M. Pilling, A. Burns, and K. Raymond. Formal specifications and proofs of inheritance protocols for real-time scheduling. *IEE/BCS Software Engineering Journal*, 5(5):263–279, September 1990.

390. P. R. H. Place and K. C. Kang. Safety-critical software: Status report and annotated bibliography. Technical Report CMU/SEI-92-TR-5 & ESC-TR-93-182, Software Engineering Institute, Carnegie-Mellon University, Pittsburgh, Pennsylvania 15213, USA, June 1993.

391. P. R. H. Place and W. Wood. Survey of formal specification techniques for reactive systems. CMU Technical Report CMU/SEI-90-TR-5, ADA22374, Software Engineering Institute, Carnegie-Mellon University, Pittsburgh, Pennsylvania 15213, USA, 1990.

392. P. R. H. Place and W. Wood. Formal development of Ada programs using Z and Anna: A case study. CMU Technical Report CMU/SEI-91-TR-1, ADA235698, Software Engineering Institute, Carnegie-Mellon University, Pittsburgh, Pennsylvania 15213, USA, February 1991.
Copies available from: Research Access Inc., 3400 Forbes Avenue, Suite 302, Pittsburgh, PA 15213, USA.

393. F. Polack and K. C. Mander. Software quality assurance using the SAZ method. In Bowen and Hall [67], pages 230–249.

394. F. Polack, M. Whiston, and P. Hitchcock. Structured analysis – a draft method for writing Z specifications. In Nicholls [375], pages 261–286.

395. F. Polack, M. Whiston, and K. C. Mander. The SAZ project: Integrating SSADM and Z. In Woodcock and Larsen [527], pages 541–557.

396. B. F. Potter, J. E. Sinclair, and D. Till. *An Introduction to Formal Specification and Z*. Prentice Hall International Series in Computer Science, 1991.

Contents: Formal specification in the context of software engineering; An informal introduction to logic and set theory; A first specification; The Z notation: the mathematical language, relations and functions, schemas and specification structure; A first specification revisited; Formal reasoning; From specification to program: data and operation refinement, operation decomposition; From theory to practice.

397. B. F. Potter and D. Till. The specification in Z of gateway functions within a communications network. In *Proc. IFIP WG10.3 Conference on Distributed Processing*. Elsevier Science Publishers (North-Holland), October 1987.

398. S. Prehn and W. J. Toetenel, editors. *VDM'91: Formal Software Development Methods*, volume 551 of *Lecture Notes in Computer Science*. Springer-Verlag, 1991. Volume 1: Conference Contributions.

The 4th VDM-Europe Symposium was held at Noordwijkerhout, The Netherlands, 21–25 October 1991. Papers with relevance to Z include [17, 35, 128, 154, 193, 249, 501, 509, 540]. See also [399].

399. S. Prehn and W. J. Toetenel, editors. *VDM'91: Formal Software Development Methods*, volume 552 of *Lecture Notes in Computer Science*. Springer-Verlag, 1991. Volume 2: Tutorials.

Papers with relevance to Z include [6, 523]. See also [398].

400. G-H. B. Rafsanjani and S. J. Colwill. From Object-Z to C++: A structural mapping. In Bowen and Nicholls [75], pages 166–179.

401. RAISE Language Group. *The RAISE Specification Language*. BCS Practitioner Series. Prentice Hall International, 1992.

402. G. P. Randell. Translating data flow diagrams into Z (and vice versa). Report no. 90019, RSRE, Ministry of Defence, Malvern, Worcestershire, UK, October 1990.

403. G. P. Randell. Data flow diagrams and Z. In Nicholls [373], pages 216–227.

404. G. P. Randell. Improving the translation from data flow diagrams into Z by incorporating the data dictionary. Report no. 92004, RSRE, Ministry of Defence, Malvern, Worcestershire, UK, January 1992.

405. D. Rann, J. Turner, and J. Whitworth. *Z: A Beginner's Guide*. Chapman & Hall, London, 1994.

406. B. Ratcliff. *Introducing Specification Using Z: A Practical Case Study Approach*. International Series in Software Engineering. McGraw-Hill, 1994.

407. A. P. Ravn, H. Rischel, and V. Stavridou. Provably correct safety critical software. In *Proc. IFAC Safety of Computer Controlled Systems 1990 (SAFECOMP'90)*. Pergamon Press, 1990.

Also available as Technical Report CSD-TR-625 from Department of Computer Science, Royal Holloway, University of London, Egham, Surrey TW20 0EX, UK.

408. M. Rawson. OOPSLA'93: Workshop on formal specification of object-oriented systems – position paper. In H. Kilov and W. Harvey, editors, *Proc. Workshop on Specification of Behavioral Semantics in Object-Oriented Information Modeling*, pages 125–135, Institute for Information Management and Department of Computer and Information Systems, Robert Morris College, Coraopolos and Pittsburgh, Pennsylvania, USA, 1993. OOPSLA.

409. K. Raymond, P. Stocks, and D. Carrington. Using Z to specify distributed systems. Technical Report 181, Key Centre for Software Technology, University of Queensland, St. Lucia 4072, Australia, 1990.

410. T. J. Read. Formal specification of reusable Ada software packages. In A. Burns, editor, *Towards Ada 9X Conference Proceedings*, pages 98–117, 1991.

411. J. N. Reed. Semantics-based tools for a specification support environment. In *Mathematical Foundations of Programming Language Semantics*, volume 298 of *Lecture Notes in Computer Science*. Springer-Verlag, 1988.

412. J. N. Reed and J. E. Sinclair. An algorithm for type-checking Z: A Z specification. Technical Monograph PRG-81, Oxford University Computing Laboratory, Wolfson Building, Parks Road, Oxford, UK, March 1990.

413. N. R. Reizer, G. D. Abowd, B. C. Meyers, and P. R. H. Place. Using formal methods for requirements specification of a proposed POSIX standard. In *IEEE International Conference on Requirements Engineering (ICRE'94)*, April 1994.

414. G. J. Reynolds. Yet another approach to the formal specification of a configurable graphics system. In *Proc. Eurographics Association Formal Methods in Computer Graphics*, June 1991.

415. B. Ritchie, J. Bicarregui, and H. P. Haughton. Experiences in using the abstract machine notation in a GKS case study. In Naftalin et al. [361], pages 93–104.

416. K. A. Robinson. Refining Z specifications to programs. In *Proc. Australian Software Engineering Conference*, pages 87–97, 1987.

417. G. A. Rose. Object-Z. In Stepney et al. [462], pages 59–77.

418. G. A. Rose and P. Robinson. A case study in formal specifications. In *Proc. First Australian Software Engineering Conference*, May 1986.

419. K. J. Ross and P. A. Lindsay. Maintaining consistency under changes to formal specifications. Technical Report 93-3, Department of Computer Science, University of Queensland, St. Lucia 4072, Australia, March 1993.

420. A. R. Ruddle. Formal methods in the specification of real-time, safety-critical control systems. In Bowen and Nicholls [75], pages 131–146.

421. P. Rudkin. Modelling information objects in Z. In J. de Meer, editor, *Proc. International Workshop on ODP*. Elsevier Science Publishers (North-Holland), 1992.

422. M. Saaltink. Z and Eves. In Nicholls [375], pages 223–242.

423. H. Saiedian. The mathematics of computing. *Journal of Computer Science Education*, 3(3):203–221, 1992.

424. A. C. A. Sampaio and S. L. Meira. Modular extensions to Z. In Bjørner et al. [39], pages 211–232.

425. P. Sanders, M. Johnson, and R. Tinker. From Z specifications to functional implementations. *British Telecom Technology Journal*, 7(4), October 1989.

426. S. A. Schuman and D. H. Pitt. Object-oriented subsystem specification. In L. G. L. T. Meertens, editor, *Program Specification and Transformation*, pages 313–341. Elsevier Science Publishers (North-Holland), 1987.

427. S. A. Schuman, D. H. Pitt, and P. J. Byers. Object-oriented process specification. In C. Rattray, editor, *Specification and Verification of Concurrent Systems*, Workshops in Computing, pages 21–70. Springer-Verlag, 1990.

428. L. T. Semmens and P. M. Allen. Using entity relationship models as a basis for Z specifications. Technical Report IES1/90, Leeds Polytechnic, Faculty of Information and Engineering Systems, Leeds, UK, 1990.

429. L. T. Semmens and P. M. Allen. Using Yourdon and Z to specify computer security: A case study. Technical Report IES4/90, Leeds Polytechnic, Faculty of Information and Engineering Systems, Leeds, UK, 1990.

430. L. T. Semmens and P. M. Allen. Using Yourdon and Z: An approach to formal specification. In Nicholls [373], pages 228–253.

431. L. T. Semmens, R. B. France, and T. W. G. Docker. Integrated structured analysis and formal specification techniques. *The Computer Journal*, 35(6):600–610, December 1992.

432. C. T. Sennett. Review of type checking and scope rules of the specification language Z. Report no. 87017, RSRE, Ministry of Defence, Malvern, Worcestershire, UK, November 1987.

433. C. T. Sennett. Formal specification and implementation. In C. T. Sennett, editor, *High-Integrity Software*, Computer Systems Series. Pitman, 1989.

434. C. T. Sennett. Using refinement to convince: Lessons learned from a case study. In Morgan and Woodcock [358], pages 172–197.

435. C. T. Sennett. Demonstrating the compliance of Ada programs with Z specifications. In Jones et al. [272].

436. C. T. Sennett and R. Macdonald. Separability and security models. Report no. 87020, RSRE, Ministry of Defence, Malvern, Worcestershire, UK, November 1987.

437. D. E. Shepherd. Verified microcode design. *Microprocessors and Microsystems*, 14(10):623–630, December 1990.
    This article is part of a special issue on *Formal aspects of microprocessor design*, edited by H. S. M. Zedan. See also [52].

438. D. E. Shepherd and G. Wilson. Making chips that work. *New Scientist*, 1664:61–64, May 1989.
    A general article containing information on the formal development of the T800 floating-point unit for the transputer including the use of Z.

439. D. Sheppard. *An Introduction to Formal Specification with Z and VDM*. International Series in Software Engineering. McGraw Hill, 1995.

440. L. B. Sherrell and D. L. Carver. Z meets Haskell: A case study. In *COMPSAC '93: 17th Annual International Computer Software and Applications Conference*, pages 320–326. IEEE Computer Society Press, November 1993.
    The paper traces the development of a simple system, the class manager's assistant, from an existing Z specification, through design in Z, to a Haskell implementation.

441. L. N. Simcox. The application of Z to the specification of air traffic control systems: 1. Memorandum no. 4280, RSRE, Ministry of Defence, Malvern, Worcestershire, UK, April 1989.

442. R. Sinnott and K. J. Turner. Modeling ODP viewpoints. In H. Kilov, W. Harvey, and H. Mili, editors, *Proc. Workshop on Precise Behavioral Specifications in Object-Oriented Information Modeling, OOPSLA 1994*, pages 121–128, Robert Morris College, Coraopolos and Pittsburgh, Pennsylvania 15108-1189, USA, 1994. OOPSLA.

443. A. Smith. The Knuth-Bendix completion algorithm and its specification in Z. In Nicholls [371], pages 195–220.

444. A. Smith. On recursive free types in Z. In Nicholls [375], pages 3–39.

445. G. Smith. *An Object-Oriented Approach to Formal Specification*. PhD thesis, Department of Computer Science, University of Queensland, St. Lucia 4072, Australia, October 1992.
    A detailed description of a version of Object-Z similar to (but not identical to) that in [167]. The thesis also includes a formalization of temporal logic history invariants and a fully-abstract model of classes in Object-Z.

446. G. Smith. A object-oriented development framework for Z. In Bowen and Hall [67], pages 89–107.

447. G. Smith and R. Duke. Specification and verification of a cache coherence protocol. Technical Report 126, Department of Computer Science, University of Queensland, St. Lucia 4072, Australia, 1989.

448. G. Smith and R. Duke. Modelling a cache coherence protocol using Object-Z. In *Proc. 13th Australian Computer Science Conference (ACSC-13)*, pages 352–361, 1990.

449. P. Smith and R. Keighley. The formal development of a secure transaction mechanism. In Prehn and Toetenel [398], pages 457–476.

450. I. Sommerville. *Software Engineering*, chapter 9, pages 153–168. Addison-Wesley, 4th edition, 1992.
    A chapter entitled *Model-Based Specification* including examples using Z.

451. I. H. Sørensen. A specification language. In J. Staunstrup, editor, *Program Specification: Proceedings of a Workshop*, volume 134 of *Lecture Notes in Computer Science*, pages 381–401. Springer-Verlag, 1981.

452. J. M. Spivey. *Understanding Z: A Specification Language and its Formal Semantics*, volume 3 of *Cambridge Tracts in Theoretical Computer Science*. Cambridge University Press, January 1988.
Published version of 1985 DPhil thesis.

453. J. M. Spivey. An introduction to Z and formal specifications. *IEE/BCS Software Engineering Journal*, 4(1):40–50, January 1989.

454. J. M. Spivey. A guide to the zed style option. Oxford University Computing Laboratory, December 1990.
A description of the Z style option 'zed.sty' for use with the LaTeX document preparation system [297].

455. J. M. Spivey. Specifying a real-time kernel. *IEEE Software*, 7(5):21–28, September 1990.
This case study of an embedded real-time kernel shows that mathematical techniques have an important role to play in documenting systems and avoiding design flaws.

456. J. M. Spivey. *The ƒUZZ Manual*. Computing Science Consultancy, 34 Westlands Grove, Stockton Lane, York YO3 0EF, UK, 2nd edition, July 1992.
The manual describes a Z type-checker and 'fuzz.sty' style option for LaTeX documents [297]. The package is compatible with the book, *The Z Notation: A Reference Manual* by the same author [457].

457. J. M. Spivey. *The Z Notation: A Reference Manual*. Prentice Hall International Series in Computer Science, 2nd edition, 1992.
This is a revised edition of the first widely available reference manual on Z originally published in 1989. The book provides a complete and definitive guide to the use of Z in specifying information systems, writing specifications and designing implementations. See also the draft Z standard [83].
Contents: Tutorial introduction; Background; The Z language; The mathematical tool-kit; Sequential systems; Syntax summary; Changes from the first edition; Glossary.

458. J. M. Spivey and B. A. Sufrin. Type inference in Z. In Bjørner et al. [39], pages 426–438.

459. J. M. Spivey and B. A. Sufrin. Type inference in Z. In Nicholls [371], pages 6–31.
Also published as [458].

460. S. Stepney. *High Integrity Compilation: A Case Study*. Prentice Hall, 1993.

461. S. Stepney and R. Barden. Annotated Z bibliography. *Bulletin of the European Association of Theoretical Computer Science*, 50:280–313, June 1993.

462. S. Stepney, R. Barden, and D. Cooper, editors. *Object Orientation in Z*. Workshops in Computing. Springer-Verlag, 1992.
This is a collection of papers describing various OOZ approaches – Hall, ZERO, MooZ, Object-Z, OOZE, Schuman & Pitt, Z^{++}, ZEST and Fresco (an object-oriented VDM method) – in the main written by the methods' inventors, and all specifying the same two examples. The collection is a revised and expanded version of a ZIP report distributed at the 1991 Z User Meeting at York.

463. S. Stepney, R. Barden, and D. Cooper. A survey of object orientation in Z. *IEE/BCS Software Engineering Journal*, 7(2):150–160, March 1992.

464. S. Stepney and S. P. Lord. Formal specification of an access control system. *Software – Practice and Experience*, 17(9):575–593, September 1987.

465. P. Stocks. *Applying formal methods to software testing*. PhD thesis, Department of Computer Science, University of Queensland, St. Lucia 4072, Australia, 1993.

466. P. Stocks and D. A. Carrington. Deriving software test cases from formal specifications. In *6th Australian Software Engineering Conference*, pages 327–340, July 1991.

467. P. Stocks and D. A. Carrington. Test template framework: A specification-based testing case study. In *Proc. International Symposium on Software Testing and Analysis (ISSTA '93)*, pages 11–18, June 1993.

Also available in a longer form as Technical Report UQCS-255, Department of Computer Science, University of Queensland.

468. P. Stocks and D. A. Carrington. Test templates: A specification-based testing framework. In *Proc. 15th International Conference on Software Engineering*, pages 405–414, May 1993.

Also available in a longer form as Technical Report UQCS-243, Department of Computer Science, University of Queensland.

469. P. Stocks, K. Raymond, and D. Carrington. Representing distributed system concepts in Z. Technical Report 180, Key Centre for Software Technology, University of Queensland, St. Lucia 4072, Australia, 1990.

470. P. Stocks, K. Raymond, D. Carrington, and A. Lister. Modelling open distributed systems in Z. *Computer Communications*, 15(2):103–113, March 1992.

In a special issue on the practical use of FDTs (Formal Description Techniques) in communications and distributed systems, edited by Dr. Gordon S. Blair.

471. B. Stoddart and P. Knaggs. The event calculus (formal specification of real time systems by means of diagrams and Z schemas). In *5th International Conference on putting into practice methods and tools for information system design*, University of Nantes, Institute Universitaire de Technologie, 3 Rue du Maréchal Joffre, 44041 Nantes Cedex 01, France, September 1992.

472. A. C. Storey and H. P. Haughton. A strategy for the production of verifiable code using the B method. In Naftalin et al. [361], pages 346–365.

473. B. A. Sufrin. Formal system specification: Notation and examples. In D. Neel, editor, *Tools and Notations for Program Construction*. Cambridge University Press, 1982.

An example of a filing system specification, this was the first published use of the schema notation to put together states.

474. B. A. Sufrin. Towards formal specification of the ICL data dictionary. *ICL Technical Journal*, August 1984.

475. B. A. Sufrin. Formal methods and the design of effective user interfaces. In M. D. Harrison and A. F. Monk, editors, *People and Computers: Designing for Usability*. Cambridge University Press, 1986.

476. B. A. Sufrin. Formal specification of a display-oriented editor. In N. Gehani and A. D. McGettrick, editors, *Software Specification Techniques*, International Computer Science Series, pages 223–267. Addison-Wesley Publishing Company, 1986.

Originally published in Science of Computer Programming, 1:157–202, 1982.

477. B. A. Sufrin. A formal framework for classifying interactive information systems. In *IEE Colloquium on Formal Methods and Human-Computer Interaction*, number 09 in IEE Digest, pages 4/1–14, London, UK, 1987. The Institution of Electrical Engineers.

478. B. A. Sufrin. Effective industrial application of formal methods. In G. X. Ritter, editor, *Information Processing 89, Proc. 11th IFIP Computer Congress*, pages 61–69. Elsevier Science Publishers (North-Holland), 1989.

This paper presents a Z model of the Unix *make* utility.

479. B. A. Sufrin and He Jifeng. Specification, analysis and refinement of interactive processes. In M. D. Harrison and H. Thimbleby, editors, *Formal Methods in Human-Computer Interaction*, volume 2 of *Cambridge Series on Human-Computer Interaction*, chapter 6, pages 153–200. Cambridge University Press, 1990.

A case study on using Z for process modelling.

480. B. A. Sufrin and J. C. P. Woodcock. Towards the formal specification of a simple programming support environment. *IEE/BCS Software Engineering Journal*, 2(4):86–94, July 1987.

481. P. A. Swatman. *Increasing Formality in the Specification of High-Quality Information Systems in a Commercial Context*. Phd thesis, Curtin University of Technology, School of Computing, Perth, Western Australia, July 1992.

482. P. A. Swatman. Using formal specification in the acquisition of information systems: Educating information systems professionals. In Bowen and Nicholls [75], pages 205–239.

483. P. A. Swatman, D. Fowler, and C. Y. M. Gan. Extending the useful application domain for formal methods. In Nicholls [375], pages 125–144.

484. P. A. Swatman and P. M. C. Swatman. Formal specification: An analytic tool for (management) information systems. *Journal of Information Systems*, 2(2):121–160, April 1992.

485. P. A. Swatman and P. M. C. Swatman. Is the information systems community wrong to ignore formal specification methods? In R. Clarke and J. Cameron, editors, *Managing Information Technology's Organisational Impact*. Elsevier Science Publishers (North-Holland), October 1992.

486. P. A. Swatman and P. M. C. Swatman. Managing the formal specification of information systems. In *Proc. International Conference on Organization and Information Systems*, September 1992.

487. P. A. Swatman, P. M. C. Swatman, and R. Duke. Electronic data interchange: A high-level formal specification in Object-Z. In *Proc. 6th Australian Software Engineering Conference (ASWEC'91)*, 1991.

488. P. F. Terry and S. R. Wiseman. On the design and implementation of a secure computer system. Memorandum no. 4188, RSRE, Ministry of Defence, Malvern, Worcestershire, UK, June 1988.

489. S. Thompson. Specification techniques [9004-0316]. *ACM Computing Reviews*, 31(4):213, April 1990.
    A review of *Formal methods applied to a floating-point number system* by Geoff Barrett [30].

490. D. Till, editor. *6th Refinement Workshop*, Workshop in Computing. Springer-Verlag, 1994. The workshop was held at City University, London, UK, 5–7 January 1994. See [181, 300].

491. B. S. Todd. A model-based diagnostic program. *IEE/BCS Software Engineering Journal*, 2(3):54–63, May 1987.

492. R. Took. The presenter – a formal design for an autonomous display manager. In I. Sommerville, editor, *Software Engineering Environments*, pages 151–169. Peter Peregrinus, London, 1986.

493. I. Toyn. *CADiZ Quick Reference Guide*. York Software Engineering Ltd, University of York, York YO1 5DD, UK, 1990.
    A guide to the CADiZ (Computer Aided Design in Z) toolkit. This makes use of the Unix *troff* family of text formatting tools. Contact David Jordan at the address above or on yse@minster.york.ac.uk via e-mail for further information on CADiZ. See also [274] for a paper introducing CADiZ. Support for LaTeX [297] is now available.

494. I. Toyn and A. J. Dix. Efficient binary transfer of pointer structures. Technical document, Computer Science Department, University of York, York YO1 5DD, UK, November 1993.

495. I. Toyn and J. A. McDermid. CADiZ: An architecture for Z tools and its implementation. Technical document, Computer Science Department, University of York, York YO1 5DD, UK, November 1993.

496. O. Traynor, P. Kearney, E. Kazmierczak, Li Wang, and E. Karlsen. Extending Z with modules. *Australian Computer Science Communications*, 17(1), 1995. Proc. ACSC'95.

497. S. Valentine. The programming language Z^{--}. *Information and Software Technology*, 37(5):293–301, May 1995.

498. S. H. Valentine. Z^{--}, an executable subset of Z. In Nicholls [375], pages 157–187.

499. S. H. Valentine. Putting numbers into the mathematical toolkit. In Bowen and Nicholls [75], pages 9–36.

500. M. J. van Diepen and K. M. van Hee. A formal semantics for Z and the link between Z and the relational algebra. In Bjørner et al. [39], pages 526–551.

501. K. M. van Hee, L. J. Somers, and M. Voorhoeve. Z and high level Petri nets. In Prehn and Toetenel [398], pages 204–219.

502. H. J. van Zuylen, editor. *The REDO Compendium: Reverse Engineering for Software Maintenance*. John Wiley & Sons, 1993.

An overview of the results of the ESPRIT REDO project, including the use of Z and Z^{++}. See in particular Chapter 16, also published in a longer form as [302].

503. M. M. West and B. M. Eaglestone. Software development: Two approaches to animation of Z specifications using Prolog. *IEE/BCS Software Engineering Journal*, 7(4):264–276, July 1992.

504. C. Wezeman and A. Judge. Z for managed objects. In Bowen and Hall [67], pages 108–119.

505. R. W. Whitty. Structural metrics for Z specifications. In Nicholls [371], pages 186–191.

506. P. J. Whysall and J. A. McDermid. An approach to object-oriented specification using Z. In Nicholls [373], pages 193–215.

507. P. J. Whysall and J. A. McDermid. Object-oriented specification and refinement. In Morris and Shaw [359], pages 151–184.

508. J. M. Wing. A specifier's introduction to formal methods. *IEEE Computer*, 23(9):8–24, September 1990.

509. J. M. Wing and A. M. Zaremski. Unintrusive ways to integrate formal specifications in practice. In Prehn and Toetenel [398], pages 545–570.

510. S. R. Wiseman and C. L. Harrold. A security model and its implementation. Memorandum no. 4222, RSRE, Ministry of Defence, Malvern, Worcestershire, UK, September 1988.

511. A. W. Wood. A Z specification of the MaCHO interface editor. Memorandum no. 4247, RSRE, Ministry of Defence, Malvern, Worcestershire, UK, November 1988.

512. K. R. Wood. The elusive software refinery: a case study in program development. In Morris and Shaw [359], pages 281–325.

513. K. R. Wood. A practical approach to software engineering using Z and the refinement calculus. *ACM Software Engineering Notes*, 18(5):79–88, December 1993.

514. W. G. Wood. Application of formal methods to system and software specification. *ACM SIGSOFT Software Engineering Notes*, 15(4):144–146, September 1990.

515. J. C. P. Woodcock. Teaching how to use mathematics for large-scale software development. *Bulletin of BCS-FACS*, July 1988.

516. J. C. P. Woodcock. Calculating properties of Z specifications. *ACM SIGSOFT Software Engineering Notes*, 14(4):43–54, 1989.

517. J. C. P. Woodcock. Mathematics as a management tool: Proof rules for promotion. In *Proc. 6th Annual CSR Conference on Large Software Systems*, Bristol, UK, September 1989.

518. J. C. P. Woodcock. Parallel refinement in Z. In *Proc. Workshop on Refinement*, January 1989.

519. J. C. P. Woodcock. Structuring specifications in Z. *IEE/BCS Software Engineering Journal*, 4(1):51–66, January 1989.

520. J. C. P. Woodcock. Transaction refinement in Z. In *Proc. Workshop on Refinement*, January 1989.

521. J. C. P. Woodcock. Z. In D. Craigen and K. Summerskill, editors, *Formal Methods for Trustworthy Computer Systems (FM89)*, Workshops in Computing, pages 57–62. Springer-Verlag, 1990.

522. J. C. P. Woodcock. Implementing promoted operations in Z. In Morris and Shaw [359], pages 366–378.

523. J. C. P. Woodcock. A tutorial on the refinement calculus. In Prehn and Toetenel [399], pages 79–140.

524. J. C. P. Woodcock. The rudiments of algorithm design. *The Computer Journal*, 35(5):441–450, October 1992.

525. J. C. P. Woodcock and S. M. Brien. W: A logic for Z. In Nicholls [375], pages 77–96.

526. J. C. P. Woodcock, P. H. B. Gardiner, and J. R. Hulance. The formal specification in Z of Defence Standard 00-56. In Bowen and Hall [67], pages 9–28.

527. J. C. P. Woodcock and P. G. Larsen, editors. *FME'93: Industrial-Strength Formal Methods*, volume 670 of *Lecture Notes in Computer Science*. Formal Methods Europe, Springer-Verlag, 1993.
    The 1st FME Symposium was held at Odense, Denmark, 19–23 April 1993. Z-related papers include [76, 125, 179, 262, 327, 395].

528. J. C. P. Woodcock and P. G. Larsen. Guest editorial. *IEEE Transactions on Software Engineering*, 21(2):61–62, 1995.
    Best papers of FME'93 [527]. See [43, 37, 127, 263].

529. J. C. P. Woodcock and M. Loomes. *Software Engineering Mathematics: Formal Methods Demystified*. Pitman, 1988.
    Also published as: *Software Engineering Mathematics*, Addison-Wesley, 1989.

530. J. C. P. Woodcock and C. C. Morgan. Refinement of state-based concurrent systems. In Bjørner et al. [39], pages 340–351.
    Work on combining Z and CSP.

531. R. Worden. Fermenting and distilling. In Bowen and Hall [67], pages 1–6.

532. J. B. Wordsworth. Teaching formal specification methods in an industrial environment. In *Proc. Software Engineering '86*, London, 1986. IEE/BCS, Peter Peregrinus.

533. J. B. Wordsworth. Specifying and refining programs with Z. In *Proc. Second IEE/BCS Conference on Software Engineering*, number 290 in Conference Publication, pages 8–16. IEE/BCS, July 1988.
    A tutorial summary.

534. J. B. Wordsworth. Refinement tutorial: A storage manager. In *Proc. Workshop on Refinement*, January 1989.

535. J. B. Wordsworth. A Z development method. In *Proc. Workshop on Refinement*, January 1989.

536. J. B. Wordsworth. The CICS application programming interface definition. In Nicholls [373], pages 285–294.

537. J. B. Wordsworth. *Software Development with Z: A A Practical Approach to Formal Methods in Software Engineering*. Addison-Wesley, 1993.
    This book provides a guide to developing software from specification to code, and is based in part on work done at IBM's UK Laboratory that won the UK Queen's Award for Technological Achievement in 1992.
    Contents: Introduction; A simple Z specification; Sets and predicates; Relations and functions; Schemas and specifications; Data design; Algorithm design; Specification of an oil terminal control system.

538. Xiaoping Jia. *ZTC: A Type Checker for Z – User's Guide*. Institute for Software Engineering, Department of Computer Science and Information Systems, DePaul University, Chicago, IL 60604, USA, 1994.

ZTC is a type checker for the Z specification language. ZTC accepts two forms of input: LATEX with zed style option and ZSL, an ASCII version of Z. ZTC can also perform translations between the two input forms. This document is intended to serve as both a user's guide and a reference manual for ZTC.

539. W. D. Young. Comparing specifications paradigms: Gypsy and Z. Technical Report 45, Computational Logic Inc., 1717 W. 6th St., Suite 290, Austin, Texas 78703, USA, 1989.

540. P. Zave and M. Jackson. Techniques for partial specification and specification of switching systems. In Prehn and Toetenel [398], pages 511–525.
Also published as [541].

541. P. Zave and M. Jackson. Techniques for partial specification and specification of switching systems. In Nicholls [375], pages 205–219.

542. Y. Zhang and P. Hitchcock. EMS: Case study in methodology for designing knowledge-based systems and information systems. *Information and Software Technology*, 33(7):518–526, September 1991.

543. Z archive. Oxford University Computing Laboratory, 1993.
A computer-based archive server at the Programming Research Group in Oxford is available for use by anyone with World-Wide Web (WWW) access, anonymous FTP access or an electronic mail address. This allows people interested in Z (and other things) to access various archived files. In particular, messages from the Z FORUM electronic mailing list [544] and a Z bibliography [53] are available.
The preferred method of access to the on-line Z archive is via the World-Wide Web (WWW) under the following 'URL' (Uniform Resource Locator):

$$\texttt{http://www.comlab.ox.ac.uk/archive/z.html}$$

Simply follow the hyperlinks of interest.
Much of the Z archive is also available via anonymous FTP on the Internet. Type the command 'ftp ftp.comlab.ox.ac.uk' (or alternatively if this does not work, 'ftp 163.1.27.2') and use 'anonymous' as the login id and your e-mail address as the password when prompted. The FTP command 'cd pub/Zforum' will get you into the Z archive directory. The file README gives some general information and 00index gives a list of the available files. The Z bibliography may be retrieved using the FTP command 'get z.bib', for example. If you wish to access any of the compressed POSTSCRIPT files in the archive, please issue the 'binary' command first.
For users without on-line access on the Internet, it is possible to access parts of the Z archive using electronic mail, send a message to archive-server@comlab.ox.ac.uk with the 'Subject:' line and/or the body of the message containing commands such as the following:

help	help on using the PRG archive server
index	general index of categories (e.g., 'z')
index z	index of Z-related files
send z z.bib	send Z bibliography in BIBTEX format
send z *file1 file2* ...	send multiple files
path *name@site*	optionally specify return e-mail address

If you have serious problems accessing the Z archive using WWW, anonymous FTP access or the electronic mail server and thus need human help, or if you wish to submit an item for the archive, please send electronic mail to archive-management@comlab.ox.ac.uk.

544. Z FORUM. Oxford University Computing Laboratory, 1986 onwards. Electronic mailing list: vol. 1.1–9 (1986), vol. 2.1–4 (1987), vol. 3.1–7 (1988), vol. 4.1–4 (1989), vol. 5.1–3 (1990).

Z FORUM is an electronic mailing list. It was initiated as an edited newsletter by R. Macdonald of DRA (formerly RSRE), Malvern, Worcestershire, UK, and is now maintained by J. P. Bowen at the Oxford University Computing Laboratory. Contributions should be sent to zforum@comlab.ox.ac.uk. Requests to join or leave the list should be sent to zforum-request@comlab.ox.ac.uk. Messages are now forwarded to the list directly to ensure timeliness. The list is also gatewayed to the USENET newsgroup comp.specification.z at Oxford and messages posted on either will automatically appear on both. A message answering some frequently asked questions is maintained and sent to the list once a month. A list of back issues of newsletters and other Z-related material is available electronically via anonymous FTP from ftp.comlab.ox.ac.uk:/pub/Zforum in the file 00index or via e-mail from the OUCL archive server [543]. For messages from a particular month (e.g., May 1995), access a file such as zforum95-05; for the most recent messages, see the file zforum.

Comp.specification.z and Z FORUM
Frequently Asked Questions

Jonathan P. Bowen

Oxford University Computing Laboratory
Wolfson Building, Parks Road, Oxford OX1 3QD, UK
Email: Jonathan.Bowen@comlab.ox.ac.uk
URL: http://www.comlab.ox.ac.uk/oucl/people/jonathan.bowen.html

Abstract. This appendix provides some details on how to access information on Z, particularly electronically. It has been generated from a message that is updated and sent out monthly on international computer networks.

This on-line information is available on-line on the following World Wide Web (WWW) hypertext page where it is split into convenient sections:

http://www.cis.ohio-state.edu/hypertext/faq/usenet/z-faq/faq.html

1 What is it?

The comp.specification.z electronic USENET newsgroup was established in June 1991 and is intended to handle messages concerned with the formal specification notation Z (pronounced 'zed'). It has an estimated readership of around 30,000 people worldwide. Z, based on set theory and first order predicate logic, has been developed at the Programming Research Group at the Oxford University Computing Laboratory (OUCL) and elsewhere since the late 1970s. It is now used by industry as part of the software (and hardware) development process in the UK, USA and elsewhere. It is currently undergoing international ISO standardization. Comp.specification.z provides a convenient forum for messages concerned with recent developments and the use of Z. Pointers to and reviews of recent books and articles are particularly encouraged. These will be included in the Z bibliography (see later) if they appear in comp.specification.z.

2 What if I do not have access to USENET news?

There is an associated Z FORUM electronic mailing list that was initiated in January 1986 by Ruaridh Macdonald, RSRE, UK. Articles are now automatically cross-posted between comp.specification.z and the mailing list for those whose do not have access to USENET news. This may apply especially to industrial Z users who are particularly encouraged to subscribe and post their experiences to the list. Please contact zforum-request@comlab.ox.ac.uk with your name, address and email address to join the mailing list (or if you change your email address or wish to be removed from the list). Readers are strongly urged to read the comp.specification.z newsgroup rather than the Z FORUM mailing list if possible. Messages for submission to the Z FORUM mailing list and the `comp.specification.z` newsgroup may be emailed to `zforum@comlab.ox.ac.uk`. This method of posting is particularly recommended for important messages like announcements of meetings since not all messages posted on `comp.specification.z` reach the OUCL.

A mailing list for the Z User Meeting educational issues session has been set by Neville Dean, Anglia Polytechnic University, UK. Anyone interested may join by emailing `zugeis-request@comlab.ox.ac.uk` with your contact details.

A specialist electronic mailing for discussion of SAZ, a combination of the structured method SSADM and Z existed for a while, but is now closed.

3 What if I do not have access to email?

If you wish to join the postal Z mailing list, please send your address to Amanda Kingscote, Praxis plc, 20 Manvers Street, Bath BA1 1PX, UK (tel +44-1225-444700, fax +44-1225-465205, email `ark@praxis.co.uk`). This will ensure you receive details of Z meetings, etc., particularly for people without access to electronic mail.

4 How can I join in?

If you are currently using Z, you are welcome to introduce yourself to the newsgroup and Z FORUM list by describing your work with Z or raising any questions you might have about Z which are not answered here. You may also advertize publications concerning Z which you or your colleagues produce. These may then be added to the master Z bibliography maintained at the OUCL (see later).

5 Where are Z-related files archived?

Information on the World Wide Web (WWW) is available under the
    `http://www.comlab.ox.ac.uk/archive/z.html`
page. See also the
    `http://www.comlab.ox.ac.uk/archive/formal-methods.html`
page on formal methods in general. The WWW global hypertext system is accessible using the 'netscape', 'mosaic' or 'lynx' programs for example. Contact your system manager if WWW access is not available on your system.

Some of the archive is also available via anonymous FTP on the Internet under the following directory:
    `ftp://ftp.comlab.ox.ac.uk/pub/Zforum`
Type the command 'ftp ftp.comlab.ox.ac.uk' (or 'ftp 163.1.27.2' if this does not work) and use 'anonymous' as the login id and your email address as the password when prompted. The FTP command 'cd pub/Zforum' will get you into the Z archive directory. The file README gives some general information and 00index gives a list of the files. (Retrieve these using the FTP command 'get README', for example.)

There is an automatic electronic mail-based electronic archive server which allows access to some of the archive such as most messages on comp.specification.z and Z FORUM, as well as a selection of other Z-related text files. Send an email message containing the command 'help' to archive-server@comlab.ox.ac.uk for further information on how to use the server. A command of 'index z' will list the Z-related files. To receive files via email, send a message containing the command 'send z *file1 file2* ...' to archive-server@comlab.ox.ac.uk. If you have serious trouble accessing the archive server, please contact the address archive-management@comlab.ox.ac.uk.

6   What tools are available?

Various tools for formatting, type-checking and aiding proofs in Z are available. A free LaTeX style file and documentation can be obtained from the OUCL archive. Access the

> ftp://ftp.comlab.ox.ac.uk/pub/Zforum/zed.sty

and

> ftp://ftp.comlab.ox.ac.uk/pub/Zforum/zguide.tex

files. A newer style 'csp_zed.sty' is available in the same location, which uses the new font selection scheme and covers CSP and Z symbols. A style for Object-Z 'oz.sty' with a guide 'oz.tex' is also accessible. LaTeX2e users may find 'zed-csp.sty' and 'zed2e.tex' useful.

The *f*UZZ package, a syntax and type checker with a LaTeX style option and fonts, is available from the Spivey Partnership, 10 Warneford Road, Oxford OX4 1LU, UK. It is compatible with the second edition of Spivey's Z Reference Manual. Access ftp:ftp.comlab.ox.ac.uk/pub/Zforum/fuzz for brief information and an order form. Contact Mike Spivey (email Mike.Spivey@comlab.oxford.ac.uk) for further information.

CADiZ is a UNIX-based suite of tools for checking and typesetting Z specifications which also supports previewing and interactive investigation of specifications. It is available from York Software Engineering, University of York, York YO1 5DD, UK (tel +44-1904-433741, fax +44-1904-433744). CADiZ supports a language like that of the Z Base Standard (Version 1.0). A particular extension allows one specification document to import another, including the mathematical toolkit as one such document. Typesetting support is available for both *troff* and for LaTeX. Browsing operations include display of information deduced by the type-checker (e.g. types of expressions and uses of variables), expansion of schemas, pre- and post-condition calculation, and simplification by the one-point rule. Work is on-going to provide support for refinement of Z specifications to Ada programs through a literate program development method and integrated proof facilities. Further information is available from David Jordan on yse@minster.york.ac.uk at York.

ProofPower is a suite of tools supporting specification and proof in Higher Order Logic (HOL) and in Z. Short courses on ProofPower-Z are available as demand arises. Information about ProofPower can be obtained automatically by sending email to ProofPower-server@win.icl.co.uk. Contact Roger Jones, International Computers Ltd, Eskdale Road, Winnersh, Wokingham, Berkshire RG11 5TT, UK (tel

+44-1734-693131 ext 6536, fax +44-1734-697636, email `rbj@win.icl.co.uk`) for further details.

Zola is a tool that supports the production and typesetting of Z specifications, including a type-checker and a Tactical Proof System. The tool is sold commercially and available to academic users at a special discount. For further information, contact K. Ashoo, Imperial Software Technology, 62–74 Burleigh Street, Cambridge CB1 1DJ, UK (tel +44-1223-462400, fax +44-1223-462500, email `ka@ist.co.uk`).

ZTC is a Z type-checker available free of charge for educational and non-profit uses. It is intended to be compliant with the 2nd edition of Spivey's Z Reference Manual. It accepts LATEX with 'zed' or 'oz' styles, and ZSL – an ASCII version of Z. ZANS is a Z animator. It is a research prototype that is still very crude. Both ZTC and ZANS run on Linux, SunOS 4.x, Solaris 2.x, HP-UX 9.0, DOS, and extended DOS. They are available via anonymous FTP under `ftp://ise.cs.depaul.edu/pub` in the directories ZANS-*x.xx* and ZTC-*x.xx*, where *x.xx* are version numbers. Contact Xiaoping Jia `jia@cs.depaul.edu` for further information.

Formaliser is a syntax-directed Z editor and type checker, running under Microsoft Windows, available from Logica Cambridge. Contact Susan Stepney, Logica Cambridge Limited, Betjeman House, 104 Hills Road, Cambridge CB2 1LQ, UK (tel +44-1223-66343, email `susan@logcam.co.uk`) for further information.

DST-*fuzz* is a set of tools based on the *fuzz* package by Mike Spivey, supplying a Motif based user interface for LATEX based pretty printing, syntax and type checking. A CASE tool interface allows basic functionality for combined application of Z together with structured specifications. The tools are integrated into SoftBench. For further information contact Hans-Martin Hoercher, DST Deutsche System-Techik GmbH, Edisonstr. 3, D-24145 Kiel, Germany (tel +49-431-7109-478, fax +49-431-7109-503, email `hmh@informatik.uni-kiel.d400.de`).

The B-Tool can be used to check proofs concerning parts of Z specifications. This is licensed by Edinburgh Portable Compilers Ltd, 17 Alva Street, Edinburgh EH2 4PH, UK (tel +44-131-225-6262, fax +44-131-225-6644). Contact the Distribution Manager (email `support@epc.ed.ac.uk`) for further information.

The B-Toolkit is a set of integrated tools which fully supports the B-Method for formal software development and is available from B-Core (UK) Limited, Magdalen Centre, The Oxford Science Park, Oxford OX4 4GA, UK. For further details, contact Ib Sørensen (email `Ib.Sorensen@comlab.ox.ac.uk`, tel +44-1865-784520, fax +44-1865-784518).

Z fonts for MS Windows and Macintosh are available on-line. For hyperlinks to these and other Z tool resources see the WWW Z page:

`http://www.comlab.ox.ac.uk/archive/z.html#tools`

A survey of Z tools (produced in 1991) may be obtained from Colin Parker, Systems Process Department, W376C, British Aerospace, Warton Aerodrome, Warton, Preston PR4 1AX, UK.

7 How can I learn about Z?

There are a number of courses on Z run by industry and academia. Oxford University offers industrial short courses in the use Z. As well as introductory courses, recent newly developed material includes advanced Z-based courses on proof and refinement, partly based around the B-Tool. Courses are held in Oxford, or elsewhere (e.g., on a company's premises) if there is enough demand. For further information, contact Jim Woodcock (tel +44-1865-283514, fax +44-1865-273839, email Jim.Woodcock@comlab.ox.ac.uk).

Logica offer a five day course on Z at company sites. Contact Rosalind Barden (email rosalind@logcam.co.uk, tel +44-1223-366343 ext 4860, fax +44-1223-322315) at Logica UK Limited, Betjeman House, 104 Hills Road, Cambridge CB2 1LQ, UK.

Praxis Systems plc runs a range of Z (and other formal methods) courses. For details contact Anthony Hall on +44-1225-444700 or jah@praxis.co.uk.

Formal Systems (Europe) Ltd run a range of Z, CSP and other formal methods courses, primarily in the US and with such lecturers as Jim Woodcock and Bill Roscoe (both lecturers at the OUCL). For dates and prices contact Kate Pearson (tel +44-1865-728460, fax +44-1865-201114) at Formal Systems (Europe) Limited, 3 Alfred Street, Oxford OX1 4EH, UK.

DST Deutsche System-Technik runs a collection of courses for either Z or CSP, mainly in Germany. These courses range from half day introductions to formal methods and Z to one week introductory or advanced courses, held either at DST, or elsewhere. For further information contact Hans-Martin Hoercher, DST Deutsche System-Techik GmbH, Edisonstr. 3, D-24145 Kiel, Germany (tel +49-431-7109-478, fax +49-431-7109-503, email hmh@informatik.uni-kiel.d400.de).

8 What has been published about Z?

A searchable on-line Z bibliography is available on the World Wide Web under

http://www.comlab.ox.ac.uk/archive/z/bib.html

in BIBTEX format. For those without WWW access, an older compressed version is available under ftp://ftp.comlab.ox.ac.uk/pub/Zforum/z.bib.Z (and also ftp://ftp.comlab.ox.ac.uk/pub/Zforum/z.ps.Z in compressed POSTSCRIPT format). Information on OUCL Technical Monographs and Reports, including many on Z, is available from the librarian (tel +44-1865-273837, fax +44-1865-273839, email library@comlab.ox.ac.uk).

Formal Methods: A Survey by S. Austin & G. I. Parkin, March 1993 includes information on the use and teaching of Z in industry and academia. Contact DITC Office, Formal Methods Survey, National Laboratory, Teddington, Middlesex TW11 0LW, UK (tel +44-181-943-7002, fax +44-181-977-7091) for a copy.

The following books largely concerning Z have been or are due to be published (in approximate chronological order):

• I. Hayes (ed.), Specification Case Studies, Prentice Hall International Series in Computer Science, 1987. (2nd ed., 1993)

- J. M. Spivey, Understanding Z: A specification language and its formal semantics, Cambridge University Press, 1988.
- D. Ince, An Introduction to Discrete Mathematics, Formal System Specification and Z, Oxford University Press, 1988. (2nd ed., 1993)
- J. C. P. Woodcock & M. Loomes, Software Engineering Mathematics: Formal Methods Demystified, Pitman, 1998. (Also Addision-Wesley, 1989)
- J. M. Spivey, The Z Notation: A reference manual, Prentice Hall International Series in Computer Science, 1989. (2nd ed., 1992) [Widely used as a de facto standard for Z. Often known as ZRM2.]
- A. Diller, Z: An introduction to formal methods, Wiley, 1990.
- J. E. Nicholls (ed.), Z user workshop, Oxford 1989, Springer-Verlag, Workshops in Computing, 1990.
- B. Potter, J. Sinclair & D. Till, An Introduction to Formal Specification and Z, Prentice Hall International Series in Computer Science, 1991.
- D. Lightfoot, Formal Specification using Z, MacMillan, 1991.
- A. Norcliffe & G. Slater, Mathematics for Software Construction, Ellis Horwood, 1991.
- J. E. Nicholls (ed.), Z User Workshop, Oxford 1990, Springer-Verlag, Workshops in Computing, 1991.
- I. Craig, The Formal Specification of Advanced AI Architectures, Ellis Horwood, 1991.
- M. Imperato, An Introduction to Z, Chartwell-Bratt, 1991.
- J. B. Wordsworth, Software Development with Z, Addison-Wesley, 1992.
- S. Stepney, R. Barden & D. Cooper (eds.), Object Orientation in Z, Springer-Verlag, Workshops in Computing, August 1992.
- J. E. Nicholls (ed.), Z User Workshop, York 1991, Springer-Verlag, Workshops in Computing, 1992.
- D. Edmond, Information Modeling: Specification and implementation, Prentice Hall, 1992.
- J. P. Bowen & J. E. Nicholls (eds.), Z User Workshop, London 1992, Springer-Verlag, Workshops in Computing, 1993.
- S. Stepney, High Integrity Compilation: A case study, Prentice Hall, 1993.
- M. McMorran & S. Powell, Z Guide for Beginners, Blackwell Scientific, 1993.
- K. C. Lano & H. Haughton (eds.), Object-oriented Specification Case Studies, Prentice Hall International Object-Oriented Series, 1993.
- B. Ratcliff, Introducing Specification using Z: A practical case study approach, McGraw-Hill, 1994.
- A. Diller, Z: An introduction to formal methods, 2nd ed., Wiley, 1994.
- J. P. Bowen & J. A. Hall (eds.), Z User Workshop, Cambridge 1994, Springer-Verlag, Workshops in Computing, 1994.
- R. Barden, S. Stepney & D. Cooper, Z in Practice, Prentice Hall BCS Practitioner Series, 1994.
- D. Rann, J. Turner & J. Whitworth, Z: A beginner's guide. Chapman & Hall, 1994.
- D. Heath, D. Allum & L. Dunckley, Introductory Logic and Formal Methods. A. Waller, Henley-on-Thames, 1994.

- L. Bottaci and J. Jones, Formal Specification using Z: A modelling approach. International Thomson Publishing, 1995.
- D. Sheppard, An Introduction to Formal Specification with Z and VDM. McGraw Hill International Series in Software Engineering, 1995.
- J. P. Bowen, Formal Specification and Documentation using Z: A Case Study Approach, International Thomson Publishing, 1995.

9 What is object-oriented Z?

Several object-oriented extensions to or versions of Z have been proposed. The book *Object orientation in Z*, listed above, is a collection of papers describing various OOZ approaches – Hall, ZERO, MooZ, Object-Z, OOZE, Schuman&Pitt, Z^{++}, ZEST and Fresco (an OO VDM method) – in the main written by the methods' inventors, and all specifying the same two examples. A more recent book entitled *Object-oriented specification case studies* surveys the principal methods and languages for formal object-oriented specification, including Z-based approaches.

10 How can I run Z?

Z is a (non-executable in general) specification language, so there is no such thing as a Z compiler/linker/etc. as you would expect for a programming language. Some people have looked at animating subsets of Z for rapid prototyping purposes, using logic and functional programming for example, but this work is preliminary and is not really the major point of Z, which is to increase human understandability of the specified system and allow the possibility of formal reasoning and development. However, Prolog seems to be the main favoured language for Z prototyping and some references may be found in the Z bibliography (see above).

11 Where can I meet other Z people?

The 8th Z User Meeting (ZUM'94) was held on 29–30 June 1994 at St. John's College, University of Cambridge, UK in association with BCS FACS. The 9th Z User Meeting is planned for 7–8 September 1995 in Limerick, Ireland. A number of tutorials will be held on 4–6 September 1995, also at Limerick. For general enquiries, contact the Conference Chair, Jonathan Bowen (tel +44-1865-283512, fax +44-1865-273839, email Jonathan.Bowen@comlab.ox.ac.uk). Further details will be issued on comp.specification.z in due course. The proceedings for Z User Meetings have been published in the Springer-Verlag Workshops in Computing series since the 4th meeting in 1989. Information on ZUM'95 is available from

http://www.comlab.ox.ac.uk/archive/z/zum95.html

via WWW or

ftp://ftp.comlab.ox.ac.uk/pub/Zforum/zum95

via anonymous FTP. The proceedings this year will be published in the Springer-Verlag LNCS series.

The 6th Refinement Workshop was held at City University, London, UK, 5–7 January 1994. The Programme Chair was David Till, Dept of Computer Science, City University, Northampton Square, London, EC1V 0HB, UK (tel +44-171-477-8552, email `till@cs.city.ac.uk`). The proceedings for these workshops are currently published in the Springer-Verlag Workshops in Computing series.

WIFT'95 (Workshop on Industrial-strength Formal specification Techniques) took place at Boca Raton, Florida, USA, 5–7 April 1995. See

`http://www.cse.fau.edu/WIFT/`

for details.

FORTE addresses formal techniques and testing methodologies applicable to distributed systems such as Estelle, Lotos, SDL, ASN.1, Z, etc. The IFIP WG6.1 7th International Conference on Formal Description Techniques for Distributed Systems and Communications Protocols (FORTE'94) was held at Berne, Switzerland, 4–7 October 1994. The 8th conference (FORTE'95) will be held in Montreal (Quebec), Canada, 17–20 October 1995. For further information, see the WWW page:

`http://www.comlab.ox.ac.uk/archive/formal-methods/conf/FORTE95.html`

Z2B – Z and its Future (Putting into Practice, Methods and Tools for Information System Design) will be held at Nantes, 10–12 October 1995. See:

`http://www.comlab.ox.ac.uk/archive/formal-methods/conf/Z2B95.html`

ICECCS'95 (IEEE International Conference on Engineering of Complex Computer Systems) includes a formal methods track and is to be held in Southern Florida, USA, 6–10 November 1995. For further information, see:

`http://rtlab12.njit.edu:8000/rtcl_pub_html/Mosaic/act/iceccs95.html`

The second FME (Formal Methods Europe) Symposium was held in Barcelona, Spain, 24–28 October 1994. The proceedings are available as Springer LNCS 873. The next FME Symposium will be held at St. Hugh's College, Oxford, UK, 18–22 March 1996. Deadline for submission is 11 September 1995. For further information, see:

`http://www.lri.fr/conferences/FME-96/`

The chairman of the umbrella organization, Formal Methods Europe, is Prof. Peter Lucas, TU Graz, Austria (email `lucas@ist.tu-graz.ac.at`).

Details of Z-related meetings may be advertized on `comp.specification.z` if desired. All the above meetings are likely to be repeated in some form. For a fuller list of meetings with a formal methods content, see:

`http://www.comlab.ox.ac.uk/archive/formal-methods/meetings.html`

12 What is the Z User Group?

The Z User Group was set up in 1992 to oversee Z-related activities, and the Z User Meetings in particular. As a subscriber to either `comp.specification.z`, ZFORUM or the postal mailing list, you may consider yourself a member of the Z User Group. There are currently no charges for membership, although this is subject to review if necessary. Contact `zforum-request@comlab.ox.ac.uk` for further information.

13 How can I obtain the draft Z standard?

The proposed Z standard under ISO/IEC JTC1/SC22 is available electronically via anonymous FTP *only* (not via the mail server since it is too large) from the Z archive at Oxford in compressed POSTSCRIPT format. Version 1.0 of the draft standard is accessible as the file

        ftp://ftp.comlab.ox.ac.uk/pub/Zforum/zstandard1.0.ps.Z

together with an annex in

        ftp://ftp.comlab.ox.ac.uk/pub/Zforum/zstandard-annex1.0.ps.Z

It is also available in printed form from the OUCL librarian (tel +44-1865-273837, fax +44-1865-273839, email `library@comlab.ox.ac.uk`) by requesting Technical Monograph number PRG-107.

14 Where else is Z discussed?

The BCS-FACS (British Computer Society Formal Aspects of Computer Science special interest group) and FME (Formal Methods Europe) are two organizations interested in formal methods in general. Contact BCS FACS, Dept of Computer Studies, Loughborough University of Technology, Loughborough, Leicester LE11 3TU, UK (tel +44-1509-222676, fax +44-1509-211586, email `FACS@lut.ac.uk`) for further information.

A *FACS Europe* newsletter is issued to members of FACS and FME. Please send suitable Z-related material to the Z column editor, David Till, Dept of Computer Science, City University, Northampton Square, London, EC1V 0HB, UK (tel +44-171-477-8552, email `till@cs.city.ac.uk`) for possible publication. Material from articles appearing on the `comp.specification.z` newsgroup may be included if considered of sufficient interest (with permission from the originator if possible). It would be helpful for posters of articles on `comp.specification.z` to indicate if they do not want further distribution for any reason.

15 How does VDM compare with Z?

See I. J. Hayes, C. B. Jones & J. E. Nicholls, Understanding the differences between VDM and Z, *FACS Europe*, Series I, 1(1):7–30, Autumn 1993, available on-line under `ftp://ftp.cs.man.ac.uk/pub/TR/UMCS-93-8-1.ps.Z` as a Technical Report, and I. J. Hayes, VDM and Z: A comparative case study, *Formal Aspects of Computing*, 4(1):76–99, 1992. VDM is discussed on the (unmoderated) VDM FORUM mailing list. Send a message containing the command 'join vdm-forum *name*' where *name* is your real name to `mailbase@mailbase.ac.uk`. To contact the list administrator, email John Fitzgerald on `vdm-forum-request@mailbase.ac.uk`.

16 What if I have spotted a mistake or an omission?

Please send corrections or new relevant information about meetings, books, tools, etc., to `zforum-request@comlab.ox.ac.uk`. New questions and model answers are also gratefully received!

Author Index

Springer-Verlag
and the Environment

Lecture Notes in Computer Science

For information about Vols. 1–903

please contact your bookseller or Springer-Verlag

Vol. 904: P. Vitányi (Ed.), Computational Learning Theory. EuroCOLT'95. Proceedings, 1995. XVII, 415 pages. 1995. (Subseries LNAI).

Vol. 905: N. Ayache (Ed.), Computer Vision, Virtual Reality and Robotics in Medicine. Proceedings, 1995. XIV,

Vol. 906: E. Astesiano, G. Reggio, A. Tarlecki (Eds.), Recent Trends in Data Type Specification. Proceedings, 1995. VIII, 523 pages. 1995.

Vol. 907: T. Ito, A. Yonezawa (Eds.), Theory and Practice of Parallel Programming. Proceedings, 1995. VIII, 485 pages. 1995.

Vol. 908: J. R. Rao Extensions of the UNITY Methodology: Compositionality, Fairness and Probability in Parallelism. XI, 178 pages. 1995.

Vol. 909: H. Comon, J.-P. Jouannaud (Eds.), Term Rewriting. Proceedings, 1993. VIII, 221 pages. 1995.

Vol. 910: A. Podelski (Ed.), Constraint Programming: Basics and Trends. Proceedings, 1995. XI, 315 pages. 1995.

Vol. 911: R. Baeza-Yates, E. Goles, P. V. Poblete (Eds.), LATIN '95: Theoretical Informatics. Proceedings, 1995. IX, 525 pages. 1995.

Vol. 912: N. Lavrac, S. Wrobel (Eds.), Machine Learning: ECML – 95. Proceedings, 1995. XI, 370 pages. 1995. (Subseries LNAI).

Vol. 913: W. Schäfer (Ed.), Software Process Technology. Proceedings, 1995. IX, 261 pages. 1995.

Vol. 914: J. Hsiang (Ed.), Rewriting Techniques and Applications. Proceedings, 1995. XII, 473 pages. 1995.

Vol. 915: P. D. Mosses, M. Nielsen, M. I. Schwartzbach (Eds.), TAPSOFT '95: Theory and Practice of Software Development. Proceedings, 1995. XV, 810 pages. 1995.

Vol. 916: N. R. Adam, B. K. Bhargava, Y. Yesha (Eds.), Digital Libraries. Proceedings, 1994. XIII, 321 pages. 1995.

Vol. 917: J. Pieprzyk, R. Safavi-Naini (Eds.), Advances in Cryptology - ASIACRYPT '94. Proceedings, 1994. XII, 431 pages. 1995.

Vol. 918: P. Baumgartner, R. Hähnle, J. Posegga (Eds.), Theorem Proving with Analytic Tableaux and Related Methods. Proceedings, 1995. X, 352 pages. 1995. (Subseries LNAI).

Vol. 919: B. Hertzberger, G. Serazzi (Eds.), High-Performance Computing and Networking. Proceedings, 1995. XXIV, 957 pages. 1995.

Vol. 920: E. Balas, J. Clausen (Eds.), Integer Programming and Combinatorial Optimization. Proceedings, 1995. IX, 436 pages. 1995.

Vol. 921: L. C. Guillou, J.-J. Quisquater (Eds.), Advances in Cryptology – EUROCRYPT '95. Proceedings, 1995. XIV, 417 pages. 1995.

Vol. 922: H. Dörr, Efficient Graph Rewriting and Its Implementation. IX, 266 pages. 1995.

Vol. 923: M. Meyer (Ed.), Constraint Processing. IV, 289 pages. 1995.

Vol. 924: P. Ciancarini, O. Nierstrasz, A. Yonezawa (Eds.), Object-Based Models and Languages for Concurrent Systems. Proceedings, 1994. VII, 193 pages. 1995.

Vol. 925: J. Jeuring, E. Meijer (Eds.), Advanced Functional Programming. Proceedings, 1995. VII, 331 pages. 1995.

Vol. 926: P. Nesi (Ed.), Objective Software Quality. Proceedings, 1995. VIII, 249 pages. 1995.

Vol. 927: J. Dix, L. Moniz Pereira, T. C. Przymusinski (Eds.), Non-Monotonic Extensions of Logic Programming. Proceedings, 1994. IX, 229 pages. 1995. (Subseries LNAI).

Vol. 928: V.W. Marek, A. Nerode, M. Truszczynski (Eds.), Logic Programming and Nonmonotonic Reasoning. Proceedings, 1995. VIII, 417 pages. 1995. (Subseries LNAI).

Vol. 929: F. Morán, A. Moreno, J.J. Merelo, P. Chacón (Eds.), Advances in Artificial Life. Proceedings, 1995. XIII, 960 pages. 1995 (Subseries LNAI).

Vol. 930: J. Mira, F. Sandoval (Eds.), From Natural to Artificial Neural Computation. Proceedings, 1995. XVIII, 1150 pages. 1995.

Vol. 931: P.J. Braspenning, F. Thuijsman, A.J.M.M. Weijters (Eds.), Artificial Neural Networks. IX, 295 pages. 1995.

Vol. 932: J. Iivari, K. Lyytinen, M. Rossi (Eds.), Advanced Information Systems Engineering. Proceedings, 1995. XI, 388 pages. 1995.

Vol. 933: L. Pacholski, J. Tiuryn (Eds.), Computer Science Logic. Proceedings, 1994. IX, 543 pages. 1995.

Vol. 934: P. Barahona, M. Stefanelli, J. Wyatt (Eds.), Artificial Intelligence in Medicine. Proceedings, 1995. XI, 449 pages. 1995. (Subseries LNAI).

Vol. 935: G. De Michelis, M. Diaz (Eds.), Application and Theory of Petri Nets 1995. Proceedings, 1995. VIII, 511 pages. 1995.

Vol. 936: V.S. Alagar, M. Nivat (Eds.), Algebraic Methodology and Software Technology. Proceedings, 1995. XIV, 591 pages. 1995.

Vol. 937: Z. Galil, E. Ukkonen (Eds.), Combinatorial Pattern Matching. Proceedings, 1995. VIII, 409 pages. 1995.